Smart Inorganic Polymers

Smart Inorganic Polymers

Synthesis, Properties, and Emerging Applications in Materials and Life Sciences

Edited by Evamarie Hey-Hawkins and Muriel Hissler

WILEY-VCH

Editors

Professor Evamarie Hey-Hawkins
Universität Leipzig
Institut für Anorganische Chemie
Chemie und Mineralogie
Johannisallee 29
D-04103 Leipzig
Germany

Professor Muriel Hissler
Université de Rennes 1
Campus de Beaulieu
Bat 10a, Bur 219
263, Av. du Général Leclerc
35042 Rennes Cedex
France

Cover Courtesy of Dr. Christoph Selg, Leipzig University, Germany

Library of Congress Card No.:
applied for

British Library Cataloguing-in-Publication Data
A catalogue record for this book is available from the British Library.

Bibliographic information published by the Deutsche Nationalbibliothek
The Deutsche Nationalbibliothek lists this publication in the Deutsche Nationalbibliografie; detailed bibliographic data are available on the Internet at <http://dnb.d-nb.de>.

Print ISBN: 978-3-527-34484-0
ePDF ISBN: 978-3-527-81911-9
ePub ISBN: 978-3-527-81913-3
oBook ISBN: 978-3-527-81914-0

Typesetting SPi Global, Chennai, India
Printing and Binding C.O.S. Printers Pte Ltd, Singapore
Printed on acid-free paper

10 9 8 7 6 5 4 3 2 1

Contents

Preface

Research in functional materials science is one of the strategic top-priority areas of development in science and technology. Environmentally friendly and cost-efficient multifunctional materials are indispensable for our modern society. Owing to their broad range of properties, polymers are used as thermoplastics, thermosets, and elastomers, and after being processed into fibers and films or used in the design of structural materials, they play an essential and ubiquitous role in everyday life. In this area, inorganic polymeric materials are growing in importance as a result of increased demand for new materials with specific properties. Inorganic polymers have remarkable properties that can be superior to those of their organic counterparts in a variety of ways including, for example, thermal stability (polysilazanes) and controlled degradability in drug-delivery systems. This facilitates their use as water repellents, flame retardants, and metal-ion conductors in batteries (polyphosphazenes), as well as precursors for ceramics, microlithography, and electroluminescent devices (polysilanes). Furthermore, the discovery of new synthetic avenues to metallopolymers has led to the rapid growth of this field and to the wide range of applications for these functional materials. Inorganic polymers offer an innovative approach to sustainability by switching from hydrocarbon/oil-based polymers to heteroelement-based polymers and lessen the burden on scarce nonrenewable resources or even provide full replacements. In terms of applications, inorganic polymers are now found in many consumer products, demonstrating their technological importance; investigation of new applications continues in laboratories worldwide.

At the European level, many academic and industrial research groups have a long-standing interest in inorganic polymers or smart materials. The idea to join their forces in the framework of a COST Action raised European competitiveness in this area to new levels. Effectively, with the COST Action CM1302: European Network on Smart Inorganic Polymers (SIPs), it was possible to give this exciting research area a new forum and a new voice. Funding priorities and opportunities are very different throughout Europe. Bringing scientists from such diverse backgrounds together has a far-reaching impact that will reverberate for many years to come. New collaborations involving inclusiveness target countries (ITCs) have emerged, for example between Portugal and Spain, Croatia and Germany, as well as Romania and France. A generation of young researchers from all over Europe has been trained in European training schools and in Short-Term Scientific Missions to receive state-of-the-art skills and expertise. Working together has been inspirational, many fundamental questions have been addressed, and for many more the ground has been prepared. Different dissemination tools have been used to increase the interest, awareness, and the visibility of this COST Action (web pages, press releases, international conferences, etc.). The final dissemination of the COST Action is the present book entitled *Smart Inorganic Polymers: Synthesis, Properties and Emerging Applications in Materials and Life Sciences.*

This book showcases some of the research highlights that emerged during the COST Action and at the same time hopes to entice the reader into the world of smart inorganic polymers and its vibrant, pan-European community. The book is aimed at academics and industrial researchers in this field, and also at scientists who want to get a better overview on the state of the art of this rapidly advancing area.

Chapter 1 by *Rudolf Pietschnig* provides an overview of the use of smart inorganic materials in daily life. For example, some polymers have the ability to undergo facile electronic excitation giving rise to materials with tunable color (electro/thermochromism) or switchable surface polarity (hydrophobicity/hydrophilicity). Such properties enable adaptive windows for privacy, security, and heat management. *Rudolf Pietschnig* has also been the dissemination manager of the COST Action SIPs. His group is recognized for the development of silanol-based surfactants.

The following chapters present the synthesis, properties, and applications of polymers incorporating different heteroelements such as boron, phosphorus, silicon, germanium, and tin.

Anne Staubitz, the equal opportunity manager of SIPs and an expert in the development of Group 13/15 element analogs of organic materials, describes together with her coworkers *Jonas Hoffmann* and *Philipp Gliese* the synthesis and properties of polymers incorporating Group 13 and Group 15 elements in Chapter 2.1. The properties of these polymers are remarkably different from those of the corresponding organic polymers: their higher polarity (boron–nitrogen vs. carbon–carbon [BN vs. CC]) or greater flexibility (boron–phosphorus vs. carbon–carbon [BP vs. CC]) leads to different solubilities or glass transition temperatures. They may be used as preceramic polymers, which is unthinkable for organic polymers. Many more exciting properties are yet to be discovered. In Chapter 2.2, *Clara Viñas,* leader of the working group in SIPs dealing with advanced applications of inorganic polymers, and her colleagues *Rosario Núñez, Isabel Romero,* and *Francesc Teixidor* emphasize the role of hollow spherical icosahedral boron clusters in the properties of polymeric and nanohybrid structures. The incorporation of such clusters into these structures not only improves their solubility, stability, and processability, but also boosts their electronic, optical, and thermal properties. All these properties make them promising materials for technological and biomedical applications.

With the rising interest in applying Group 14-based materials in the field of electronics and energy-related materials, Chapter 3 by *Ana Torvisco, David Scheschkewitz,* and *Frank Uhlig* presents the preparation of stable Group 14 metal-containing linear polymers, focusing on metal hydrides (R_nEH_{4-n}, E = Ge, Sn) as building blocks. In addition, the application of trihydrides REH_3 toward the preparation of novel branched polymeric networks is described.

David Scheschkewitz and *Frank Uhlig* have been leaders of the working group in SIPs dealing with the development of inorganic molecular building blocks.

Chapter 4 by *Andreas Orthaber* and *Alejandro Presa Soto* summarizes the synthesis of polymers of phosphorus and the heavier pnictogens comprising unsaturated motifs, such as phosphazene or phosphaalkene, as well as saturated motifs, e.g. phosphane and phosphole or P(=O)O units. They detail the synthesis of suitable precursors and specific polymerization methods that lead to functional and smart materials.

In Chapter 5, *Anne-Marie Caminade* (specialist in inorganic dendrimers and leader of the working group in SIPs dealing with the synthesis of smart inorganic polymers) describes the synthesis of "inorganic" dendrimers comprising either silicon or phosphorus branching points with emphasis on carbosilane and phosphorhydrazone dendrimers. A few miscellaneous examples of dendrimers based on germanium, tin, or bismuth branching points are also presented. The properties of these dendrimers, presented in Chapter 10, are relevant to three main fields: (i) catalysis, with emphasis on the "dendritic effect" and the possibility of recovering and reusing the dendritic catalysts; (ii) nanomaterials exclusively composed of dendrimers, or materials incorporating inorganic dendrimers in their structure, or surfaces of materials modified with inorganic dendrimers; and (iii) biology/nanomedicine with applications in bioimaging, gene therapy, and for treatment of viruses, brain diseases, cancers, and inflammatory diseases.

In Chapter 6, *Jiří Vohlídal*, an expert in metallosupramolecular polymers, and *Muriel Hissler* focus on constitutionally dynamic polymers (dynamers) whose chains consist of alternating molecular units with two or more chelating end-groups (unimers) and metal ions linked by coordination. This chapter provides in brief the classification of dynamers and related terms, synthetic approaches to unimers, methods of monitoring the self-assembly and characterization of metallosupramolecular dynamers, as well as their functional properties and potential applications.

Chapter 7 highlights the different kinds of heteroatom-based materials that have been used in electronic devices, such as organic light-emitting diodes (OLEDs), organic photovoltaic (OPV) cells, dye-sensitized solar cells (DSSCs), organic field-effect transistors (OFETs), and electrochromic cells. This chapter has been written by *Muriel Hissler*, co-chair and STSM coordinator of COST Action SIPs, and her coworkers *Matthew P. Duffy* and *Pierre-Antoine Bouit*.

Owing to their intrinsic structures, most smart inorganic polymers show excellent thermal stability and fire retardancy. *De-Yi Wang* and *Tarik Eren*, having extensive expertise in fire-retardant materials, detail the synthesis and fire-retardant properties of different classes of polymers in Chapter 8 together with their coworker *Raghvendra Kumar Mishra*.

Inorganic polymers have also potential medical applications. In Chapter 9, *Andreia Valente*, *Rafaella L. M. Precker*, and *Evamarie Hey-Hawkins*, experts on molecules and polymers for medicinal applications, present an overview of the use of inorganic polymers, metal-containing polymers and metal-organic frameworks in medicinal applications, mainly tissue engineering and drug delivery. *Evamarie Hey-Hawkins* was Chair of COST Action CM1302 SIPs.

Finally, *Walter Caseri* and *Irene Weymuth* present other examples of inorganic polymers for the development of functional materials. For instance, substituted 1,2,4-triazole ligands can establish coordination polymers with a number of metal ions, e.g. Ag^{I}, Co^{II}, Cu^{II}, and Fe^{II}. These polymers can exhibit not only strictly linear structures but also complicated network structures. Depending on the system, thermochromism, diamagnetism–paramagnetism transitions, and color sensitivity to small amounts of water can arise. Soluble polymers can be processed into lyogels, xerogels, films, and blended fibers. Application in areas such as humidity or temperature sensors, thermochromic fibers, and anticounterfeiting measures can be envisaged for such polymers.

We would like to thank the COST Office, especially the Science Officer, Dr. Lucia Forzi, and the Administrative Officer, Ms. Svetlana Voinova, for their great support and help in organizing the COST Action SIPs.

We are very grateful to all authors for their contributions and their patience. Last but not least, we would like to thank the Wiley team, especially Martin Preuss and Shirly Samuel, for their continuous support in planning and compiling this book, which gives a timely overview of the evolving potential and emerging applications of smart inorganic polymers.

7 September 2018

Evamarie Hey-Hawkins
Leipzig, Germany

Muriel Hissler
Rennes, France

1

Current Status and Future Perspectives of Functional and Smart Materials in Daily Life

Rudolf Pietschnig

University of Kassel, Department of Chemistry and Center for Interdisciplinary Nanostructure Science and Technology, Heinrich-Plett-Straße 40, 34132 Kassel, Germany

1.1 Introduction

Smart materials in general, are defined as materials with adaptive properties that undergo change upon exposure to an external stimulus. These changing properties can be very diverse including, but not limited to, reversible changes in volume, shape, opacity, or color, which may be triggered by changes in pH, temperature, magnetic/electric fields, or light. Because of their adaptive nature, these properties play a key role in the manufacture of many technical devices used in daily life, and the ability to switch them is the basis for the surging interest in smart materials for emerging technologies and applications. Besides the responsive features outlined above, materials with optoelectronic and thermoelectric properties that allow conversion of light or heat to electricity at different temperature levels are also sometimes termed as smart materials, but will not be covered in this chapter. Self-healing materials that have the intrinsic ability to repair damage have also not been included. On the other hand, shape memory materials are included. In this chapter, a brief outline on the current status and future perspectives of smart materials in daily life will be given, with a special focus on polymeric materials and inorganic polymeric materials in particular.

1.2 Properties and Applications

The properties and applications outlined in this section rely mainly on the features of smart materials related to their mechanical and rheological properties, electronic excitation, optical features, supramolecular recognition, and chemical reactivity, and further aspects such as switchable polarity.

1.2.1 Applications Based on Mechanical and Rheological Properties

Shape memory polymers (SMPs) are capable of changing shape and dimensions upon stimuli such as heat or upon exposure to light of suitable wavelength.

Smart Inorganic Polymers: Synthesis, Properties, and Emerging Applications in Materials and Life Sciences, First Edition. Edited by Evamarie Hey-Hawkins and Muriel Hissler.

The behavior is comparable with that of shape memory metals and alloys that are already in use for medical purposes but may sometimes entail biological side effects owing to their metallic constituents. Besides their superior biocompatibility, SMPs show a faster switching time owing to their lower heat conductivity and capacity that make them attractive candidates for medical implants in minimal invasive surgery [1]. Recently, hybrid inorganic–organic SMPs have been developed that show improved cycling behavior for a series of switching events with response times (<0.5 seconds) and a transformation temperature near body temperature (40 °C) [2].

Polyhedral silsesquioxane (POSS) cages are nanostructures with diameters in the range of 1–3 nm and, hence, may be considered as the smallest existing silica particles. POSS cages are compatible with many polymer matrices and improve their viscosity, strength, modulus, rigidity, and flammability [3–6]. Moreover, they may affect gas permeation through membranes, which is the basis for application of POSS cages in food packaging. Similar applications may be envisaged for polymers and polymer blends involving carboranes based on the intriguing properties of the latter (cf. Chapter 2) [7–9].

Aiming at further improvements in medical implant and handling technology, antimicrobial properties have been endowed to cellulosic fibers, by coating with hydrophilic TiO_2, and subsequently equipped with a fluorinated silanol coating. The resulting amphiphobic material featured excellent water- and hexane-repellent properties, originating from both surface chemistry and morphology. It was further demonstrated that *Escherichia coli* cells did not adhere to such surfaces owing to the rather low surface energies (Figure 1.1) [10].

The sizing of natural fibers is an important step in paper or textile manufacturing by which the surface properties can be modified to control the wear and sorption characteristics. While reactive organic components (e.g. alkyl ketene dimer (AKD) and alkenyl succinic anhydride (ASA)) are used for the sizing of bulk cellulosic fibers, surface sizing uses mainly modified starch to form a water-repellent film on the paper surface. Owing to cost restrictions, improved processes are limited to advanced purposes such as nanopapers and applications in electronics where highly hydrophobic surfaces with well-defined surface properties can be obtained using vapor phase processing of cellulosic fibers with alkoxysilanes [11]. Similarly, modification of wool fibers via coating with a siloxane-based polymer film equipped the fibers with anti-felting properties and significantly improved the shrinkage behavior of the wool upon heat treatment [12]. Silicon-based precursors and their polymeric condensation products have found application in hair care as well. A variety of patents and products reported silanetriols as important binder materials for improved hair dyeing, or to incorporate permanent fragrances [13, 14]. Moreover, it has been reported that conditioning agents containing silanetriols may reduce the stress on hair during dyeing processes as was demonstrated for Caucasian and Afroethnic hair [15, 16].

1.2.2 Applications Based on Electronic Excitation

A salient feature of the compounds of heavier main group elements is the smaller energetic separation between bonding and anti-bonding states compared with

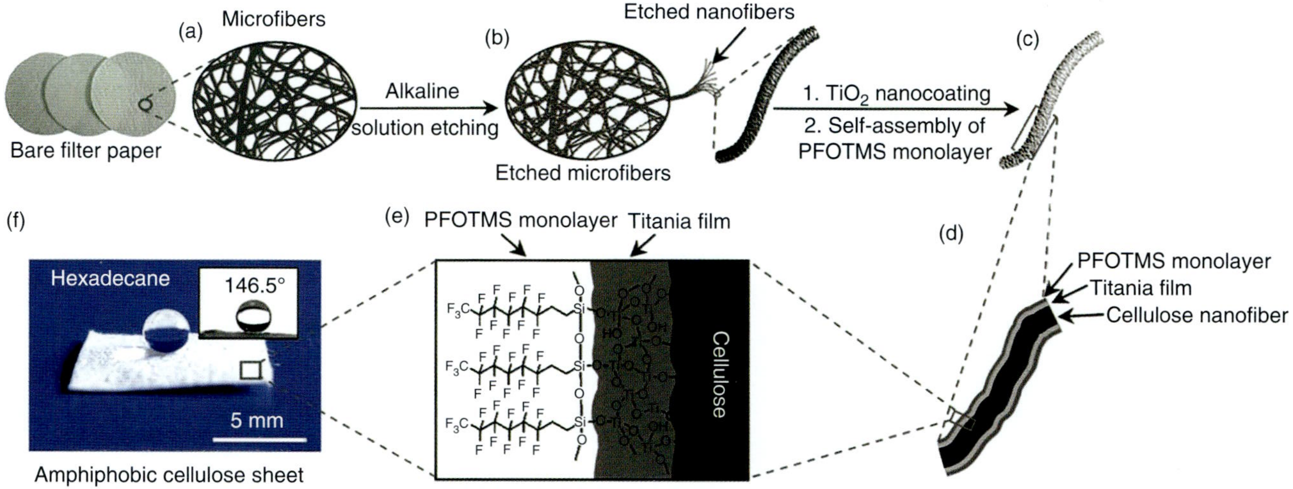

Figure 1.1 Sketch of the fabrication process of amphiphobic cellulose sheets. (a) Cellulose microfibers of commercial filter paper; (b) rougher cellulose fibers of filter paper resulting from alkaline solution etching; (c) cellulose nanofiber of etched filter paper deposited with titania–PFOTMS (perfluorooctyl trimethoxysilane) composite ultrathin films; (d, e) structure of titania–PFOTMS ultrathin film-modified cellulose nanofiber; (f) resulting amphiphobic cellulose sheet. Source: Jin et al. 2012 [10]. Reproduced with permission of Royal Society of Chemistry.

their organic counterparts. This aspect is not only relevant for π-bonding, but also for σ-bonding especially in the homonuclear bonding situations enabling σ-conjugation in polysilanes and polystannanes and conventional π-conjugation in multiple-bonded organopnictogen and especially organophosphorus compounds [17–21]. Based on these phenomena, electronic excitation is facilitated by incorporation of heavier main group elements into (in)organic hybrid materials giving rise to attractive properties (cf. Chapter 7). Owing to their desirable properties, such low band gap materials are the preferred choice for applications relevant for daily life such as light-emitting diodes, solar cells, and molecular electronics in general. The use of phospholes in organic light-emitting diodes (OLEDs) evolves from their emissive properties (fluorescence, phosphorescence), which makes them well suited as emissive layers in such devices (Figure 1.2) for which a brightness of >44 000 cd/m^2 with a turn on voltage of <3 V can be achieved [22]. By combining different components with complementary emissive colors it was possible to build white organic light-emitting diodes (WOLEDs) in which the phosphole oligomer contributes a more orange color tone. Besides the use of phosphole-based materials in the emissive layer, phospholes have been employed in electron-transporting layers as well. The principle of an OLED where charge recombination is used to generate photon emission can be reversed by using photon absorption to generate electron excitation followed by charge separation for instance in photovoltaic devices. In view of the close relationship between both types of devices phosphole-based materials have been implemented successfully in solar cells, e.g. in dye sensitized solar cells (DSSC) or bulk heterojunction solar cells (BHJ-SC).

By employing phospholes, a power conversion efficiency of 7.1% has been achieved in a BHJ-SC, whereas to date a power conversion efficiency of 5.6% is the highest η value reported for such materials in a DSSC [22]. With respect to organic electronics, silole-based materials are even more established than their phosphole analogs. The attractive properties of materials incorporating this group 14 heterocycle in part evolve around their large hole mobilities [23, 24]. Using silole-based materials a power conversion efficiency of 7.3% has been achieved in a BHJ-SC device [25]. Further details on the role and beneficial properties of organic inorganic hybrid structures for optoelectronic applications can be found in Chapter 7.

In general, homo-catenated polymers of group 14 are prominent examples that display the unique feature of σ-conjugation, which entails electronic excitation of the saturated backbone by light in the visible or long-wavelength UV region.

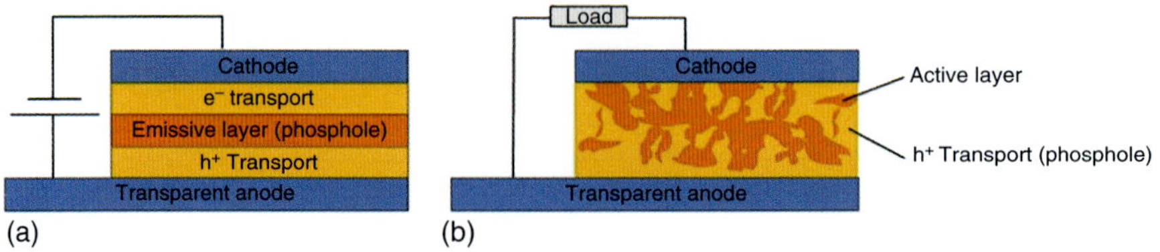

Figure 1.2 Simplified schematic sketch of phosphole-based optoelectronic devices: (a) OLED and (b) BHJ-SC.

Since anti-bonding states are populated during excitation, these polymers are then prone to fragmentation, yielding group 14-based radicals or tetrylenes among which silylenes are the most prominent ones [21]. In the case of polysilanes, applications have emerged where polysilanes serve as photoinitiators (PIs) for polymerizations behaving as a safe and stable polymeric radical precursor before initiation. Furthermore, the silylated by-products obtained from the PI's reactivity, which will be retained in the resulting polymer matrix, are nontoxic. Based on a similar principle, germanium-based photoinitiators have emerged as radical sources for advanced applications such as dental fillings. Such acylgermanes have a redshifted absorption characteristic, which is advantageous for use in restorative dental medicine as exposure of living tissue to harmful UV-radiation can be avoided. Moreover, photocuring at 500 nm is important for the development of bulk-curable restorative composites with increased curing depth and can be performed using an LED. Members of the COST action SIPs have developed acylgermanes that have improved radical efficiency and curing behavior [26]. Based on the excellent PI properties these materials have great potential as PIs for dental fillings and for additional applications such as photopolymerizable adhesives, coatings, or 3D-lithography techniques (Figure 1.3). Moreover, they extended this methodology to acylstannanes, which reveal a significant redshift of the longest wavelength absorption compared with that of the corresponding germanium compounds [27]. In contrast to the well-known toxicity of some trialkylstannanes, both the Ames-test, a biological assay to assess the mutagenic potential of chemical compounds, and cytotoxicity studies reveal intriguingly low toxicity for acylstannanes. The excellent performance of the latter as photoinitiators was demonstrated, which opened new avenues for polymer-based restorative dental medicine.

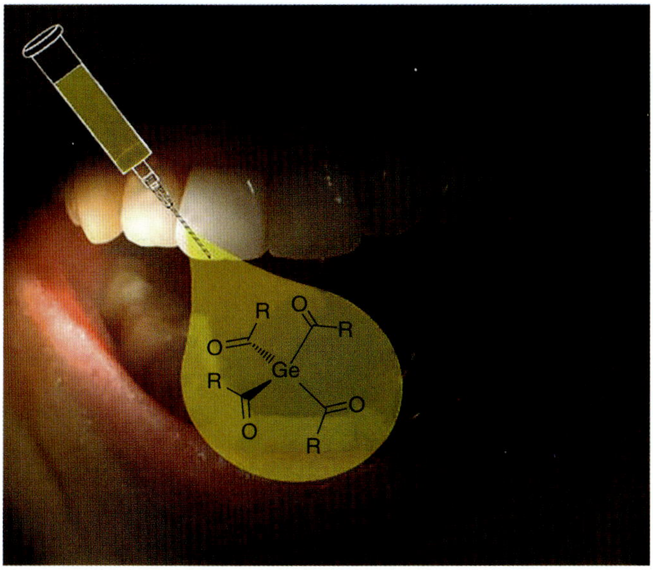

Figure 1.3 Sketch of a germanium-based PI developed as initiator for the curing of dental fillings. Source: Radebner et al. 2017 [26]. Reproduced with permission of John Wiley & Sons.

1.2.3 Applications Based on Optical Features

Smart materials changing optical features such as opacity and color upon exposure to an external stimulus are very attractive for applications in display technology and energy management in smart buildings. Such materials show thermochromic or electrochromic behavior changing color with temperature or the voltage applied. While smart materials show reversible color changes, there are other applications where a temperature seal indicates if a maximum temperature was maintained by inducing an irreversible change to ensure a cooling chain during storage and transport of temperature sensitive products (e.g. food, biological samples or tissue, medication). Materials undergoing reversible color change do not necessarily have a complex composition. For instance, simple inorganic compounds such as HgI_2 may show such thermochromic behavior. In the 1970s, thermochromic flat thermometers became quite popular using materials that are less problematic in terms of toxicity during the product lifecycle including production and disposal. Similarly, tea or coffee mugs changing color or motifs depending on the temperature of the liquid inside taking advantage of thermochromism are well-known examples from daily life as well. Heat flow in daily life applications is usually a rather slow diffusion-based process and given the limited temperature differences involved switching times are usually rather long.

Fast switching is usually required for electrochromic materials where the color change is usually effected by the voltage applied. There are smart materials that can be switched between two states such as colorless and dark for use as shades or blinds in smart windows in buildings (Figure 1.4) or in vehicles that cover a full range of colors relevant for display technology. Generally, switching properties via a voltage difference is favorable for integration into IT-based control systems, which is the reason for the high interest in electrochromic materials. Again, materials with simple chemical constitution such as WO_3 are capable of electrochromic behavior. In such thin films a cathodic current leads to injection of electrons into the pale yellow WO_3 lattice leading to partially reduced tungsten ions, which entails a color change [29, 30]. The ability to switch off vision through a window is essential for applications related to privacy and security. Thus, electrochromic windows are installed in the driver cabin of ICE 3 trains and in cockpits on airplanes. Alternatively, simple on/off switching of smart windows can be achieved by an array of micro-mirrors as well, which require very sophisticated production processes as compared with redox-active thin films outlined above [31].

Figure 1.4 Switching sequence of an electrochromic laminated glass. Source: Baetens et al. 2010 [28]. Reproduced with permission of Elsevier.

To overcome the limitations of switching between only two states in mechanically inflexible solid state materials, ferrocene has been used to induce electrochromic behavior to smart polymeric materials [32]. By adjusting an appropriate voltage difference the redox-active ferrocene unit can be switched reversibly from the neutral ferrocene to the cationic ferrocenium form. These changes are accompanied by characteristic color changes between the reduced and oxidized forms and more than two colored states may be accessible by overlaying the color effects of the whole ensemble of ferrocene/ferrocenium units at the polymer chains with those of the conjugated polymer showing redox activity (doping) as well. EDOT-Fc based polymers with pendant ferrocenyl groups can be switched between purple and blue, for example [33]. Different numbers of achievable colored states can be achieved by substitution of the ferrocenyl unit with other units leading the way to multicolored electrochromic materials. It is worth mentioning that, for practical considerations, incorporation of the ferrocenyl substituent enhances the optical contrast in the visible region resulting in better transparency for the oxidized state [34]. Besides the composition of such materials, the method of preparation also impacts the properties and durability. For instance, ferrocene-based conjugated polymers prepared by electrochemical polymerization on transparent indium tin oxide (ITO) electrodes using potential cycling are superior to those produced potentiostatically [35]. For polymer films obtained using the potentiodynamic technique the electroactivity is retained even after hundreds of cycles between their redox states associated with distinctive color changes between neutral and reduced/oxidized forms involving metal-to-ligand charge transfer (MLCT) and/or intraligand charge transfer (ILCT) from Fc moieties to the polymer skeleton.

A challenge in the field of electrochromic materials has been the realization of displays providing a full range of colors for applications in electronic books, billboards, shelf-edge labels, and charge gauges for batteries, to name just a few. A class of materials, which meets these requirements, is based on the inorganic polymer polyferrocenylsilane (PFS) as a cross-linked matrix in which silica spheres are embedded in an opal-like arrangement acting as a photonic structure [36]. Reversible expansion and contraction of the photonic lattice is achieved by expansion or contraction of the polymeric PFS matrix, which can be tuned by either chemical or electrochemical means. Thus, a solvent leads to the swelling of the matrix and ultimately a change in reflected color [37]. Similarly, the degree of swelling of the metallopolymer is dependent on the oxidation state of the metals in its backbone, providing a material whose structural color can be electrically tuned through (Figure 1.5). This material, for which the name "photonic ink" (P-ink) has been coined, enables a variety of optical functions that are attractive for display technology [36]. While the first demonstration of full-color tuning, based on the electroactive PFS polymer–gel/silica opal composite outlined above, was still limited by slow switching times and high drive voltage requirements, improved varieties have been developed [38]. Electrolyte permeation through the contiguous space-filling opal lattice embedded within the polymer–gel matrix was identified as a key parameter that impeded electron and charge transport via ion migration thus limiting the overall performance of the device. In variation of the same working principle, a high-performance

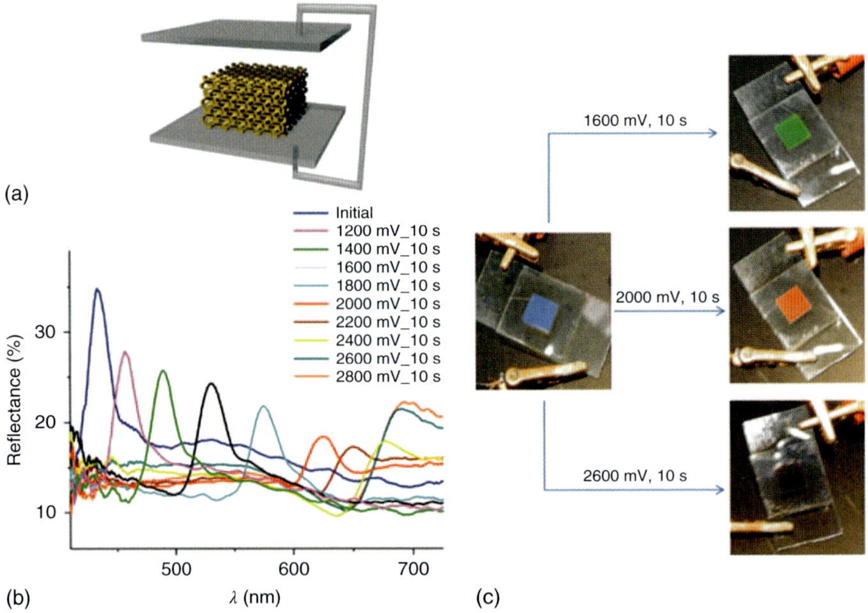

Figure 1.5 PFS-based P-ink at different charging levels. Source: Puzzo et al. 2009 [38]. Reproduced with permission of John Wiley & Sons.

electroactive photonic material has been obtained using an inverse polymer–gel opal in which the silica spheres are replaced by spherical voids through which the electrolyte freely infuses the nanoporous lattice.

The polymers chosen for the inverse polymer–gel opal are a blend of inorganic polymers derived from PFS, namely polyferrocenylmethylvinylsilane (PFMVS) and polyferrocenyldivinylsilane (PFDVS). A narrow molecular weight distribution (polydispersity, PDI < 1.1) is achieved through anionic ring-opening polymerization from the appropriate silaferrocenophanes where ring opening is favored over addition to the C=C double bonds present in each monomer unit leaving the latter intact for subsequent cross-linking. Based on this inverse opal approach full color tuning at very low drive voltages (~2 V) with unprecedented wavelength shifts traversing the ultraviolet, visible, and near infrared spectral ranges has been accomplished. Future technological hurdles that need to be overcome include enhanced reflectivity, reverse scan speed, and cycle lifetime. The PFS-based approach relies on changing the lattice dimension by an electrical stimulus. One might speculate that future developments could take advantage of changing the lattice dimension by an optical or magnetic stimulus or by contrast, continuous, reversible, and rapid altering of the refractive index by electrical, optical, or magnetic stimuli. Regardless of the physical concept, the potential for practical application will depend on the ability to prepare any such material with high structural and optical quality, and on a large scale and at low cost.

A metal-free electrochromic material is the phosphol-based phosphaviologen, which is the primary example of a phosphaviologen that can switch between colors for all three electrochemically accessible states [39]. Although this principle has been successfully implemented in a device, demonstrating the potential of these π-conjugated phosphorus species for applications such as dimming devices and multicolored displays, this technology is still in an experimental stage.

1.2.4 Applications Based on Supramolecular Recognition

Larger supramolecular entities consisting of distinguishable subunits are capable of forming molecular patterns that offer the potential to serve as information storage and release carriers similar to genetic encoding in biological systems. The complexity of current systems imposes limitations on the synthetic reproduction of sets of identical patterns. This complexity is, however, the key to applications such as anti-counterfeiting protection for product safety and consumer protection. For SMPs the use as switchable QR-code carriers has been successfully demonstrated [40]. The underlying principle is to encode an information (e.g. a QR-code), which may be stored and erased for single use validation or allows decoding only with special equipment for instance as safety certificate or tag. This switching process can further be used as a means for tracking the exposure to temperature or radiation. Current research is aiming at improving the contrast of the encoded features for instance via implementation of thermochromic pigments and the extension to data matrix codes, other codes and alphanumeric characters [41]. While the pioneering examples of SMP-based codes used organic polymers where the encoding process can be realized via locally constricted guest diffusion of surface layers of two structurally different SMPs, inorganic components offer an even larger structural variety, resulting, in turn, in improved storage capacity.

A breakthrough to the third dimension is the supramolecular assembly of multi-block copolymers containing a metallopolymer sequence such as PFS in addition to other block sequences as pioneered by Manners and coworkers [42]. These block-copolymers phase-separate on the nanoscale due to the immiscibility of the PFS block with a wide range of organic or inorganic co-blocks, allowing the formation of a variety of interesting nanostructured materials with crystalline iron-rich domains and various shapes and morphologies. Such polymers are accessible via sequential living anionic polymerization of several block units in a controlled fashion. Following this methodology, cylindrical micelles can be prepared that can act as nanoscale pixels with fully tunable color emissions when equipped with fluorescent tags [43]. Seeded growth of different ratios of block-copolymer unimers enabled the preparation of monodisperse co-assembled cylindrical micelles that emitted different colors depending on the amount of each fluorescent unimer added, for which the luminescence was uniform throughout the micelle (Figure 1.6).

Such systems provide new possibilities for the preparation of encoded nanomaterials, bearing a defined pattern that is invisible to standard methods but may be identified or decoded with other techniques. Ultimately, these materials may serve as tools for invisibly identifying valuable objects or conveying sensitive information. As yet, micelles with 18 distinct colors as well as non-centrosymmetric structures with up to seven blocks have been achieved. Based on these colors, this platform allows in principle to fabricate micelles

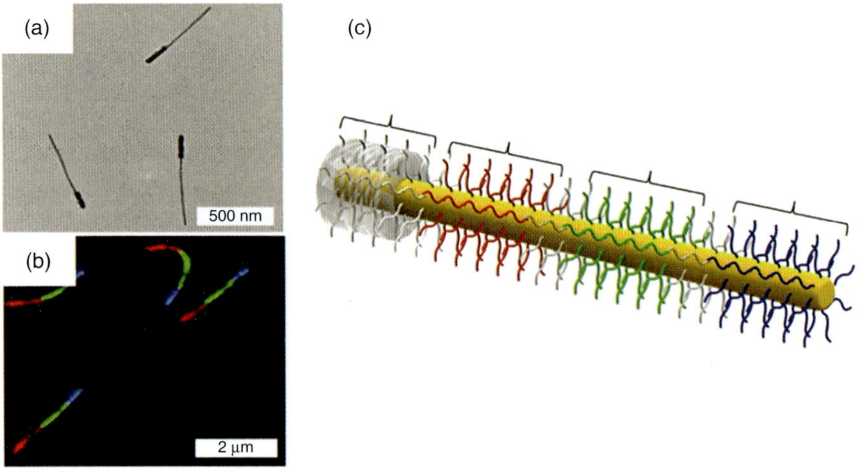

Figure 1.6 (a) TEM image, (b) laser scanning confocal microscopy (LSCM) image, and (c) schematic representation of non-centrosymmetric RGB "nanopixel" block comicelles. Scale bars are (a) 500 nm and (b) 2 μm. Source: Reproduced with permission of Hudson et al. 2014 [43]. Copyright 2014, Springer Nature.

with 18^7 or 6.1×10^8 combinations of fluorescent color. The number of combinations not only exceeds the data density of binary encoding, but facilitates the encoding of nanostructures with more complex patterns as well. A further advantage is the fact that such block comicelles are stable over long periods with no detected rearrangement of the fluorescent unimers along the length of the cylinders, which indicated that they exist as static, kinetically trapped structures.

1.2.5 Applications Based on Chemical Reactivity

The redox activity of metal-containing hybrid polymers has been employed to prepare responsive coatings that are able to switch between hydrophobic and hydrophilic behavior upon exposure to an external redox stimulus. Such switchable polarity can be achieved elegantly using ferrocene-based polymers. The application of this concept to surfaces dates back to Whitesides et al. who, two decades ago, reported the change of surface wettability of self-assembled monolayers of a ferrocene compound on a gold surface upon oxidation [44]. An instructive example is a study in which ferrocene polymers have been attached to a silicon-based surface allowing for switchable surface wettability (Scheme 1.1) [45]. In the case of PFcMA (poly(2-(methacryloyloxy)ethyl ferrocenecarboxylate) attached to a silicon substrate via surface-initiated atom transfer radical polymerization, a remarkable drop of the contact angle (CA) for water of approximately 70° could be observed upon oxidation. The reversible nature of switching the surface wettability has been demonstrated and different oxidants have been used to trigger the redox stimuli. Moreover, polymers with longer chain length show a more significant CA drop than those with shorter ones [45]. From an industrial point of view, the hydrophobization, either of fillers or other hydrophilic materials, is an important feature used in the building and

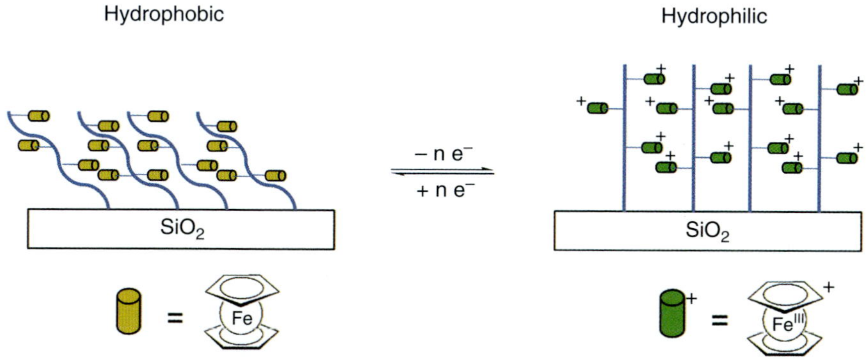

Scheme 1.1 Sketch of a surface with tunable wettability. Source: Pietschnig 2016 [32]. http://pubs.rsc.org/ru/content/articlehtml/2016/cs/c6cs00196c. Licensed under CC by 3.0.

construction industry for providing a water barrier for the construction of roads or buildings [46–49]. A switchable version of such behavior would be of great interest with respect to humidity management in smart buildings or in transport systems.

The redox activity of cross-linkable alkoxysilane PFS derivatives can be used to endow electrospun microfibers with properties that allow for electroactuation [50]. These materials display interesting strain-induced buckling behavior on electroactuation at low voltages (1.5–2.0 V). In contrast to fibers from organic polymers such as polyaniline (PANI) and polypyrrole (PPy), which require a polymer blend for electrospinning, inorganic hybrid polymers such as PFS do not have such limitations and may be spun as neat compounds. Compared with organic polyacrylonitrile (PAN) fibers that are not inherently redox active requiring large potentials to actuate electronically, PFS-based polymer fibers show electroactuation at low voltages (1.5–2.0 V). Future developments will aim at improving the mechanical strength of PFS yarns by combining hundreds of electrospun fibers in a rope-like manner while maintaining their rapid response times in order to be competitive with the existing bilayer actuators.

In general, highly charged polymer chains are prone to interact with biomolecules. Polyanionic heparin is a particularly important coating agent in the fabrication of medical materials in contact with blood due to its rather high charge density and specific inhibition of blood coagulation. In a similar vein, a polyanionic inorganic polymer with a polyphosphazene backbone has been demonstrated to act as adjuvant for intradermal immunization [51]. Current vaccine administration approaches rely strongly on immune enhancing additives (i.e. immunoadjuvants) to induce the desirable protective immune responses. Unfortunately, many of the currently used vaccine adjuvants such as the most commonly used alum are not compatible with intradermal delivery approaches owing to adverse side effects. By contrast poly[di(carboxylatophenoxy) phosphazene] (PCPP) is a potent adjuvant for intradermal immunization and its excellent immunomodulating potential has been demonstrated in multiple animal models with both viral and bacterial antigens. In addition, use of PCPP has progressed into clinical trials and PCPP formulated vaccines were found to be safe and immunogenic in humans.

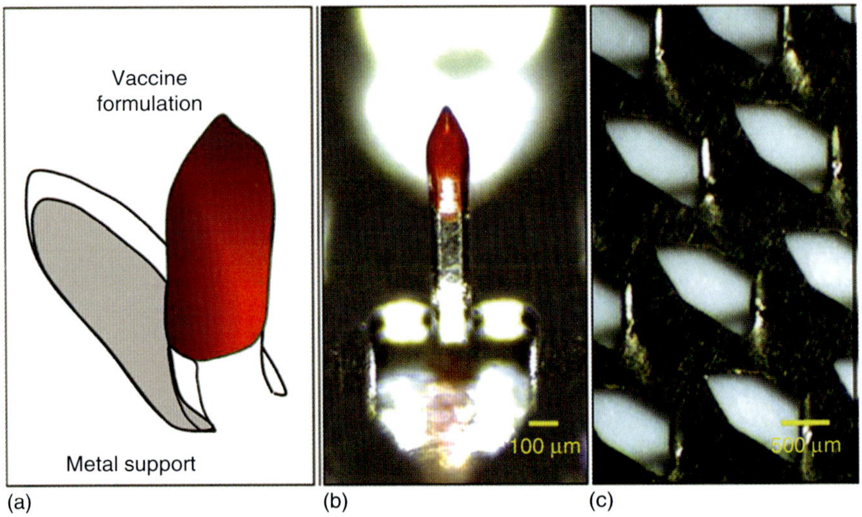

Figure 1.7 Microneedles and microneedle array with PCPP-coated tip (schematic representation (a), optical microscopy images of a coated microneedle (b), and an array of microneedles (c)). Source: Reproduced with permission of Andrianov et al. 2009 [51].

Its dual function as an immunoadjuvant and as a film forming/microfabrication material makes further macromolecular excipients obsolete. These smart properties have paved the way to PCPP-coated microneedles and microneedle arrays (Figure 1.7). Microneedles and microinjection devices can provide a convenient alternative potentially offering ease of application and distribution, with the possibility of self-administration and pain-free delivery of a vaccine. Further achievements in the field of biomedical applications that have evolved from closely related dendritic polymers are discussed in detail in Chapters 9 and 10 [52, 53].

1.2.6 Further Applications

Flame retardants are compounds that are present in consumer products, as for instance in clothing, toys, furniture, and seats in public transportation and also in construction materials. Flame retardants are in fact necessary additives if products are to meet legal fire-safety requirements. In recent years growing body of evidence has indicated that brominated flame-retardant chemicals used in upholstered furniture may, in some cases, be endocrine disrupters or have neurological and other health effects that make them unacceptable for use in daily life [54]. In line with these findings, since 2016, the *European Union* has prohibited the manufacture, placing on the market, and use of hexabromocyclododecane (HBCDD), an organic flame retardant used in a variety of consumer products. Therefore, finding replacements for these established brominated flame retardants is a surging field of investigation. Certain inorganic components and polymers were demonstrated to have great potential as replacements for the halogenated organic retardants and the detailed achievements in this area will be outlined in Chapter 8.

1.3 Perspective

Smart inorganic polymers are an emerging field of molecular materials with a bright perspective owing to their multifaceted properties. Other than for petrol-based organic polymers, the variety of fundamental building blocks is much larger for inorganic polymers, which helps avoiding limited resources in the production cycle. The nature of the elements involved is key to the versatile properties of this compound class. Based on their unique bonding situation, the energy barriers for torsional and rotational mobility of the backbone and the adjacent substituents are low, allowing increased flexibility with lower glass transition temperatures, higher gas permeation, and adaptive rheological properties as compared with their organic counterparts. The electronic features of heavier atoms endow such materials with optoelectronic properties relevant to LEDs, photovoltaic devices, applications such as flat-panel displays, and solid-state lighting. Charging and discharging of inorganic polymers can be used to switch polarity and wettability from hydrophobic to hydrophilic and vice versa. Such dynamic surface modification has potential for future applications in window screens, for transportation, and in the building sector. In a similar vein, dynamic switching of windows for privacy, security, or heat management is a strategy that has already entered the commercial market. For consumer safety and anti-counterfeiting purposes, molecular bar codes and encoded supramolecular entities have started to emerge on a high level of sophistication, and are almost ready for specialized applications. To summarize, smart inorganic polymers show high potential for numerous applications ranging from convenience and health care products to optoelectronic and information storage devices including security and safety purposes. Although the field has just started to emerge, a variety of applications has already passed the proof-of-principle state and future developments will help produce fascinating new materials based on smart inorganic polymers.

Acknowledgments

The author would like to acknowledge funding by the German Research Foundation (DFG) and COST Action CM1302 (SIPS): "European Network on Smart Inorganic Polymers."

References

1 Lendlein, A., Jiang, H., Jünger, O., and Langer, R. (2005). Light-induced shape-memory polymers. *Nature* 434: 879.
2 Chan, B.Q.Y., Liow, S.S., and Loh, X.J. (2016). Organic–inorganic shape memory thermoplastic polyurethane based on polycaprolactone and polydimethylsiloxane. *RSC Adv.* 6: 34946–34954.
3 Hartmann-Thompson, C. (2011). *Applications of Polyhedral Oligomeric Silsesquioxanes*, vol. 3. Heidelberg: Springer.

4 Hurkes, N., Bruhn, C., Belaj, F., and Pietschnig, R. (2014). Silanetriols as powerful starting materials for selective condensation to bulky POSS cages. *Organometallics* 33: 7299–7306.
5 Lickiss, P.D. and Rataboul, F. (2008). Fully condensed polyhedral oligosilsesquioxanes (POSS): from synthesis to application. *Adv. Organomet. Chem.* 57: 1–116.
6 Spirk, S., Nieger, M., Belaj, F., and Pietschnig, R. (2009). Formation and hydrogen bonding of a novel POSS-trisilanol. *Dalton Trans.* 163–167.
7 Frank, R., Ahrens, V.M., Boehnke, S. et al. (2015). Charge-compensated metallacarborane building blocks for conjugation with peptides. *ChemBioChem* 17: 308–317.
8 Nuñez, R., Romero, I., Teixidor, F., and Viñas, C. (2016). Icosahedral boron clusters: a perfect tool for the enhancement of polymer features. *Chem. Soc. Rev.* 45: 5147–5173.
9 Teixidor, F., Sillanpää, R., Pepiol, A. et al. (2015). Synthesis of globular precursors. *Chem. Eur. J.* 21: 12778–12786.
10 Jin, C., Jiang, Y., Niu, T., and Huang, J. (2012). Cellulose-based material with amphiphobicity to inhibit bacterial adhesion by surface modification. *J. Mater. Chem.* 22: 12562–12567.
11 Oh, M.-J., Lee, S.-Y., and Paik, K.-H. (2011). Preparation of hydrophobic self-assembled monolayers on paper surface with silanes. *J. Ind. Eng. Chem.* 17: 149–153.
12 Hu, Y., Liu, J., and Xu, C. (2011). Antifelting-coating for wool from $PPD[Si(OH)_3]_2$ agent via a sol–gel process. *Adv. Mater. Res.* 233–235: 151–154.
13 Savaides, A., Tasker, R., Hier, J., Raymond, D., and Russell, J.L. (2007). Method and composition for reducing malodor in permanently waved hair comprising hydrolyzed vegetable protein, polyquaternium-7, dimethicone and fragrance. Patent US20070190007A1, 1–9.
14 Yamaki, S. (2010). Hair dye pretreatment preparations containing amino acid derivatives and plant-originated protein hydrolyzates (Japan). Patent JP2010235572A, 10.
15 Dias, T.C., Baby, A.R., Kaneko, T.M., and Velasco, M.V.R. (2008). Protective effect of conditioning agents on afro-ethnic hair chemically treated with thioglycolate-based straightening emulsion. *J. Cosmet. Dermatol.* 7: 120–126.
16 Gama, R.M., Balogh, T.S., Franca, S. et al. (2011). Thermal analysis of hair treated with oxidative hair dye under influence of conditioners agents. *J. Therm. Anal. Calorim.* 106: 399–405.
17 Baumgartner, T. and Réau, R. (2006). Organophosphorus π-conjugated materials. *Chem. Rev.* 106: 4681–4727.
18 Caseri, W. (2016). Polystannanes: processible molecular metals with defined chemical structures. *Chem. Soc. Rev.* 45: 5187–5199.
19 Hissler, M., Dyer, P.W., and Reau, R. (2003). Linear organic π-conjugated systems featuring the heavy group 14 and 15 elements. *Coord. Chem. Rev.* 244: 1–44.
20 Hissler, M., Dyer, P.W., and Reau, R. (2005). The rise of organophosphorus derivatives in π-conjugated materials chemistry. *Top. Curr. Chem.* 250: 127–163.

21 Miller, R.D. and Michl, J. (1989). Polysilane high polymers. *Chem. Rev.* 89: 1359–1410.

22 Duffy, M.P., Delaunay, W., Bouit, P.A., and Hissler, M. (2016). π-Conjugated phospholes and their incorporation into devices: components with a great deal of potential. *Chem. Soc. Rev.* 45: 5296–5310.

23 Chen, J. and Cao, Y. (2007). Silole-containing polymers: chemistry and optoelectronic properties. *Macromol. Rapid Commun.* 28: 1714–1742.

24 He, X. and Baumgartner, T. (2013). Conjugated main-group polymers for optoelectronics. *RSC Adv.* 3: 11334–11350.

25 Pron, A., Berrouard, P., and Leclerc, M. (2012). Thieno[3,4-*c*]pyrrole-4,6-dione-based polymers for optoelectronic applications. *Macromol. Chem. Phys.* 214: 7–16.

26 Radebner, J., Eibel, A., Leypold, M. et al. (2017). Tetraacylgermanes: highly efficient photoinitiators for visible-light-induced free-radical polymerization. *Angew. Chem. Int. Ed.* 56: 3103–3107.

27 Radebner, J., Eibel, A., Leypold, M. et al. (2018). Tetraacylstannanes as long-wavelength visible-light photoinitiators with intriguing low toxicity. *Chem. Eur. J.* 24 (33): 8281–8285.

28 Baetens, R., Jelle, B.P., and Gustavsen, A. (2010). Properties, requirements and possibilities of smart windows for dynamic daylight and solar energy control in buildings: a state-of-the-art review. *Sol. Energy Mater. Sol. Cells* 94: 87–105.

29 Crandall, R.S. and Faughnan, B.W. (1976). Dynamics of coloration of amorphous electrochromic films of WO_3 at low voltages. *Appl. Phys. Lett.* 28: 95–97.

30 Faughnan, B.W., Crandall, R.S., and Lampert, M.A. (1975). Model for the bleaching of WO_3 electrochromic films by an electric field. *Appl. Phys. Lett.* 27: 275–277.

31 Li, Q., Jäkel, A., Viereck, V., and Hillmer, H. (2010). Design of self-assembling micromirror arrays for light guiding applications. *Microsyst. Technol.* 16: 895–899.

32 Pietschnig, R. (2016). Polymers with pendant ferrocenes. *Chem. Soc. Rev.* 45: 5216–5231.

33 Akpinar, H., Balan, A., Baran, D. et al. (2010). Donor–acceptor–donor type conjugated polymers for electrochromic applications: benzimidazole as the acceptor unit. *Polymer* 51: 6123–6131.

34 Özdemir, S., Balan, A., Baran, D. et al. (2011). Green to highly transmissive switching multicolored electrochromes: ferrocene pendant group effect on electrochromic properties. *React. Funct. Polym.* 71: 168–174.

35 Gulce, H., Yetkin, A., Akgul, E., and Gulce, A. (2013). A ferrocene functionalized polymer: poly[*N*-(ferrocenylmethyl)-*o*-phenylenediamine]. Electrochemical production and spectroelectrochemical investigation in acetonitrile medium. *Thin Solid Films* 545: 81–88.

36 Arsenault, A.C., Míguez, H., Kitaev, V. et al. (2003). A polychromic, fast response metallopolymer gel photonic crystal with solvent and redox tunability: a step towards photonic ink (P-ink). *Adv. Mater.* 15: 503–507.

37 Arsenault, A.C., Kitaev, V., Manners, I. et al. (2005). Vapor swellable colloidal photonic crystals with pressure tunability. *J. Mater. Chem.* 15: 133–138.

38 Puzzo, D.P., Arsenault, A.C., Manners, I., and Ozin, G.A. (2009). Electroactive inverse opal: a single material for all colors. *Angew. Chem. Int. Ed.* 48: 943–947.
39 Reus, C., Stolar, M., Vanderkley, J. et al. (2015). A convenient N-arylation route for electron-deficient pyridines: the case of π-extended electrochromic phosphaviologens. *J. Am. Chem. Soc.* 137: 11710–11717.
40 Pretsch, T., Ecker, M., Schildhauer, M., and Maskos, M. (2012). Switchable information carriers based on shape memory polymer. *J. Mater. Chem.* 22: 7757–7766.
41 Ecker, M. and Pretsch, T. (2014). Novel design approaches for multifunctional information carriers. *RSC Adv.* 4: 46680–46688.
42 Hailes, R.L.N., Oliver, A.M., Gwyther, J. et al. (2016). Polyferrocenylsilanes: synthesis, properties, and applications. *Chem. Soc. Rev.* 45: 5358–5407.
43 Hudson, Z.M., Lunn, D.J., Winnik, M.A., and Manners, I. (2014). Colour-tunable fluorescent multiblock micelles. *Nat. Commun.* 5: 3372.
44 Abbott, N.L. and Whitesides, G.M. (1994). Potential-dependent wetting of aqueous solutions on self-assembled monolayers formed from 15-(ferrocenylcarbonyl)pentadecanethiol on gold. *Langmuir* 10: 1493–1497.
45 Elbert, J., Gallei, M., Ruettiger, C. et al. (2013). Ferrocene polymers for switchable surface wettability. *Organometallics* 32: 5873–5878.
46 Friedel, M., Ljesic, S., Wassmer, C., and Zoelitz, S. (2012). Preparation of solutions comprising alkali metal propylsilanolates and orthosilicates as protective sealing agents for hydrophobization of artificial and natural building materials. Patent DE102011007137A1, 15.
47 Fukazu, S., and Murase, M. (2000). Organosiloxane-based primer sealing composition for building. Patent JP2000086990A, 4.
48 Guilbault, L.J., and Rider, C. (2012). Trialkoxysilane or alkyl siliconate waterproofing agents and acrylate ester polymer emulsion binders for stabilizing soil and road base materials. Patent US20120207551A1, 7.
49 Schildbach, D., Auer, D., Felix, K.-H., and Stepp, M. (2012). Water-soluble organo-siliconate powder used for hydrophobicing of mineral building materials. Patent WO2012022544A1, 68.
50 McDowell, J.J., Zacharia, N.S., Puzzo, D. et al. (2010). Electroactuation of alkoxysilane-functionalized polyferrocenylsilane microfibers. *J. Am. Chem. Soc.* 132: 3236–3237.
51 Andrianov, A.K., DeCollibus, D.P., Gillis, H.A. et al. (2009). Poly[di(carboxylatophenoxy)phosphazene] is a potent adjuvant for intradermal immunization. *Proc. Natl. Acad. Sci. U.S.A.* 106: 18936.
52 Caminade, A.M. (2016). Inorganic dendrimers: recent advances for catalysis, nanomaterials, and nanomedicine. *Chem. Soc. Rev.* 45: 5174–5186.
53 Mignani, S., El Brahmi, N., Eloy, L. et al. (2017). Anticancer copper(II) phosphorus dendrimers are potent proapoptotic bax activators. *Eur. J. Med. Chem.* 132: 142–156.
54 Schulz, W.G. (2012). Ablaze over furniture fires. *Chem. Eng. News* 90: 27–37.

2

Boron-Containing Polymers

2.1

Group 13–Group 15 Element Bonds Replacing Carbon–Carbon Bonds in Main Group Polyolefin Analogs

Anne Staubitz[1,2], *Jonas Hoffmann*[1,2,3], *and Philipp Gliese*[1,2]

[1] *University of Bremen, Institute for Organic and Analytical Chemistry, Leobener Straße 7, 28359 Bremen, Germany*
[2] *University of Bremen, MAPEX Center for Materials and Processes, Bibliothekstraße 1, 28359 Bremen, Germany*
[3] *UMR 6226 CNRS-Université de Rennes 1, Institut des Sciences Chimiques de Rennes, Campus de Beaulieu, 263 Avenue du Général Leclerc, Rennes 35042, France*

2.1.1 Introduction

The combination of one element from group 13 and another from group 15 leads to bonds that are isoelectronic to carbon–carbon bonds, which are the backbone of all polyolefins. However, the electronegativity differences between group 13–group 15 elements are either large (for BN, the difference in electronegativity is 1.0), or the bond is relatively long with a smaller difference in electronegativity (for BP, the difference in electronegativity is 0.1, but the bond length is 190–200 pm). This means that these bonds are either very polar or at least have some dipole character. The dipole moment affects the properties of any polymer in which these bonds are incorporated. The difference in polarity changes the reactivity of the hydrogen substituents on the elements B, N, and P as opposed to C and thus leads to completely different chemical properties for amine–boranes and phosphine–boranes [1], as well as the resulting polymers. This chapter describes how main chain BN and BP analogs of polyolefins can be prepared, the mechanisms that are invoked, and the polymer properties that result.

In addition to CC/BN or CC/BP single bond substitutions, a replacement of a C=C bond by a B=N group also significantly affects the polymer, if it occurs in the substituents in the side chain.[1] Of such compounds, only very few examples exist to date, which will be a second focus of this chapter.

The following sections discuss BN and BP polymers (Figure 2.1.1); mixed BN–BP oligomers and polymers have also been described computationally [2], but they are much less researched and are beyond the scope of this chapter.

1 B=P-containing polymers have not been reported to date.

Smart Inorganic Polymers: Synthesis, Properties, and Emerging Applications in Materials and Life Sciences, First Edition. Edited by Evamarie Hey-Hawkins and Muriel Hissler.

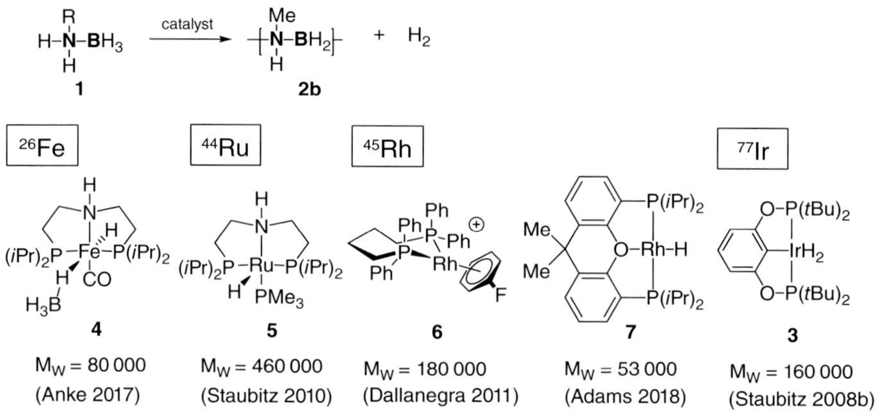

Figure 2.1.1 Types of group 13–group 15 analogs of polyolefins **1–3** that are discussed in this chapter.

2.1.2 Group 13–Group 15 Element-Containing Polyolefin Analogs with the Heteroatoms in the Main Chain

2.1.2.1 Poly(phosphinoboranes)

The first well-defined group 13–group 15 bond-containing polyolefins were the poly(phosphinoboranes) [3, 4]. The key for their preparation was the formation of the element–element bonds: while carbon-based polyolefins are typically prepared by addition to the double bond of olefinic monomers, double bonds to the element phosphorus are much less common and free phosphaborenes (compounds with P=B double bonds) are unknown.[2] Since no simple B=P analogs of olefins are available up to now, direct polymerizations by standard polyolefination methods, analogously to polymerizations with C=C monomers is therefore seen as impossible. A breakthrough was achieved only when the analogy was not searched for in olefins, but instead in the dehydrocoupling processes of silanes, R_2SiH_2, that could be transformed into polysilanes under the evolution of dihydrogen [7]. This thought was first implemented in 1999 by Manners and coworkers [3]. They performed the dehydropolymerization of *P*-phenylphosphine-borane (**4a**) with a rhodium(I) catalyst at 110 °C (Scheme 2.1.1). This reaction led to gas evolution of dihydrogen and a white material that proved to be the corresponding poly(*P*-phenylphosphinoborane) (**1a**). The ^{31}P NMR signal was very broad at $\delta = -48.9$ ppm and split into a doublet with a coupling constant of 360 Hz in the ^{31}P NMR spectrum. The ^{11}B NMR spectrum showed a similarly broadened characteristic signal at $\delta = -34.7$ ppm. The molecular weight was estimated to be around 31 kDa by static light scattering. It emerged that in order to obtain high molecular weights, it was essential to heat the reaction mixture without solvent.

2 However, it is possible to stabilize such motifs with Lewis bases [5] or metal complexes [6].

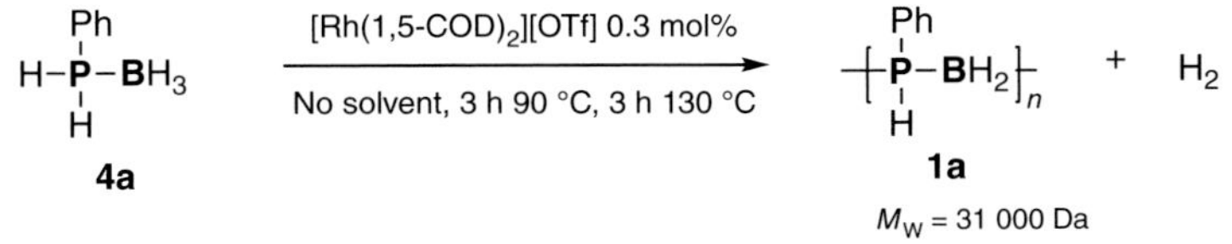

Scheme 2.1.1 Reaction of a rhodium(I) catalyst with a primary phosphine–borane **4a**, leading to the first high molecular weight poly(phosphinoborane) **1a** by dehydrocoupling. Source: Dorn et al. 1999 [3]. Reproduced with permission of John Wiley and Sons.

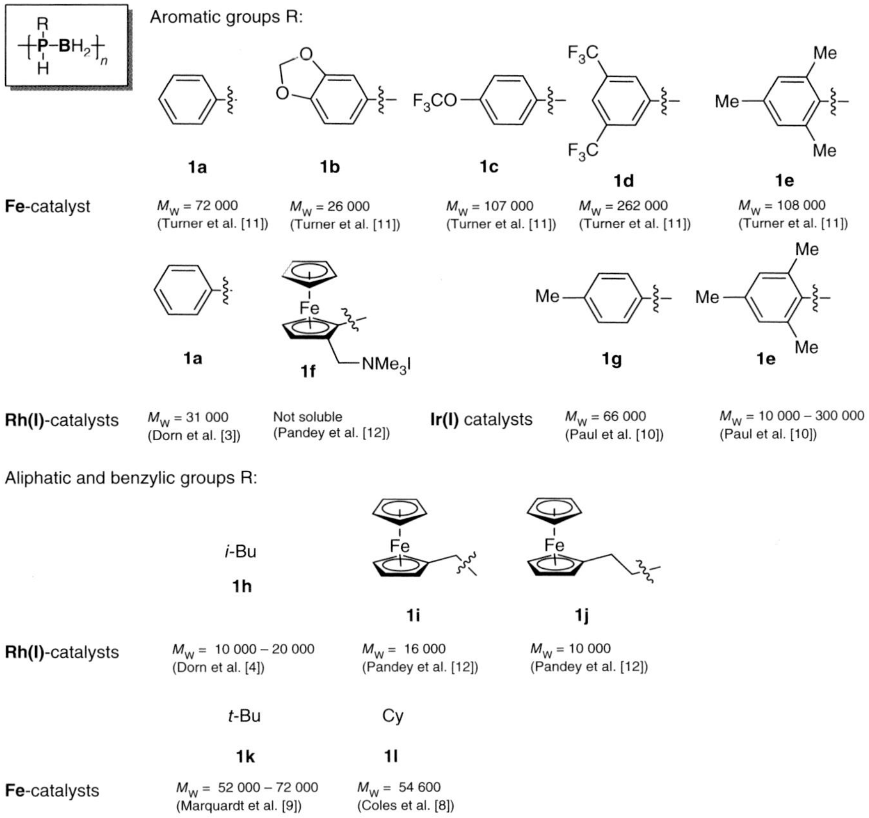

Figure 2.1.2 Summary of the poly(phosphinoboranes) **1a-l** known today. The molecular weights were determined either by light scattering or by conventional gel permeation chromatography (GPC). All molecular weights are given in Dalton.

After this initial result, other primary phosphine–boranes were polymerized, with an increasingly long list of active catalysts and polymerizable substrates (Figure 2.1.2).

2.1.2.1.1 Metal Complexes as Catalysts for the Dehydrocoupling of Phosphine–Boranes

As rhodium(I)-based catalysts were the first successful complexes for catalytic dehydrocoupling of phosphine–boranes, investigations into mechanistic pathways were centered on this metal [13–15]. These studies were experimentally difficult because for primary phosphine–boranes, intermediates could not be isolated so that secondary phosphine–boranes were used as substrates. However, these do not polymerize but dimerize to give linear or cyclic dimers. At present it is not possible to determine whether this is for steric or electronic reasons or both. Nevertheless, a plausible reaction mechanism has been suggested (Scheme 2.1.2). It assumes that the rhodium catalyst **5** coordinates the phosphine–borane **6** via two hydridic H-atoms on boron (species 7), followed by complexation of another equivalent of phosphine–borane **6** that undergoes P–H activation at the Rh center to give species **8**. This is followed by an isomerization reaction to give **9**, which is the direct precursor for the rate-determining step, which is the formation of dihydrogen, still complexed to the Rh(III) center in **10**. The release of dihydrogen from the complex is directly accompanied by B—P bond formation, but the actual turnover-limiting step appears to be the release of the linear dimer **12**, because the equilibrium of this step lies on **11**, which can be removed from the cycle by reacting to **13**. This may account for the fact that the catalysis is extremely sensitive to the substitution pattern on the phosphorus atom of the phosphine–borane.

In 2008, a piano-stool iron-based catalyst was found to dehydrocouple *P,P*-diphenylphosphine–borane (**6**) to its dimer **12** [16]. However, it was only in 2015 that iron-based catalysts made their entry for the dehydrocoupling of primary phosphine–boranes [17]. In this work, the iron-based precatalyst

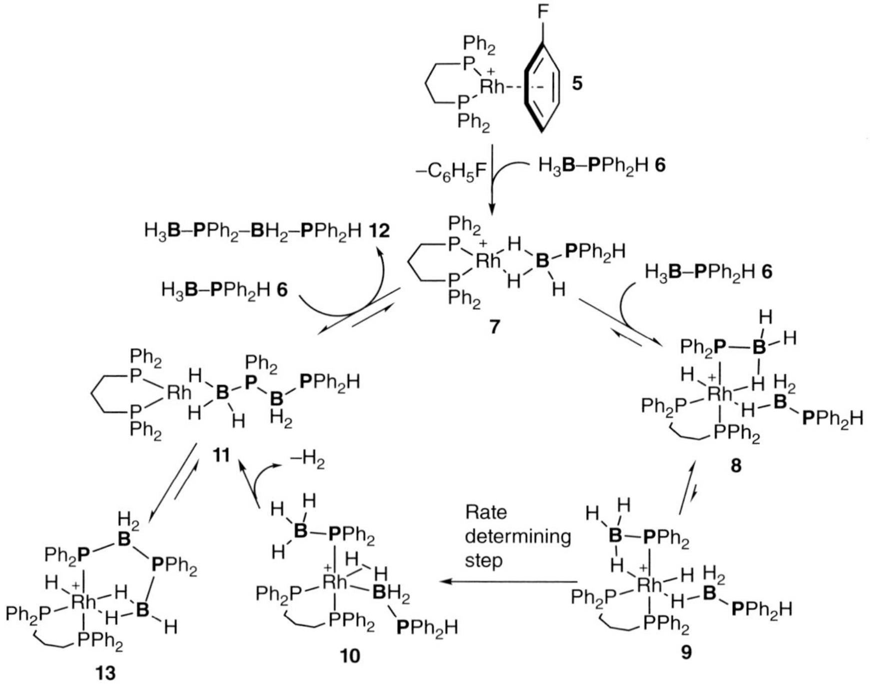

Scheme 2.1.2 Proposed mechanism for the formation of linear dimers from the secondary *P,P*-diphenylphosphine–borane (**6**). Counter ions are omitted for clarity. Source: Huertos and Weller 2013 [15]. Reproduced with permission of Royal Society of Chemistry.

[Cp(CO)$_2$FeOTf] **14** was used with *P*-phenylphosphine-borane (**4a**). Although the catalytic activity was not very high and a comparatively low molecular weight (M_w = 18 000 Da) was obtained, the study gave unique insights into the mechanism. It appears that the catalyst forms an intermediate phospido-metal complex **15**, which reacts with a further equivalent of **4a** to initiate the growing chain (Scheme 2.1.3).

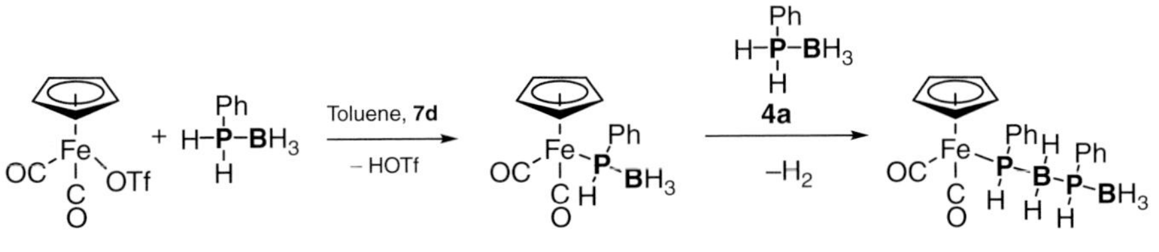

14 **4a** **15** **16**

Scheme 2.1.3 In situ formation of the active initiator **16** and addition of another monomer unit **4a**; **15** could be isolated and crystallized whereas **16** was assumed based on MS data and density functional theory (DFT) calculations. Source: Schäfer et al. 2015 [17]. Reproduced with permission of John Wiley and Sons.

Supportive evidence of the chain growth mechanism was the decreased chain length upon increased catalyst loadings [17]. A different iron-catalyzed dehydropolymerization was reported by Webster and coworkers [8]. They made use of a three-coordinated iron(II) β-diketimine and performed a dehydropolymerization on phenylphosphine–borane (**4a**). The resulting polymer was analyzed by gel permeation chromatography (GPC) and showed high molecular weight (M_w = 55 000 Da) and a polydispersity index (PDI) of 1.9. However, *P*-cyclohexylphosphine-borane (**4l**), was only poorly converted with a yield lower than 10% (M_w = 54 600 Da, PDI = 1.3).

2.1.2.1.2 Lewis Acid Promoted Dehydrocoupling of Phosphine–Boranes

Besides transition-metal-catalyzed dehydrogenation, a novel approach toward poly(phosphinoboranes) was possible by a catalytic dehydrocondensation reaction with the use of the strong Lewis acid B(C$_6$F$_5$)$_3$ as a catalyst [18] (Scheme 2.1.4).

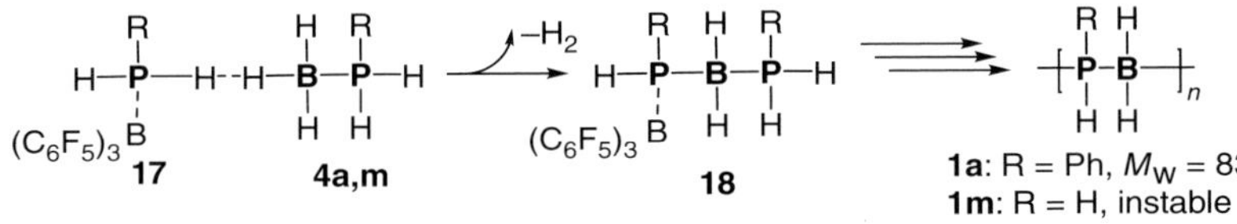

17 **4a,m** **18**

1a: R = Ph, M_W = 830–3900
1m: R = H, instable

Scheme 2.1.4 Preparation of poly(phosphinoboranes) **4a,m** via catalytic dehydrogenation reaction in the presence of a strong Lewis acid. Counter ions are omitted for clarity. Source: Denis et al. 2003 [18]. Reproduced with permission of Royal Society of Chemistry.

The strongly Lewis acidic B(C$_6$F$_5$)$_3$ supposedly exchanges with the borane part of the phosphine–borane to give **17**. This polarizes the P—H bond in **17** so strongly that the H atom becomes protic enough to react with the hydridic B—H of a phosphine–borane **4a,m** with loss of H$_2$ to give **18**. However, the molecular weight of the polymer **1a** was low. With this methodology, the first purely hydrogen-substituted poly(phosphinoborane) **1m** could be obtained. Unfortunately, this polymer was instable toward oxygen or moisture, precluding further analysis.

2.1.2.1.3 Lewis Base Promoted Dehydrocoupling of Phosphine–Boranes

Not only strong Lewis acids, but also Lewis bases can lead to polymerization: A Lewis base protected *P*-alkylphosphanylborane **19** underwent gentle dissociative thermolysis and liberated the highly reactive RPH=BH_2 **20** species, which then polymerized (Scheme 2.1.5) [9, 19]. This elegant concept of a masked monomer was later also employed for polyaminoboranes (see below) [20].

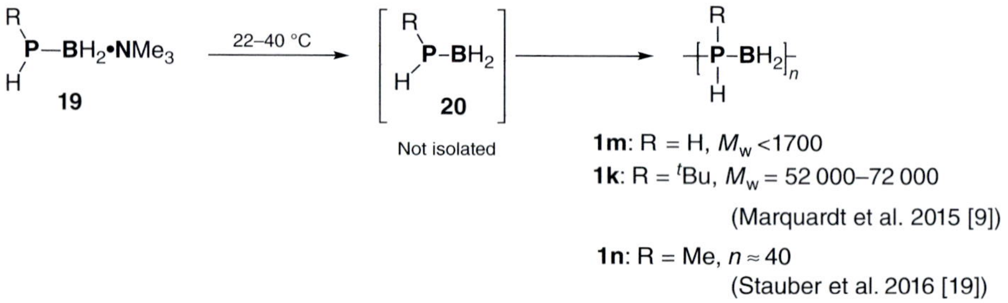

Scheme 2.1.5 Thermolysis of Lewis–base adducts of *P*-alkylphosphanylborane (**19**) via the phosphaborene **20** to give the poly(phosphinoboranes) **1k**, **1m**, and **1n**. All molecular weights are given in Dalton. Source: From Marquardt et al. 2015 [9] and Stauber et al. 2016 [19].

With sterically hindered monomers, this method led to high molar mass polymers **1k** (M_W = 52 000–72 000 Da) with low PDIs (1.6–1.9) (Figure 2.1.3) [9]. However, also the elusive poly(phosphinoborane) (**1m**) and poly(methylphosphinoborane) (**1n**), could be synthesized, which had not been possible by previous methods [19]. **1m** was largely insoluble, but the soluble fraction could be characterized by NMR, dynamic light scattering (DLS) and mass spectrometry. In this report, also the secondary *P*,*P*-diphenylphosphine-borane could be oligomerized with c. six repeating units. This novel metal-free synthetic

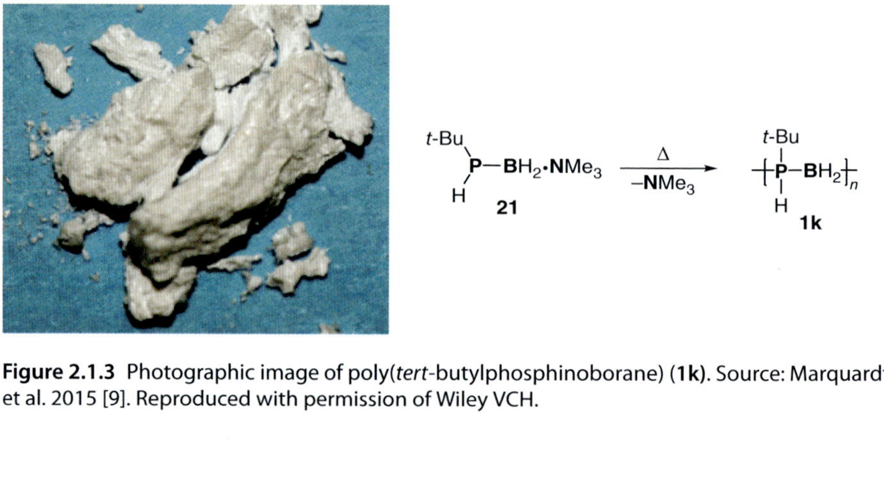

Figure 2.1.3 Photographic image of poly(*tert*-butylphosphinoborane) (**1k**). Source: Marquardt et al. 2015 [9]. Reproduced with permission of Wiley VCH.

route might therefore enable the preparation of various primary and secondary poly(phosphinoboranes).

2.1.2.1.4 Poly(phosphinoborane)-Based Materials

Although the unsubstituted poly(phosphinoborane) **1m** is instable toward oxygen and moisture and insoluble, some high molecular weight poly(*P*-alkyl-phosphinoborane)s can be handled under ambient conditions. This stability is a decisive feature for any real-life materials application (see further) as opposed to being merely of academic interest.

Poly(phosphinoboranes) are generally soft, pliable materials, with glass transitions (T_g) that can vary from $T_g = -68\,°C$ up to >135 °C. Interestingly, in the case of polymer **1h** ($T_g = 5\,°C$) the analogous organic polymer poly(4-methylpent-1-ene) exhibits a significantly higher glass transition temperature ($T_g = 45–50\,°C$). Also, for the $[(p\text{-}n\text{BuC}_6\text{H}_4)\text{PH-BH}_2]_n$ polymer the glass temperature is lower compared to the organic analog poly(*p*-*n*-butylstyrene) ($T_g = 8\,°C$ compared to $T_g = 33\,°C$) [23]. This might be explicable by the increased flexibility of the BP polymers compared to the CC analogs, due to the relatively longer BP-bonds (1.90–2.00 Å vs. 1.54 Å in polyolefins). Thermogravimetric analysis (TGA) revealed that upon heating, additional dihydrogen was removed from the polymeric network. Contrary to poly(aminoboranes), the observed ceramic yield was relatively high and could reach up to 80%, but this value was strongly affected by the nature of the substituent on phosphorus: aryl substituents gave higher ceramic yields than alkyl substituents.

A variation of the thermophysical properties can be achieved by varying the substituents on the phosphorus atom. The effect of changing an aryl group [11] or an alkyl group [24] on the phosphorus atom upon the T_g and decomposition were investigated. The variation of the aryl group in the polymers **1a–e** led to T_gs ranging from 29 °C (**1c**) to >135 °C (**1e**); long alkyl chains on the phosphorus atom resulted in a low T_g from −68 to −58 °C whereas neither the length of the polymer nor the length of the alkyl side chain appeared to impart a substantial impact on the T_g.

In some cases, poly(phosphinoboranes) were found to be swellable gels, which was attributed to intramolecular cross-linking of the polymer chains under release of hydrogen gas. This occurred for poly(phosphinoboranes) that were synthesized either by iron or ruthenium dehydrogenation reactions. The substituent on phosphorus appears to affect this property, as well as the catalyst used: it was assumed that the iron catalyst **14** generates less cross-linking [11].

2.1.2.1.5 Potenial Applications of Poly(phosphinoboranes)

Poly(phosphinoboranes) can be used for thin film patterning (Figure 2.1.4) [17].

The relative hydrophobicity of the polymers, which depends on the substituent was determined by the measurement of the respective contact angles (Figure 2.1.5) [11].

1a was compared with its organic analog polystyrene (PS), and contact angles of 70° vs. 87° were found for **1a** and PS, respectively. The lower contact angle for **1a** confirmed the more polar nature of polymer **1a**.

Figure 2.1.4 Use of poly(phenylphosphinoborane)s **1a** for soft lithography. (a) The polymer after isolation. (b)–(d) are SEM images of the prepared polymer used in soft lithography on silicon wafers. Source: Schäfer et al. 2015 [17]. Reproduced with permission of Wiley VCH.

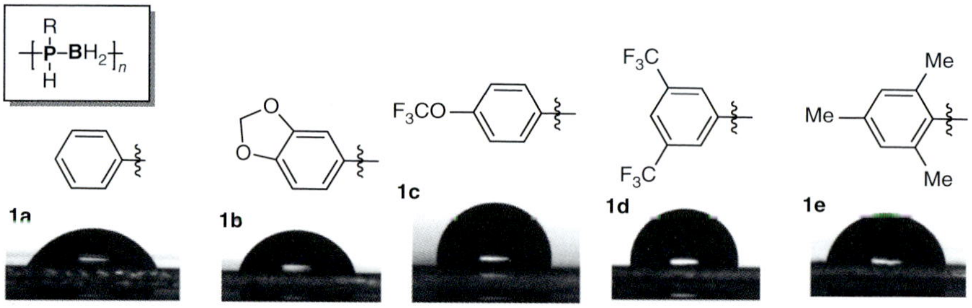

Figure 2.1.5 Water droplets on functional poly(phosphinoboranes). For **1c** and **1d** hydrophobic and for **1a**, **1b**, and **1e** a hydrophilic surface, indicated by their contact angles, were found. Source: Turner et al. 2017 [11]. Reproduced with permission of Wiley VCH.

In general, poly(phosphinoboranes) are considered as precursors for several phosphorus- or boron-containing materials. Poly(phosphinoboranes) can serve as preceramic polymers for the luminescent boron phosphide [23], which exhibits semiconducting properties [25]. Such polymers have also successfully been employed as electron beam resist for metal-free lithography [26]. Even properties such as flame-retardancy have been suggested for poly(phosphinoboranes) [9], but this has not yet been thoroughly investigated and remains an active area of research.

Besides applications that have already been investigated in practice, additional properties might emerge that thus far have only been investigated on a computational basis. From these computational studies, it appears likely that a nonlinear optical (NLO) response can be expected due to a high longitudinal dipolar first hyperpolarizability (β) [27]. However, due to its more conductive character, the dehydrogenated poly(phosphinoborene) $(R^1P{=}BR^2)_n$, as an analog of polyacetylene, would display a much higher potential for NLO applications than the hydrogenated poly(phosphinoborane), because of its predicted semiconducting nature. However, this material remains to be prepared in practice and is one of the main challenges in the field.

2.1.2.2 Poly(aminoboranes)

To access poly(aminoboranes) by catalytic dehydrocoupling routes may seem trivial at first because this route had been successfully applied to the related poly(phosphinoboranes) [3]. However, transferring this synthesis to poly(aminoboranes) remained a challenge for many years, because the amine–borane monomers, $RNH_2{-}BH_3$, did not only release just 1 equiv of H_2, but more, leading to intractable, presumably cross-linked materials. A breakthrough was achieved in 2008: Goldberg, Heinekey, and coworkers had reported an iridium-based pincer complex catalyst for the dehydrocoupling of ammonia-borane, $H_3N{-}BH_3$, which released only 1 equiv of H_2 [28]. The resulting product was insoluble. However, when Manners and coworkers then used the same catalyst with $MeNH_2{-}BH_3$, a soluble high molecular weight polymer was formed [29] (Scheme 2.1.6). Analysis by conventional GPC in tetrahydrofuran (THF) with 0.1 wt/wt% $n\mathrm{Bu_4NBr}$), calibrated against PS, revealed a polymer with a weight of $M_\mathrm{W} = 160\,000$ Da with a broad PDI of 2.9. However, the actual molecular weight may be somewhat below this value, because the polymer–solvent interactions for this solvent system would have been very different for the PS-standards and poly(*N*-methylaminoborane) (**2b**). In the $^{11}B\{^1H\}$ NMR spectrum, a broad peak at $\delta = -6.7$ ppm appeared that showed no peak splitting in the proton-coupled ^{11}B NMR spectrum.

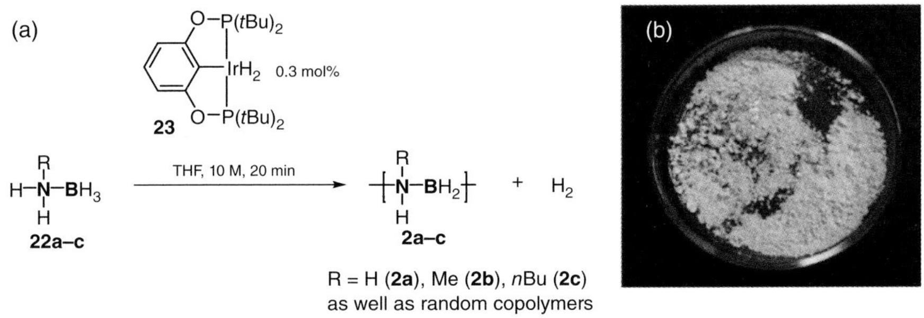

Scheme 2.1.6 (a) Catalytic dehydrocoupling of *N*-alkylamine-boranes **22a–c** leading to high molecular weight, well-defined polyaminoboranes; (b) photographic image of poly(*N*-methylaminoborane) (**2b**). Source: Staubitz et al. 2008 [29]. Reproduced with permission of Wiley VCH.

Following this lead, it emerged that there was actually a large number of catalysts that could be used for this transformation. Initially, they were found by screening of transition metal complexes that often serve as catalysts in

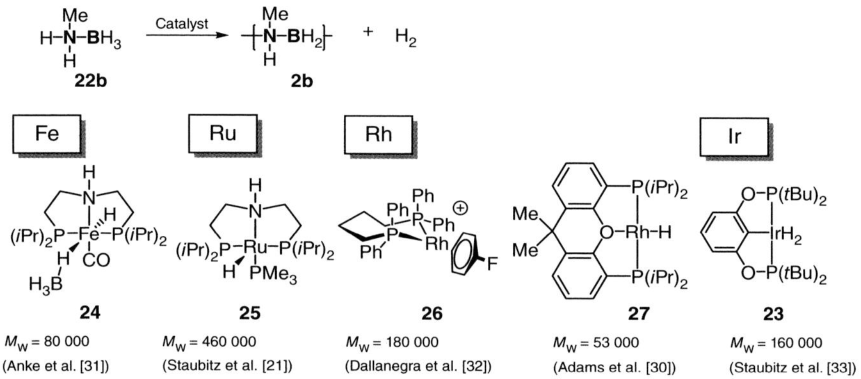

Figure 2.1.6 Catalysts **23–27** that have been successfully used for dehydrocoupling of *N*-methylamine-borane (**22b**). All molecular weights are given in Dalton.

other transformations [21]. This was augmented by other strategies in which mechanistic analogies with similar chemical reactions were used as a basis for catalyst identification and design. The catalysts that are known today are typically late transition metals (For examples see Figure 2.1.6) [21, 30–32], although early transition metal catalysts have been used for dehydrocoupling of secondary amine–boranes to give cyclic species.

To develop catalysts based on analogies with other reactions or to design highly active catalysts, the operating reaction mechanisms need to be understood in detail. For the dehydrocoupling of ammonia-borane, a number of mechanistic studies have emerged, both experimentally and computationally. Much fewer reports exist on the polymerization of *N*-alkyl-amine-boranes, but it is likely that analogies exist. One of the best understood catalytic polymerization reactions is the dehydrocoupling of *N*-methylamine-borane (**22b**) with the rhodium-based catalyst **27** [30] or the iridium-based catalyst **23** [21, 34]. The dehydrocoupling of the amine–borane cannot be easily generalized and is highly dependent on the metal center and the ligands. However, it seems to be a common denominator that $RNH{=}BH_2$ occurs as an intermediate in such polymerizations. While this species is instable and able to polymerize on its own (see the work by Alcaraz and coworkers further [20]), it is assumed that in the presence of a metal, the metal center can react as a Lewis acid, generating an aminoborohydride **28** that has a lone pair on nitrogen (Scheme 2.1.7a). This can then attack further $RNH{=}BH_2$ **22**, leading to a chain growth mechanism that is also observed by kinetic measurements. In addition, even for nonmetallic catalysts, a similar mechanism may operate as was demonstrated with a frustrated Lewis pair chain growth dehydro-oligomerization (Scheme 2.1.7b) [35]. In this case, the frustrated pair is generated in species **30** by separating the Lewis acidic and Lewis basic parts of the molecule. This arrangement can then heterolytically cleave the B—H bond in **22b**, leading to the zwitterionic **31**. The N—H bond in **31** will become relatively protic, resulting in a further B—H polarization of another molecule **22b**. Under evolution of H_2, the new N—B bond is formed to give **32** and so forth.

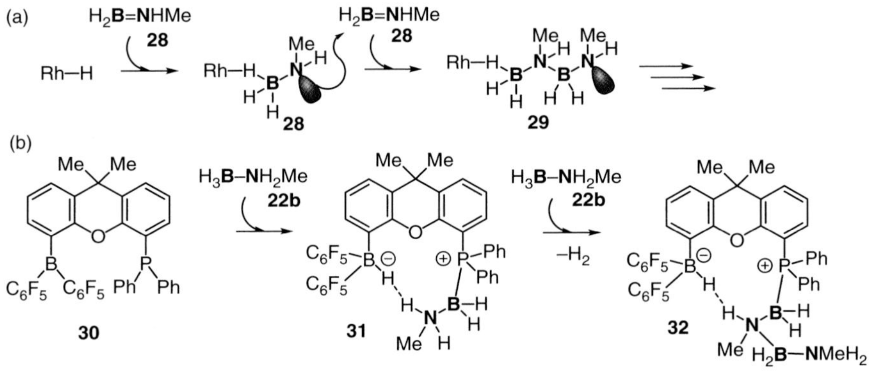

Scheme 2.1.7 Polymerization mechanisms of amino-boranes **22** on (a) metal centers [30] and (b) with frustrated Lewis pairs. Source: Mo et al. 2016 [35]. Reproduced with permission of American Chemical Society.

At the time of writing, all of the monomers used for this reaction were *N*-functionalized amine–boranes **22**. Such monomers can be accessed relatively easily: the free amine is dissolved in an appropriate solvent such as THF, dichloromethane (DCM), or ether and BH_3·THF is added. For *N*-alkylamineboranes that are not substituted at boron, the dative bond between the amine and the borane forms with a formation enthalpy of c. 30 kcal/mol, depending on the substituents on the amine [36]. High purities of the monomers are certainly important and therefore, the monomers are often sublimed or crystallized.

The catalytic dehydrocoupling route has been proved to be robust and versatile, but it is often intrinsically not possible to use monomers that are functionalized with carbon–carbon double bonds: in the reaction, hydrogen is liberated while at the same time, a transition metal catalyst is present, many of which can catalyze the addition of hydrogen to double bonds [37]. However, this issue can be entirely circumvented if the newly discovered polymer synthesis by amine–aminoborane exchange polymerization is followed (Scheme 2.1.8) [20]. In this reaction, the highly sterically hindered *N*,*N*-di-isoproylamino-borane (**33**) was used as a monomer precursor. If this compound is mixed with the amines **34a–f**, the amine exchanges against the iPr$_2$N moiety. It is formally oxidized whereas the iPr$_2$N moiety is reduced to the amine. The amino-boranes **37a–f**, which are thus generated in situ, are much less sterically hindered and can polymerize immediately to give high molecular weight polymers **2a–f**. Contrary to all catalyzed dehydrocoupling reactions of amine–boranes, in this reaction, no H_2 is liberated and therefore, double bonds are easily tolerated in this approach (see for example polymer **2f**).

Interestingly, in hindsight, there have been indications that such a process might be feasible [38]. In this work, a linear diborazane, MeNH_2–BH_2–NHMe–BH_3 **38**, was dehydrocoupled catalytically at 20 °C or thermally at 70 °C. The proposed reaction mechanism (Scheme 2.1.9) was a metal-free redistribution reaction to give amine–borane **22b** and amino-borane **37b**. The latter could then polymerize directly as a free monomer in solution as also assumed in the work

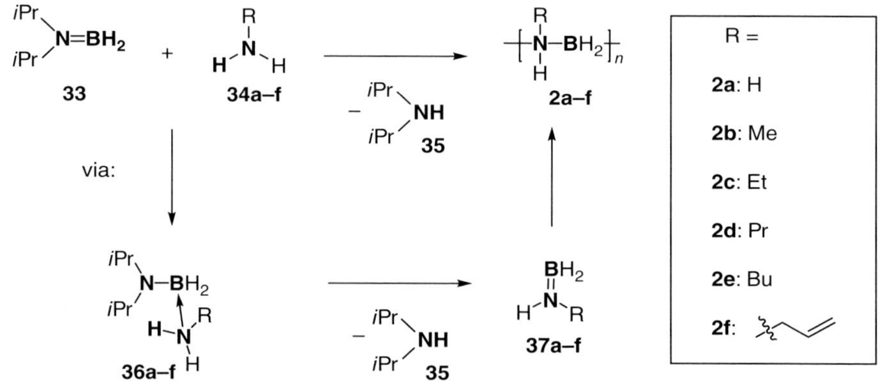

Scheme 2.1.8 Amine–aminoborane exchange polymerization for the generation of poly(aminoboranes) **2a–f**. Olefinic substituents on nitrogen are tolerated.

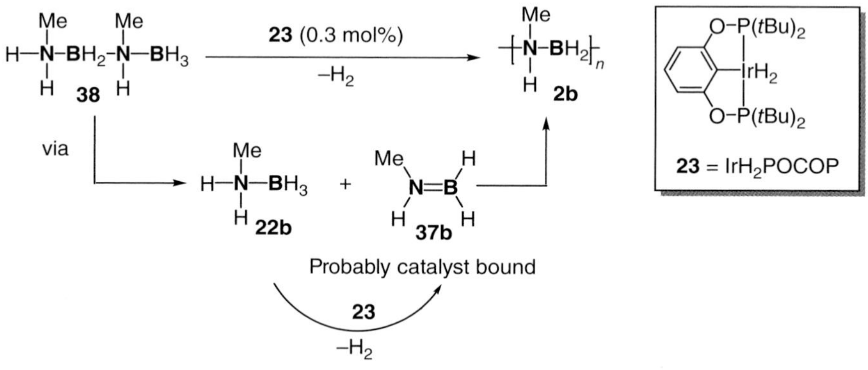

Scheme 2.1.9 Proposed polymerization mechanism of diborazanes. A redistribution reaction and catalytic dehydrocoupling leads to poly(*N*-methylaminoborane) (**2b**). Source: Robertson et al. 2011 [38]. Reproduced with permission of American Chemical Society.

by Alcaraz. The amine–borane **22b** on the other hand would be either reduced or directly polymerized with the catalyst, which was deemed more likely. The reaction temperature seems to play a deciding role; in both the work by Manners and coworkers [38] and Alcaraz and coworkers [20], high temperatures led to a large number of side products. This could be entirely avoided when low temperatures (−40 °C) were used; in that case, only the polymeric product was found [20].

The properties of poly(aminoboranes) differ significantly from the analogous polyolefins (for an overview see Table 2.1.1) [22]. Poly(*N*-methylaminoborane) (**2b**) is the best investigated poly(aminoborane), but even for this material, there are many characteristics that are still unknown [33]. Both poly(*N*-methylaminoborane) (**2b**) and the corresponding carbon analog, polypropylene (PP) are colorless powders, but in their physical appearance, this is almost the only similarity. Poly(*N*-methylaminoborane) (**2b**) is soluble in solvents with an intermediate polarity such as THF, DCM, chloroform, or dimethyl

Table 2.1.1 Comparison between selected properties of polypropylene as an all-carbon polyolefin and its BN analog poly(*N*-methylaminoborane) (**2b**).

Entry	Property	Polypropylene (PP)	Poly(*N*-methylaminoborane)
1	Structure	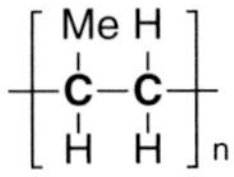	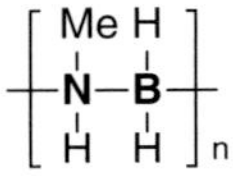
2	Solubility	Xylene (elevated *T*)	THF, CH_2Cl_2, $CHCl_3$, DMSO
3	Thermal decomposition	<400 °C	<160 °C
4	Appearance	Colorless powder	Colorless powder

Source: Staubitz 2018 [22]. Reproduced with permission of John Wiley and Sons.

sulfoxide (DMSO), but insoluble in very nonpolar solvents such as alkanes. Polypropylene, on the other hand, is only soluble in the nonpolar xylene, and even then only at elevated temperatures. This highlights the electronic difference between these isoelectronic polymers. Poly(*N*-methylaminoborane) (**2b**) can be detected by many mass-spectrometric technologies, ranging from nano-spray, electrospray-ionization, or matrix-assisted-laser-desorption-ionization (MALDI). What is unusual, however, is that the polymer can be detected both in positive and negative mode, which illustrates the ease with which both a hydridic H or a protic H may be removed from the polymer. Multiple ionizations can also occur [21].

In poly(*N*-methylaminoborane) (**2b**), there is still a hydrogen atom that is protic in nature (on nitrogen) and another that is hydridic in nature (on boron). Therefore, it may release a further equivalent of H_2. This explains the much decreased thermal stability of this polymer to some extent; however, TGA reveals that not only hydrogen is liberated upon heating but that larger fragments appear. Therefore, the decreased bond strength in the formally dative N → B bond is most likely another factor that decreases the comparatively low thermal stability compared to polypropylene that is stable up to 400 °C depending on the heating conditions.

Now, many avenues to poly(aminoboranes) are available, and in most cases these materials can be prepared on a gram scale. It can be explored how their properties can be exploited for future applications. One of the initial driving forces in this field was that poly(aminoboranes) may serve as preceramic polymers for boron nitride. Boron nitride exists in three modifications, from hexagonal boron nitride (α-BN), to cubic boron nitride (β-BN) and the hexagonal Wurzite boron nitride (γ-BN). In α-BN, planar BN layers are arranged similarly to graphite. This material is colorless, chemically very inert at temperatures below 700 °C. It is very soft and therefore can be used, for example, as a lubricant [39]. Preceramic polymers have the advantage that they can be processed into almost any desired shape before heating. To prevent shrinkage and crack formation during the ceramization process, it is essential that the ceramic yield (the mass percentage of the ceramic compared to the preceramic polymer) is high. In unsubstituted poly(aminoborane), $(NH_2{-}BH_2)_n$ **2a**, the maximal theoretical ceramic yield is 86%. This polymer was therefore analyzed thermogravimetrically

Figure 2.1.7 Pellet of poly(aminoborane) $(NH_2–BH_2)_n$ **2a** after (a) and before (b) pyrolysis. The Euro coin was shown to give a sense of the length scale of the samples. Source: Staubitz et al. 2008 [36]. Reproduced with permission of Wiley VCH.

and on a larger scale in a tube furnace. In the latter case, the material was pressed into a pellet and heated at 10 K/min, which resulted in a ceramic yield of only 36% at 900 °C (Figure 2.1.7) [33]. The polymer was white in appearance whereas the pyrolyzed materials were gray, and showed cracks. The relatively low ceramic yield is most likely a consequence of the formation of volatile $B_xN_yH_z$ molecules during heating. The remaining ceramic material was identified as α-BN by infrared (IR) spectroscopy and powder X-ray diffraction (XRD).

The most simple amine–borane, ammonia-borane, has extensively been discussed as a potential hydrogen storage material [1, 40]. As the poly(amino-boranes) can be prepared from amine–boranes by an oxidation reaction but are still capable of releasing an additional equivalent of H_2, it is conceivable that such polymers can serve as hydrogen storage materials. Although they may be discounted for the transportation sector, due to the too low hydrogen weight percentage, they may still serve in more specialized applications in which weight is not a problem, but the exact dosage of H_2 is relevant.

2.1.3 Group 13–Group 15 Element-Containing Polyolefin Analogs with the Heteroatoms in the Side Chain

PS and its analogs are thermoplastics, which are produced worldwide on an annual million-ton scale due to their facile processing chemical inertness, and robustness. They are widely used in everyday life in molded products e.g. foams, packaging, or sheet materials [41, 42]. Inorganic analogs of PS on the other hand are still very rare. Especially, analogs that are structurally very similar but electronically different may show very interesting but complementary properties. In this context, the formal replacement of C—C by B—N bonds in the aromatic ring are of special interest, since the average size of the bonds and the atoms are virtually the same, whereas a B-N bond introduces a high dipole moment.

2.1.3.1 Borazine-Containing PS Analogs

Synthetic efforts to prepare poly(vinylborazine) date back to the 1980s. Borazine **39** is the all BN analog of benzene; therefore, a logical approach to form the

BN analog of PS was to functionalize borazine with a vinyl group to form monomer **41** in a yield of 72% (Scheme 2.1.10). This was made possible by a rhodium-catalyzed hydroboration of acetylene [43]. The monomer polymerized under radical reaction conditions to form polymer **41** with a molecular weight of M_w = 18 000 Da (GPC and low angle laser light scattering [LALLS]/UV with PS standard) in a quantitative yield [43–45].

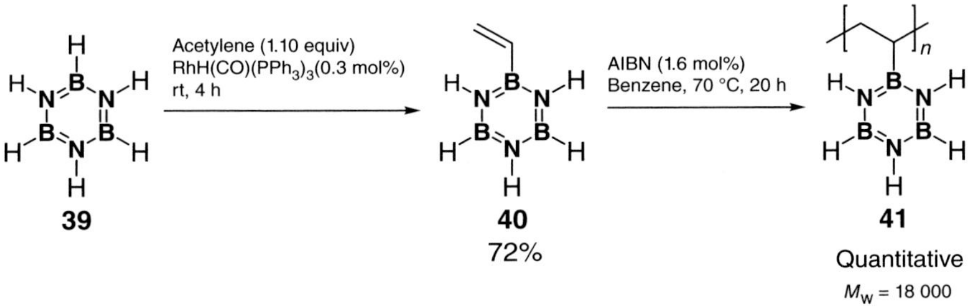

Scheme 2.1.10 Polymerization of *B*-vinyl borazine **40** and the synthesis from borazine **39**. Source: From Lynch and Sneddon 1987 [43], Lynch and Sneddon 1989 [44] and Paine and Sneddon 1994 [45].

The molecular weight distribution of poly(*B*-vinylborazine) **41** was skewed for higher molecular weights. This was interpreted as an indication of cross-linking or branching by the formation of BN linkages. In addition, a random copolymer with styrene was also obtained and analyzed. The composition ranged from 10 to 90 wt/wt% styrene with molecular weight ranging from M_w = 1000 to over 1 000 000 Da. The authors report that the high molecular weight polymers form as a result of branching, which starts at 23 wt% of the *B*-vinyl borazine **40** [45]

Polymer **41** has also been explored as a ceramic precursor.[3] By pyrolyzing it at 1000 °C, boron nitride formed basal planes with several unaligned units (turbostratic); however, the carbon content was relatively high unless a flow of ammonia was used during pyrolysis. Only heating such a pyrolyzed sample to even higher temperatures (1450 °C) under argon led to α-BN which was analyzed by powder XRD. The ceramic yield was reported with 73.6% under these conditions [44].

2.1.3.2 Azaborinine-Containing PS Analogs

Borazine is – despite formal and structural analogies to benzene – not aromatic. However, there are BN-containing heterocycles that show a large extent of aromaticity, the azaborinines. So far, only very few polymers-containing this motif in the side chain of a polyolefin have been reported (Figure 2.1.8).

In 2016, the first syntheses of azaborinine-containing PS analogs **3a** and **3b** were reported (Scheme 2.1.11) [46]. To prepare monomer **46**, the *B*-butoxy

3 The elemental analysis of this polymer suggested that the structure did not entirely have the structure as depicted but that additional borazinyl rings were incorporated. The elemental analysis was $B_3N_3C_{1.6}H_{7.5}$.

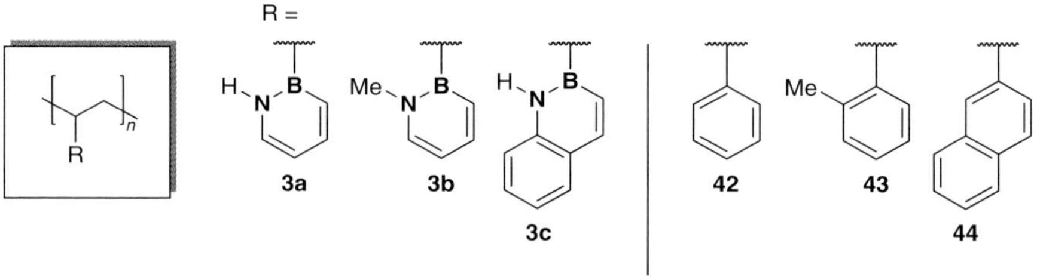

Figure 2.1.8 Polymers **3a–c** with BN motifs in the side chain and their all-carbon analogs **42–44**.

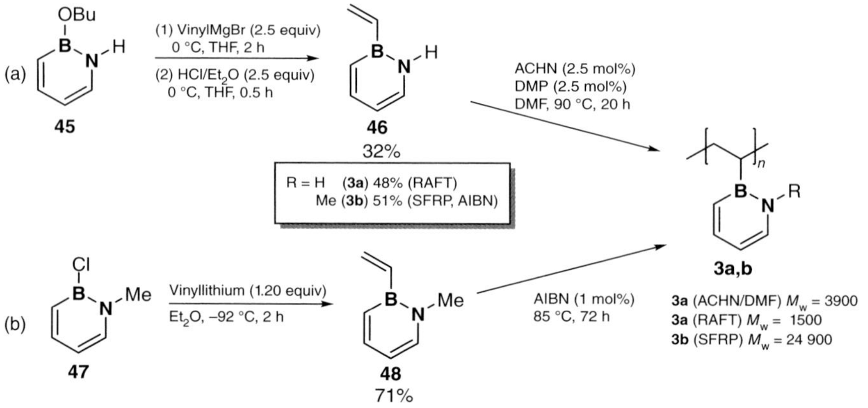

Scheme 2.1.11 Syntheses and polymerization of vinylazaborines **46** (a) and **48** (b) [46, 47]. All molecular weights are given in Dalton. Source: Thiedemann et al. 2017 [47]. Reproduced with permission of Royal Society of Chemistry.

substituted 1-hydro-2-vinyl-1,2-azaborinine (**45**) was vinylated at the boron atom in a nucleophilic substitution reaction.

The polymerization with azobis(cyclohexanecarbonitrile) (ACHN) in *N,N*-dimethylformamide (DMF) provided an oligomer with a M_w of 3900 Da and a PDI of 2.07 (GPC/PS standard). Encouraged by the results, the authors used a reversible addition–fragmentation chain transfer (RAFT) polymerization with ACHN as an initiator in DMF. It was demonstrated that the polymer **3a** could be prepared with a molecular weight of $M_w = 1500$ Da and a PDI of 1.38 (GPC/PS-standard). Moreover, **3a** was soluble in alcohols and could be dispersed in water.

Coinciding with this work, the polymerization of *B*-vinyl-*N*-methylazaborine (**48**) was demonstrated to give poly(*B*-vinyl-*N*-methylazaborine) (**3b**) (Scheme 2.1.12b) [47]. In this work, the monomer was obtained from *B*-chloro-*N*-methyl-1,2-azaborine (**47**) by vinylation and polymerized under solvent-free condition, using azobis(isobutyronitrile) (AIBN). Polymer **3b** showed a low PDI (1.51) and a high M_w (24 900 Da; GPC/PS standard) with a conversion of 51%. A comparison with the all carbon analog ($M_w = 86\,000$ da; PDI = 1.84; GPC/PS-standard; conversion: 87%) and the respective copolymer ($M_w = 36\,900$ Da; PDI = 1.69; GPC/PS-standard; conversion: 86%) showed that the radical

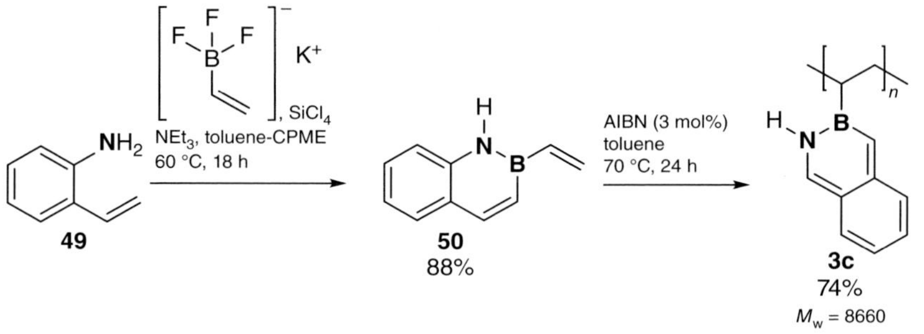

Scheme 2.1.12 Gram scale approach for the synthesis of a BN-naphthyl-substituted polyethylene derivative **50**. Source: van de Wouw et al. 2017 [48]. Reproduced with permission of Royal Society of Chemistry.

polymerization of **48** was much slower than *ortho*-methylstyrene with a decreased overall conversion.

In 2017, Klausen and coworkers demonstrated the first gram scale synthesis of 2-vinyl-1,2-dihydrobenzo[*e*][1,2]azaborinine (**50**) and their polymerization (Scheme 2.1.12) [48]. It was observed that 2-vinylaniline (**49**) could be cyclized with potassium vinyltrifluoroboate to give the monomer **50**. A radical polymerization with AIBN then led to polymer **3c** with a M_w of 8660 Da and a PDI of 1.43 (GPC/PS standard). Besides the homopolymer **3c**, it was possible to synthesize the copolymer with the respective CC analog.

2.1.4 Conclusion and Outlook

In the last decade, several synthetic breakthroughs have enabled the synthesis of poly(aminoboranes) and poly(phosphinoboranes). Catalytic methods have allowed to form well-defined polymers in high yields, high purity, and high molecular weight. In particular, for metal complex catalyzed reactions, our understanding of the mechanisms has helped progress in large strides and this has also led to a more thorough understanding of the electronic structure of the starting materials as well as the resulting polymers.

Nevertheless, large swathes of research remain underexplored or even entirely unexplored. For example, although there are numerous methods for living polymerizations for organic polyolefins, ranging from anionic to cationic to radical or coordination polymerization, there is not a single example of a controlled polymerization for poly(aminoboranes) or poly(phosphinoboranes). Likewise, while there are many methods for controlling the tacticity of polyolefins, in the inorganic main chain BN and BP-analogs, there has been no report of this and even the analysis of tacticity has not progressed yet beyond the observation of split peaks in NMR spectra and hypotheses.

At present, all BN and BP analogs of polyolefins are linear or assumed to be linear or cross-linked or ill-defined. Controlled access to more complex polymer architectures such as block copolymers, telechelic polymers, or brush-polymers have never been realized and offer enormous scope for scientific discovery.

In terms of chemical variety, we have only a glimpse of what might become possible. Catalytic methods have so far been restricted to very simple substituents such as alkyl groups; in some cases, aryl or even potentially redox-active aryl groups have been used. But it is only with the advent of more selective and efficient catalysts or by using the newly invented amine–aminoborane exchange polymerization reaction that other functional groups such as double bonds can be introduced in the side chains of such polymers. The latter is a fact that sets these inorganic analogs of polyolefins apart from their organic counterparts where this is intrinsically not possible as olefins are in fact the functional groups that polymerize.

Because the area of well-defined BN and BP analogs of polyolefins is still nascent, it is unsurprising that at the time of writing, no products based on these polymers have been commercialized. However, as they do show new and complementary properties to polyolefins and we are just at the beginning of exploiting these for applications, continued and increased research efforts in both chemistry and materials science of BN and BP polyolefin analogs are urgently required.

Acknowledgments

The authors have been supported by the Institutional Strategy of the University of Bremen, funded by the German Excellence Initiative. A. Staubitz acknowledges support from the German Research Foundation (DFG) for an Emmy-Noether-Fellowship (STA1195/2-1). J. Hoffmann acknowledges funding for a Short Term Scientific Mission from the COST action COST 1302 (SIPS): "European Network on Smart Inorganic Polymers." The authors thank the COST Action CM1302 (SIPs): "European Network on Smart Inorganic Polymers."

References

1 Staubitz, A., Robertson, A.P.M., Sloan, M.E., and Manners, I. (2010). Amine- and phosphine-borane adducts: new interest in old molecules. *Chem. Rev.* 110: 4023–4078.
2 Jacquemin, D. and Perpète, E.A. (2005). Ab initio assessment of the first hyperpolarizability of saturated and unsaturated polyaminoborane/polyphosphinoborane copolymers. *J. Phys. Chem. A* 109: 6380–6386.
3 Dorn, H., Singh, R.A., Massey, J.A. et al. (1999). Rhodium-catalyzed formation of phosphorus–boron bonds: synthesis of the first high molecular weight poly(phosphinoborane). *Angew. Chem. Int. Ed.* 38: 3321–3323.
4 Dorn, H., Singh, R.A., Massey, J.A. et al. (2000). Transition metal-catalyzed formation of phosphorus–boron bonds: a new route to phosphinoborane rings, chains, and macromolecules. *J. Am. Chem. Soc.* 122: 6669–6678.
5 Rivard, E., Merrill, W.A., Fettinger, J.C., and Power, P.P. (2006). A donor-stabilization strategy for the preparation of compounds featuring P:B and As:B double bonds. *Chem. Commun. (Cambridge, UK)* 3800–3802.

6 Kubo, K., Kawanaka, T., Tomioka, M., and Mizuta, T. (2012). Synthesis and crystal structures of P-iron-substituted phosphinoborane monomers. *Organometallics* 31: 2026–2034.
7 Corey, J.Y. (2004). Dehydrocoupling of hydrosilanes to polysilanes and silicon oligomers: a 30 year overview. *Adv. Organomet. Chem.* 51: 1–52.
8 Coles, N.T., Mahon, M.F., and Webster, R.L. (2017). Phosphine- and amine-borane dehydrocoupling using a three-coordinate iron(II) β-diketiminate precatalyst. *Organometallics* 36: 2262–2268.
9 Marquardt, C., Jurca, T., Schwan, K.-C. et al. (2015). Metal-free addition/head-to-tail polymerization of transient phosphinoboranes, $RPH{-}BH_2$: a route to poly(alkylphosphinoboranes). *Angew. Chem. Int. Ed.* 54: 13782–13786.
10 Paul, U.S.D., Braunschweig, H., and Radius, U. (2016). Iridium-catalysed dehydrocoupling of aryl phosphine–borane adducts: synthesis and characterisation of high molecular weight poly(phosphinoboranes). *Chem. Commun.* 52: 8573–8576.
11 Turner, J.R., Resendiz-Lara, D.A., Jurca, T. et al. (2017). Synthesis, characterization, and properties of poly(aryl)phosphinoboranes formed via iron-catalyzed dehydropolymerization. *Macromol. Chem. Phys.* 218: 1700120. (1700121–1700128).
12 Pandey, S., Lönnecke, P., and Hey-Hawkins, E. (2014). Phosphorus–boron-based polymers obtained by dehydrocoupling of ferrocenylphosphine–borane adducts. *Eur. J. Inorg. Chem.* 2014: 2456–2465. https://doi.org/10.1002/ejic.201402021.
13 Hooper, T.N., Huertos, M.A., Jurca, T. et al. (2014). Effect of the phosphine steric and electronic profile on the Rh-promoted dehydrocoupling of phosphine–boranes. *Inorg. Chem.* 53: 3716–3729.
14 Huertos, M.A. and Weller, A.S. (2012). Intermediates in the Rh-catalysed dehydrocoupling of phosphine–borane. *Chem. Commun.* 48: 7185–7187.
15 Huertos, M.A. and Weller, A.S. (2013). Revealing the P–B coupling event in the rhodium catalysed dehydrocoupling of phosphine–boranes $H_3B{-}PR_2H$ ($R = t$Bu, Ph). *Chem. Sci.* 4: 1881–1888.
16 Lee, K., Clark, T.J., Lough, A.J., and Manners, I. (2008). Synthesis and dehydrocoupling reactivity of iron and ruthenium phosphine–borane complexes. *Dalton Trans.* 2732–2740.
17 Schäfer, A., Jurca, T., Turner, J. et al. (2015). Iron-catalyzed dehydropolymerization: a convenient route to poly(phosphinoboranes) with molecular-weight control. *Angew. Chem. Int. Ed.* 54: 4836–4841.
18 Denis, J.-M., Forintos, H., Szelke, H. et al. (2003). $B(C_6F_5)_3$-catalyzed formation of B—P bonds by dehydrocoupling of phosphine–boranes. *Chem. Commun.* 54–55.
19 Stauber, A., Jurca, T., Marquardt, C. et al. (2016). A convenient route to monoalkyl-substituted phosphanylboranes ($HRP{-}BH_2{-}NMe_3$): prospective precursors to poly[(alkylphosphino)boranes]. *Eur. J. Inorg. Chem.* 2016: 2684–2687.

20 De Albuquerque Pinheiro, C.A., Roiland, C., Jehan, P., and Alcaraz, G. (2018). Solventless and metal-free synthesis of high-molecular-mass polyaminoboranes from diisopropylaminoborane and primary amines. *Angew. Chem. Int. Ed.* 57: 1519–1522.
21 Staubitz, A., Sloan, M.E., Robertson, A.P.M. et al. (2010). Catalytic dehydrocoupling/dehydrogenation of *N*-methylamine-borane and ammonia–borane: synthesis and characterization of high molecular weight polyaminoboranes. *J. Am. Chem. Soc.* 132: 13332–13345.
22 Staubitz, A. (2018). Generation of high-molecular-weight polymers with diverse substituents: an unusual metal-free synthesis of poly(aminoboranes). *Angew. Chem. Int. Ed.* 57: 5990–5992.
23 Dorn, H., Rodezno, J.M., Brunnhoefer, B. et al. (2003). Synthesis, characterization, and properties of the polyphosphinoboranes $[RPH{-}BH_2]_n$ (R = Ph, *i*Bu, *p*-*n*BuC_6H_4, *p*-dodecylC_6H_4): inorganic polymers with a phosphorus–boron backbone. *Macromolecules* 36: 291–297.
24 Cavaye, H., Clegg, F., Gould, P.J. et al. (2017). Primary alkylphosphine–borane polymers: synthesis, low glass transition temperature, and a predictive capability thereof. *Macromolecules* 50: 9239–9248.
25 Kumashiro, Y. (2011). Refractory semiconductor of boron phosphide. *J. Mater. Res.* 5: 2933–2947.
26 Clark Timothy, J., Rodezno José, M., Clendenning Scott, B. et al. (2005). Rhodium-catalyzed dehydrocoupling of fluorinated phosphine–borane adducts: synthesis, characterization, and properties of cyclic and polymeric phosphinoboranes with electron-withdrawing substituents at phosphorus. *Chem. Eur. J.* 11: 4526–4534.
27 Jacquemin, D. (2004). Theoretical study of dehydrogenation effects upon the first hyperpolarizability of polyphosphinoborane. *J. Phys. Chem. A* 108: 500–506.
28 Denney, M.C., Pons, V., Hebden, T.J. et al. (2006). Efficient catalysis of ammonia borane dehydrogenation. *J. Am. Chem. Soc.* 128: 12048–12049.
29 Staubitz, A., Soto, A.P., and Manners, I. (2008). Iridium-catalyzed dehydrocoupling of primary amine–borane adducts: a route to high molecular weight polyaminoboranes, boron–nitrogen analogues of polyolefins. *Angew. Chem. Int. Ed.* 47: 6212–6215.
30 Adams, G.M., Colebatch, A.L., Skornia, J.T. et al. (2018). Dehydropolymerization of $H_3B{\cdot}NMeH_2$ to form polyaminoboranes using [Rh(Xantphos–alkyl)] catalysts. *J. Am. Chem. Soc.* 140: 1481–1495.
31 Anke, F., Han, D., Klahn, M. et al. (2017). Formation of high-molecular weight polyaminoborane by Fe hydride catalyzed dehydrocoupling of methylamine borane. *Dalton Trans.* 46: 6843–6847.
32 Dallanegra, R., Robertson, A.P.M., Chaplin, A.B. et al. (2011). Tuning the $[L_2Rh{\cdots}H_3B{\cdot}NR_3]^+$ interaction using phosphine bite angle. Demonstration by the catalytic formation of polyaminoboranes. *Chem. Commun.* 47: 3763–3765.
33 Staubitz, A., Presa Soto, A., and Manners, I. (2008). Iridium-catalyzed dehydrocoupling of primary amine–borane adducts: a route to high molecular weight polyaminoboranes, boron–nitrogen analogues of polyolefins. *Angew. Chem. Int. Ed.* 120: 6308–6311.

34 Bhunya, S., Malakar, T., and Paul, A. (2014). Unfolding the crucial role of a nucleophile in Ziegler–Natta type Ir catalyzed polyaminoborane formation. *Chem. Commun. (Cambridge, UK)* 50: 5919–5922.

35 Mo, Z., Rit, A., Campos, J. et al. (2016). Catalytic B–N dehydrogenation using frustrated Lewis pairs: evidence for a chain-growth coupling mechanism. *J. Am. Chem. Soc.* 138: 3306–3309.

36 Staubitz, A., Besora, M., Harvey, J.N., and Manners, I. (2008). Computational analysis of amine–borane adducts as potential hydrogen storage materials with reversible hydrogen uptake. *Inorg. Chem.* 47: 5910–5918.

37 Sloan, M.E., Staubitz, A., Lee, K., and Manners, I. (2011). Scope and selectivity of heterogeneous Rh^0-catalyzed tandem dehydrocoupling/hydrogenation using $Me_2NH{\cdot}BH_3$ as a stoichiometric H_2 source. *Eur. J. Org. Chem.* 2011: 672–675. S672/671-S672/615.

38 Robertson, A.P.M., Leitao, E.M., and Manners, I. (2011). Catalytic redistribution and polymerization of diborazanes: unexpected observation of metal-free hydrogen transfer between aminoboranes and amine–boranes. *J. Am. Chem. Soc.* 133: 19322–19325.

39 Chen, B., Bi, Q., Yang, J. et al. (2008). Tribological properties of solid lubricants (graphite, h-BN) for Cu-based P/M friction composites. *Tribol. Int.* 41: 1145–1152.

40 Hamilton, C.W., Baker, R.T., Staubitz, A., and Manners, I. (2009). B–N compounds for chemical hydrogen storage. *Chem. Soc. Rev.* 38: 279–293.

41 PlasticsEurope (2017). Plastics – The Facts 2017.

42 Scheirs, J. and Priddy, D. (2003). *Modern Styrenic Polymers: Polystyrenes and Styrenic Copolymers*. Wiley.

43 Lynch, A.T. and Sneddon, L.G. (1987). Transition-metal-promoted reactions of boron hydrides. 10. Rhodium-catalyzed syntheses of *B*-alkenyiborazines. *J. Am. Chem. Soc.* 109: 5867–5868.

44 Lynch, A.T. and Sneddon, L.G. (1989). Transition-metal-promoted reactions of boron hydrides. 12.1 syntheses, polymerizations, and ceramic conversion reactions of *B*-alkenylborazines. *J. Am. Chem. Soc.* 111: 6201–6209.

45 Paine, R.T. and Sneddon, L.G. (1994). Recent developments in borazine-based polymers. In: *Inorganic and Organometallic Polymers II*, ACS Symposium Series, Chapter 27, vol. 572, 358–374.

46 Wan, W.-M., Baggett, A.W., Cheng, F. et al. (2016). Synthesis by free radical polymerization and properties of BN-polystyrene and BN-poly(vinylbiphenyl). *Chem. Commun.* 52: 13616–13619.

47 Thiedemann, B., Gliese, P.J., Hoffmann, J. et al. (2017). High molecular weight poly(*N*-methyl-*B*-vinylazaborine) – a semi-inorganic B–N polystyrene analogue. *Chem. Commun.* 53: 7258–7261.

48 van de Wouw, H.L., Lee, J.Y., and Klausen, R.S. (2017). Gram-scale free radical polymerization of an azaborine vinyl monomer. *Chem. Commun.* 53: 7262–7265.

2.2

Highlighting the Binding Behavior of Icosahedral Boron Clusters Incorporated into Polymers: Synthons, Polymers Preparation, and Relevant Properties

Clara Viñas[4], Rosario Núñez[4], Isabel Romero[5], and Francesc Teixidor[4]

[4] *Institut de Ciència de Materials de Barcelona (ICMAB-CSIC), Campus UAB, Bellaterra, 08193, Barcelona, Spain*
[5] *Universitat de Girona, Facultat de Ciències, Departament de Química, C/Maria Aurelia Campmany 69, Girona 17003, Spain*

2.2.1 Introduction

The preparation of carbon-based π-conjugated chains, is fundamental to the science and technology of conducting and photoluminescent polymers [1]. Extraordinary efforts have been deployed in the development of conducting organic polymers (COPs) and fluorescent π-conjugated polymers for practical applications, which require appropriate industrial processing techniques such as spin coating, layer-by-layer assembly, fiber spinning, and the use of printing technology. Accordingly, these materials, which are seen as playing a key role in future growth areas [2], have been investigated for use in rechargeable batteries [3], electrochromic displays [4], nonlinear optics [5], light emission [6], memory materials [7], antistatic materials [8], anticorrosive coatings [9], electro catalysis [10], sensors [11], electromechanical devices [12], and photovoltaics (solar cells) [13], and in applications directed to biomedicine including biosensors [14] and scaffolds for tissue engineering [15].

Icosahedral boron clusters display many specific characteristics that do not find a parallel in their organic counterparts. These are: unique stabilities and 3D geometrical properties of the isomeric *closo* $C_2B_{10}H_{12}$ carboranes and metallabisdicarbollides $[3,3'\text{-M}(1,2\text{-}C_2B_9H_{11})_2]^-$ (M = Co^{3+}, **1**; Fe^{3+}, **2**); unparalleled chemical inertness; both electron-withdrawing and electron-donating characteristics depending on whether the substituent is bonded at the carbon or the boron vertexes; and high molecular volume combined with an adaptable hydrophilicity. In addition, $[3,3'\text{-M}(1,2\text{-}C_2B_9H_{11})_2]^-$, which is amphiphilic [16], displays a low charge density because the negative charge is distributed among 45 atoms [17].

Carborane clusters are ligands of great interest in supramolecular chemistry because of their particular properties that may induce unconventional properties in the supramolecular structures in which they are inserted. Although boron clusters are genuinely considered as weak coordinating compounds, there are many examples where such a vision is a matter of dispute. Recently, a review on 1D coordination polymers (CPs) has been published in which icosahedral carborane clusters are the major participants that interact with main group metals,

Smart Inorganic Polymers: Synthesis, Properties, and Emerging Applications in Materials and Life Sciences, First Edition. Edited by Evamarie Hey-Hawkins and Muriel Hissler.

and d- and f-block elements [18]. This is complemented by another review by CE. Housecroft [19], also on carborane coordination polymers, which mentions that 1D polymeric structures with B—H···M interactions are required in their formation.

Four polymer categories that incorporate icosahedral boron clusters are presented in this chapter: COPs; fluorescent π-conjugated polymers, thermally resistant polymers, and coordination polymers.

2.2.2 Conducting Organic Polymers Containing Icosahedral Boron Clusters

COPs are also known as "synthetic metals" because one of their potential applications is as substitutes for metals or inorganic semiconductors. There are currently over 25 reported COPs classes with one of the most popular being polypyrrole (PPy) followed by polythiophene (PTh) and poly(3,4-ethylenedioxythiophene) (PEDOT). The most important aspect of COPs is their ability to act as electronic semiconductors controlled by redox switching at a specific potential supplemented by the movement of dopant ions into or out of the material depending on net polymer charge. Conducting and reducing states of COPs can be easily electrochemically generated from the corresponding monomers. During electropolymerization, anions from the solution are introduced into the polymer matrix to neutralize the positive charges generated in the oxidized polymer threads, maintaining the material's electroneutrality. These anions, which are called doping agents, play a significant role for the chemical and physical properties of the polymer both in its stability and morphology. Scheme 2.2.1 shows the electropolymerization process of the monomer EDOT to produce PEDOT(**Anion A**) along with the PEDOT polymer reduction. The COP's conducting state is the one requiring a doping agent and is produced by oxidizing the insulating material. The COPs chemical stability limit is known to be related to the over-oxidation resistance limit (ORL) of the material, beyond which the conducting material properties are definitely lost. COPs have been made more

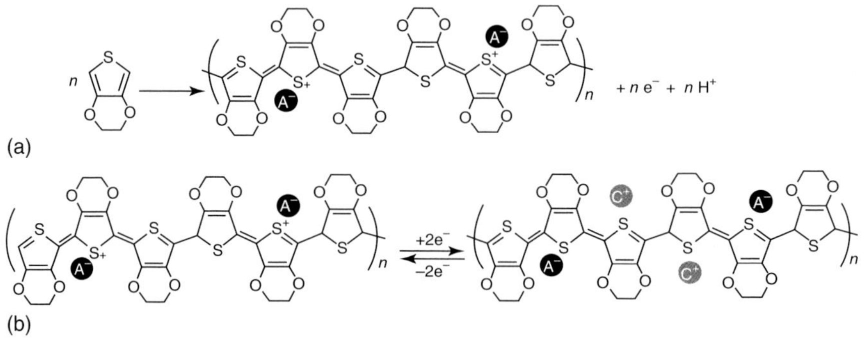

Scheme 2.2.1 (a) Electropolymerization of EDOT to form PEDOT(**A**). (b) PEDOT(**A**) oxidation/reduction reaction that occur via de-insertion and insertion of cations.

resistant to over-oxidation by doping with low nucleophilic, low electron density and low coordinating hydrophobic anions. Boron clusters, which possess these properties, have been incorporated into the COPs with the following objectives: (i) to play the role of the doping agent; (ii) to modify the chemical composition as well as the architecture of the COPs by bonding the 3D boron cluster doping agent at the branches of the polymer main chain; and (iii) to modify the chemical composition as well as the structure of the COPs by incorporating the 3D boron clusters into the polymer main chain.

2.2.2.1 Icosahedral Boron Clusters as Doping Agents in COPs

The electrochemical formation of conductive PPy films takes place by the anodic oxidation of pyrrole in CH_3CN at a moderate positive potential ($E_{1/2}$ = 1 to 1.3 V vs. Saturated Calomel Electrode [SCE]). The first described PPy that incorporated anionic icosahedral boron clusters as doping agent was reported in 2000 [20], when electrochemical polymerization of pyrrole (0.1 M), in the presence of 3.5×10^{-2} M in Cs[3,3′-Co(1,2-$C_2B_9H_{11}$)$_2$],Cs[**1**] with 1% (v/v) water, in CH_3CN was performed [20]. Later, the surface functionalization of a Pt electrode with PPy doped with Na[3,3′-Fe(1,2-$C_2B_9H_{11}$)$_2$] or with polyTHF derivatives (**2**–**4**) (Chart 2.2.1) by electro polymerization of pyrrole was reported [21].

The preparation of PPy materials doped with aryl, **5**–**8**, polyether **9**–**10** derivatives of **1** [22], as well as with the anionic boranes **11**–**15** (Chart 2.2.1) was

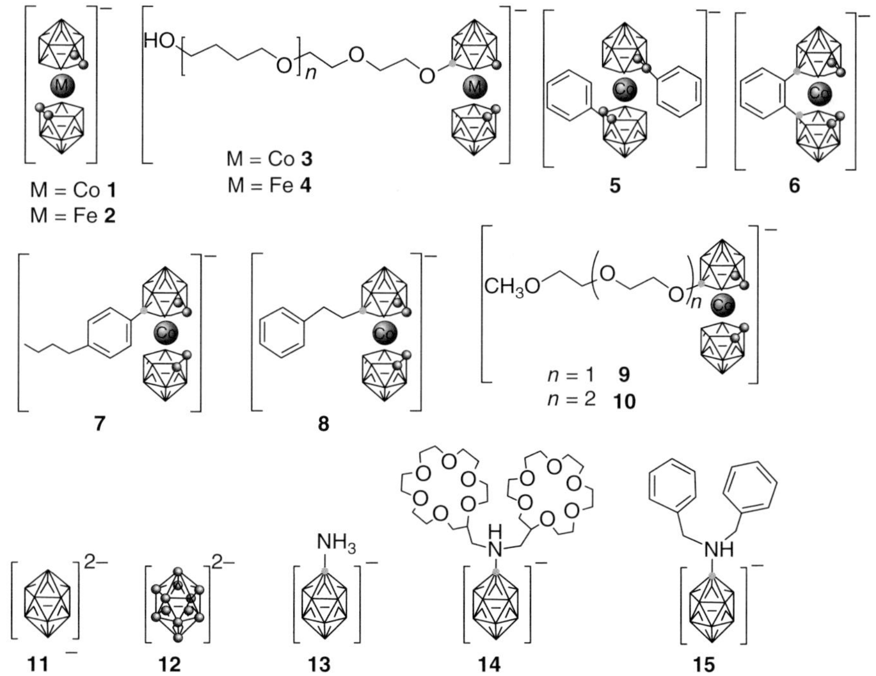

Chart 2.2.1 Schematic structure of anionic boron clusters 1–15 doping agents.

Figure 2.2.1 SEM images of electropolymerized PPy films: (a) PPy(**6**), (b) PPy(**10**), and (c) PPy($\mathbf{ClO_4}$).

accomplished to study the influence of the different icosahedral boron cluster on the PPy film growth and its properties [22]. Different morphologies (Figure 2.2.1) were observed by scanning electron microscopy (SEM) between PPy membranes doped with **1**, PPy(**1**), and common anion PPy($\mathbf{ClO_4}$) and PPy($\mathbf{PF_6}$) and the ORL and thermal resistance of these films were studied by linear sweep voltammetry (LSV) [20]. A higher ORL value of PPy (**1**) material, approx. 300 mV, as compared to commonly used doping anions, (**DBS**) and ($\mathbf{PF_6}$), was obtained.

Among COPs, PEDOT polymers are the ones with the highest prospects in the field of bioelectronics due to their outstanding conductivity, stability, flexibility, transparency, easy processability, and biocompatibility. Despite the exceptional physicochemical applications of doped PEDOT polymers as COPs, only one example on PEDOT doped with nonbonding icosahedral boron clusters has been reported [23]. The PEDOT(**1**) membrane was prepared by electropolymerization of monomer EDOT on indium tin oxide (ITO) from Cs[**1**] in CH_3CN [23]. The reversible oxidation/reduction process of PEDOT(**1**) took place by cation exchange due to the non-extrudable character of the metallabisdicarbollide $[\mathbf{1}]^-$doping anion.

2.2.2.2 Icosahedral Boron Clusters in COPs Side Chains to Modify the Chemical Composition and Act as Doping Agent

The first step in the preparation of these COP films is to synthesize the modified pyrrole and thiophene unit that possess the boron cluster covalently linked at one or both β carbon atoms of the ring, 3-, 3,4-, or at the N-vertex of the pyrrole or thiophene synthon. The reported pyrrole and thiophene derivatives that contain neutral *closo*- (synthons **16–23**) or anionic *nido*-carborane (synthons **24–25**) clusters [24] as well as pyrrole rings covalently bonded at the N- by a poly(ethyleneglycol) (PEG) chain to the anionic **1** (synthons **26–27**) [25], are schematized in Chart 2.2.2. The PPy(carboranyl)($\mathbf{PF_6}$) and PTh(carboranyl)($\mathbf{PF_6}$) films were obtained with these synthons by anodic oxidation of the corresponding substituted pyrrol or thiophene monomer in 10^{-1} M $[NBu_4][PF_6]$ in anhydrous CH_3CN [26]. All the PPy(carboranyl)($\mathbf{PF_6}$) and PTh(carboranyl)($\mathbf{PF_6}$) exhibit strongly increased ORL.

The polymerization of the anionic monomer (**26**) as well as its copolymerization with pyrrole units led to the formation of a new group of self-doped

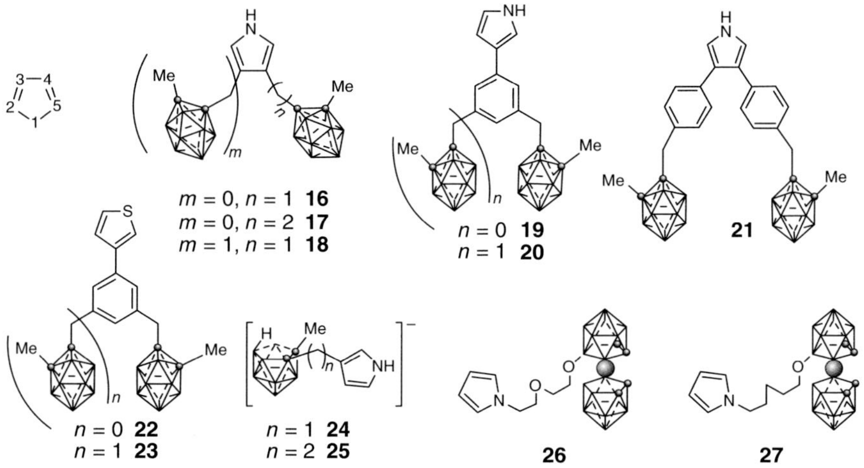

Chart 2.2.2 Pyrrole and thiophene synthons containing carboranyl clusters.

polymers, PPy(**26**) and PPy(**27**), as well as copolymer PPy(**27**):PPy [17], improving the already outstanding properties previously described by PPy(**1**) [20].

2.2.2.3 Icosahedral Boron Clusters Incorporated into the Polymer Main Chain of the COPs

Only PTh films that incorporate boron clusters in the main chain, have been reported to date [27]. The first step to achieve the preparation of these novel COPs is the synthesis of the starting thiophene monomers (**28**–**33**, Chart 2.2.3) that have one or the two α ring carbons directly bonded to the carbon of the boron cluster (C_c). The PThs28($\mathbf{PF_6}$)–PThs32($\mathbf{PF_6}$) COPs that contain the carborane clusters in the polymer backbone were produced by electropolymerization. These polymers show high thermal and electrochemical stabilities compared with the parent PTh($\mathbf{PF_6}$) [27].

In an attempt to promote a cooperative π–π intermolecular dimerization with the prospect of enhanced conductivity that stabilizes the radical cations formed during the electropolimerization process, electroactive carboranyl-functionalized thiophene oligomers (synthons **34**–**35**) have been synthesized (Chart 2.2.3) and their PTh prepared. The electrochemical

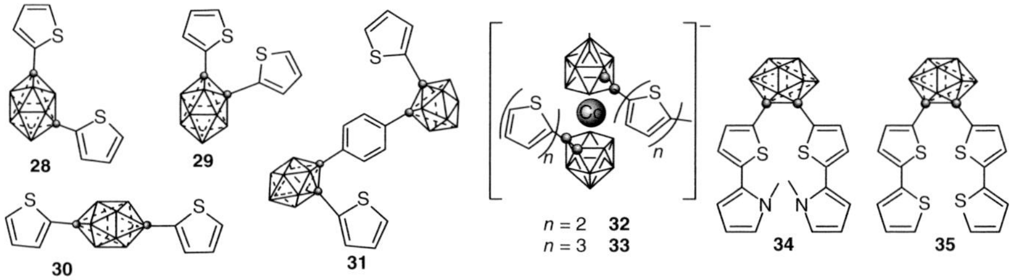

Chart 2.2.3 Thiophene synthons directly bonded to the carbon of the boron cluster.

characterization of these PPy(**34**) and PTh(**35**) polymers determined a low doping level with enhanced chemical and thermal stabilities, which support them as promising candidates for the development of carborane-functionalized conducting materials [27d].

2.2.3 Fluorescent Carborane-Containing Polymers

In π-conjugated polymers, the extended conjugation through aromatic moieties can cause a tendency to bundle and π-stack, which influences their processability and stability. This can pose a problem for their utilization in optoelectronic devices. The incorporation of carborane clusters into the polymers aids in conferring solubility and stability, while also altering their luminescent properties.

In 2009, Carter and coworkers reported the synthesis of *o*- and *p*-carborane-containing polyfluorene polymers from **36**, **37**, and **38** as the starting synthons (Scheme 2.2.2) [28, 29]. Polymer **39** was prepared using a Yamamoto type Ni(0) dehalogenative reaction in the presence of [Ni(COD)$_2$], whereas polymer **40** was prepared via a Stille coupling reaction (Scheme 2.2.2) [30]. Another type of polyfluorene polymers bearing pendant silylcarboranes, **41**, was prepared by homopolymerization or copolymerization of the monomer **38** by using similar microwave-assisted Yamamoto coupling reaction [31].

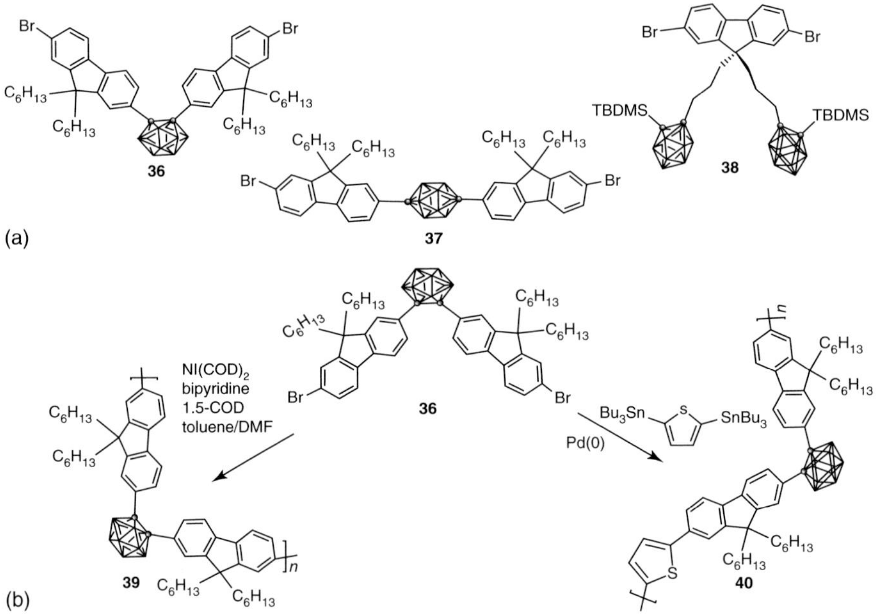

Scheme 2.2.2 (a) Main synthons to prepare carboranyl-containing polyfluorenes (**36–38**). (b) Synthesis of two different carboranyl-containing polyfluorenes (**39–40**).

Regarding the photoluminescence properties of these polymers, the polyfluorene polymer bearing *p*-carborane exhibited blue emission at λ = 400 nm in

$CHCl_3$, whereas its homolog with *o*-carborane **39** showed a maximum emission at 423 nm, an additional band around 565 nm, suggesting that the energy absorbed by the fluorene units is transferred to the *o*-carborane and emitted in the orange region. Polymer **40** showed blueshifted emission in the solid state with a maximum at 570 nm compared to the $CHCl_3$ solution (602 nm), which was attributed to forced aggregation of adjacent fluorene segments. Polymer **39** displayed a green emission centered at 520 nm in the solid state. This seemed to be caused by excimer formation due to aggregation. Emission spectra of the thin films of polymers prepared from monomer **38** before and after annealing for three hours at 160 °C indicated that the fluorescence emission maximum (535 nm) was blueshifted with respect to conventional fluorenes, with a decrease in intensity when increasing the carborane content.

Independently of this research, Chujo and coworkers reported the synthesis of linear π-conjugate aryl-ethynyl-carborane polymers. The reaction of synthons **41**, **42** [32–34], or **43** [35] (Scheme 2.2.3a), with different *p*-phenylene–ethynylene by a Sonogashira–Hagihara coupling polycondensation reaction in tetrahydrofuran (THF) and triethylamine solutions, led to the formation of the corresponding polymers (i.e. polymers **45** in Scheme 2.2.3b), in which the carboranyl units are linked to the aryl-ethynyl backbone through the C_c. Following the same type of reaction, the monomer 9,12-bis-(*p*-phenylene–ethynylene)-*o*-carborane (**44**) reacted with the corresponding di-iodophenyl derivative, to give a polymer in which the *o*-carborane is linked to the *p*-phenylene–ethynylene units through the B_c atoms (Scheme 2.2.3b) [36]. The same authors prepared a set of polymers (**46**, which is not shown in the Scheme) with different M_n (4600–3200) by the reaction of *nido*-decaborane ($B_{10}H_{14}$) with the poly(*p*-phenylene–ethynylene) polymer in various ratios (0.05, 0.10, 0.20, 0.40) [37]. These polymers showed high thermal resistance and an enhancement in ceramic yields upon thermolysis to 1050 °C.

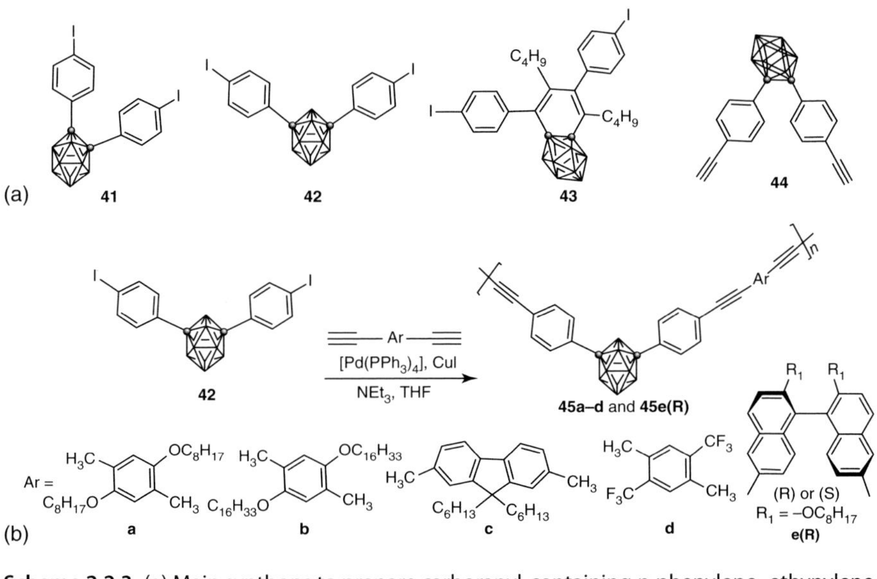

Scheme 2.2.3 (a) Main synthons to prepare carboranyl-containing *p*-phenylene–ethynylene polymers (**41–44**). (b) Synthesis of polymers (**45a–d** and **45e(R)**).

Polymers **45a–d** and **45e(R)** and polymers obtained from **44** showed intense blue emission in $CHCl_3$ with a maximum at $\lambda = 388$ nm; polymers containing the benzocarborane **43** also exhibited intense blue luminescence in THF solution. On the other hand, those polymers bearing *o*-carboranyl units were not fluorescent in THF solutions due to an intramolecular charge-transfer (CT) process from the *p*-phenylene–ethynylene units to the antibonding orbital of the C_c—C_c bond in the *o*-carborane. Nevertheless, the latter exhibited orange emission in suspensions of THF/H_2O, 1/99 (v/v) with emissions in the range of 550–583 nm, originated from aggregation-induced emission (AIE). Polymers **46** also exhibited blue emission in THF (471 nm), whereas in suspensions the emission maximum was redshifted to 529 nm, suggesting a shortening of the effective π-conjugation length due to the introduction of the carboranyl unit.

2.2.4 Thermally Resistant Carborane-Based Polymers

Since the 1960s, a large number of methods to produce carboranyl-containing hybrid organic–inorganic silane and siloxane-based polymers have been developed [38]. The main interest in these polymers is because of their high thermal stability and enhanced mechanical properties, usually improved by the introduction of carboranyl units.

According to the literature, compound **47** (Chart 2.2.4) is one of the more useful synthons to achieve carborane–siloxane–acetylene polymers through diverse reaction conditions, as is exemplified in Scheme 2.2.4 [39, 40]. Polymers **50**–**52** were synthesized by polycondensation reaction of 1,4-dilithio-1,3-butadiyne with **47** and the corresponding reagents (Scheme 2.2.4a). *m*-Carborane–siloxane–diacetylene copolymers with different calorimetric glass transitions were prepared by using different siloxane:carborane:diacetylene molar ratios (10 : 9 : 1, 5 : 4 : 1, and 3 : 2 : 1) [41]. Polymer **51** exhibited high thermal stability when compared with other ferrocenyl-siloxane polymers, with weight retention of 78% at 1000 °C. Polymers **52** exhibited excellent thermal and oxidative stability at elevated temperatures when compared to other poly-acetylene polymers, giving ceramic yields in the range of 79–86% by heating at 1000 °C under N_2. Poly(carborane–siloxane–arylacetylene) polymers **53** and **54** were synthesized from **47** using Grignard derivatives (Scheme 2.2.4b) [42]. Their thermal treatment resulted in a fully cross-linked thermoset at 500 °C. The developed polymers have exceptional thermo-oxidative properties similar to those of their diacetylene counterpart poly(carborane–siloxane–acetylene) with better mechanical properties for the thermoset samples than those from poly(carborane–siloxane–acetylene) polymers.

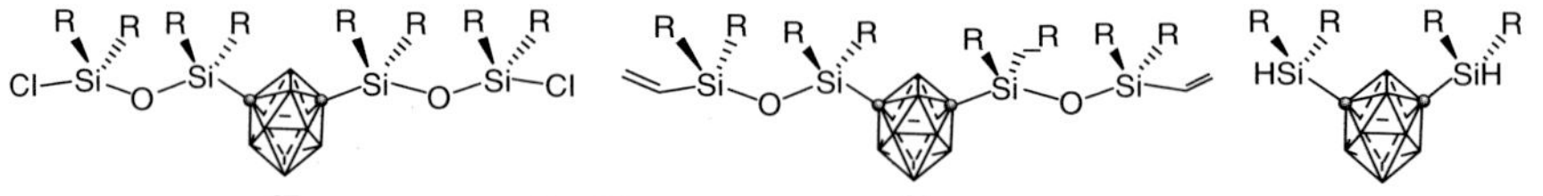

47 R = CH_3 **48** R = CH_3 (**49a**), C_6H_5 (**49b**)

Chart 2.2.4 Synthons for carborane–siloxane and carborane–silane polymers.

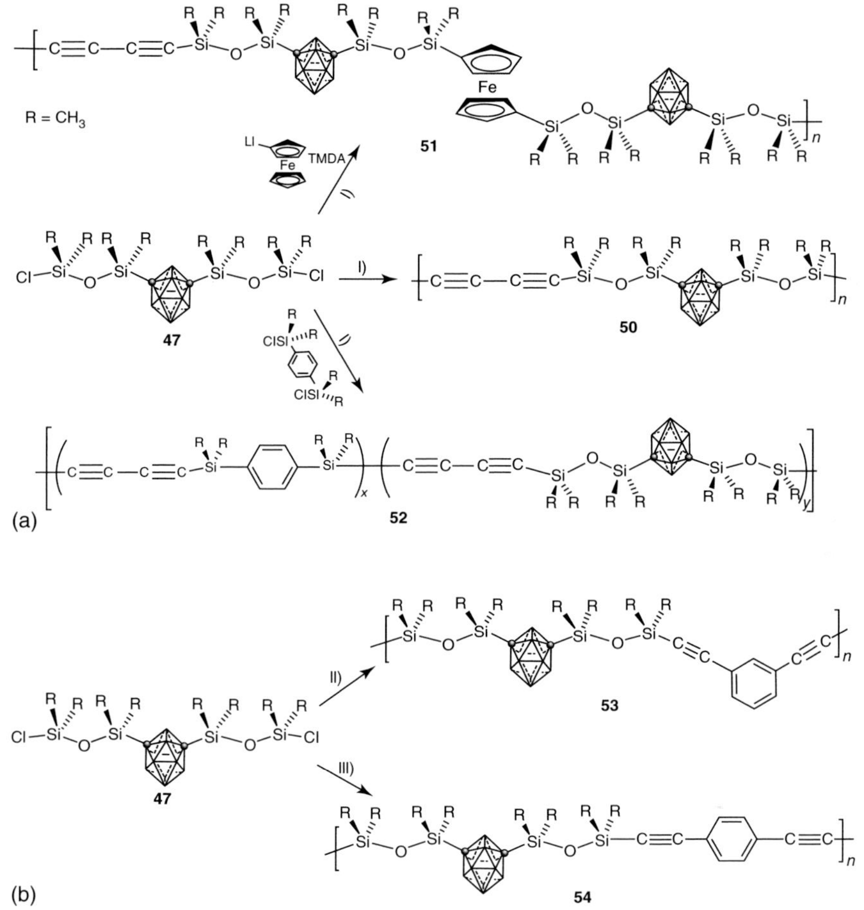

Scheme 2.2.4 Synthesis of poly(carborane–siloxane–diacetylene) polymers **50–52** (a) (i) 1,4-dilithio-1,3-butadiyne; and **53–54** (b) (ii) magnesium bromide salt of 1,3-diethynylbenzene (*m*-DEB); and (iii) magnesium bromide salt of 1,4-diethynylbenzene (*p*-DEB).

Owing to the difficulty in obtaining poly(carborane–siloxane–diacetylene) polymers **50–54**, research has been focused on the development of alternative methodologies using other class of synthons (Chart 2.2.4). *m*-Carborane–siloxane polymers were prepared via rapid Karstedt-catalyzed hydrosilylation reaction of synthon **48** (Chart 2.2.4) with poly(methylhydroxilosane) [43] or a branched siloxane cross-linker monomer [44]. The latter networks showed an oxidative stability up to 300 °C and their Td_5 (temperature at which the polymer losses 5% of its weight) are in the range 500–550 °C. Carborane–silane–acetylene polymers were also obtained from the hydrosilylation reaction of diethynylbenzene-silene polymers with synthons **49a** or **49b** (Chart 2.2.4) [45]. These polymers exhibited high thermal stability in air, and no weight loss up to 600 °C occurred with ceramic yields of around 95% at 800 °C, whereas in N_2 the ceramic yields were higher than 90% at the same temperature. Poly(silylene–arylacetylene)s **55** containing *o*-carborane were prepared by the coupling reaction between poly(silylene–arylacetylene) and $B_{10}H_{14}$ with different molar ratios [46]. Another type of *m*-carborane-containing poly(carborane–silane–arylacetylene) polymer showed very good solubility in organic solvents offering good processability to fabricate composites [47]. The obtained thermosets exhibited excellent thermal and thermo-oxidative stabilities with ceramic yields of around 96% in air and 92% in N_2. In general, poly(carborane–silane–acetylene) polymers possess better mechanical and thermal stability than poly(carborane–siloxane–diacetylene) polymers, making them suitable candidates as a matrix resin for advanced composites and as a precursor for ceramics.

Apart from the previously described polymers, carborane-containing polyesters [48], poly(hydroxy ethers) [49], polyphenylene [50], polyaryl(ether-ketone) [51], polyimide [52], and polystyrene [53] polymers, among others, with high thermostability and ceramics yields were successfully synthesized [18].

2.2.5 Coordination Polymers and Nanoparticles Incorporating *closo*-Carborane Clusters

Coordination polymers (CPs), are infinite multidimensional 1D linear or twisted chains, 2D squares and polygons, and 3D cubes and polyhedra materials with highly ordered structures [54]. These structures are highly conditioned by the nature of multifunctional ligands and the coordination requirements of the metallic centers or the counter ions. A slight modification in any of these factors can originate new structures and consequently new properties. The presence of metal–ligand association can lead to materials suitable for diverse applications; the metal can provide redox and magnetic properties among others and the appropriate ligand may determine the functionality of the material [55].

The incorporation of carborane clusters as linkers in 1D coordination polymers and nanohybrid materials was performed in two steps: (i) by introducing the metal-binding group directly bonded to the cage and (ii) coordination to the corresponding metal.

2.2.5.1 Carboxylate-Functionalized Carboranes

Carboranylcarboxylate ligands have been used to introduce C_2B_{10} cages into the backbone of 1D coordination polymers. The reaction of two different carboranylcarboxylate ligands **56** and **57**, with $MnCO_3$ in water leads to the polymeric compounds $[Mn(\mu\text{-}H_2O)(\mu\text{-}1\text{-}CH_3\text{-}2\text{-}CO_2\text{-}1,2\text{-}closo\text{-}C_2B_{10}H_{10})_2]_n{\cdot}(H_2O)_n$ **58** [56] or $[Mn(H_2O)(\mu\text{-}1\text{-}CO_2\text{-}1,2\text{-}closo\text{-}C_2B_{10}H_{11})_2]_n\ {\cdot}2H_2O$ **59** [57], respectively (Scheme 2.2.5). X-ray analysis revealed substantial differences between both compounds, which was attributed to the existence of different substituents, $-CH_3$ or $-H$, on one of the C_c of the carboranylcarboxylate ligand. While a six-coordinated Mn^{II} compound with water molecules bridging each two Mn^{II} centers was observed in **58**, a square pyramidal geometry around each Mn^{II} ion with terminal water molecules coordinated to each Mn^{II} center was found in **59**.

The polymeric nature of both compounds can be easily fragmented by coordinating solvents, e.g. diethyl ether, generating oligomer species. Moreover, the

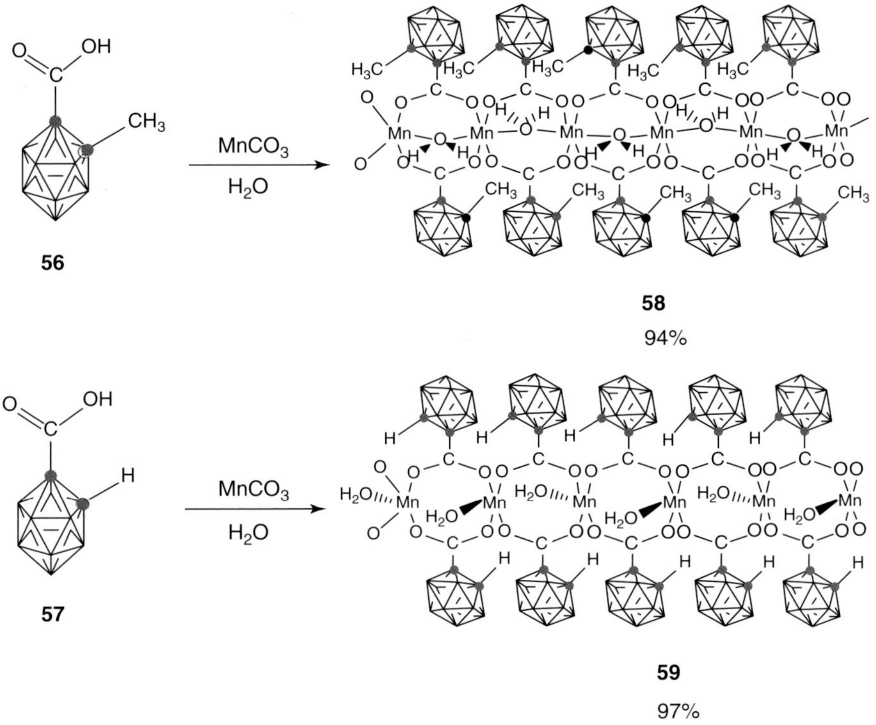

Scheme 2.2.5 Synthetic strategy for the preparation of polymers **58** and **59**.

reactivity of polymer **58** with Lewis bases as 2,2′-bipyrimidine (bpm), leads to the formation of a new high hybrid polymer [Mn(1-CH_3-2-CO_2-1,2-*closo*-$C_2B_{10}H_{10}$)$_2$(bpm)]$_n$ **60**. Magnetic measurements show weak antiferromagnetic interactions between the Mn atoms in all cases. Polymers **58** and **60** have proven to be active in the catalytic epoxidation of alkenes, showing high conversion and selectivity values in short reaction time [58]. The reaction of **56** with $CoCO_3$ in water allows to isolate the polymer compound [Co(μ-H_2O)(μ-1-CH_3-2-CO_2-1,2-*closo*-$C_2B_{10}H_{10}$)$_2$]$_n$·(H_2O)$_n$ **61** [59], with a structure similar to polymer **58**.

2.2.5.2 Phosphinate- and Phosphino-Functionalized Carboranes

Organophosphinate CPs with transition metals have not been available until recently, because of the reducing power of the ligands. Not long ago, the first examples of 1D manganese polymers with a carboranylphosphinic acid have been described [60]. The reaction of Na[1-OPH(O)-1,7-*closo*-$C_2B_{10}H_{11}$] **62** salt with $MnCl_2$·4H_2O in methanol leads to the 1D [Mn(CH_3OH)$_2$(1-OPH(O)-1,7-*closo*-$C_2B_{10}H_{11}$)$_2$]$_n$, **63** polymer (Scheme 2.2.6). When **63** is dissolved in an ethanol solution of 2,2′-bipyridine (bpy), a new 1D polymer [Mn(bpy)(1-OPH(O)-1,7-*closo*-$C_2B_{10}H_{11}$)$_2$]$_n$, **64** was formed. These polymers are not only the first examples of all-inorganic-made inorganic CPs; they are also the first described

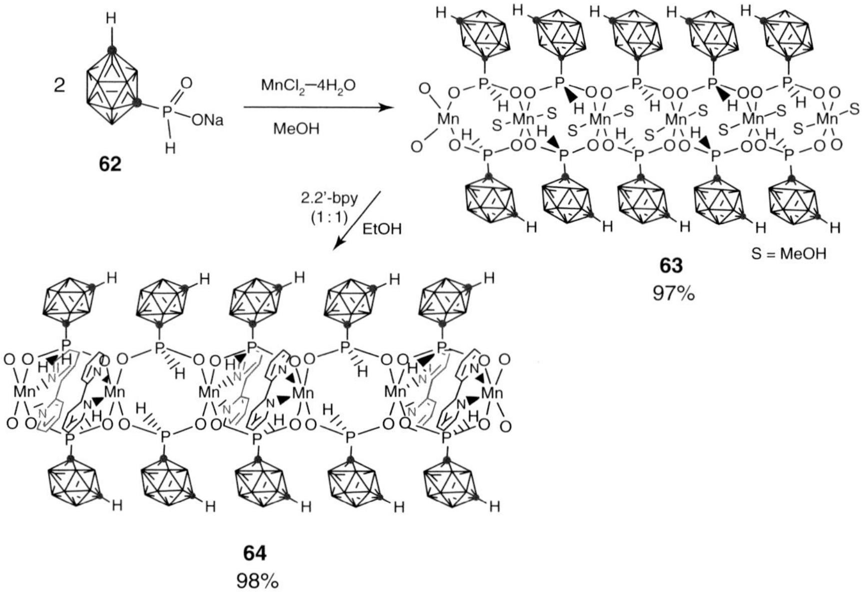

Scheme 2.2.6 Synthetic strategy for the preparation of polymers **63** and **64**.

coordination polynuclear Mn^{II} polymers with phosphinate ligands bridging each of the two manganese centers. The backbone is strictly made of manganese and the all-inorganic-made carboranylphosphinate runs in a zigzag 1D chain.

On the other hand, the reaction of the *closo*-carborane–diphenylphosphine, 1,2-$(PPh_2)_2$-1,2-*closo*-$C_2B_{10}H_{10}$ **65** with $HgCl_2$ in CH_2Cl_2 produced a 1D chain polymer via bridged Hg—Cl bonds, $[Hg_2\{1,2\text{-}(PPh_2)_2\text{-}1,2\text{-}closo\text{-}C_2B_{10}H_{10}\}_2(\mu\text{-}Cl_2)_2(\text{-}HgCl_2)_3]_n\cdot(H_2O)_n$ **66** [61].

2.2.5.3 Nanohybrid Materials Based on Functionalized Carboranes

Phosphinate-functionalized carboranes have also been used to achieve multifunctional nanohybrid materials, as magnetic nanoparticles (MNPs) and quantum dots (QDs), with unexpected properties for a wide range of potential applications.

Magnetite/maghemite nanoparticles (MNPs) decorated with the phosphinate carboranyl ligand **62** have been prepared via coordination of phosphinate bidentated bridging group to the iron atoms [62] (Figure 2.2.2) with the aim to develop new purely inorganic hybrid nanomaterials that could act as an anticancer drug for dual therapy (boron neutron capture therapy [BNCT] and thermotherapy). The stability of the MNPs under the autoclave sterilization process that is required for "in vitro" and "in vivo" biological tests was studied. The phosphinate carboranyl ligands were neither oxidized nor altered through the sterilization process (steam heated to 121 °C under pressure for one hour). The results suggest that these new nanohybrid icosahedral boron cluster MNPs might offer a broad scope for exciting research and future biomedical applications.

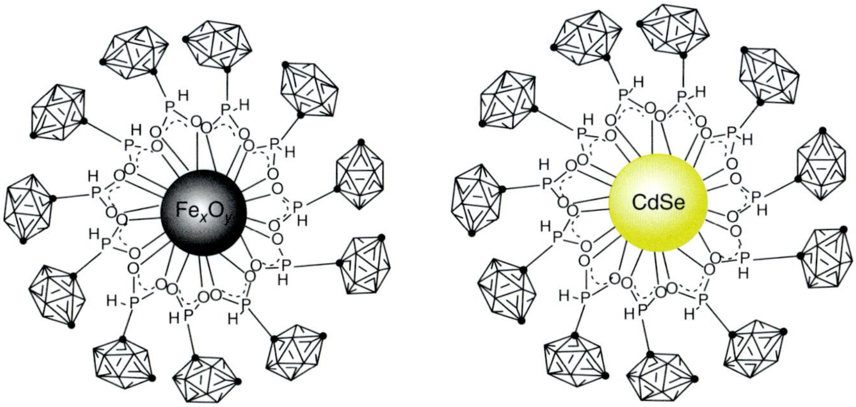

Figure 2.2.2 Magnetite/maghemite nanoparticles (MNPs) and quantum dots (QDs).

The phosphinate carboranyl ligands have been used to create a new type of architecture in quantum dots named "core-canopy QDs," which produces an unprecedented phenomenon, namely kinetic fluorescence switching (KFS) [63]. Carboranes can be visualized as hollow spheres. Very few carbon-based molecules are hollow spheres and they are quite difficult to obtain, hence necessitating the use of carboranes for this purpose. These hollow spheres were linked to the QDs through the phosphinate binding site of the phosphinate carboranyl for the first time and this fact led to the introduction of the new "KFS" concept that arises from the development of this novel "core-canopy" architecture in QDs. This is based on a QD core from which hollow spheres emanate out with the spheres being connected to the core via pillars. This architecture leaves channels between the hollow spheres and the core and it is susceptible to distinct properties. The foremost among these properties is the aforementioned KFS. Owing to this KFS property, these "core-canopy QDs," which are initially fluorescent, lose their fluorescence when left to aggregate, but upon being provided mild kinetic energy (like being shaken) again shine back to their maximum fluorescent intensity. Upon being left to aggregate they again lose their fluorescence and regain it when shaken (provided with kinetic energy). This sequential process of luminescence: fading, back to luminescence, fading, and so on, can be continuously run showing minimal loss of intensity, even after a period of five months [63]. This property makes these "core-canopy QDs" attractive for commercial applications, particularly those taking advantage of their high extinction coefficient and photobleaching stability such as solar light harvesting, light-emitting diodes, photocatalysis, and others.

Furthermore, gold nanoparticles capped with mercaptocarborane ligands lead to monolayer protected clusters (MPCs) [64] (Scheme 2.2.7) that exhibit the uniqueness to store and release cationic charge that is counterbalanced by electrons in the metal core. The rigid and spherical structure of the carborane cluster plus its capability for self-assembly by hydrogen or dihydrogen intramolecular

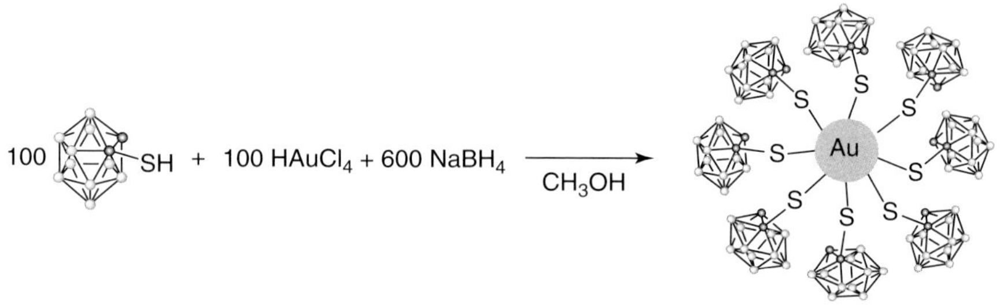

Scheme 2.2.7 Synthetic strategy for the preparation of gold nanoparticles.

interactions make the difference between organic thiols and mercaptocarborane. Very few carbon-based molecules are hollow spheres and they are quite difficult to obtain, hence necessitating the use of carboranes for this purpose. These gold nanoparticles capped with mercaptocarborane ligands, which are hollow spheres ready to link to gold, lead to a monolayer-protected clusters that exhibit the uniqueness to store and release cationic charge that is counterbalanced by electrons in the metal core. Unexpectedly, the gold nanoparticles are highly water soluble. However, the most interesting property is the reversible phase transfer of the particles from water to diethyl ether that is achieved by acidification of the aqueous phase with dilute hydrochloric acid. In a two-phase system, direct extraction from the aqueous to the organic phase is observed. Re-extracting the particles into an aqueous phase is possible by addition of sodium borohydride but cannot be achieved simply by a base solution. This demonstrates that it is not an acid–base reaction but a redox reaction. The process is fully reversible. The same particles can be isolated as a solid and re-dissolved and moved back and forth between diethyl ether and water phases without any evidence of aging effects. To explain such an unusual property, it was postulated that the particles act as both electron- and ion-storage devices. We thought that when preparing the gold nanoparticles Na^+ cations are retained close to the gold surface that would be negatively charged. To demonstrate the ability of the gold nanoparticles to store ionic charge, the sample was dissolved in water and passed through a cation (Li, K, Mg, and Ca) exchange columns. These MPCs showed toxicity in front of HeLa cells, a common human fibroblast cell line, and led to cancer cell death, after 24 hours of incubation with them [64].

Later [65], it was reported that these 2–3 nm MPCs protected by hollow sphere mercaptocarborane ligands (i) readily penetrate phospholipid bilayer membranes and establish a membrane potential as partitioning anions and (ii) affect the transport of alkali cations across the membrane and hence represent a class of membrane alkali-ion specific transporters. Two complementary techniques have been used to study these phenomena: voltage-dependent fluorescence spectroscopy of dispersions of vesicles and potentiometric and potential step experiments across a free-standing bilayer membrane. The gold nanoparticles act as alkali-ion-specific transporters across free-standing membranes under potentiostatic control. Magnesium ions are not transported [65].

2.2.6 Conclusion and Outlook

The incorporation of icosahedral boron clusters into π-conjugated polymers improves their stability and solubility and clearly alters their electrochemical and photoluminescent properties. Further, metallacarborane doping agents produce the highest ORL films known to date and carboranes provide exceptional thermal, thermo-oxidative, and elastic properties for high-temperature elastomers as hybrid polymers. After pyrolysis at very high temperatures, these polymers form highly thermally stable ceramics that resist oxidative degradation. Icosahedral boron clusters induce exceptional properties in the supramolecular structure of CPs, gold NPs, MNPs, and QDs due to the electron-withdrawing characteristic of the rigid and spherical carborane structure.

Polymers and nanohybrids containing icosahedral boron clusters are promising materials for technological and biomedical applications due to their enhanced electronic, optical, and thermal properties provided by the uniqueness of the clusters.

Acknowledgments

The authors have been supported by the Spanish Ministerio de Economía y Competitividad (MINECO, project CTQ2016-75150-R and CTQ2015-66143-P) and Generalitat de Catalunya (2017 SGR 1720). C. V. acknowledges the COST Action CM1302 (SIPS): "European Network on Smart Inorganic Polymers."

References

1 Skotheim, T.A. and Reynolds, J.R. (eds.) (2007). *Handbook of Conducting Polymers*, 3e. Boca Raton, FL: CRC Press, Taylor & Francis Group.
2 Naarmann, H. (2000). Polymers, electrically conducting. In: *Ullmann's Encyclopedia of Industrial Chemistry*. Wiley-VCH.
3 Xie, J., Gu, P.Y., and Zhang, Q.C. (2017). Nanostructured conjugated polymers: toward high-performance organic electrodes for rechargeable batteries. *ACS Energy Lett.* 2 (9): 1985–1996.
4 Neo, W.T., Ye, Q., Chua, S.J., and Xu, J.W. (2016). Conjugated polymer-based electrochromics: materials, device fabrication and application prospects. *J. Mater. Chem. C* 4 (31): 7364–7376.
5 Oikawa, H., Onodera, T., Masuhara, A. et al. (2010). New class materials of organic–inorganic hybridized nanocrystals/nanoparticles, and their assembled micro- and nano-structure toward photonics. *Adv. Polym. Sci.* 231: 147–190.
6 Tao, S., Zhu, S.J., Feng, T.L. et al. (2017). The polymeric characteristics and photoluminescence mechanism in polymer carbon dots: a review. *Mater. Today Chem.* 6: 13–25.
7 Yen, H.-J., Shan, C.S., Wang, L. et al. (2017). Development of conjugated polymers for memory device applications. *Polymers* 9 (1): 25.

8 Advincula, R. (2009). Nanostructured conjugated polymer network ultrathin films and coatings using the precursor polymer approach. In: *Nanotechnology Applications in Coatings*, ACS Symposium Series, vol. 1008 (ed. R.H. Fernando and L.P. Sung), 51–61. Washington, DC: American Chemical Society.

9 Iqbal, S. and Ahmad, S. (2018). Recent development in hybrid conducting polymers: synthesis, applications and future prospects. *J. Ind. Eng. Chem.* 60: 53–84.

10 Janaky, C. and Rajeshwar, K. (2015). The role of (photo)electrochemistry in the rational design of hybrid conducting polymer/semiconductor assemblies: from fundamental concepts to practical applications. *Prog. Polym. Sci.* 43: 96–135.

11 (a) Thomas, S.W. III, Joly, G.D., and Swager, T.M. (2007). Chemical sensors based on amplifying fluorescent conjugated polymers. *Chem. Rev.* 107 (4): 1339–1386. (b) McQuade, D.T., Pullen, A.E., and Swager, T.M. (2000). Conjugated polymer-based chemical sensors. *Chem. Rev.* 100 (7): 2537–2574.

12 (a) Otero, T.F., Martinez, J.G., and Arias-Pardilla, J. (2012). Biomimetic electrochemistry from conducting polymers. A review. Artificial muscles, smart membranes, smart drug delivery and computer/neuron interfaces. *Electrochim. Acta* 84: 112–128. (b) Jager, E.W.H., Smela, E., and Inganas, O. (2000). Microfabricating conjugated polymer actuators. *Science* 290 (5496): 1540–1545.

13 (a) Lin, Y., Li, Y., and Zhan, X. (2012). Small molecule semiconductors for high-efficiency organic photovoltaics. *Chem. Soc. Rev.* 41 (11): 4245–4272. (b) Jorgensen, M., Norrman, K., and Krebs, F.C. (2008). Stability/degradation of polymer solar cells. *Sol. Energy Mater. Sol. Cells* 92 (7): 686–714.

14 (a) Lin, P. and Yan, F. (2012). Organic thin-film transistors for chemical and biological sensing. *Adv. Mater.* 24 (1): 34–51. (b) Feng, X., Liu, L., Wang, S., and Zhu, D. (2010). Water-soluble fluorescent conjugated polymers and their interactions with biomacromolecules for sensitive biosensors. *Chem. Soc. Rev.* 39 (7): 2411–2419.

15 (a) Guimard, N.K., Gomez, N., and Schmidt, C.E. (2007). Conducting polymers in biomedical engineering. *Prog. Polym. Sci.* 32 (8–9): 876–921. (b) Tian, H., Tang, Z., Zhuang, X., and Chen, X. (2012). Biodegradable synthetic polymers: preparation, functionalization and biomedical application. *Prog. Polym. Sci.* 37 (2): 237–280.

16 (a) Bauduin, P., Prevost, S., Farràs, P. et al. (2011). A theta-shaped amphiphilic cobaltabisdicarbollide anion: transition from monolayer vesicles to micelles. *Angew. Chem. Int. Ed.* 50: 5298. (b) Fernandez-Alvarez, R., Ďord'ovič, V., Uchman, M., and Matelǐcek, P. (2018). Amphiphiles without head-and-tail design: nanostructures based on the self-assembly of anionic boron cluster compounds. *Langmuir* 34 (12): 3541–3554.

17 Masalles, C., Llop, J., Viñas, C., and Teixidor, F. (2002). Extraordinary overoxidation resistance increase in self-doped polypyrroles by using non-conventional low charge-density anions. *Adv. Mater.* 14: 826.

18 Núñez, R., Romero, I., Teixidor, F., and Viñas, C. (2016). Icosahedral boron clusters: a perfect tool for the enhancement of polymer features. *Chem. Soc. Rev.* 45: 5147–5173.

19 Housecroft, C.E. (2015). Carboranes as guest, counterions and linkers in coordination polymers and networks. *J. Organomet. Chem.* 798: 218–228.

20 Masalles, C., Borrós, S., Viñas, C., and Teixidor, F. (2000). Are low-coordinating anions of interest as doping agents in organic conducting polymers? *Adv. Mater.* 12: 1199–1202.

21 Tarrés, M., Viñas, C., Cioran, A.M. et al. (2014). Towards multifunctional materials incorporating elastomers and reversible redox-active fragments. *Chem. Eur. J.* 20: 15808–15815.

22 Gentil, S., Crespo, E., Rojo, I. et al. (2005). Polypyrrole materials doped with weakly coordinating anions: influence of substituents and the fate of the doping anion during the overoxidation process. *Polymer* 46: 12218–12225.

23 David, V., Viñas, C., and Teixidor, F. (2006). Poly(3,4-ethylenedioxythiophene) doped with a non-extrudable metallacarborane anion electroactive during synthesis. *Polymer* 47: 4694–4702.

24 (a) Fabre, B., Caleb Clark, J., and Vicente, M.G. (2006). Synthesis and electrochemistry of carboranylpyrroles. Toward the preparation of electrochemically and thermally resistant conjugated polymers. *Macromolecules* 39: 112–119; (b) Hao, E., Fabre, B., Fronczek, F.R., and Vicente, M.G. (2007). Syntheses and electropolymerization of carboranyl-functionalized pyrroles and thiophenes. *Chem. Mater.* 19: 6195–6105. (c) Chayer, S., Jaquinod, L., Smith, K.M., and Vicente, M.G. (2001). Syntheses of carboranylpyrroles. *Tetrahedron Lett.* 42: 7759–7761.

25 Llop, J., Masalles, C., Viñas, C. et al. (2003). The $[3,3'\text{-Co}(1,2\text{-C}_2\text{B}_9\text{H}_{11})_2]^-$ anion as a platform for new materials: synthesis of its functionalized monosubstituted derivatives incorporating synthons for conducting organic polymers. *Dalton Trans.* 556–561.

26 Fabre, B., Chayer, S., and Vicente, M.G. (2003). Polythiophenes containing in-chain cobaltabisdicarbollide centers. *Electrochem. Commun.* 5: 431–434.

27 (a) Hao, E., Fabre, B., Fronczek, F.R., and Vicente, M.G.H. (2007). Poly[di(2-thiophenyl)carborane]s: conducting polymers with high electrochemical and thermal resistance. *Chem. Commun.* 4387–4389; (b) Barriere, F., Fabre, B., Hao, E. et al. (2009). Electropolymerizable 2,2′-Carboranyldithiophenes. Structure–property investigations of the corresponding conducting polymer films by electrochemistry, UV–visible spectroscopy and conducting probe atomic force microscopy. *Macromolecules* 42: 2981–2987. (c) Cansu-Ergun, E.M. and Cihaner, A. (2013). A new carborane based polymeric electrochrome. *J. Electroanal. Chem.* 707: 78–84. (d) Zhao, N., Fabre, B., Bobadova-Parvanova, P. et al. (2017). Synthesis and electropolymerization of a series of 2,2′-(*ortho*-carboranyl)bisthiophenes. *J. Organomet. Chem.* 828: 157–165.

28 Peterson, J.J., Werre, M., Simon, Y.C. et al. (2009). Carborane-containing polyfluorene: *o*-carborane in the main chain. *Macromolecules* 42: 8594–8598.

29 Peterson, J.J., Simon, Y.C., Coughlin, E.B., and Carter, K.R. (2009). Polyfluorene with *p*-carborane in the backbone. *Chem. Commun.* 4950–4952.

30 Davis, A.R., Peterson, J.J., and Carter, K.R. (2012). Effect of *o*-carborane on the optoelectronic and device-level properties of poly(fluorene)s. *ACS Macro Lett.* 1: 469–472.

31 Simon, Y.C., Peterson, J.J., Mangold, C. et al. (2009). Synthesis of polyfluorenes with pendant silylcarboranes. *Macromolecules* 42: 512–516.
32 Kokado, K. and Chujo, Y. (2009). Emission via aggregation of alternating polymers with *o*-carborane and *p*-phenylene–ethynylene sequences. *Macromolecules* 42: 1418–1420.
33 Kokado, K., Tokoro, Y., and Chujo, Y. (2009). Luminescent *m*-carborane-based π-conjugated polymer. *Macromolecules* 42: 2925–2930.
34 Kokado, K., Tokoro, Y., and Chujo, Y. (2009). Luminescent and axially chiral π-conjugated polymers linked by carboranes in the main chain. *Macromolecules* 42: 9238–9242.
35 Kokado, K., Tominaga, M., and Chujo, Y. (2010). Aromatic ring-fused carborane-based luminescent *p*-conjugated polymers. *Macromol. Rapid Commun.* 31: 1389–1394.
36 Tominaga, M., Morisaki, Y., and Chujo, Y. (2013). Luminescent polymer consisting of 9,12-linked *o*-carborane. *Macromol. Rapid Commun.* 34: 1357–1362.
37 Kokado, K. and Chujo, Y. (2010). Polymer reaction of poly(*p*-phenylene–ethynylene) by addition of decaborane: modulation of luminescence and heat resistance. *Polym. J.* 42: 363–367.
38 Grimes, R.N. (2016). *Carboranes*, 3e. Academic Press.
39 Sundar, R.A. and Keller, T.M. (1997). Linear diacetylene polymers containing bis(dimethylsilyl)phenyl and/or bis-(tetramethyldisiloxane)carborane residues: their synthesis, characterization and thermal and oxidative properties. *J. Polym. Sci., Part A: Polym. Chem.* 35: 2387–2394.
40 Kolel-Veetil, M.K. and Keller, T.M. (2003). The effects of concentration dilution of cross-linkable diacetylenes on the plasticity of poly(*m*-carborane–disiloxane–diacetylene)s. *J. Mater. Chem.* 13: 1652–1656.
41 Kolel-Veetil, M.K., Beckham, H.W., and Keller, T.M. (2004). Dependence of thermal properties on the copolymer sequence in diacetylene-containing polycarboranylenesiloxanes. *Chem. Mater.* 16: 3162–3167.
42 Kolel-Veetil, M.K., Dominguez, D.D., Klug, C.A. et al. (2013). Hybrid inorganic–organic poly(carborane–siloxane–arylacetylene) structural isomers with in-chain aromatics: synthesis and properties. *J. Polym. Sci., Part A: Polym. Chem.* 51: 2638–2650.
43 Houser, E.J. and Keller, T.M. (1998). Hydrosilation routes to materials with high thermal and oxidative stabilities. *J. Polym. Sci., Part A: Polym. Chem.* 36: 1969.
44 Kolel-Veetil, M.K. and Keller, T.M. (2006). Formation of elastomeric network polymers from ambient heterogeneous hydrosilations of carboranylenesiloxane and branched siloxane monomers. *J. Polym. Sci., Part A: Polym. Chem.* 44: 147–155.
45 Kimura, H., Okita, K., Ichitani, M. et al. (2003). Structural study of silyl-carborane hybrid diethynylbenzene–silylene polymers by high-resolution solid-state ^{11}B, ^{13}C, and ^{29}Si NMR spectroscopy. *Chem. Mater.* 15: 355–362.
46 Jiang, Y., Li, X., Huang, F. et al. (2015). *o*-Carborane-containing poly(siloxane–arylacetylene)s with thermal and thermo-oxidative stabilities. *J. Macromol. Sci. Part A: Pure Appl. Chem.* 52: 476–484.

47 Zhou, Q., Mao, Z., Ni, L., and Chen, J. (2007). Novel phenyl acetylene terminated poly(carborane–silane): synthesis, characterization, and thermal property. *J. Appl. Polym. Sci.* 104: 2498–2503.

48 Li, N., Zeng, F., Qu, D. et al. (2016). Synthesis and characterization of carborane-containing polyester with excellent thermal and ultrahigh char yield. *J. Appl. Polym. Sci.* 133: 44202.

49 Yanga, Z., Lib, N., Yanga, X.-X. et al. (2017). Synthesis and characterization of carborane-containing poly(hydroxy ethers) with excellent thermal stability. *Chin. J. Polym. Sci.* 35 (12): 1463–1473.

50 Cheng, S., Yan, K., Wang, X. et al. (2017). Poly(phenylene-carborane) for boron-carbide/carbon ceramic precursor synthesized via nickel catalysis. *Polymer* 115: 224–231.

51 Colquhoun, H.M., Lewis, D.F., Daniels, J.A. et al. (1997). Polyetherketones from diarylcarboranes: a new approach to semi-inorganic polymers. *Polymer* 2447–2453. and references therein.

52 Xing, T. and Zhang, K. (2015). Synthesis of novel soluble carborane polyimides with ultrahigh thermal stability. *Polym. Int.* 64: 1715–1721.

53 Liang, L., Rapakousiou, A., Salmon, L. et al. (2011). "Click" assembly of carborane-appended polymers and stabilization of gold and palladium nanoparticles. *Eur. J. Inorg. Chem.* 3043–3049.

54 Férey, G. (2008). Hybrid porous solids: past, present, future. *Chem. Soc. Rev.* 37: 191–214.

55 (a) Verdaguer, M. and Robert, V. (2013). Vol. 8: coordination and organometallic chemistry. In: *Comprehensive Inorganic Chemistry II: From Elements to Applications*, vol. 8 (ed. V.W.W. Yam), 131–189. Amsterdam: Elsevier. (b) Xu, L.J., Xu, G.-T., and Chen, Z.-N. (2014). Recent advances in lanthanide luminescence with metal-organic chromophores as sensitizers. *Coord. Chem. Rev.* 273–274: 47–62.

56 Fontanet, M., Rodríguez, M., Romero, I. et al. (2013). A water soluble Mn(II) polymer with aqua metal bridges. *Dalton Trans.* 42: 7838–7841.

57 Fontanet, M., Rodríguez, M., Fontrodona, X. et al. (2014). Water-soluble manganese inorganic polymers: the role of carborane clusters and producing large structural adjustments from minor molecular changes. *Chem. Eur. J.* 20: 13993–14003.

58 Fontanet, M., Rodríguez, M., Viñas, C. et al. (2017). Carboranycarboxylate complexes as efficient catalysts in epoxidation reactions. *Eur. J. Inorg. Chem.* 4425–4429.

59 Fontanet, M., Rodríguez, M., Fontrodona, X. et al. (2016). Carving a 1D Co^{II}-carboranylcarboxylate system by using organic solvents to create stable trinuclear molecular analogues: complete structural and magnetic studies. *Dalton Trans.* 45: 10916–10927.

60 Oleshkevich, E., Viñas, C., Romero, I. et al. (2017). *m*-Carboranylphosphinate as versatile building blocks to design all inorganic coordination polymers. *Inorg. Chem.* 56: 5502–5505.

61 Kong, L., Zhang, D., Su, F. et al. (2011). The novel 1D chain and 3D network of mercury carborane–diphenylphosphine complexes constructed by covalent bond or hydrogen bond interactions. *Inorg. Chim. Acta* 370: 1–6.

62 Oleshkevich, E., Teixidor, F., Rosell, A., and Viñas, C. (2018). Merging icosahedral boron clusters and magnetic nanoparticles: aiming toward multi-functional nanohybrid materials. *Inorg. Chem.* 57: 462–470.
63 Saha, A., Oleshkevich, E., Viñas, C., and Teixidor, F. (2017). Biomimetic inspired core-canopy quantum dots: ions trapped in voids induce kinetic fluorescence switching. *Adv. Mater.* 29: 1704238.
64 Cioran, A.M., Musteti, A.D., Teixidor, F. et al. (2012). Mercaptocarborane-capped gold nanoparticles: electron pools and ion traps with switchable hydrophilicity. *J. Am. Chem. Soc.* 134: 212–221.
65 Grzelczak, M.P., Danks, S.P., Klipp, R.C. et al. (2017). Ion transport across biological membranes by carborane-capped gold nanoparticles. *ACS Nano* 11: 12492–12499.

3

Synthesis of Group 14 Metal-Containing Polymers

Ana Torvisco[1], Frank Uhlig[1], and David Scheschkewitz[2]

[1] *Graz University of Technology, Institute for Inorganic Chemistry, Stremayrgasse 9, 8010 Graz, Austria*
[2] *Saarland University, Krupp-Chair of General and Inorganic Chemistry, Campus C4.1, 66123 Saarbrücken, Germany*

3.1 Introduction

Group 14 metal inorganic oligomers and polymers have gained interest in recent years due to their applicability in a variety of fields. Polysilanes have been well reviewed and are materials exhibiting interesting electronic and optical properties [1–5]. Similar to polysilanes, linear oligo- and polygermanes and stannanes are compounds with the general formula $-[R_2E]_n-$, with a covalently linked (semi-)metal backbone and thus a molecular wire embedded in an organic jacket (R) (Figure 3.1) [6–20].

Similar to the structurally analogous saturated polymeric hydrocarbons, the polymers of group 14 elements can be approximated as linear chains of E—E single bonds disregarding side chains and cross-linking [21]. With increasing number of covalent E—E single bonds, the delocalization of σ-electrons along the backbone progressively transforms discrete frontier orbitals into a σ–σ* band structure (σ-conjugation) [22–25] with strong σ–σ* transitions [22, 26–38]. This leads to characteristic yellow or orange-yellow colors due to absorption maxima in UV–vis spectra at about 375–410 nm [21, 39–41]. Similar to σ-conjugated organic polymers, the bandgap is decreased with increasing degree of catenation and the longest wavelength absorption of the σ–σ* transitions therefore shifted to the red [6, 42]. Even lower band gaps can be achieved if the π-electrons of pending aromatic groups confer additional σ–π delocalization [22, 25, 37].

Heavier Ge and Sn atoms in the backbone of σ-conjugated polymers are expected to give stronger σ-conjugation, exhibit narrower bandgaps, and induce more metallic character to the resulting material as compared to their Si counterparts [1, 5, 7, 23, 28, 43–46]. Such highly conductive metallic materials could be promising for applications in charge-transport, photoresists in microlithography, nonlinear optical materials, semiconductors in doping, and electronic devices [22, 26, 28, 29, 35, 37, 47–53] for instance as one dimensional quantum wires in printed electronic applications [26, 28, 30, 33, 51, 54]. These features are highly dependent on the degree of σ-delocalization and as a consequence on the chain

Smart Inorganic Polymers: Synthesis, Properties, and Emerging Applications in Materials and Life Sciences, First Edition. Edited by Evamarie Hey-Hawkins and Muriel Hissler.

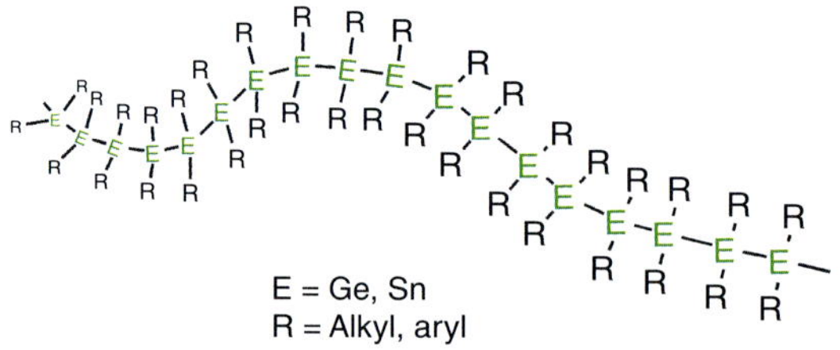

Figure 3.1 Schematic representation of a linear group 14 polymer.

length and the nature of the substituents attached to the main chain, as well as backbone conformation and other secondary structural features in the solid state.

Over the years, different polymerization methods have been established in order to synthesize polystannanes, which were summarized in a number of review articles [12, 19, 20, 55–57]. Classically, Wurtz-type coupling reactions involving group 14 metal chloride (R_2ECl_2) precursors with alkali/alkaline earth metals in organic solvents such as toluene have been widely used to obtain polygermanes and polystannanes of high molar mass. However, these reactions are often accompanied by the formation of side products including shorter oligomeric chains and/or cyclic derivatives. Other routes involving group 14 metal chloride (R_2ECl_2) precursors include electrochemical synthesis leading to products with undesired residual metal–halide bonds and unknown fractions of cyclic oligomers in variable yields [58]. In the case of germanium, a demethanative coupling reaction has been developed [59, 60]. Recently, dehydropolymerization routes involving group 14 metal dihydride (R_2EH_2) precursors have become more prolific. These routes include dehydropolymerization [22, 30–32, 34, 37–41, 47, 61–70] of group 14 metal dihydrides catalyzed by transition metal complexes, lanthanides, as well as "nontraditional" iron and molybdenum alkyls. Moreover, due to the thermal sensitivity of group 14 metal hydrogen bonds, solvent- and catalyst-free thermal based dehydrogenation reactions also yield polystannanes that exhibit only modest molecular weight and broad polydispersity [71]. In addition, amine bases such as pyridine [72–75] and TMEDA (*N*,*N*,*N*′,*N*′-tetramethylethylenediamine) have been reported to efficiently catalyze the polymerization reaction of organotin dihydrides leading to poly(diarylstannane)s [76].

With the rising interest in these materials in the field of electronics, this chapter will focus on the present pursuits toward the preparation of stable group 14 metal-containing linear polymers, focusing on group 14 metal hydrides as building blocks (R_nEH_{4-n}). Improved synthetic methods as well as the recent advancements in improving solubility and light and moisture stability of these promising materials will be highlighted. In addition, the application of alternative group 14 metal hydride precursors, namely group 14 metal trihydrides REH_3, toward the preparation of novel group 14 metal-containing polymeric networks is presented.

3.2 Organohydrides of Group 14, R_nEH_{4-n}

The most recent (and most convenient) route toward generating group 14 metal hydrides of germanium and tin (R_nEH_{4-n}, E = Ge, Sn) is the reaction of the corresponding group 14 metal halides (R_nEX_{4-n}, X = Cl, Br) with lithium aluminum hydride ($LiAlH_4$) in a polar solvent (Scheme 3.1). This route allowed for the synthesis of a variety of organohydrides of group 14 metals (Table 3.1).

$$R_nEX_{4-n} \xrightarrow[\text{Polar solvent}]{LiAlH_4} R_nEH_{4-n}$$

Scheme 3.1 Synthesis of organogermanium and tin hydrides.

Organohydrides of the heavier group 14 metals readily decompose even at room temperature as a function of the number of organic substituents. The parent metal hydrides (EH_4) are more prone to the elimination of dihydrogen going from Ge to Sn. In accordance with the dissociation energy of the M—H bond [98], the formation enthalpy of EH_4 becomes increasingly endothermic and thus the deposition of the element favorable. This temperature sensitivity can be lowered by the substitution of a hydrogen atom with an alkyl or aryl moiety. Furthermore, the reactivity toward atmospheric oxygen increases with the number of residual hydride functionalities leading to poorly soluble oxidized precipitates [99]. Despite these inherent issues with compound stability, the solid state structures of several of group 14 organohydrides with aryl residues have been elucidated by single crystal X-ray diffraction (Table 3.2).

Table 3.1 Group 14 metal organohydrides prepared with $LiAlH_4$.

R_2GeH_2	$RGeH_3$	R_2SnH_2	$RSnH_3$
		R =	
Cyclopropyl [77]	Cyclopropyl [77]	Methyl [89]	Methyl [89]
Cyclobutyl [77]	Cyclobutyl [77]	Ethyl [67]	Phenyl [92]
Cyclopentyl [77]	Cyclopentyl [77]	Propyl [67]	*o*-Tolyl[92]
Cyclohexyl [78]	Adamantyl [84]	Butyl [90, 91]	2,4-Xylyl [92]
Benzyl [78]	Phenyl [80]	Pentyl [67]	2,6-Xylyl [92]
Phenyl [79]	*p*-Tolyl [80]	Hexyl [67]	*p*-^{n}*Butylphenyl* [94]
p-Anisyl [80]	*p*-Anisyl [80]	Octyl [67]	1-Naphthyl [92]
2,4,6-Me_3-C_6H_2 [81, 82]	9-MeO-1-Naphthyl [85]	Dodecyl [67]	2,4,6-Me_3-C_6H_2 [95]
2,4,6-$^i Pr_3$-C_6H_2 [83]	2,4,6-$^i Pr_3$-C_6H_2 [86]	^{n}Bu(phenyl) [76]	2,4,6-$^i Pr_3$-C_6H_2 [93, 96]
	2,4,6-$^t Bu_3$-C_6H_2 [87]	Phenyl [18, 92]	2,6-(2,6-$^i Pr_2$-C_6H_3)$_2$-C_6H_3 [93, 97]
	2,6-(2,6-$^i Pr_2$-C_6H_3)$_2$-C_6H_3 [88]	*p*-^{n}Bu-C_6H_4 [76]	
		p-^{t}Bu-C_6H_4 [37]	
		p-^{n}Hex-C_6H_4 [37]	
		o-Et-C_6H_4 [37]	
		p-^{n}BuO-C_6H_4 [37]	
		o-Et-*p*^{n}BuO-C_6H_4 [37]	
		p-$^n Bu_2$N-C_6H_4 [37]	
		p-(Me_3Si)$_2$N-C_6H_4 [37]	
		1-Naphthyl [92]	
		2-Naphthyl [92]	
		p-Biphenyl [92]	
		2,4,6-Me_3-C_6H_2 [63, 66]	
		2,4,6-$^i Pr_3$-C_6H_2 [93]	

Table 3.2 Single crystal X-ray diffraction structurally characterized group14 metal organohydrides.

R_2GeH_2	$RGeH_3$	R_2SnH_2	$RSnH_3$
	R =		
2,4,6-Me_3-C_6H_2 [82] 2,6-(2,4,6-^{i}Pr_3-C_6H_2)$_2$-C_6H_3 [100]	*p*-Anisyl [80] 2,4,6-^{t}Bu_3-C_6H_2 [87] 2,6-(2,6-^{i}Pr_2-C_6H_3)$_2$-C_6H_3 [88]	2,4,6-Me_3-C_6H_2 [66]	2,4,6-Me_3-C_6H_2 [95] 2,4,6-^{i}Pr_3-C_6H_2 [93] 2,6-(2,4,6-^{i}Pr_3-C_6H_2)$_2$-C_6H_3 [101]

Unsurprisingly, due to the high reactivity of the E—H bonds, very few group 14 organodi- and trihydrides have been structurally characterized. In the majority of cases, sterically encumbering residues are necessary to stabilize these species. Specifically, aryl residues with substituents at both ortho positions are capable of stabilizing the metal center, for example, the mesityl group (mesityl = 2,4,6-$Me_3C_6H_2$) is quite successful in providing access to solid group 14 organohydrides, (2,4,6-Me_3-C_6H_2)$_2$$GeH_2$ [82], (2,4,6-Me_3-C_6H_2)$_2$$SnH_2$ [66] (Figure 3.2), and (2,4,6-Me_3-C_6H_2)SnH_3 [95] (Figure 3.2). In the case of (*p*-anisyl)GeH_3 [80], in situ crystallization was used in order to provide X-ray quality crystals. The germanium-bonded hydrogen atoms were not found in the difference map and were thus placed at fixed positions and distances (Ge—H = 1.40 Å). The same issue was encountered for (2,4,6-^{t}Bu_3-C_6H_2)GeH_3 (Ge—H = 1.20 Å) [87]. In the case of (2,4,6-Me_3-C_6H_2)SnH_3 [95], the compound was stable at slightly below room temperature, and all hydrogen positions were found on the difference map (average Sn—H = 1.65(8) Å). For (2,6-(2,4,6-^{i}Pr_3-C_6H_2)$_2$-C_6H_3)SnH_3 [101] with the most sterically encumbering residue, all Sn—H positions were located in the difference map and were found to be disordered (average Sn—H = 1.51(2) Å). Table 3.3 summarizes all determined E—H bond lengths in these species.

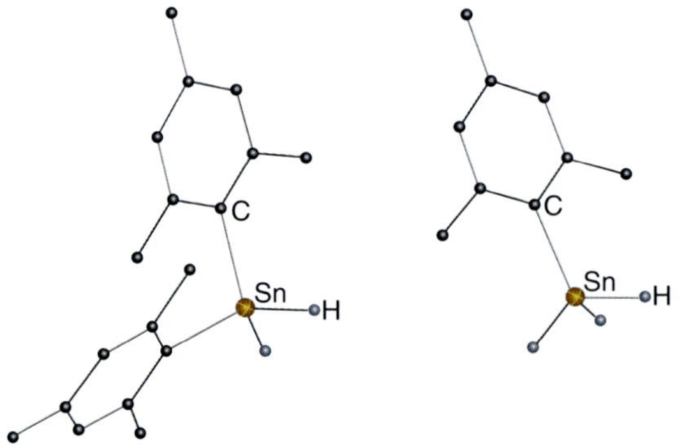

Figure 3.2 Crystal structure of (2,4,6-Me_3-C_6H_2)$_2$$SnH_2$ [66] and (2,4,6-Me_3-C_6H_2)SnH_3 [95]. All non-carbon atoms shown as 30% shaded ellipsoids. Hydrogen atoms (except Sn—H) removed for clarity.

Table 3.3 Summary of average E—H bond lengths in group 14 organodi- and trihydrides.

R_2EH_2	E—H (Å) (average)	REH_3	E—H (Å) (average)
$(2,4,6\text{-}Me_3\text{-}C_6H_2)_2GeH_2$ [82]	1.43(3)	(p-Anisyl)GeH_3 [80]	1.400
$(2,6\text{-}(2,4,6\text{-}^iPr_3\text{-}C_6H_2)_2\text{-}C_6H_3)_2GeH_2$ [100]	1.30(4)	$(2,4,6\text{-}^tBu_3\text{-}C_6H_2)GeH_3$ [87]	1.200
$(2,4,6\text{-}Me_3\text{-}C_6H_2)_2SnH_2$ [66]	1.67(2)	$(2,6\text{-}(2,6\text{-}^iPr_2\text{-}C_6H_3)_2\text{-}C_6H_3)GeH_3$ [88]	1.45(2)
		$(2,4,6\text{-}Me_3\text{-}C_6H_2)SnH_3$ [95]	1.66(2)
		$(2,4,6\text{-}^iPr_3\text{-}C_6H_2)SnH_3$ [93]	1.63(2)
		$(2,6\text{-}(2,4,6\text{-}^iPr_3\text{-}C_6H_2)_2\text{-}C_6H_3)SnH_3$ [101]	1.51(2)

3.3 Diorganodihydrides of Group 14, R_2EH_2, as Building Blocks for Chain-Type Polymers

3.3.1 Metal-Catalyzed Dehydropolymerization

The most common and widely applied method of generating tin-based organopolymers from corresponding dihydrides is by catalytic dehydropolymerization by a wide range of transition metal and lanthanide complexes [22, 30–32, 34, 37–41, 47, 61–69]. A variety of alkyl and aryl residues at the group 14 center is tolerated under these conditions (Scheme 3.2). In the case of tin, this synthetic route has resulted in polystannanes with a range of molecular weights ($M_w = (4.2–97) \times 10^3$ Da) and a range of polydispersities (PDI = 1.1–6.7) as summarized in a number of review articles [20, 56]. In contrast, only one example of catalytic dehydropolymerization of an organogermanium hydride to a polygermane has been reported (Table 3.4). The reaction of diphenyl germane (Ph_2GeH_2) in the presence of Cp_2TiMe_2 resulted in the formation of a hydride-terminated linear tetramer, $H(Ph_2Ge)_4H$ [69].

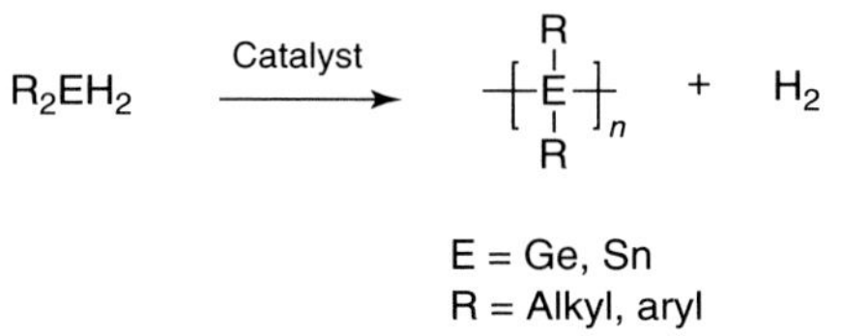

Scheme 3.2 Metal-catalyzed dehydrogenative coupling reaction of dioorgano group 14 dihydrides.

3.3.2 Dehydrogenative Coupling Using an Amine Base

Adding a nitrogen base such as pyridine to a tin hydride solution in order to afford the formation of a Sn—Sn bond was first described by Neumann and

Table 3.4 Group 14 metal-containing polymers synthesized from organodihydride (R_2EH_2) building blocks.

Building block	Method	Conditions	Polymer	M_w ($\times 10^3$) (Da)	Absorption maximum UV–vis (nm)	Repeat units	PDI
Ph_2GeH_2 [69]	Dehydropolymerization (3.3.1)	Cp_2TiMe_2, 70 °C, toluene	$H(Ph_2Ge)_4H$	—	300	—	—
$^nBu_2SnH_2$ [71]	Dehydropolymerization (3.3.3)	200 °C, low pressure, inert atm	$-[(^nBu)_2Sn]_n-$	18.0	370	—	6.9
$^nBu_2SnH_2$ [18]	Condensation (3.3.4)	$^nBu_2Sn(NEt_2)_2$, 0 °C, Et_2O, inert atm	$-[(^nBu)_2Sn]_n-$	18.1	379	77	2.3
Ph_2SnH_2 [76]	Dehydropolymerization (3.3.2)	TMEDA, RT, Et_2O, inert atm	$-[(Ph)_2Sn]_n-$	—	470	—	—
$(p\text{-}^nBuPh)_2SnH_2$ [76]	Dehydropolymerization (3.3.2)	TMEDA, RT, Et_2O, inert atm	$-[(p\text{-}^nBuPh)_2Sn]_n-$	46.0	420	120	3.2
$^nBu(Ph)\ SnH_2$ [76]	Dehydropolymerization (3.3.2)	TMEDA, RT, Et_2O, inert atm	$-[(^nBu)PhSn]_n-$	13.0	410	51	2.0
Ph_2SnH_2 [18]	Condensation (3.3.4)	$^nBu_2Sn(NEt_2)_2$, 0 °C, Et_2O, inert atm	$-[Ph_2Sn\text{-}alt\text{-}\ Sn(^nBu)_2]_n-$	18.8	390	74	2.9
Ph_2SnH_2 [18]	Condensation (3.3.4)	$Me_2Sn(NEt_2)_2$, 0 °C, Et_2O, inert atm	$-[Ph_2Sn\text{-}alt\text{-}SnMe_2]_n-$	242.5	395	1150	2.0
$^nBu_2SnH_2$ [18]	Condensation (3.3.4)	$Me_2Sn(NEt_2)_2$, 0 °C, Et_2O, inert atm	$-[Me_2Sn\text{-}alt\text{-}Sn(^nBu)_2]_n-$	82.1	361	180	2.9

atm, atmosphere; M_w, weight-average molecular weight; PDI, polydispersity; RT, room temperature.

Koenig [72, 73] and further developed by Davies, Mathiasch, and coworkers [74, 75]. Although further investigations on the reaction mechanism have been neglected thus far, it is proposed to occur via either a polar process or a radical chain mechanism. It is possible that the polar mechanism proceeds through a four-membered transition state involving an R_3SnX moiety [102, 103]. In the alternative radical pathway, an R_3Sn·radical would displace a hydrogen atom from another tin hydride resulting in a H·species, which subsequently leads to the production of H_2 through abstracting a hydrogen from another tin center.

In 2011, Uhlig and coworkers described the application of TMEDA as a convenient amine base generating poly(dialkylstannane)s as well as poly(diarylstannane)s (Scheme 3.3) [76]. Reaction of Ph_2SnH_2 with TMEDA led to a yellow precipitate. Elemental analysis revealed few impurities in the poly(diphenylstannane). As this polymer was insoluble in all solvents tested, the molar mass could not be determined. By attaching a butyl group in the para position of the phenyl substituent in ((p-^{n}BuPh)$_2$SnH$_2$), the increased solubility of the polyarylstannanes allowed for standard polymer characterization techniques, such as ^{119}Sn NMR and gel permeation chromatography (GPC). In this manner, (p-^{n}BuPh)$_2$SnH$_2$ was converted with TMEDA in Et_2O to result in –[(p-^{n}BuPh)$_2$Sn]$_n$– exhibiting a molar mass of 46×10^3 Da, which corresponds to approximately 120 (p-^{n}BuPh)$_2$Sn repeat units. Increased solubility was also achieved by use of a mixed residue stannane, ^{n}Bu(Ph)SnH$_2$. Both ^{n}Bu(Ph)SnH$_2$ and (p-^{n}BuPh)$_2$SnH$_2$, polymerized to highly viscous oils that were soluble in CH_2Cl_2, toluene, hexane, tetrahydrofuran (THF), and diethyl ether.

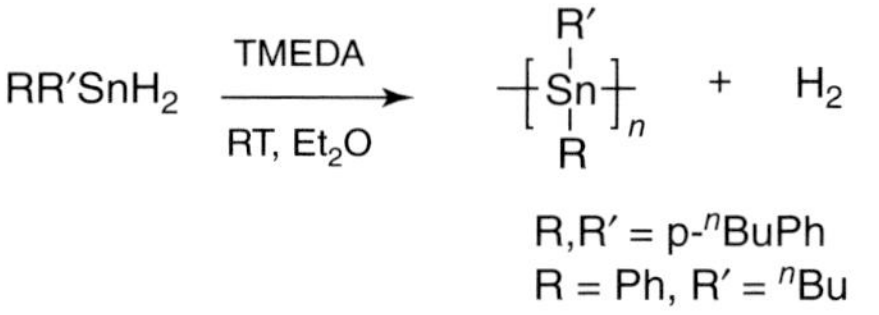

Scheme 3.3 Dehydrogenative coupling reaction of dioorganotin dihydrides in the presence of TMEDA.

3.3.3 Solvent- and Catalyst-Free Dehydrogenative Coupling

In 2010, Foucher and coworkers reported the successful preparation of oligo- and polystannanes by thermal dehydrocoupling under inert conditions (Scheme 3.4). ^{n}Bu$_2$SnH$_2$ was heated for six hours yielding a hydride-terminated distannane, H(^{n}Bu)$_2$SnSn(^{n}Bu)$_2$H, various cyclic stannanes, ((^{n}Bu)$_2$Sn)$_{n=5,6}$, and the polystannane, –[(^{n}Bu)$_2$Sn]$_n$–. The generated polystannane showed modest molecular weight ($M_w = 18 \times 10^3$ Da), broad PDI (PDI = 6.9), and a characteristic UV–vis absorbance (λ_{max}) of ≈370 nm [71].

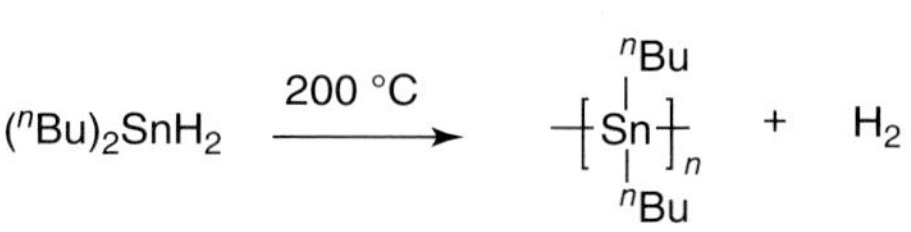

Scheme 3.4 High-temperature dehydrogenative coupling reaction of dibutyltin dihydride.

3.3.4 Condensation

While condensation reactions involving stannylamines with organotin hydrides have been reported to form cyclostannanes by Sommer et al. [104] and Uhlig and coworkers [105], the route has also been used to isolate alternating oligostannanes. Stepwise condensation of $(^n\mathrm{Bu})_2\mathrm{SnH}_2$ with $(^n\mathrm{Bu})_2\mathrm{Sn(OCH_2CH_2OEt)(NMe_2)}$ to produce linear oligostannanes containing 3–15 tin atoms in the presence of a strong base was reported by Sita et al. [30]. Recently, Foucher and coworkers reported on a synthetic protocol for the generation of alternating polystannanes, which involved the stoichiometric reaction of organotin dihydrides and organotin diamides in diethyl ether or toluene under mild reaction conditions (Scheme 3.5). This procedure afforded three novel moderate molecular weight alternating polystannanes in yields up to 86%: $-[\mathrm{Ph_2Sn}\text{-}alt\text{-}\mathrm{Sn}(^n\mathrm{Bu})_2]_n-$, $-[\mathrm{Ph_2Sn}\text{-}alt\text{-}\mathrm{SnMe_2}]_n-$, and $-[\mathrm{Me_2Sn}\text{-}alt\text{-}\mathrm{Sn}(^n\mathrm{Bu})_2]_n-$ [18]. By introducing both, diaryl and dialkyl substituents into the tin backbone in an alternating fashion, solubility as well as light stability of the product improved in contrast to homopolymers such as $-[(\mathrm{Ph})_2\mathrm{Sn}]_n-$ [18, 25].

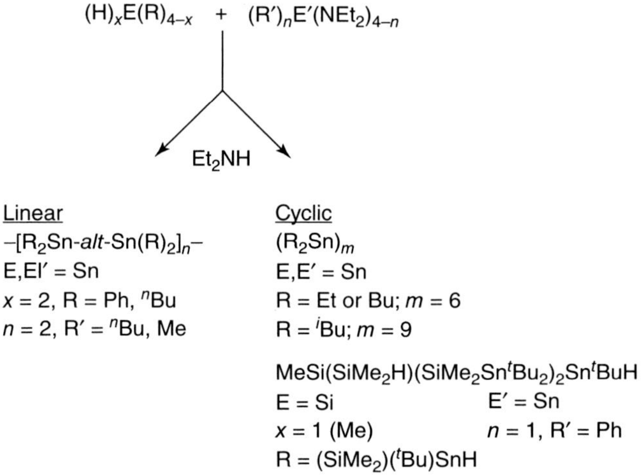

Scheme 3.5 Preparation of alternating polystannanes by condensation [18, 104, 105].

3.4 Monoorganotrihydrides of Group 14, REH_3, as Building Blocks for 3D Polymers

The conversion of monoorganotrihydrides of group 14 (REH_3) as trifunctionalized educts toward 3D polymers, which had been neglected thus far, presents a promising starting point to the formation of nanostructured organometallic particles.

3.4.1 Metal-Catalyzed Dehydropolymerization

In contrast to the well-studied and well-reviewed metal-catalyzed dehydropolymerization reaction of tin dihydrides, there are no equivalent studies employing tin trihydrides. In the case of germanium, $PhGeH_3$ was shown to be catalytically converted to partially cross-linked polyphenylgermanes of high molecular weights (M_w = 40–70 × 10^3) by Cp_2ZrCl_2 or $Cp^*CpZrCl_2$ at low temperature (−78 °C) in hexanes, albeit in moderate yields (Table 3.5) [106]. These polyphenylgermanes were found to be soluble in common solvents. Unlike heavily cross-linked polygermynes prepared by Wurtz coupling [108], the resulting polymer appears to behave, at least electronically, more like regular polygermanes.

3.4.2 Dehydrogenative Coupling Using an Amine Base

Sindlinger and coworkers reported on the base-catalyzed reductive dihydrogen elimination from the sterically very encumbered aryltin trihydride Ar^*SnH_3 (Ar^* = 2,6-$trip_2(C_6H_3)$, trip = 2,4,6-triisopropylphenyl). The resulting aryltin (II) hydride species could be obtained with a variety of amine bases such as pyridine, DMAP (4-dimethylaminopyridine) and TMEDA. With catalytic amounts of base, the dehydrocoupling product diorganodistannane $Ar^*H_2Sn{-}SnH_2Ar^*$ was obtained quantitatively. In contrast, a large excess of the base (>4 equiv) led to almost exclusive formation of the thermally quite robust monomeric base adduct Ar^*SnH. Owing to the considerable bulk of the aryl residue of the converted aryltin trihydride Ar^*SnH_3, the formation of larger oligomeric Sn species by full hydrogen release was not observed. Apparently, neither the distannane nor the base-stabilized stannylene species allows for further hydrogen abstraction. Conversely, the addition of an amine base to group 14 monoaryltrihydrides with less encumbering aryl residues (*o*-tolyl) leads to dehydrogenative coupling to novel, 3D polymeric materials (Scheme 3.6), which can be described as aryl decorated nanoparticles (E@aryl) [95, 107].

The coordination of TMEDA to the Sn atom of the starting material and the ensuing Sn—H bond elongation and weakening facilitates the formation of H_2 and therefore initializes the dehydrogenative coupling reaction of *o*-$tolylSnH_3$. The reaction of *o*-$tolylSnH_3$ with TMEDA proceeds from a colorless solution to a distinct color change, consistent with that exhibited by tin compounds in an oxidation state +IV, to a yellow and then an orange precipitate with the evolution of hydrogen (Figure 3.3) indicating tin in an oxidation state +II. A similar color change due to the generation of Sn—Sn bonds that allow for σ–σ^* and σ–π transitions, as well as the generation of hydrogen, had previously been reported for the dehydrogenative coupling reaction of Ph_2SnH_2 to $-[(Ph_2)Sn]_n-$ applying pyridine or TMEDA [72–76]. The yellow color had been taken as an indication for the formation of linear polystannanes. The presence of remaining Sn—H bonds allows for branching and or cross-linking at the metal sites thus propagating 3D networks. The loss of both H_2 and the cleavage of Sn—C(aryl) bonds results in particles containing Sn(0) metal centers (Scheme 3.7). The particles are isolated as highly air- and moisture-sensitive black polymeric and insoluble materials (Figure 3.3).

As the black precipitate is not soluble in common organic solvents, standard polymer characterization techniques (e.g. liquid state ^{1}H, ^{13}C, ^{119}Sn NMR, GC–MS, GPC) could not be applied. According to GC–MS, toluene (arylH)

Table 3.5 Group 14 metal-containing polymers synthesized from organomonohydride (REH_3) building blocks.

Building block	Method	Conditions	Polymer	M_w ($\times 10^3$) (Da)	UV–vis (nm)
$PhGeH_3$ [106]	Dehydropolymerization (3.4.1)	$[(PhGeH)_1(PhGe)_m]_n$	$-[(PhGeH)_1(PhGe)_m]_n-$	73	329
$PhGeH_3$ [106]	Dehydropolymerization (3.4.1)	Cp_2ZrCl_2 (0.7 mol%, 24 hours), ^{n}BuLi −78 °C, hexanes	$-[(PhGeH)_1(PhGe)_m]_n-$	35	329
$PhGeH_3$ [106]	Dehydropolymerization (3.4.1)	Cp_2ZrCl_2 (0.7 mol%, 72 hours), ^{n}BuLi −78 °C, hexanes	$-[(PhGeH)_1(PhGe)_m]_n-$	41	329
$PhGeH_3$ [106]	Dehydropolymerization (3.4.1)	$CpZrCl_2$ (1 mol%, 120), ^{n}BuLi −78 °C, hexanes	$-[(PhGeH)_1(PhGe)_m]_n-$	44	329
$PhGeH_3$ [106]	Dehydropolymerization (3.4.1)	$Cp^*CpZrCl_2$ (1 mol%), ^{n}BuLi −78 °C, hexanes	$-[(PhGeH)_1(PhGe)_m]_n-$	37	329
$PhGeH_3$ [106]	Dehydropolymerization (3.4.1)	$Cp^*CpZrCl_2$ (1 mol%, 72 hours), ^{n}BuLi −78 °C, hexanes	$-[(PhGeH)_1(PhGe)_m]_n-$	43	329
o-TolylSnH_3 [107]	Dehydropolymerization (3.4.2)	TMEDA, RT, Et_2O, inert atm	Sn@*o*-tolyl		459–527

atm, atmosphere; M_w, weight-average molecular weight; RT, room temperature.

Scheme 3.6 Dehydrogenative coupling reaction of monoorgano group 14 trihydrides in the presence of an amine base.

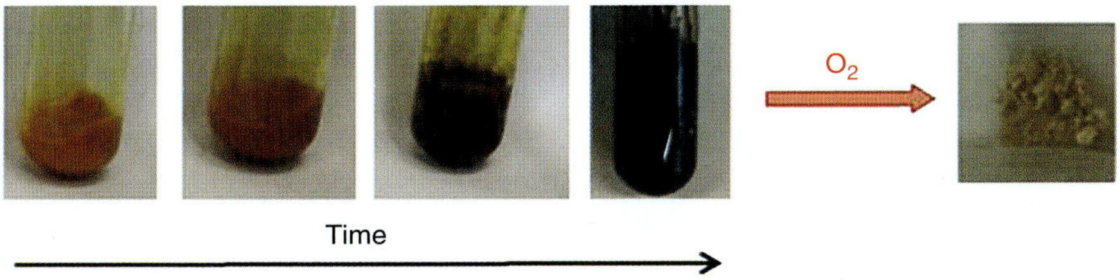

Figure 3.3 Visible reaction progress in the dehydrogenative coupling reaction of *o*-tolylSnH_3 forming Sn@*o*-tolyl. Subsequent color change of black material upon exposure to oxygen.

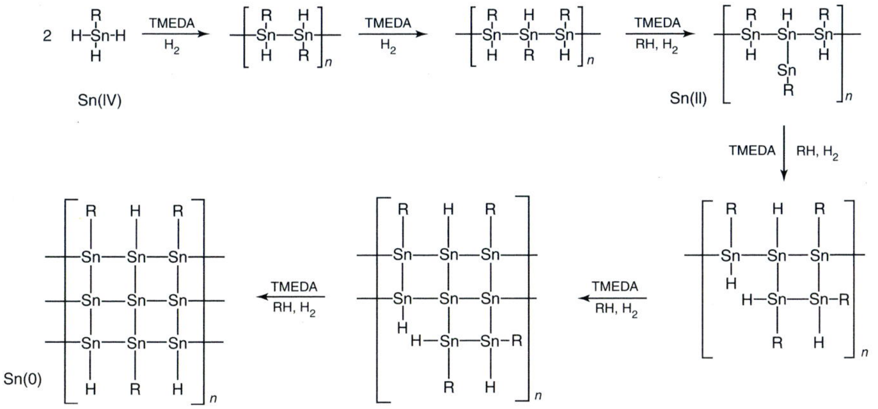

Scheme 3.7 Polymer propagation from linear (Sn(II)) to branched 3D polymeric systems (Sn(0)).

is detected as a side product, readily explained by Sn—C bond cleavage. Upon exposure to air, the product easily undergoes oxidation, changing color from black to yellowish white in the process (Figure 3.3). This oxidation process to tin oxides is noticeably exothermic, which suggests the presence of a high number of reaction sites due to an increased surface area. In situ UV measurements of the dehydrogenative coupling reaction of *o*-tolylSnH_3 in diethyl ether showed three absorptions at lower energies at 459, 480, and 527 nm (Table 3.5).

Field Emission Scanning Electron Microscopy (FESEM) measurements of the black, insoluble powder show a discrete spherical morphology. The particles seen in FESEM exhibit a diameter of approximately 1 μm (Figure 3.4a,b). Increased magnification displays that these spheres (superstructures) are composed of smaller subunits, which seem to also be of spherical nature in a size range of 7–30 nm (Figure 3.4c). In situ small-angle X-ray scattering (SAXS) measurements revealed a discrete nanostructure in the material with a correlation length

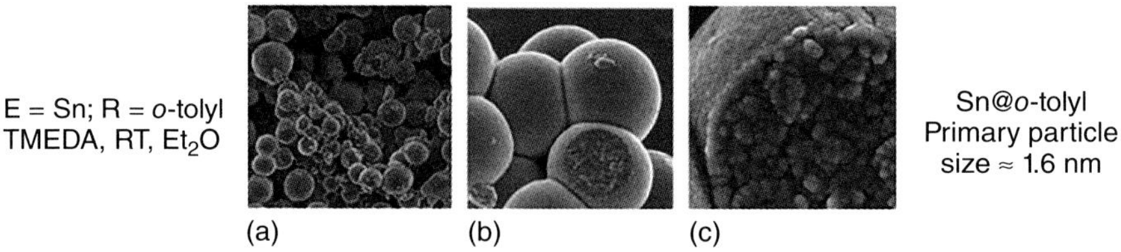

Figure 3.4 FESEM images of Sn@*o*-tolyl. (a, b) Spherical superstructures (≈1 μm). (c) Spherical substructures (≈7–30 nm).

of ≈1.6 nm. The nanosubstructures agglomerate over reaction time under the formation of superstructures in the 1 μm range (Figure 3.4a).

3.5 Applications

The quest for novel electrode materials is one of the key research topics since lithium ion batteries (LIBs) have come into focus as power sources of vehicles and portable electronics [109, 110]. Owing to the increasing energy consumption of today's world and the demand for novel energy storage systems, the quest for novel materials for this purpose has become a central task [111–117]. In the past, germanium has been considered as Li-allyoing electrode material for next generation lithium ion batteries, for possibly improving the capacity compared to standard graphite anodes [112, 118–121]. Graphite itself has a relatively low theoretical capacity of 372 mAh/g. Incorporating Ge as a Li- alloying anode leads to an increase of theoretical capacity about four times higher than that of the commercial graphite anode (to 1623 mAh/g). Although the theoretical capacity of the Li–Ge system is by far lower compared to the theoretical capacity of Li–Si, Ge has the big advantage that the diffusion rate of Li^+ in Ge is significantly higher and furthermore shows a higher electrical conductivity (10^4 times higher than that of silicon) [120].

Tin has recently gained considerable attention among various interesting candidates in the research field of novel anode materials due to its high theoretical capacity of 944 mAh/g, which is in accordance with the formation of $Li_{22}Sn_{59}$ [122]. Therefore, a series of tin-based materials, such as tin dioxide (SnO_2) [123], tin disulfide (SnS_2) [124], tin diselenide ($SnSe_2$) [125] as well as intermetallic compounds such as nickel (Ni–Sn) [126] and cooper (Cu–Sn) [127] have been investigated as potential high-energy density anode material since all of them show higher capacity than the commercialized graphite anodes (~372 mAh/g). Despite the high abundance and low price of tin, its application has not been commercialized yet, since one of the main drawbacks of Sn-based anode materials is the severe capacity fading upon cycling caused by the enormous volume change of the metal (about 300%) [126, 128, 129]. Thus, the structure of these promising Sn-based materials ought to be tuned to effectively accommodate the volume expansion [111, 130].

This issue is proposed to be elegantly circumvented by utilizing electrode materials at the nanoscale, which are reported to provide significantly improved performance of LIBs [109, 110, 129, 131, 132]. This is due to an increased number of electrochemically active sites as well as to a high ability to buffer the stress, which is induced upon lithiation and delithiation. In particular, nanostructured composites of carbon and tin have been proposed as a high capacity substitute for the carbon anode of lithium ion batteries. A variety of Sn–carbon hybrids such as Sn nanoparticles embedded in porous carbon [126, 133–141], Sn or Sn@C nanoparticles encapsulated with carbon nanotubes [142–146], and carbon-encapsulated Sn nanostructures [147–150] have been designed. Deliberately manufactured composites of tin and carbon on the nanoscale may solve the poor cyclability of the tin electrodes as they provide space to buffer mechanical stress.

Incorporation of novel E@aryl nanoparticles into carbon-based systems could be achieved through either mixing the already formed nanoparticles into a graphite layer or from the E@aryl particles in situ in the presence of graphite (Scheme 3.8).

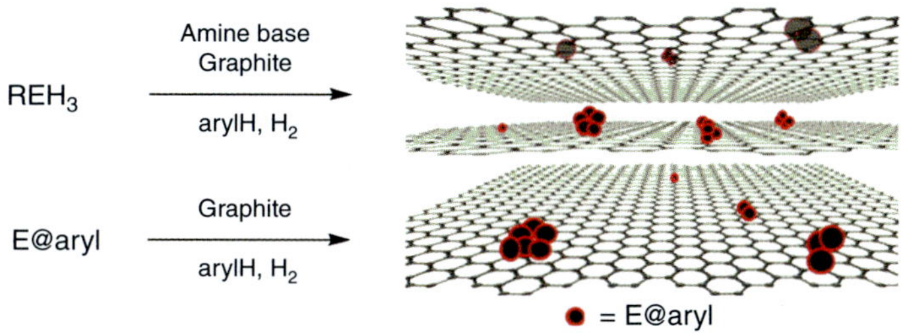

Scheme 3.8 Formation of novel nanostructured composites of carbon and group 14 metal-based nanoparticles.

Efforts to incorporate Sn@C nanoparticles into graphite surfaces produced via the dehydrogenative coupling reaction in the presence of an amine base have produced promising results. In both incorporation methods, elemental tin could be detected in CV experiments (Figure 3.5) [151]. Owing to the presence of tin, the graphite redox properties were slightly affected by the resulting broadening of the respective peaks (Figure 3.5a). The CV measurements of Sn@aryl particles in Li-ion battery half-cells show reduction and oxidation processes. These are most pronounced during the first cycle and fade quickly over the course of a few cycles, caused by the swelling of tin within the electrodes leading to contact loss or by the incomplete delithiation of the tin particles. The swelling was shown to be mitigated by the use of graphite. When Sn@aryl particles were used without the addition of graphite (Figure 3.5b), it was observed that all intercalation steps occur over an enormously broadened potential range during the first cycle. However, the intercalation of lithium into the Sn particles seems to occur only during the first cycle as, in the second cycle, the oxidation and reduction curves are minimized to one peak each, lying at 0.2 V for the reduction and at approximately 0.5 V for the oxidation process. These peaks indicate that after the first cycle only the carbon in the samples, be it the additive carbon black or the aryl substituents "decorating" the tin particles, are electrochemically active.

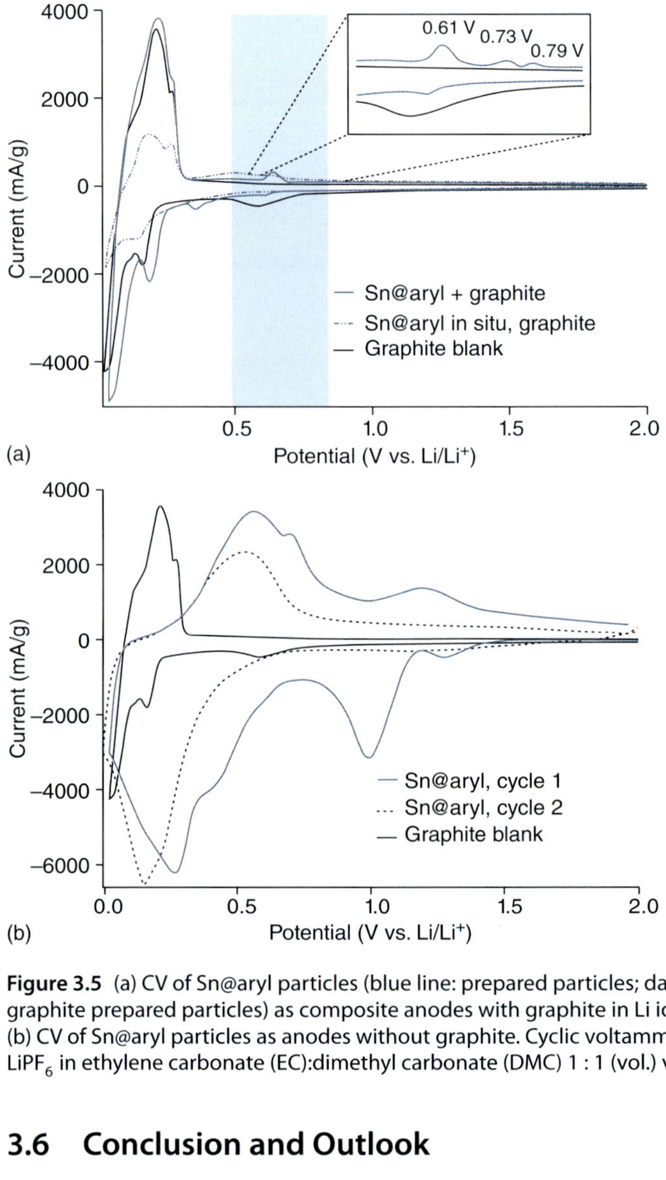

Figure 3.5 (a) CV of Sn@aryl particles (blue line: prepared particles; dashed line: in situ in graphite prepared particles) as composite anodes with graphite in Li ion batteries at 1 mV/s. (b) CV of Sn@aryl particles as anodes without graphite. Cyclic voltammetry was measured in $LiPF_6$ in ethylene carbonate (EC):dimethyl carbonate (DMC) 1 : 1 (vol.) vs. Li in Li ion half-cells.

3.6 Conclusion and Outlook

While being structurally analogous to saturated polymeric hydrocarbons and polysilanes, group 14 metal-containing linear polymers have the capacity to catenate. Heavier germanium and tin atoms in the backbone of σ-conjugated polymers give stronger σ-conjugation and narrower bandgaps and induce more metallic character to the resulting material as compared to their silicon counterparts and can be described as molecular wires embedded in an organic jacket. A number of group 14 aliphatic and aromatic high molecular weight metal-containing linear polymers have been prepared in a variety of synthetic methods involving group 14 metal chloride precursors (Wurtz coupling, electrochemical synthesis). However, these synthetic methods have typically resulted in the formation of a cyclic byproduct, low yields, or remaining metal–halide bonds in the polymer. In addition, issues including polymer solubility and light and moisture sensitivity have hindered their application in the field of electronics.

More recently, the dehydropolymerization synthetic approach employing group 14 metal dihydrides has become more attractive. Whether in the presence of a metal-based catalyst or amine base, reactions from group 14 metal dihydride precursors have resulted in a number of aliphatic and aromatic linear polymers with modest to high molecular weight and a range of polydispersities under mild reaction conditions. Owing to the sensitivity toward elevated temperatures of the E—H bond, solvent- and catalyst-free dehydrogenative coupling reactions can also be employed. Condensation routes involving reacting stannylamines with organotin hydrides have been used to isolate moderate molecular weight alternating oligostannanes, which show increased solubility as well as light stability by introducing both diaryl and dialkyl substituents alternatively into the metal backbone.

Contrary to the well-studied and well-reviewed metal dehydropolymerization reaction of group 14 metal dihydrides, the conversion of monoorganotrihydrides of group 14 (REH_3) as trifunctionalized educts has been largely neglected, possibly due to their high sensitivity. Nevertheless, through the dehydropolymerization reaction of *o*-tolySnH_3 in the presence of TMEDA, novel, 3D polymeric materials which can be described as aryl decorated nanoparticles (E@aryl) have been obtained. These particles are found to be spherical in nature and due to the loss of both H_2 and the cleavage of Sn—C(aryl) bonds, result in particles containing Sn(0) metal centers. The increased metallic nature of these novel polymeric materials has significant implications in potential use in electronic applications. Ongoing research dedicated to incorporating these novel materials into lithium ion batteries has been promising and should serve as a starting point for research into the synthesis, characterization, and uses of this new class of group 14 metal-based materials.

Acknowledgments

The authors thank the COST Action CM1302 (SIPs): "European Network on Smart Inorganic Polymers." F. Uhlig and A. Torvisco acknowledge support from the Graz University of Technology.

References

1 Miller, R.D. and Michl, J. (1989). Polysilane high polymers. *Chem. Rev.* 89: 1359–1410.

2 West, R. (1986). The polysilane high polymers. *J. Organomet. Chem.* 300: 327–346.

3 West, R. (1989). Polysilanes. In: *The Chemistry of Organosilicon Compounds* (ed. S. Patai and Z. Rappoport), 1207–1240. New York: Wiley.

4 West, R. (1995). Comprehensive organometallic chemistry. In: *Comprehensive Organometallic Chemistry II* (ed. A.G. Davies), 77–110. Oxford, U.K.: Pergamon Press.
5 West, R. (2003). Polysilanes: conformations, chromotropism and conductivity. In: *The Chemistry of Organic Silicon Compounds* (ed. Z. Rappoport and Y. Apeloig), 541–563. John Wiley & Sons.
6 Adams, S. and Draeger, M. (1988). Quantitative relationships between molecular structure, NMR data, and electronic transitions in linear and branched organopolystannanes. *Main Group Met. Chem.* 11: 151–180.
7 Sita, L.R. (1994). Heavy-metal organic chemistry: building with tin. *Acc. Chem. Res.* 27: 191–197.
8 Sita, L.R. (1995). Structure/property relationships of polystannanes. *Adv. Organomet. Chem.* 38: 189–243.
9 Manners, I. (1996). Polymers and the periodic table: recent developments in inorganic polymer science. *Angew. Chem., Int. Ed.* 35: 1602–1621.
10 Jurkschat, K. and Mehring, M. (2002). *Organometallic Polymers of Germanium, Tin and Lead.* John Wiley & Sons.
11 Davies, A.G. (2004). Organotin hydrides. In: *Organotin Chemistry*, 244–265. Wiley-VCH Verlag.
12 Manners, I. (2004). Polymers with metal–metal bonds in the main chain. In: *Synthetic Metal-Containing Polymers*, 181–202. Wiley-VCH Verlag.
13 Carraher, C.E. Jr., Pittman, C.U. Jr., Zeldin, M., and Abd-El-Aziz, A.S. (2005). Organogermanium polymers. *Macromol. Contain. Met. Met. Like Elem.* 4: 225–261.
14 Choffat, F., Smith, P., and Caseri, W. (2008). Polystannanes: polymers of a molecular, jacketed metal-wire structure. *Adv. Mater.* 20: 2225–2229.
15 Sharma, H.K. and Pannell, K.H. (2008). Organotin polymers and related materials. In: *Tin Chemistry: Fundamentals, Frontiers, and Applications* (ed. A.G. Davies, M. Gielen, K.H. Pannell and E.R.T. Tiekink), 376–391. John Wiley & Sons.
16 Trummer, M., Choffat, F., Smith, P., and Caseri, W. (2012). Polystannanes: synthesis, properties, and outlook. *Macromol. Rapid Commun.* 33: 448–460.
17 Watanabe, A. (2014). Ge- and Sn-containing polymers. In: *Encyclopedia of Polymeric Nanomaterials* (ed. S. Kobayashi and K. Müllen), 1–10. Berlin, Heidelberg: Springer.
18 Harrypersad, S. and Foucher, D. (2015). Alternating polystannanes: syntheses and properties. *Chem. Commun.* 51: 7120–7123.
19 Caseri, W. (2016). Polystannanes: processible molecular metals with defined chemical structures. *Chem. Soc. Rev.* 45: 5187–5199.
20 Foucher, D. Catenated germanium and tin oligomers and polymers. In: *Main Group Strategies towards Functional Hybrid Materials* (ed. T. Baumgartner and F. Jäkle), 209–236. Wiley.
21 de Lima, G.M., Toupance, T., Arkis, E. et al. (2008). Materials chemistry and structural chemistry of tin compounds. In: *Tin Chemistry* (ed. A.G. Davies, M. Gielen, K.H. Pannell and E.R.T. Tiekink), 285–411. John Wiley & Sons.
22 Lu, V. and Tilley, T.D. (1996). Low-band-gap, σ-conjugated polymers: poly(diarylstannanes). *Macromolecules* 29: 5763–5764.

23 Sita, L.R. (1992). A new strategy for the synthesis of homologously pure linear polystannane oligomers. *Organometallics* 11: 1442–1444.
24 Sita, L.R., Terry, K.W., and Shibata, K. (1995). Sandorfy Hueckel molecular orbital approximation for modeling the electronic structures of long-chain polystannanes. *J. Am. Chem. Soc.* 117: 8049–8050.
25 Harrypersad, S., Liao, L., Khan, A. et al. (2015). Linear oligostannanes: a synthetic and TD-DFT study. *J. Inorg. Organomet. Polym.* 25: 515–528.
26 Devylder, N., Hill, M., Molloy, K.C., and Price, G.J. (1996). Wurtz synthesis of high molecular weight poly(dibutylstannane). *Chem. Commun.* 711–712.
27 Haeberle, K. and Draeger, M. (1987). Polygermanes. XVIII. Synthesis of the multifunctional digermanes, $Ph_{6-n}Ge_2Cl_n$ ($n = 2, 3, 4$) and of Ph_2GeCl_2 under pressure. *Z. Naturforsch. B Chem. Sci.* 42: 323–329.
28 Okano, M., Matsumoto, N., Arakawa, M. et al. (1998). Electrochemical synthesis of dialkylsubstituted polystannanes and their properties. *Chem. Commun.* 1799–1800.
29 Zou, W.K. and Yang, N.L. (1992). Synthesis and characterization of high-molecular-weight linear poly(*n*-dibutylstannane). *Polym. Prepr. (Am. Chem. Soc., Div. Polym. Chem.)* 33: 188–189.
30 Babcock, J.R. and Sita, L.R. (1996). Highly branched, high molecular weight polystannane from dibutylstannane via a novel dehydropolymerization/rearrangement process. *J. Am. Chem. Soc.* 118: 12481–12482.
31 Imori, T., Lu, V., Cai, H., and Tilley, T.D. (1995). Metal-catalyzed dehydropolymerization of secondary stannanes to high molecular weight polystannanes. *J. Am. Chem. Soc.* 117: 9931–9940.
32 Imori, T. and Tilley, T.D. (1993). High molecular mass polystannanes via dehydropolymerization of di-butylstannane. *J. Chem. Soc., Chem. Commun.* 1607–1609.
33 Miles, D., Burrow, T., Lough, A., and Foucher, D. (2010). Wurtz coupling of perfluorinated dichlorostannanes. *J. Inorg. Organomet. Polym. Mater.* 20: 544–553.
34 Tilley, T.D. and Imori, T. (1996). *High Molecular Weight Polystannanes by Metal-Catalyzed Dehydropolymerization*, 15. USA: The Regents of the University of California.
35 Trummer, M., Solenthaler, D., Smith, P., and Caseri, W. (2011). Poly(dialkylstannane) and poly(diarylstannane) homo- and random copolymers synthesized in liquid ammonia. *RSC Adv.* 1: 823–833.
36 Trummer, M., Zemp, J., Sax, C. et al. (2011). Reaction products of dichlorodiorganostannanes with sodium in liquid ammonia: in-situ investigations with ^{119}Sn NMR spectroscopy and usage as intermediates for the synthesis of tetraorganostannanes. *J. Organomet. Chem.* 696: 3041–3049.
37 Lu, V.Y. and Tilley, T.D. (2000). Poly(diaryl)stannanes: influence of substituents on the σ–σ transition energy. *Macromolecules* 33: 2403–2412.
38 Neale, N.R. and Tilley, T.D. (2004). Reactions of hafnocene stannyl complexes with stannanes: implications for the mechanism of the metal-catalyzed dehydropolymerization of secondary stannanes. *Tetrahedron* 60: 7247–7260.

39 Harada, T. (1939). Metallo-organic compounds. IV. Alkyltin compounds and their derivatives. *Sci. Pap. Inst. Phys. Chem. Res. (Jpn.)* 35: 290–329.
40 Thompson, S.M. and Schubert, U. (2003). Dehydrogenative stannane coupling by platinum complexes. *Inorg. Chim. Acta* 350: 329–338.
41 Choffat, F., Buchmueller, Y., Mensing, C. et al. (2009). Poly(di(ω-alkylphenyl)stannane)s. *J. Inorg. Organomet. Polym. Mater.* 19: 166–175.
42 Adams, S. and Draeger, M. (1987). Polystannanes. 5. Polystannanes $Ph_3Sn(tert\text{-}Bu_2Sn)_nSnPh_3$ ($n = 1$–4) – on the route to molecular metals? *Angew. Chem.* 99: 1280–1282.
43 Adams, S. and Dräger, M. (1987). Polystannanes $Ph_3Sn–SnPh_3$ ($n = 1$–4): a route to molecular metals? *Angew. Chem. Int. Ed.* 26: 1255–1256.
44 Takeda, K. and Shiraishi, K. (1992). Electronic structure of chain-like polystannane. *Chem. Phys. Lett.* 195: 121–126.
45 Drenth, W., Janssen, M.J., van der Kerk, G.J.M., and Vliegenthart, J.A. (1964). Studies in group IV organometallic chemistry: XIV. Ultraviolet absorption due to bonds between group IV metals. *J. Organomet. Chem.* 2: 265–270.
46 Drenth, W., Noltes, J.G., Bulten, E.J., and Creemers, H.M.J.C. (1969). Ultraviolet absorption spectra of perethylpolygermanes and -polystannanes. *J. Organomet. Chem.* 17: 173–174.
47 Skotheim, T.A. (1997). *Handbook of Conducting Polymers*, 2ee. New York: CRC Press.
48 Brédas, J.L. and Silbey, R. (2012). *Conjugated Polymers: The Novel Science and Technology of Highly Conducting and Nonlinear Optically Active Materials*. Netherlands: Springer.
49 Carraher, C.E., Currell, B., Pittman, C. et al. (2012). *Inorganic and Metal-Containing Polymeric Materials*. USA: Springer.
50 Mallela, S.P., Saar, Y., Hill, S., and Geanangel, R.A. (1999). Reactions of $LiE(SiMe_3)_3$, E = Si, Ge: X-ray crystal structure of the cyclotetrastannane $[ClSnSi(SiMe_3)_3]_4$. *Inorg. Chem.* 38: 2957–2960.
51 Deacon, P.R., Devylder, N., Hill, M.S. et al. (2003). Organotin compounds bearing mesogenic sidechains: synthesis, X-ray structures and polymerisation chemistry. *J. Organomet. Chem.* 687: 46–56.
52 Choffat, F., Wolfer, P., Smith, P., and Caseri, W. (2010). Light-stability of poly(dialkylstannane)s. *Macromol. Mater. Eng.* 295: 210–221.
53 Colton, R. and Dakternieks, D. (1988). Tin-119 NMR studies on adduct formation and stereochemistry of organoyltin(IV)trihalides. *Inorg. Chim. Acta* 143: 151–159.
54 Mustafa, A., Achilleos, M., Ruiz-Iban, J. et al. (2006). Synthesis and structural characterisation of various organosilane–organogermane and organosilane–organostannane statistical copolymers by the Wurtz reductive coupling polymerisation: ^{119}Sn NMR and EXAFS characterisation of the stannane copolymers. *React. Funct. Polym.* 66: 123–135.
55 Priegert, A.M., Rawe, B.W., Serin, S.C., and Gates, D.P. (2016). Polymers and the *p*-block elements. *Chem. Soc. Rev.* 45: 922–953.
56 Gielen, M., Davies, A.G., Pannell, K., and Tiekink, E. (2008). *Tin Chemistry: Fundamentals, Frontiers, and Applications*. Wiley.

57 Dehnen, S. (2009). Tin chemistry. Fundamentals, frontiers, and applications. Herausgegeben von Alwyn G. Davies, Marcel Gielen, Keith H. Pannell und Edward R. T. Tiekink. *Angew. Chem.* 121: 4164–4165.
58 Okano, M., Watanabe, K., and Totsuka, S. (2003). Electrochemical synthesis of network polystannanes. *Electrochemistry* 71: 257–259.
59 Katz, S.M., Reichl, J.A., and Berry, D.H. (1998). Catalytic synthesis of poly(arylmethylgermanes) by demethanative coupling: a mild route to σ-conjugated polymers. *J. Am. Chem. Soc.* 120: 9844–9849.
60 Reichl, J.A. (1998). The catalytic demethanative coupling of tertiary germanes: a new and efficient route to polygermanes. Doctoral thesis. Philadelphia: Univeristy of Pennsylvania, p. 335.
61 Choffat, F., Smith, P., and Caseri, W. (2005). Facile synthesis of linear poly(dibutylstannane). *J. Mater. Chem.* 15: 1789–1792.
62 Bukalov, S.S., Leites, L.A., Lu, V., and Tilley, T.D. (2002). Order–disorder phase transition in poly(di-*n*-butylstannane) observed by UV–Vis and Raman spectroscopy. *Macromolecules* 35: 1757–1761.
63 Neale, N.R. and Tilley, T.D. (2002). A new mechanism for metal-catalyzed stannane dehydrocoupling based on *α*-H-elimination in a hafnium hydrostannyl complex. *J. Am. Chem. Soc.* 124: 3802–3803.
64 Woo, H.-G., Park, J.-M., Song, S.-J. et al. (1997). Catalytic dehydropolymerization of di-*n*-butylstannane $n\text{-}Bu_2SnH_2$ by group 4 and 6 transition metal complexes. *Bull. Korean Chem. Soc.* 18: 1291–1295.
65 Woo, H.-G., Song, S.-J., and Kim, B.-H. (1998). Redistribution/dehydrocoupling of tertiary alkylstannane $n\text{-}Bu_3SnH$ catalyzed by group 4 and 6 transition metal complexes. *Bull. Korean Chem. Soc.* 19: 1161–1164.
66 Schittelkopf, K., Fischer, R.C., Meyer, S. et al. (2010). Catalytic dehydrogenative coupling of diorganotindihydrides by lanthanide diamide complexes. *Appl. Organomet. Chem.* 24: 897–901.
67 Choffat, F., Käser, S., Wolfer, P. et al. (2007). Synthesis and characterization of linear poly(dialkylstannane)s. *Macromolecules* 40: 7878–7889.
68 Thompson, S.M. and Schubert, U. (2004). Formation of cyclo- and polystannanes by dehydrogenative stannane coupling catalyzed by platinum(II) complexes. *Inorg. Chim. Acta* 357: 1959–1964.
69 Aitken, C., Harrod, J.F., Malek, A., and Samuel, E. (1988). Oligomerization of phenylgermanes by catalytic dehydrocoupling. *J. Organomet. Chem.* 349: 285–291.
70 Sharma, H.K., Arias-Ugarte, R., Metta-Magana, A.J., and Pannell, K.H. (2009). Dehydrogenative dimerization of di-tert-butyltin dihydride photochemically and thermally catalyzed by iron and molybdenum complexes. *Angew. Chem. Int. Ed.* 48: 6309–6312.
71 Khan, A., Gossage, R.A., and Foucher, D.A. (2010). A convenient route to distannanes, oligostannanes, and polystannanes. *Can. J. Chem.* 88: 1046–1052.
72 Neumann, W.P. and Koenig, K. (1962). The preparation and structure of diphenyltin. *Angew. Chem.* 74: 215.
73 Neumann, W.P. and Koenig, K. (1964). Organotin compounds. VII. Diphenyltin. *Justus Liebigs Ann. Chem.* 677: 1–11.

74 Mathiasch, B. (1977). 1,2-Dichlorotetramethyldistannane. *Inorg. Nucl. Chem. Lett.* 13: 13–15.
75 Davies, A.G. and Osei-Kissi, D.K. (1994). The formation of distannanes from tin hydrides. *J. Organomet. Chem.* 474: C8–C10.
76 Lechner, M.-L., Trummer, M., Braeunlich, I. et al. (2011). From poly(dialkylstannane)s to poly(diarylstannane)s: comparison of synthesis methods and resulting polymers. *Appl. Organomet. Chem.* 25: 769–776.
77 Dakkouri, M. and Kehrer, H. (1983). Darstellung der cyclischen Reihe $(C_nH_{2n-1})GeX_3(n = 3,4,5$ und $X = Cl, H)$. *Chem. Ber.* 116: 2041–2043.
78 Kamitani, M., Fukumoto, K., Tada, R. et al. (2012). Catalytic synthesis of cyclic and linear germoxanes mediated by an iron complex. *Organometallics* 31: 2957–2960.
79 Noda, H., Tanaka, K., Tanabe, M., and Osakada, K. (2018). Cyclic platina(borasiloxane)s and platina(siloxane)s and their chemical properties. *Organometallics* 37: 22–29.
80 Riedmiller, F., Wegner, G.L., Jockisch, A., and Schmidbaur, H. (1999). Preparation, structure, and ^{73}Ge NMR spectroscopy of arylgermanes $ArGeH_3$, Ar_2GeH_2, and Ar_3GeH. *Organometallics* 18: 4317–4324.
81 Cooke, J.A., Dixon, C.E., Netherton, M.R. et al. (1996). Improved synthesis of 1,2-dichlorotetramesityldigermane and other mesitylgermanes. *Synth. React. Inorg. Met.-Org. Chem.* 26: 1205–1217.
82 Samanamu, C.R., Anderson, C.R., Golen, J.A. et al. (2011). Syntheses and structural analysis of the sterically encumbered germanes $(o\text{-}Bu^tC_6H_4)_3GeX$ ($X = Br, H, Cl, OH$), $(o\text{-}Bu^tC_6H_4)_2GeBr_2$, and Mes_2GeH_2: distortions arising from the presence of an ortho-tert-butyl substituent. *J. Organomet. Chem.* 696: 2993–2999.
83 Nieder, D., Klemmer, L., Kaiser, Y. et al. (2018). Isolation and reactivity of a digerma analogue of vinyllithiums: a lithium digermenide. *Organometallics* 37: 632–635.
84 Gar, T.K., Chernysheva, O.N., Kisin, A.V., and Mironov, V.F. (1985). Germyladamantanes. *Zh. Obshch. Khim.* 55: 1057–1063.
85 Cosledan, F., Castel, A., Rivière, P. et al. (1998). Synthesis, structural characterization, and reactivity of (8-methoxynaphthyl)hydridogermanium triflates and iodides. *Organometallics* 17: 2222–2227.
86 Chaubon, M.A., Dittrich, B., Escudie, J. et al. (1997). New (2,4,6-triisopropylphenyl) hydro-, halo- and hydrohalogermanes. *Synth. React. Inorg. Met.-Org. Chem.* 27: 519–533.
87 Masafumi, U., Yuuki, K., and Hideyuki, M. (2001). Crystal structure of 2,4,6-tri-tert-butylphenyltrichlorogermane and 2,4,6-tri-tert-butylphenylgermane. *Heteroat. Chem.* 12: 238–243.
88 Spikes, G.H., Fettinger, J.C., and Power, P.P. (2005). Facile activation of dihydrogen by an unsaturated heavier main group compound. *J. Am. Chem. Soc.* 127: 12232–12233.
89 Finholt, A.E., Bond, A.C.,.J., Wilzbach, K.E., and Schlesinger, H.I. (1947). The preparation and some properties of hydrides of elements of the fourth group of the periodic system and of their organic derivatives. *J. Am. Chem. Soc.* 69: 2692–2696.

90 van der Kerk, G.J.M., Noltes, J.G., and Luijten, J.G.A. (1957). Investigations on organo-tin compounds. VII* the addition of organo-tin hydrides to olefinic double bonds. *J. Appl. Chem.* 7: 356–365.

91 Finholt, A.E., Bond, A.C., and Schlesinger, H.I. (1947). Lithium aluminum hydride, aluminum hydride and lithium gallium hydride, and some of their applications in organic and inorganic chemistry 1. *J. Am. Chem. Soc.* 69: 1199–1203.

92 Zeppek, C., Pichler, J., Torvisco, A. et al. (2013). Aryltin chlorides and hydrides: preparation, detailed NMR studies and DFT calculations. *J. Organomet. Chem.* 740: 41–49.

93 Sindlinger, C.P. and Wesemann, L. (2014). Hydrogen abstraction from organotin di- and trihydrides by N-heterocyclic carbenes: a new method for the preparation of NHC adducts to tin(II) species and observation of an isomer of a hexastannabenzene derivative [R_6Sn_6]. *Chem. Sci.* 5: 2739–2746.

94 Zeppek, C., Fischer, R.C., Torvisco, A., and Uhlig, F. (2014). Stabilizing, non-covalent interactions in the solid state structure of novel aryltin hydrides and halogenides. *Can. J. Chem.* 92: 556–564.

95 Zeppek, C. (2015). *Amine Base Induced Polymerization of Aryltin Hydrides: Mechanistic Insights and Nanomaterial Characterization*. Graz, Austria: Graz University of Technology.

96 Hayes, P.G., Gribble, C.W., Waterman, R., and Tilley, T.D. (2009). A hydrogen-substituted osmium stannylene complex: isomerization to a metallostannylene complex via an unusual α-hydrogen migration from tin to osmium. *J. Am. Chem. Soc.* 131: 4606–4607.

97 Saito, M., Hashimoto, H., Tajima, T., and Ikeda, M. (2007). Synthesis and structures of polychalcogenadistannabicyclo[k.l.m]alkanes. *J. Organomet. Chem.* 692: 2729–2735.

98 Voronkov, M.G. and Egorochkin, A.N. (2003). Similarities and differences of organic compounds of germanium, tin and lead. In: *The Chemistry of Organic Germanium, Tin and Lead Compounds* (ed. Z. Rappoport), 131–168. John Wiley & Sons.

99 Dillard, C.R., McNeill, E.H., Simmons, D.E., and Yeldell, J.B. (1958). Synthesis and properties of some tin alkyls. *J. Am. Chem. Soc.* 80: 3607–3609.

100 Peng, Y., Guo, J.-D., Ellis, B.D. et al. (2009). Reaction of hydrogen or ammonia with unsaturated germanium or tin molecules under ambient conditions: oxidative addition versus arene elimination. *J. Am. Chem. Soc.* 131: 16272–16282.

101 Sindlinger, C.P., Stasch, A., Bettinger, H.F., and Wesemann, L. (2015). A nitrogen-base catalyzed generation of organotin(II) hydride from an organotin trihydride under reductive dihydrogen elimination. *Chem. Sci.* 6: 4737–4751.

102 Neumann, W.P. (1970). *The Organic Chemistry of Tin*. Interscience.

103 Poller, R.C. (1970). *The Chemistry of Organotin Compounds*. Logos P.

104 Sommer, R., Neumann, W.P., and Schneider, B. (1964). Kondensation von stannyl-aminen mit organozinn-hydriden. *Tetrahedron Lett.* 5: 3875–3878.

105 Costisella, B., Englich, U., Prass, I. et al. (2000). Syntheses and reactivity of stannyloligosilanes. 2. A branched stannyloligosilanes. *Organometallics* 19: 2546–2550.

106 Choi, N. and Tanaka, M. (1998). Zirconocene-catalyzed dehydrogenative coupling of phenylgermane and properties of the resulting partially network polyphenylgermanes. *J. Organomet. Chem.* 564: 81–84.

107 Zeppek, C., Torvisco, A., Flock, M. et al. Towards aryl decorated tin nanoparticles (Sn@Aryl)-insights into an unknown reaction pathway. *J. Mater. Chem.*, in preparation.

108 Szymanski, W.J., Visscher, G.T., and Bianconi, P.A. (1993). Polygermynes: synthesis and properties of germanium–germanium bonded network polymers. *Macromolecules* 26: 869–871.

109 Armand, M. and Tarascon, J.M. (2008). Building better batteries. *Nature* 451: 652–657.

110 Tarascon, J.M. and Armand, M. (2001). Issues and challenges facing rechargeable lithium batteries. *Nature* 414: 359–367.

111 Kamali, A.R. and Fray, D.J. (2011). Tin-based materials as advanced anode materials for lithium ion batteries: a review. *Rev. Adv. Mater. Sci.* 27: 14–24.

112 Bruce, P.G., Scrosati, B., and Tarascon, J.-M. (2008). Nanomaterials for rechargeable lithium batteries. *Angew. Chem. Int. Ed.* 47: 2930–2946.

113 Li, X., Liang, J., Hou, Z. et al. (2015). The design of a high-energy Li-ion battery using germanium-based anode and $LiCoO_2$ cathode. *J. Power Sources* 293: 868–875.

114 Lim, L.Y., Fan, S., Hng, H.H., and Toney, M.F. (2015). Storage capacity and cycling stability in Ge anodes: relationship of anode structure and cycling rate. *Adv. Energy Mater.* 5: 1500599.

115 Li, W., Li, M., Yang, Z. et al. (2015). Carbon-coated germanium nanowires on carbon nanofibers as self-supported electrodes for flexible lithium-ion batteries. *Small* 11: 2762–2767.

116 Liu, Y., Zhang, S., and Zhu, T. (2014). Germanium-based electrode materials for lithium-ion batteries. *ChemElectroChem* 1: 706–713.

117 Xu, Y., Zhu, X., Zhou, X. et al. (2014). Ge nanoparticles encapsulated in nitrogen-doped reduced graphene oxide as an advanced anode material for lithium-ion batteries. *J. Phys. Chem. C* 118: 28502–28508.

118 Kim, H., Son, Y., Park, C. et al. (2013). Catalyst-free direct growth of a single to a few layers of graphene on a germanium nanowire for the anode material of a lithium battery. *Angew. Chem. Int. Ed.* 52: 5997–6001.

119 Jo, G., Choi, I., Ahn, H., and Park, M.J. (2012). Binder-free Ge nanoparticles-carbon hybrids for anode materials of advanced lithium batteries with high capacity and rate capability. *Chem. Commun.* 48: 3987–3989.

120 Park, C.-M., Kim, J.-H., Kim, H., and Sohn, H.-J. (2010). Li-alloy based anode materials for Li secondary batteries. *Chem. Soc. Rev.* 39: 3115–3141.

121 Yoon, S., Park, C.-M., and Sohn, H.-J. (2008). Electrochemical characterizations of germanium and carbon-coated germanium composite anode for lithium-ion batteries. *Electrochem. Solid-State Lett.* 11: A42–A45.

122 Courtney, I.A. and Dahn, J.R. (1997). Electrochemical and in situ X-ray diffraction studies of the reaction of lithium with tin oxide composites. *J. Electrochem. Soc. India* 144: 2045–2052.

123 Idota, Y., Kubota, T., Matsufuji, A. et al. (1997). Tin-based amorphous oxide: a high-capacity lithium-ion-storage material. *Science* 276: 1395–1397.
124 Zhi Xiang, H., Ye, W., Jen It, W., and Hui Ying, Y. (2015). Free standing SnS_2 nanosheets on 3D graphene foam: an outstanding hybrid nanostructure anode for Li-ion batteries. *2D Mater.* 2: 024010.
125 Ibarz, A., Ruiz, E., and Alvarez, S. (1998). Electronic structure of host lattices for intercalation compounds: SnS_2, $SnSe_2$, ZrS_2, and TaS_2. *Chem. Mater.* 10: 3422–3428.
126 Wang, B., Luo, B., Li, X., and Zhi, L. (2012). The dimensionality of Sn anodes in Li-ion batteries. *Mater. Today* 15: 544–552.
127 Shin, H.C. and Liu, M. (2005). Three-dimensional porous copper–tin alloy electrodes for rechargeable lithium batteries. *Adv. Funct. Mater.* 15: 582–586.
128 Hu, R., Zhu, M., Wang, H. et al. (2012). Sn buffered by shape memory effect of NiTi alloys as high-performance anodes for lithium ion batteries. *Acta Mater.* 60: 4695–4703.
129 Kravchyk, K., Protesescu, L., Bodnarchuk, M.I. et al. (2013). Monodisperse and inorganically capped Sn and Sn/SnO_2 nanocrystals for high-performance Li-ion battery anodes. *J. Am. Chem. Soc.* 135: 4199–4202.
130 Chen, J.S. and Lou, X.W. (2012). SnO_2 and TiO_2 nanosheets for lithium-ion batteries. *Mater. Today* 15: 246–254.
131 Goodenough, J.B. and Kim, Y. (2010). Challenges for rechargeable Li batteries. *Chem. Mater.* 22: 587–603.
132 Xu, L., Kim, C., Shukla, A.K. et al. (2013). Monodisperse Sn nanocrystals as a platform for the study of mechanical damage during electrochemical reactions with Li. *Nano Lett.* 13: 1800–1805.
133 Xu, Y., Liu, Q., Zhu, Y. et al. (2013). Uniform nano-Sn/C composite anodes for lithium ion batteries. *Nano Lett.* 13: 470–474.
134 Xu, Y., Guo, J., and Wang, C. (2012). Sponge-like porous carbon/tin composite anode materials for lithium ion batteries. *J. Mater. Chem.* 22: 9562–9567.
135 Deng, D. and Lee, J.Y. (2009). Reversible storage of lithium in a rambutan-like tin–carbon electrode. *Angew. Chem. Int. Ed.* 48: 1660–1663.
136 Derrien, G., Hassoun, J., Panero, S., and Scrosati, B. (2007). Nanostructured Sn–C composite as an advanced anode material in high-performance lithium-ion batteries. *Adv. Mater.* 19: 2336–2340.
137 Hassoun, J., Derrien, G., Panero, S., and Scrosati, B. (2008). A nanostructured Sn–C composite lithium battery electrode with unique stability and high electrochemical performance. *Adv. Mater.* 20: 3169–3175.
138 Hassoun, J., Lee, K.-S., Sun, Y.-K., and Scrosati, B. (2011). An advanced lithium ion battery based on high performance electrode materials. *J. Am. Chem. Soc.* 133: 3139–3143.
139 Hwang, J., Woo, S.H., Shim, J. et al. (2013). One-pot synthesis of tin-embedded carbon/silica nanocomposites for anode materials in lithium-ion batteries. *ACS Nano* 7: 1036–1044.
140 Qiu, Y., Yan, K., and Yang, S. (2010). Ultrafine tin nanocrystallites encapsulated in mesoporous carbon nanowires: scalable synthesis and excellent electrochemical properties for rechargeable lithium ion batteries. *Chem. Commun.* 46: 8359–8361.

141 Yu, Y., Gu, L., Zhu, C. et al. (2009). Tin nanoparticles encapsulated in porous multichannel carbon microtubes: preparation by single-nozzle electrospinning and application as anode material for high-performance Li-based batteries. *J. Am. Chem. Soc.* 131: 15984–15985.

142 Hou, X., Jiang, H., Hu, Y. et al. (2013). In situ deposition of hierarchical architecture assembly from Sn-filled CNTS for lithium-ion batteries. *ACS Appl. Mater. Interfaces* 5: 6672–6677.

143 Seo, S.-D., Lee, G.-H., Lim, A.-H. et al. (2012). Direct assembly of tin-MWCNT 3D-networked anode for rechargeable lithium ion batteries. *RSC Adv.* 2: 3315–3320.

144 Wang, Y., Wu, M., Jiao, Z., and Lee, J.Y. (2009). Sn@CNT and Sn@C@CNT nanostructures for superior reversible lithium ion storage. *Chem. Mater.* 21: 3210–3215.

145 Yu, Y., Gu, L., Wang, C. et al. (2009). Encapsulation of Sn@carbon nanoparticles in bamboo-like hollow carbon nanofibers as an anode material in lithium-based batteries. *Angew. Chem. Int. Ed.* 48: 6485–6489.

146 Zhang, H., Song, H., Chen, X., and Zhou, J. (2012). Enhanced lithium ion storage property of Sn nanoparticles: the confinement effect of few-walled carbon nanotubes. *J. Phys. Chem. C* 116: 22774–22779.

147 Hsu, K.-C., Liu, C.-E., Chen, P.-C. et al. (2012). One-step vapor-solid reaction growth of Sn@C core-shell nanowires as an anode material for Li-ion batteries. *J. Mater. Chem.* 22: 21533–21539.

148 Lee, K.T., Jung, Y.S., and Oh, S.M. (2003). Synthesis of tin-encapsulated spherical hollow carbon for anode material in lithium secondary batteries. *J. Am. Chem. Soc.* 125: 5652–5653.

149 Li, X., Dhanabalan, A., Gu, L., and Wang, C. (2012). Three-dimensional porous core-shell Sn@carbon composite anodes for high-performance lithium-ion battery applications. *Adv. Energy Mater.* 2: 238–244.

150 Zhang, W.-M., Hu, J.-S., Guo, Y.-G. et al. (2008). Tin-nanoparticles encapsulated in elastic hollow carbon spheres for high-performance anode material in lithium-ion batteries. *Adv. Mater.* 20: 1160–1165.

151 Biedermann, J. (2018). *Synthesis and Electrochemical Studies of Aryl Subsituted Group 14 Compounds.* Graz, Austria: Graz University of Technology.

4

Synthesis of Polymers Containing Group 15 Elements

Andreas Orthaber[1] and Alejandro P. Soto[2]

[1] *Uppsala University, Department of Chemistry – Ångström Laboratories, Synthetic Molecular Chemistry, Box 523, Uppsala 75120, Sweden*
[2] *University of Oviedo, School of Chemistry, Department of Organic and Inorganic Chemistry, Julián Clavería, 8, Oviedo 33006, Spain*

4.1 Introduction

In this chapter, we will discuss the following classes of pnictogen-containing polymers: conjugated polymers with saturated and unsaturated pnictogen motifs, homopolymers and nonconjugated tri- and penta-valent organopnictogen polymers, polyphosphazenes, and polymers containing P(=O)—O motifs in their backbone (Figure 4.1).

The pnictogen atoms in nonconjugated polymers can possess the heteroelement in its trivalent, i.e. $\sigma^3\lambda^3$, or pentavalent, i.e. $\sigma^4\lambda^5$, state. In many cases, "oxygen sensitivity" of the trivalent forms promotes oxidation in the final polymers. Polymerization procedures often take reference from protocols used to prepare organic polymers; however, occasional surprises and peculiarities of these hybrid polymers give rise to fascinating materials with unprecedented properties. Similarly, conjugated polymers containing heavier pnictogens entail fascinating optoelectronic properties that go beyond the scope of all organic congeners.

The importance of pnictogen atoms in macromolecular science can be clearly illustrated by the growth of the broad field of polyphosphazene chemistry, $[N{=}PR_2]_n$. The properties of polyphosphazenes depend on both the nature of the inorganic backbone and the structure of the organic side groups. The wide variety of polyphosphazene architectures provides multiple opportunities for these developments not only in biomedicine and energy research but also in the fields of elastomer technology, photonics, adhesion, membranes, and many other areas.

On the other hand, polyphosphoesters with repeating phosphoester bonds in the main chain are a very structurally versatile, biocompatible, and biodegradable class of polymers. They are appealing for many biological applications (e.g. drug/gene delivery, tissue engineering, etc.), as well as flame retardants.

Smart Inorganic Polymers: Synthesis, Properties, and Emerging Applications in Materials and Life Sciences, First Edition. Edited by Evamarie Hey-Hawkins and Muriel Hissler.

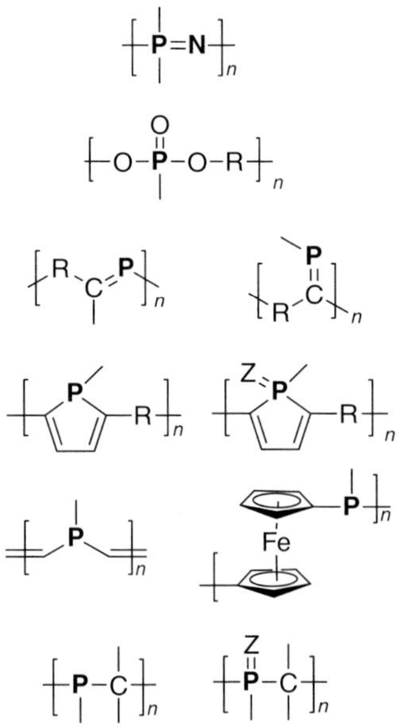

Figure 4.1 Representative examples of pnictogen-containing polymers.

4.2 Conjugated Polymers Containing Group 15 Elements

Conjugated (conductive) polymers are often major components employed in the fabrication of organic electronic devices. Tailoring the optoelectronically active materials prompts chemists to explore the diversity of the chemical space and it was already shown earlier on that incorporation of heavier pnictogens could have a profound impact on the resulting properties. In principle, one can distinguish two types of conjugated polymers. In analogy to polyolefins and related polymers, Type **I** is characterized by an unsaturated pnictogen center, bound to a group 14 – especially carbon – or group 15 element, whereas in Type **II** the heteroelement is not part of the unsaturated backbone. Noteworthy, the classification "unsaturated" is ambiguous and includes different unsaturated motifs. We use the "*lambda convention*," $\sigma^n\lambda^m$, which provides a more accurate description of the bonding situation of the pnictogen centers. This convention describes the valency (λ^m, i.e. the number of bonding valence electrons) as well as the coordination number (σ^n). The most common examples of these conjugated polymers contain diphosphene (two $\sigma^2\lambda^3$ phosphorus atoms, —P=P—) and phosphaalkene (one $\sigma^2\lambda^3$ phosphorus atom bound to an sp^2 carbon, —P=C<) and phosphole (a $\sigma^3\lambda^3$ phosphorus atom in an unsaturated five-membered heterocycle) motifs.

4.2.1 Phosphaalkenes, Arsaalkenes, and Diphosphenes

In general, two main strategies have led to oligomers and polymers possessing group 15 heteroalkene motifs. The first strategy utilizes the heteroelement – carbon bond formation in the final step(s) of polymerization, while the second strategy is based on the polymerization of building blocks containing these unsaturated motifs.

In 1976, Becker developed a procedure, creating a P=C double bond in two steps starting from an acyl chloride and a disilyl phosphane. In the first step, the P—C single bond is established by a silyl chloride condensation. The second step is carried out in situ by raising the reaction temperature to trigger the sigmatropic 1,3-silyl shift from phosphorus to the carbonyl moiety establishing the phosphaalkene double bond (Scheme 4.1, Eq. (1)). The nature of the rearrangement however implies that *E*- and *Z*-isomeric products are formed and their relative ratio is governed by the steric demand of the substituents [1]. In 2002, Wright and Gates applied this method using tetramethyl terephthaloyl chloride (**1**) and 1,4-diphosphanyl benzene (**2**). Thermolysis at 80 °C of neat solutions affords a highly viscous yellow oil of the low molecular weight polymeric *phospha*-OPV **3**. The degree of polymerization was estimated based on ^{31}P-NMR end group analysis between 5 and 21, with a *Z*/*E* isomer ratio of approx. 1.1. The moderate redshifted absorption of approximately 20 nm (λ_{max} 328–338 nm) tailing into the visible region characterizes the electronic structure of these systems (**3**) [2].

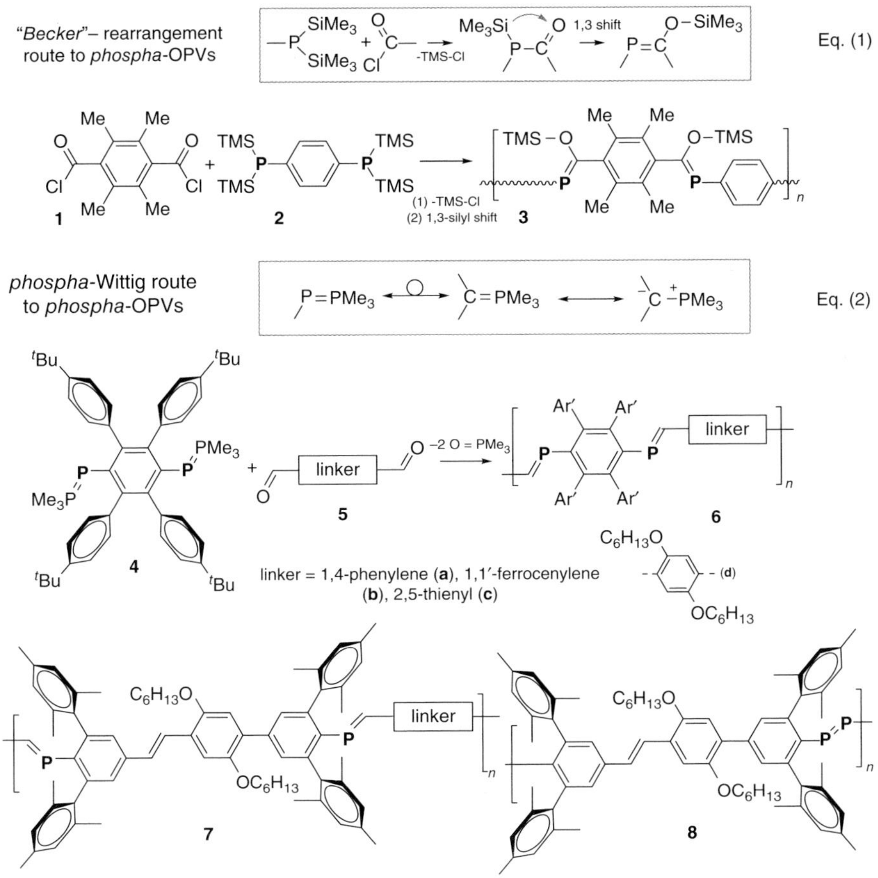

Scheme 4.1 Polymers containing unsaturated phosphaalkene and diphosphene motifs in the conjugated polymer backbone.

A synthetic strategy that takes reference from organic chemistry was used to prepare a tetraaryl phenylene bridged phosphaalkene polymer. The *phospha*-Wittig reaction was developed analogous to the "carbon" Wittig reaction replacing the phosphonium ylide with a phosphanylidene-phosphorane (**4**), which is easily obtained from the corresponding aryl dichlorophosphane and PMe_3 in the presence of Zn dust. The subsequent reaction with aldehydes, and in some cases also ketones, yields predominantly *E*-phosphaalkenes (Scheme 4.1, Eq. (2)). Protasiewicz and coworkers have studied a variety of bridging dialdehydes forming polymers featuring phosphaalkene repeating units. While polymers **6a–c** show only limited solubility, introduction of two hexyloxy side chains in **6d** indeed increases the solubility in hexanes. ^{1}H and ^{31}P-NMR end group analysis confirms a degree of polymerization of approx. 6, i.e. a polymer featuring, in average, 12 phosphaalkene units. Remarkably, this polymer shows a strongly redshifted absorption maximum at 445 nm compared to that by oligomeric model compounds and is even more redshifted than regular oligo(phenylene vinylene) (OPV) despite the modest conjugation length [3]. In addition to the moderate degree of polymerization, the effective conjugation length is also limited due to the unfavorable orientation of the P=C π-systems with respect to the adjacent sterically encumbered aromatic system.

The same authors have expanded this concept to a bis-phosphanylidene phosphorane monomer consisting of two *meta*-terphenyl units separated by an alkyloxy-decorated phenyl vinylene unit. Reaction with various aromatic dialdehydes gives photoluminescent polymers (7) with a moderate degree of polymerization (*n* approx. 6) and good polydispersity (*Đ* approx. 2). Interestingly, photolysis of the bis-phosphanylidene phosphorane monomer leads to elimination of volatile trimethyl phosphane and formation of the non-emissive diphosphene-containing polymer **8** [4].

The Becker route and the *phospha*-Wittig reactions are extensively used in the synthesis of conjugated oligomers [5, 6]. Recently, a phosphorus variant of the Horner–Wittig–Emmons reaction has been developed and promising conjugated materials could be prepared with almost exclusive *E*-configuration of the phosphaalkene [7].

The previously described phosphaalkene (diphosphene) polymers preparations utilized the P=C (P=P) bond formation also as the polymerization step. In an alternate approach, the unsaturated pnictogen center is embedded into a polymerizable conjugated framework. Orthaber and coworkers used the dithienyl fluorenylidene phosphine and arsine (**9**) as suitable precursors, which can undergo electrochemical oxidative polymerization. Under these conditions, a polymer film is deposited on the electrode surface. Both glassy carbon electrodes and transparent electrodes of conductive glass (e.g. fluorine-doped tin oxide [FTO]) could be used in the preparation of these phospha- or arsaalkene-containing polymers (**10**). This process occurs analogously to conventional (di-)thiophene polymerizations at the positions of highest radical density, i.e. the α–carbons of the thienyl substituents (Figure 4.2, Eq. (3)). Postsynthetic modification of these electrochromic polymers was achieved by coordinating Au(I)Cl (**11**) resulting in a redshift of absorption from approx. 580 to 730 nm of the weak π–π* transition [8].

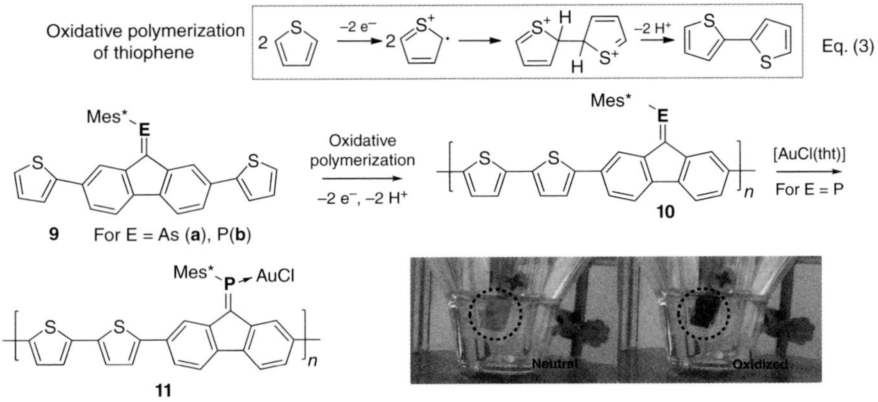

Figure 4.2 Electropolymerization of fluorenylidene phosphanes and arsines and their electrochromic behavior (dashed circles).

4.2.2 Group 15-Based Heteroles

Group 15-based heteroles are by far the most explored class of building blocks for conjugated polymers and among these $\sigma^4\lambda^5$-phospholes (e.g. phosphole sulfides) dominated this area of research. In contrast, $\sigma^3\lambda^3$-phospholes are often due to their tendency toward oxidation formed only as intermediates in preparative protocols. A common approach for this class of compounds is to design monomeric phospholes with suitable polymerizable groups. Similar to the above-described monomer **9**, thienyl substituents or annulated thieno motifs can be used for electropolymerizations. The synthesis of suitable 2,5-disubstituted heterole derivatives mostly employs the Fagan–Nugent method (Figure 4.3, Eq. (4)) using alkyl-bridged dialkynyls forming a zirconacene intermediate, which can subsequently be reacted with the corresponding pnictogen dihalide [9].

Oxidative polymerization of, for example, monomer **12** is an effective method in the preparation of phosphole-thienyl copolymers (**13**, **14**). The chemical diversity and possible modifications at the phosphorus center allow conversion of the trivalent phosphole to pentavalent oxides (**13c**) or sulfides (**13b**) as well as cationic phospholium salts (**14**). All precursors undergo effective electropolymerization [10, 11]. This polymerization method is not limited to linear phospholes and oligomeric monomers (**15**) [12], and it has also been shown that particular 1,1′-biphospholes (**16**) engage in similar polymerization [13]. The intrinsic properties of these phospholes and the possibility to functionalize the heteroelement pre- and post-polymerization make these materials extremely interesting for optoelectronics. Devices and specific applications are discussed in Chapter 7. These approaches generally afford the desired polymers as insoluble materials that can be directly deposited on the electrode surfaces.

Alternatively, the phospholes can be functionalized with suitable leaving groups and engage in chemical polymerizations via C–C coupling reactions such as Stille- or Suzuki-coupling protocols. The advantage of the procedure is that solubilizing groups can be introduced and co-monomers can be chosen at convenience. Matano et al. have in this way synthesized donor–acceptor systems

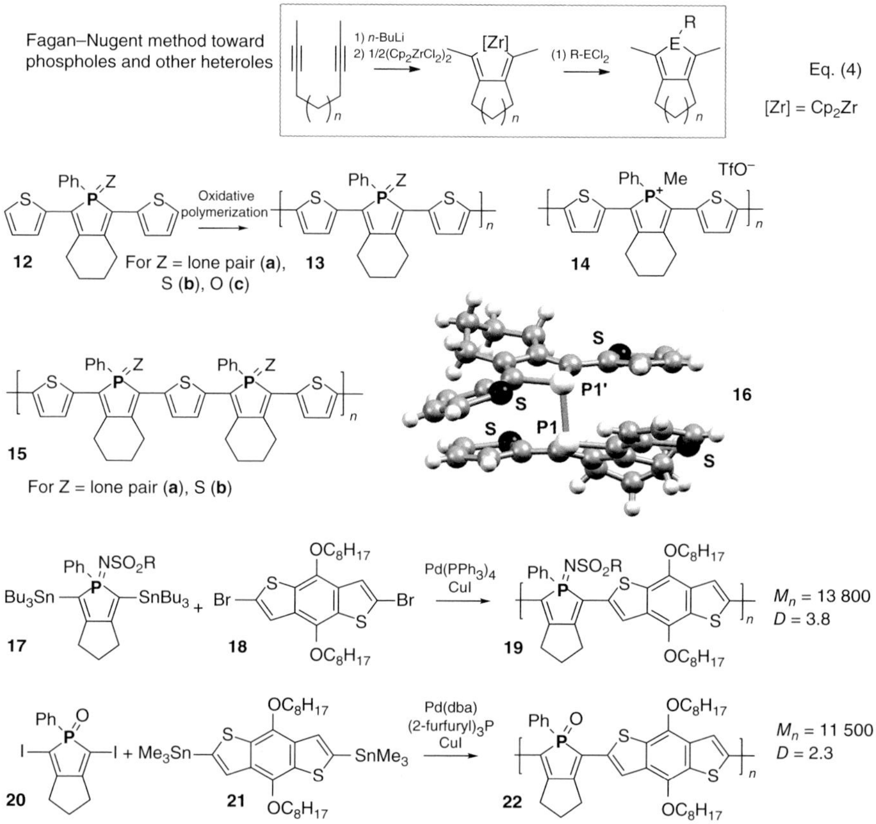

Figure 4.3 Chemical and electrochemical polymerization of phospholes-containing building blocks.

(**19**, **22**) as active solar cell materials polymerizing different $\sigma^4\lambda^5$-phospholes (**17**, **20**) with suitably functionalized benzodithienophene (**18**, **21**) monomers [14].

Annulated heterole derivatives are most commonly obtained by the reaction of dilithio-biaryls with aryl dichloro pnictogens (Scheme 4.2, Eq. (5)) [15]. In this way, the dithienophosphole building block was obtained and further functionalization was studied extensively. Reactivity at both the phosphorus center – to alter the optical and electronic properties – and at the annulated thienyl rings – to introduce solubilizing and leaving groups – led to a huge library of monomeric building blocks (**23**). Typical Stille, Suzuki, and dehydrocoupling protocols could be used to obtain polymers of moderate conjugation length (**24–26**) [16–18]. These polymers are emissive with photoluminescence maxima between 460 and 570 nm and quantum yields generally above 50%. Unprecedented optoelectronic properties have been developed using the dipyridino-annulated phosphole (**27**). The alkylated *phospha*-viologene polymers (**28**) are electrochromic materials and can be used as polymer electrolytes [19, 20].

Recently, a growing interest in stable trivalent heteroles prompted researchers to study the heavier pnictogen analogs, e.g. arsoles and bismoles. These are known to be less oxophilic, thus their trivalent form (i.e. $\sigma^3\lambda^3$) could potentially be stable enough under ambient conditions and enable advanced optoelectronic studies. Similar to the previous work on dithienophospholes, Heeney and

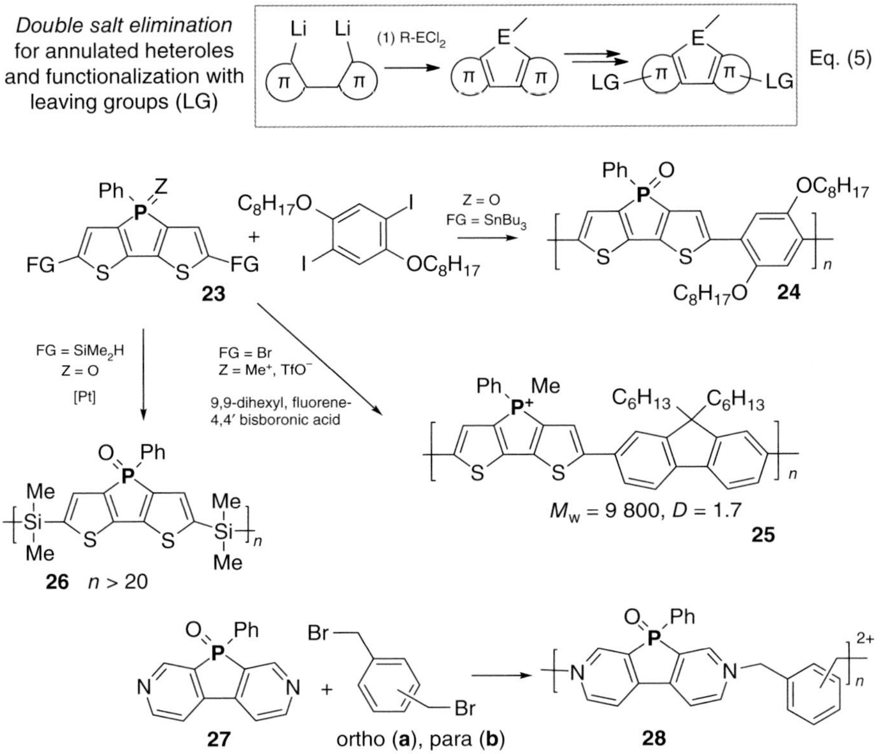

Scheme 4.2 Functionalization and polymerization of annulated phospholes.

coworkers have devised a suitable dithienoarsole building block (**29**). The anticipated stability under ambient conditions allowed the preparation of polymers with trivalent arsenic centers (**30**) using microwave-assisted Stille coupling, completely avoiding the formation of corresponding $\sigma^4\lambda^5$-arsoles [21]. Similarly, Suzuki coupling of dibromo-derivatized arsoles (**31**) with a fluorenyl bisboronate gave a yellow emitting polymer (**32**) with high quantum yields in solution (ϕ_{PL} = 0.44) and dispersed in a poly(methyl methacrylate) (PMMA) matrix (ϕ_{PL} = 0.06) [22] (Scheme 4.3).

An alternative approach toward arsole-containing polymers was followed by Tomita and coworkers, who used the formation of a titanacyclopentadienyl intermediate (**33**) as the polymerization step. This reactive polymer undergoes an exchange with diiodophenyl arsine at low temperatures resulting in a clean post-element transformation to an arsole polymer (**34**). Electrochemical studies of this polymer showed quasi-reversible oxidative and reductive processes, which is in stark contrast to the irreversible oxidation of the related phosphole derivatives. The authors also demonstrated postsynthetic functionalization of the polymer through gold chloride coordination (**35**), which induced a bathochromic shift of the absorption maxima [23]. Although there have been some recent theoretical and experimental publications highlighting the favorable luminescent properties of bismoles (e.g. dithienobismoles) [24] the only reported polymer containing a bismole repeating unit to date was featured almost 10 years ago. Interestingly, the synthetic strategy uses the Fagan–Nugent approach on the polymer backbone. However, instead of directly converting the zirconocenes

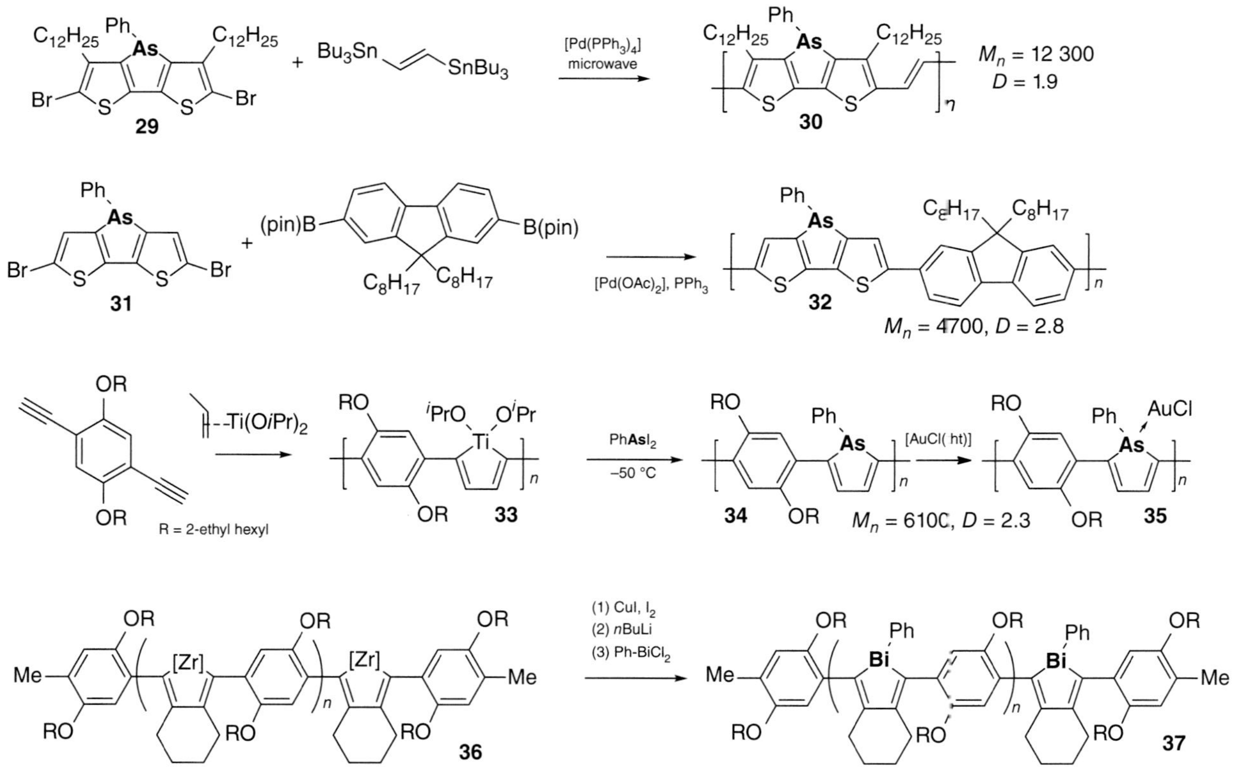

Scheme 4.3 Synthetic approaches toward heavier group 15 heteroles and their corresponding polymers.

(**36**) into the corresponding bismoles, it is initially reacted with iodine, followed by metal–halogen exchange (i.e. I to Li), and subsequent reaction with dichloro aryl bismuthane giving polymer **37** [25].

4.3 Polymers with two Unsaturated Organic Moieties Adjacent to the Heteroelement Motif

This class of polymers can be cross-, homo-, or nonconjugated polymers. Nonetheless, most examples display significant extension of their conjugation across the bridging motifs. Cross-conjugated motifs show significantly reduced delocalization compared to their linearly conjugated counterparts; however, secondary effects, such as possible switching behavior, nonlinear properties or good semiconductor characteristics have been reported for organic polymers [26]. While (negative) hyperconjugation plays only a minor role in these systems, the participation of the pnictogen lone pair is crucial to the electronic features of these systems. Therefore, switching to heavier pnictogens entails tuning of pyramidalization, delocalization, and optoelectronic properties. Best explored are polyanilines (PANIs), where the degree of oxidation determines the contribution of quinoid (oxidized), and non-quinoid (reduced) resonance structure. This optoelectronic tuning allows color tuning from a colorless to a green/blue and finally a dark blue/violet state (Figure 4.4).

4.3.1 Cross-Conjugated Group 15 Heteroalkene-Containing Materials

Organic cross-conjugated motifs are commonly found in conductive polymers greatly expanding the scope of optoelectronic materials. In contrast, their hybrid, i.e. heteroalkene-containing congeners are significantly less explored, presumably due to their decreased stability and the incompatibility of many synthetic protocols with polymer preparations. A prominent approach toward cross-conjugated phosphaalkene motif involves the reaction of propargylic chlorides with *n*-BuLi (or Mg) (Scheme 4.4, Eq. (6)) followed by the addition of an aryl dichlorophosphane [27]. Several oligomeric materials have been obtained in this way using the cross-conjugated precursors in Sonogashira couplings with various diiodo-arenes (**38**) or Ullman-type coupling reactions (**39**) [28–30]. The resulting cross-conjugated materials exhibit significantly redshifted absorption maxima and milder reduction potentials compared to related organic polymers. Recently, two groups independently demonstrated facile ways to sequentially couple alkynyl groups to Mes*P=CBr_2 (**40**) giving regio-selective syntheses of *C,C*-diacetylenic phosphaalkenes (**41**) opening new avenues toward polymeric materials [31, 32].

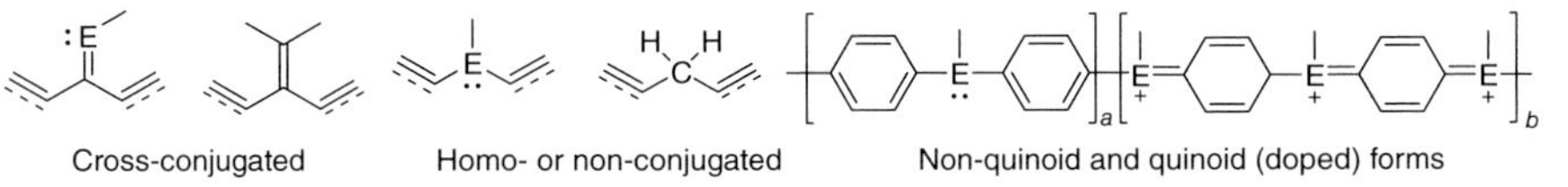

Figure 4.4 Different types of conjugation and delocalization observable in organopnictogen polymers.

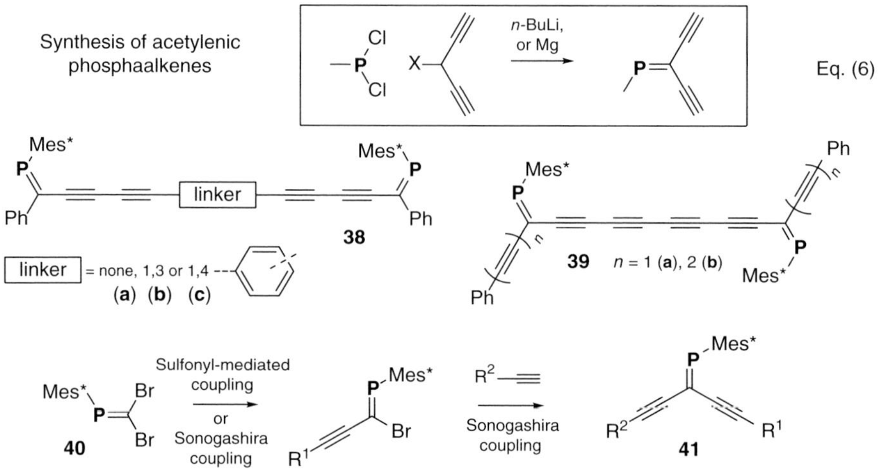

Scheme 4.4 Different building blocks containing cross-conjugated phosphaalkene motifs.

4.3.2 Group 15 Elements with two Adjacent Alkynes, Alkenes, or Arene Motifs of the Polymer Backbone

Different aspects such as significant contribution of quinoid resonance structures, delocalization via contribution of the heteroelement's lone pair, and other polarization effects are used to rationalize the effective conjugation observed in many of these systems.

The first polyarylene(arylphosphane)s were prepared by Ni-mediated cross-coupling of bis-chloroaryl phosphane oxide (**42**). Importantly, the monomeric building block must contain an oxidized, i.e. a $\sigma^4\lambda^5$ phosphorus center, to give high molecular weight polymers (**43**, M_n = 15 300) with moderate polydispersity ($Đ$ approx. 1.6). The final polymer is partially converted into the $\sigma^3\lambda^3$-P-polymer (**44**) using phenyl silane as the reducing agent shifting the onset for electronic excitation to approx. 350 nm. Oxygen sensitivity and poor solubility however precluded further studies at that time [33]. Similar derivatives could be prepared using two fundamentally different P–C coupling procedures (Scheme 4.5, Eq. (7)). In one case, primary (**45**) or secondary arylphosphanes (**47**, **50**) were coupled with dihaloarenes to give true PANI analogs (**46**, **48**). Pd-coupling procedures show large versatility giving pure phosphorus as well as mixed P/N-containing polymers (**51**) in low to moderate molecular weights. Importantly, this methodology is compatible with $\sigma^3\lambda^3$-P centers and allows postsynthetic modifications, e.g. oxidation with hydrogen peroxide, resulting in the corresponding P-oxidized polymers (**49**) [34–36].

In an alternative approach, phenyl dichloro phosphane is reacted with diacetelyenic building blocks in a Ni-catalyzed P—C bond formation (**52**). Introduction of solubilizing hexyl groups (**52b**) gives a moderate degree of polymerization (M_w = 10–12 kDa and $Đ \sim 1.9$). Strong bathochromic shifts compared to molecular model compounds indicate efficient conjugation across the pnictogen centers. Importantly, the observed "turn-on" fluorescence through oxidation is observed only in the polymer (**53**) [37].

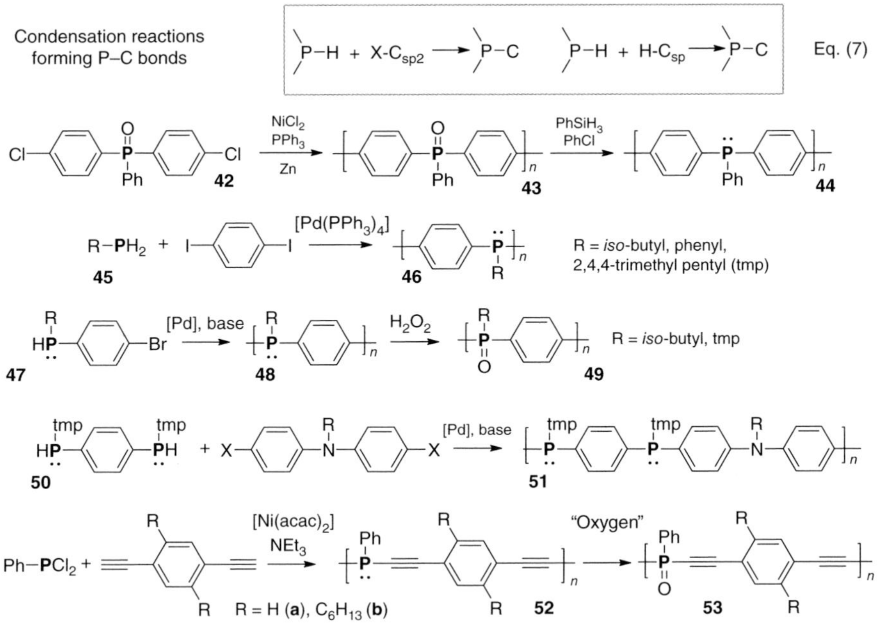

Scheme 4.5 Metal-catalyzed P—C bond formations toward polyarylene(arylphosphanes).

4.3.2.1 Ring-Opening Polymerization

Ring-opening reactions of phospha-ferrocenophanes (Scheme 4.6, Eq. (8)) or phosphirene afford polymeric materials with an unsaturated heteroelement motif in the backbone as well. Strained [1]-ferrocenophanes (**54**) with a single phosphorus atom in the bridge undergo facile living ring-opening-polymerization (ROP, **55**) using anionic (e.g. NaCp), photo-initiated anionic [38] or photo-initiated [39]

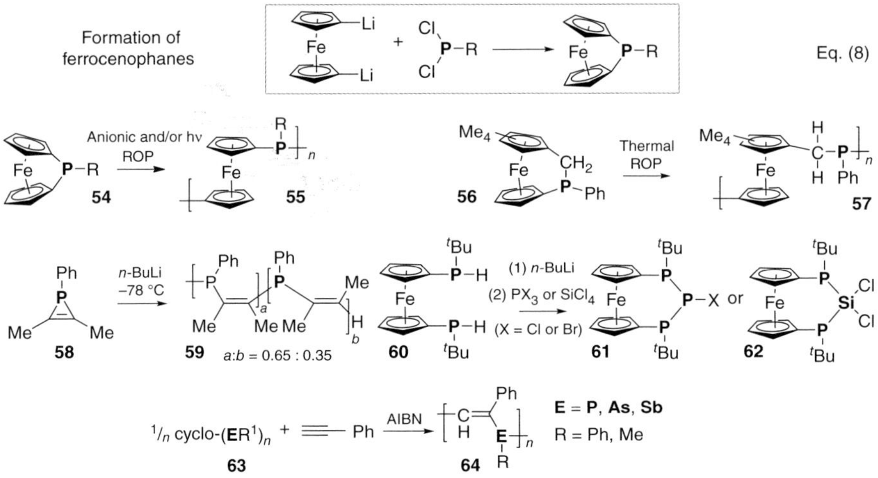

Scheme 4.6 Ring-opening polymerizations of ferrocenophanes and other (strained) cyclic precursors.

methods. Similarly, P–C-[2]-ferrocenophanes (**56**), can be used in thermal ROPs giving high molecular weight polycarbophosphaferrocene (**57**) [40]. The living anionic polymerization procedures lead to excellent polymer characteristics, such as low polydispersity (*Đ*: 1.05–1.23) and excellent control of molecular weights based on the monomer-to-initiator ratio. Furthermore, the living polymerization enabled the synthesis of block copolymers with polyferrocenyl silanes. The combination of redox-active ferrocene centers and main group elements (P and Si) creates unprecedented possibilities in postsynthetic metal coordination, redox catalysis, and other areas [38]. Anionic ROP was also reported for the simple 1-phenyl-2,3-dimethylphosphirene (**58**), i.e. a highly strained three-membered heterocycle giving regio-irregular polyvinylenephosphane (**59**). The *n*-BuLi initiated reaction yields low to high molecular weight (1.7–18 kDa) polymers with rather high polydispersity ($Đ$ = 1.23–1.58) [41]. Ferrocenophanes, such as the triphospha-[3]-ferrocenophane **61** and 1,2-diphospha-sila-[3]-ferrocenophane **62**, are significantly less strained; however, the reactive P–X or SiX_2 groups are formidable positions to create oligo- and polyphosphanes with pendant ferrocenylene functionalities [42, 43].

In a different approach, polyvinylene phosphanes (**64**) and other pnictogen analogs have been synthesized by radical initiated polymerization of the respective homocycles (**63**) in the presence of terminal acetylenes [44]. This method is particularly useful in the syntheses of heavier pnictogen derivatives illustrated by high molecular weight polyvinylene arsine and polyvinylene stibine derivatives [45–47].

4.4 Organic–Inorganic Hybrid Polymers Containing Saturated Phosphorus Centers

The most common approach toward organic polymers with a saturated backbone is the addition polymerization of the C=C bond using olefinic monomers. The analogy of alkenes and phosphaalkenes to form such polymethylene phosphanes was first explored in 2003 for **65a** using a radical initiator, but later anionic initiators (MeLi and *n*-BuLi) were also tested (Scheme 4.7, Eq. (9)). Owing to instability of the trivalent phosphorus center, the resulting polymers are often sulfurized to the more stable pentavalent derivative [48]. The following examples of living anionic polymerizations routes yielded interesting derivatives including polymethylene phosphanes (**66**) with pendant ferrocenes (**66c**) as well as block copolymers with polystyrene (derivatives) [49, 50]. Quite interestingly, the radical polymerizations predominantly take an alternative polymer propagation through a methyl radical at the P-Mes substituent giving a different micro-structure (**67**) [51]. Later, it was demonstrated that anionic initiators could also give regio-irregular polymers [52]. These polyphosphane polymers display a large variability at the phosphorus center, allowing metal–borane coordination and oxidation/sulfurization as well as quaternization reactions. 1-Phospha-isoprene (**68**) can also be polymerized using anionic initiators. This phosphorus analog of natural rubber (**69**) potentially could undergo addition polymerization at the C=C (fraction *n*) or P=C (fraction *o*) moiety, or an

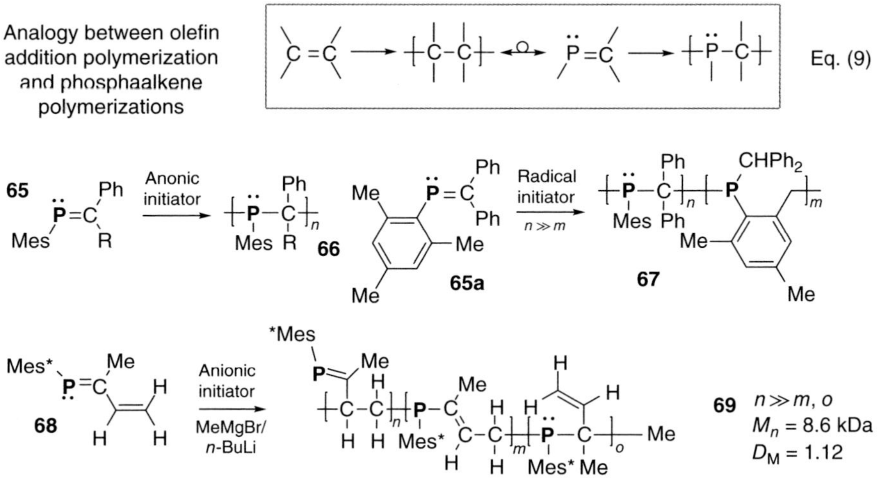

Scheme 4.7 Addition polymerization of phosphaalkene-containing precursors giving different microstructures.

addition-isomerization polymerization (fraction *m*). Based on NMR spectroscopic evidence the microstructure *n*, which maintains the integrity of the phosphaalkene, dominates this polymer, allowing for further functionalization, such as AuCl coordination [53].

4.4.1 Miscellaneous Polymers

Introducing the functional group 15-containing motif as the pendant functionality is an interesting alternative to incorporating the heteroelement unperturbed from the polymer backbone. Baumgartner and coworkers realized a random polymer using a blend of styrene and two-styrene functionalized dithienophospholes (**70**). The resulting polymer (**71**) shows favorable white light emission in the solid [54].

Incorporating highly unsaturated phosphorus building blocks, such as phosphaalkynes, is extremely challenging; however, the resulting linear conjugation is an attractive electronic feature. Recently, Crossley and coworker have synthesized a bimetallic scaffold terminated by two (P≡C) moieties (**72**) by desilylation of the bis(trimethylsilylphosphaalkyne) terminated precursor (**73**) [55].

Utilizing metal coordination and combining the phosphole building block with terpyridine-based coordination motifs gives fascinating oligomeric coordination compounds (**75**). In dilute solutions, approx. 10 unimeric units are assembled showing fast and slow exchange dynamics for Zn^{2+} and Fe^{2+} metal centers, respectively [56] (Figure 4.5).

4.5 Polyphosphazene

Polyphosphazenes (Figure 4.6a) are a class of polymers consisting of $[N{=}P(R_1R_2)]_n$ repeating units with a pentavalent phosphorus bearing two

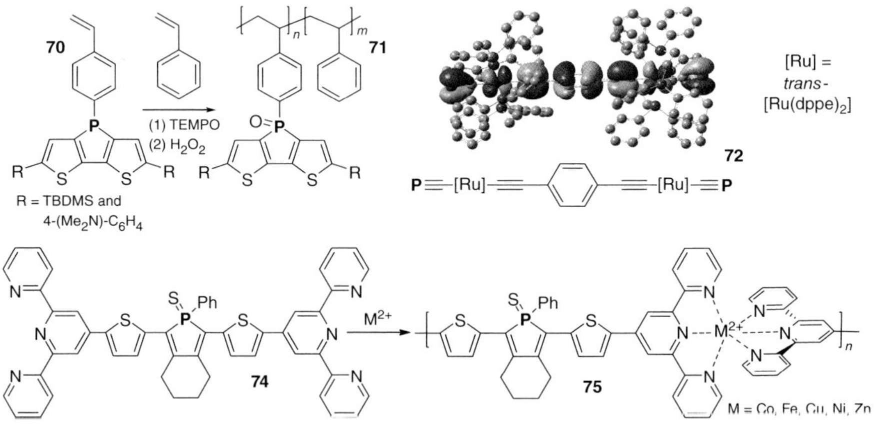

Figure 4.5 Miscellaneous polymers having side-chain phosphole groups, main-chain phospholes-coordination polymers and oligomers incorporating phosphaalkyne units.

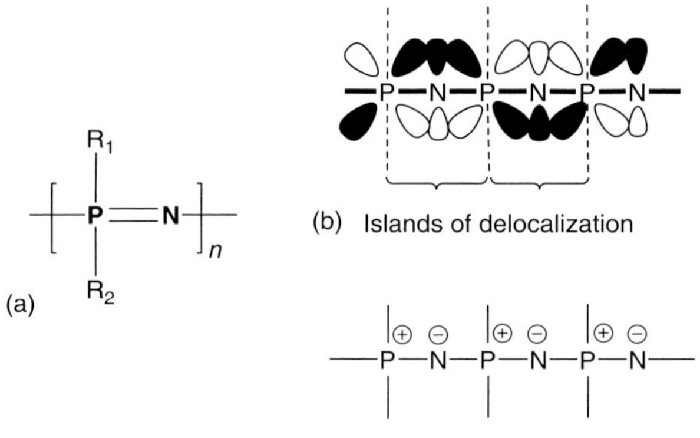

Figure 4.6 (a) General structure of polyphosphazenes. (b) Valence bond representation of d_π–p_π bonding in polyphosphazene main chain. (c) Structure of polyphosphazene main chain based on negative hyperconjugation model.

substituents, R (equal or different) and a trivalent nitrogen [57]. The bonding nature of these polymers can generate considerable debate principally due to the different properties of the alternating —P=N—P=N— main chain with respect to those of the conjugated —C=C—C=C— backbones (e.g. poly(acetylene)s). Indeed, conjugated organic polymers have a stiff backbone due to the π–π orbital overlap within the delocalized structure, which is responsible for their significant inherent properties (electronic conductivity, color, etc.). However, the alternating —P=N—P=N— main chain that is transparent in the visible region, exhibits very low torsion barriers of the backbone bonds, and no electronic conductivity. These differences can be explained on the basis of the nature of the phosphazene bonding. There are two main theories explaining the —P—N— bonding. The Allcock theory is based on the so-called *"island of delocalization,"* in which the 3d orbitals of the phosphorus atoms overlap with the 2d orbitals of the adjacent nitrogen atoms originating a node at each phosphorus (Figure 4.6b). This theory would allow delocalization within each —P—N—P— unit, interrupted at each phosphorus (no electronic conductivity). Moreover, the orbital overlap would also explain the —P—N— bond shortening in polyphosphazenes compared with single-bonded —P—N— units, and the low torsion barriers of the backbone as the nitrogen 2p orbitals might overlap at different torsional angles with several different phosphorus 3d orbitals [58]. However, latest theories suggest that the —P—N— bonding in polyphosphazenes consists of an ionic σ-bond and a π-bond induced largely by negative hyperconjugation, which involves electron donation from the lone-pair π(N) orbitals to σ*(P) bonds. In contrast to Allcock's model, the participation of d-orbitals in the π bonding is negligible. Thus, a better valence structure of polyphosphazene is that described in Figure 4.6c [59].

Phosphorus and nitrogen are the two most important elements that prevent combustion of organic materials. Moreover, the presence of both phosphorus and nitrogen in a material is synergistic. In this regard, the backbone of polyphosphazenes ensures the nonflammability of most of the poly(organophosphazene)s. In this regard, polyphosphazenes can be used as a flame retardant for other materials with which they are in contact by both suppressing the free radical process that occurs in the flame or by creating an intumescent shield that covers the material from the oxygen [60]. For instance, poly(diphenoxyphosphazene), $[N{=}P(OPh)_2]_n$ (**75**), which is a commercially available material for fire-resistant heat and sound insulation, can be mixed with polyurethane precursors to make fire-resistant foam rubbers (for fire-resistant materials see Chapter 8) [61].

Although the —P=N— backbone imparts its own properties to polyphosphazenes, R side groups that may be organic, organometallic, or inorganic, also play a critical role based on their intrinsic properties (steric demand, intermolecular interactions, solubility, optical properties, etc.) [57, 62]. For instance, most of the organic side groups on phosphorus protect the backbone from hydrolysis. However, other side groups such as amino acid esters, glyceryl moieties, sugar units, and ethoxy groups, among others, facilitate the hydrolysis of the polyphosphazene main chain [63]. Thus, polyphosphazenes can be designed to be hydrolytically unstable in a biological environment, in which the polyphosphazene main chain hydrolyzes into ammonia and phosphates, resulting in a buffered combination that prevents necrosis of living tissues. The hydrolysis of the phosphazene backbones has also been exploited in drug-delivery [64] applications (see also Chapters 1 and 9).

General synthetic routes to polyphosphazenes are shown in Scheme 4.8.

The thermal (250 °C) ROP of the cyclic phosphonitrilic chloride trimer $(N{=}PCl_2)_3$ (**76**), first reported in 1965 [65], represents the most well-established route to high molecular weight linear soluble poly(dichlorophosphazene), $[N{=}PCl_2]_n$ (**77**), which led to a wide variety of polyphosphazenes (more than 700 examples) by means of the very versatile and efficient macromolecular nucleophilic substitution reaction of the chlorine atoms, which can be carried out sequentially [57, 62]. The possibilities of this macromolecular substitution for the preparation of homopolymers and random copolymers with very different chemical compositions have been reviewed elsewhere (Scheme 4.9) [66].

The ROP of **76** can be also carried out in solution (1,3,5-trichlorobenzene) at 210 °C in the presence of a catalytic amount of $CaSO_4{\cdot}5H_2O$ and H_3NSO_3 [67], and at room temperature in the presence of carboranyl anion-stabilized trialkysilylium cations as initiators [68]. A less-employed method to $[N{=}PCl_2]_n$ (77)

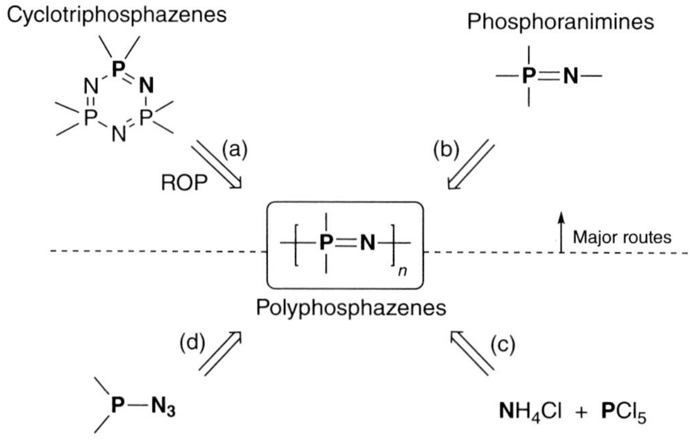

Scheme 4.8 General routes to polyphosphazenes: (a) Ring-opening polymerization (ROP) of cyclic phosphonitrilic trimer. (b) Condensation of phosphoranimines (iminophosphoranes). (c) Direct synthesis from NH_4Cl and PCl_5. (d) Condensation of azidophosphanes.

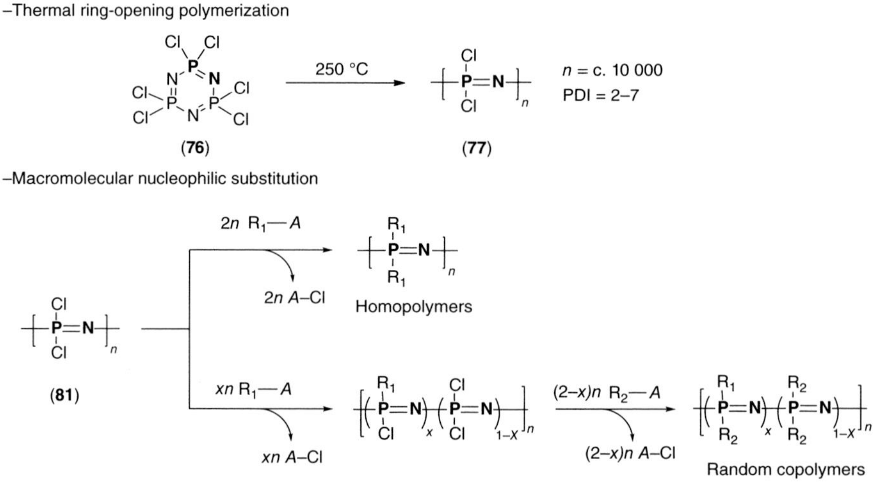

Scheme 4.9 Synthesis of high molecular weight linear $[N{=}PCl_2]_n$ (**77**) by thermal ring-opening polymerization of phosphonitrilic chloride trimer (**76**), and general synthesis to polyphosphazenes, homopolymers, and random copolymers, by nucleophilic macromolecular substitution of $[N{=}PCl_2]_n$ (**77**).

is based on the thermal condensation of $Cl_3P{=}N{-}P(O)Cl_2$ (**78**) at c. 250 °C (Figure 4.7, Eq. (10)) [69]. Polyphosphazenes bearing alkyl or aryl substituents are not accessible by macromolecular substitution from $[N{=}PCl_2]_n$ (77) with aryl or alkyl Grignard or lithium reagents, and the ROP of the related cyclotriphosphazenes having alkyl or aryl substituent requires high temperature conditions resulting in cross-linked materials with ill-define compositions. Thus, the synthesis of soluble and well-defined poly(alkyl/aryl)phosphazenes was first achieved by the thermal condensation of *N*-trimethylsilyl phosphoranimines of general formula $(CF_3CH_2O)(R_1R_2)P{=}N{-}SiMe_3$ (R = alkyl or aryl group,

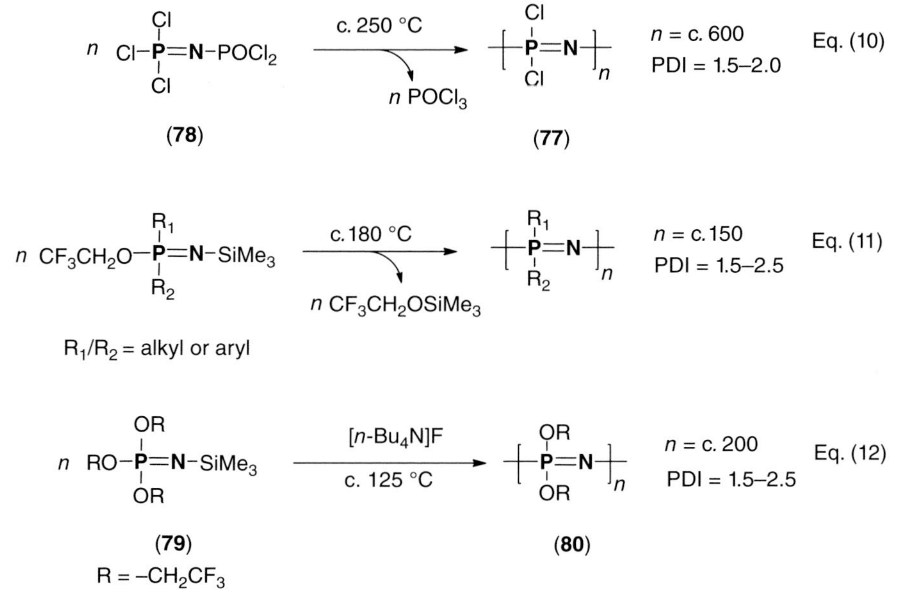

Figure 4.7 Routes to $[N{=}PCl_2]_n$ (**77**) (Eq. (10)), poly(alkyl/aryl)phosphazenes (Eq. (11)), and poly(bistrifluoroethoxy)phosphazene $[N{=}P(OCH_2CF_3)_2]_n$ (**80**, Eq. (12)).

Figure 4.7, Eq. (11)) [70]. Recently, poly(alkyl/aryl)phosphazenes were also synthesized by reacting $X(R_1R_2)P{=}N{-}SiMe_3$ (X = Cl, Br; R = alkyl or aryl group) with trimethylphosphite $P(OMe)_3$ at room temperature [71]. Also, Matyjaszewski and coworkers described the synthesis of poly(bistrifluoroethoxy)phosphazene, $[N{=}P(OCH_2CF_3)_2]_n$ (**80**), by anionic polymerization of the phosphoranimine $(CF_3CH_2O)_3P{=}N{-}SiMe_3$ (**79**) in the presence of fluoride anions (Figure 4.7, Eq. (12)) [72].

Other minor routes to polyphosphazenes include the thermal direct synthesis of high molecular weight $[N{=}PCl_2]_n$ (**77**) from NH_4Cl and PCl_5 (Route c in Scheme 4.8) [73], and the thermal condensation of phosphorus azides, which mainly led to mixtures of low molecular weight linear and cyclic polyphosphazenes [74, 75] (Route d in Scheme 4.8).

However, none of these methods allowed the synthesis of polyphosphazenes with control of the molecular weights and narrow polydispersity index (PDI ≤ 1.2). In 1995, the groups of Allcock and Manners developed the ambient temperature living cationic chain-growth condensation polymerization of *N*-trimethylsilyl phosphoranimines promoted by PCl_5 (Scheme 4.10) [76]. Thus, the polymerization of $Cl_3P{=}N{-}SiMe_3$ (**81**) is initiated by PCl_5 providing molecular weight control of $[N{=}PCl_2]_n$ (**77**) by simply altering the monomer/initiator ratio ($Cl_3P{=}N{-}SiMe_3/PCl_5$). Polyphosphazenes obtained via this route have much more narrow polydispersities (PDI ≤ 1.3) than those obtained via the thermal ROP route. The polymerization is considered "*living*" because the kinetics of the propagation steps are of first order with respect to concentration of monomer $Cl_3P{=}N{-}SiMe_3$ (**81**), the molecular weights follow

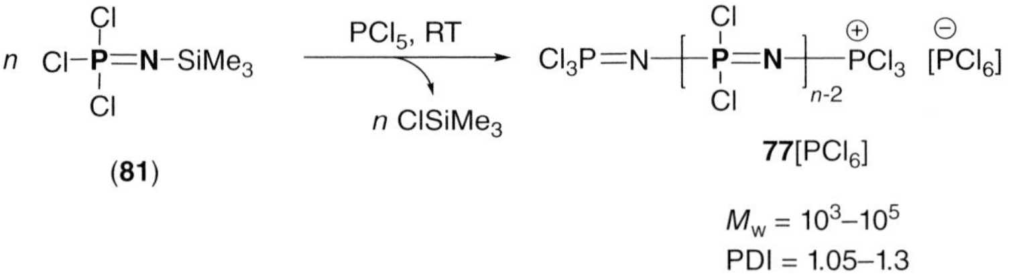

Scheme 4.10 Living cationic polymerization of $Cl_3P{=}N{—}SiMe_3$ (**81**) promoted by PCl_5.

a Poisson distribution, and the end-groups remain active at least one hour after the consumption of the initial amount of monomer [77].

The proposed mechanism of this polymerization is shown in Scheme 4.11.

This polymerization proceeds by condensation of 2 equiv of PCl_5 with $Cl_3P{=}N{—}SiMe_3$ (**81**) to yield the reactive cationic initiator $[Cl_3P{=}N{=}PCl_3][PCl_6]$ ([**82**]PCl_6 as shown in Scheme 4.11) [76, 77]. Subsequent condensation reactions (elimination of Cl-$SiMe_3$) between this initiator and $Cl_3P{=}N{—}SiMe_3$ (**81**) yield the poly(dichlorophosphazene), $Cl_2P{=}N{—}[Cl_2P{=}N]_{(n+1)}{-}PCl_3^+$ $[PCl_6]^-$ (**77**[PCl_6]) which has available reactive $-PCl_3^+$ end groups (see *propagation step* in Scheme 4.11). However, recent investigation of this process by Manners and coworkers demonstrated that the two end groups of the living $Cl_2P{=}N{—}[Cl_2P{=}N]_{(n+1)}{-}PCl_3^+$ (**77**[PCl_6]) chains have different activities depending on the chain lengths of the growing polymer [78] and therefore, sufficiently long chains will grow bi-directionally [79], which evidences that

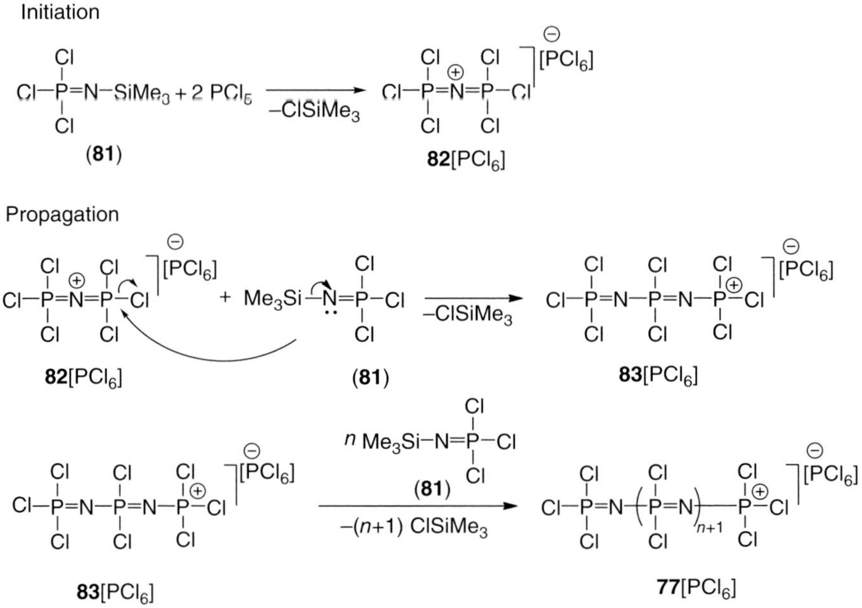

Scheme 4.11 Proposed mechanism for the living cationic polymerization of $Cl_3P{=}N{—}SiMe_3$ (**81**) promoted by PCl_5.

the mechanism of this living cationic chain-growth polycondensation is more complex than that previously proposed. This cationic polymerization was also carried out with a variety of phosphoranimines, $(X)(R_1)(R_2)P{=}N{-}SiMe_3$, combining different alkyl, aryl, alkoxy, and halogen groups on phosphorus [80]. Importantly, this living polymerization provides access to polyphosphazene block copolymers of type polyphosphazene-*b*-polyphosphazene (PP-*b*-PP′) by a sequential monomer addition. As depicted in Scheme 4.12, a variety of phosphoranimines with alkyl or aryl groups on phosphorus were polymerized from the active $-PCl_3^+$ end group of a living poly(dichlorophosphazene), **77**$[PCl_6]$, prepared by the reaction of $Cl_3P{=}N{-}SiMe_3$ (**81**) and PCl_5. The macromolecular substitution of the chlorine atoms of the $[N{=}PCl_2]_n$ block with an appropriate alkoxy or aryloxy nucleophiles led to block copolymers **84** [81].

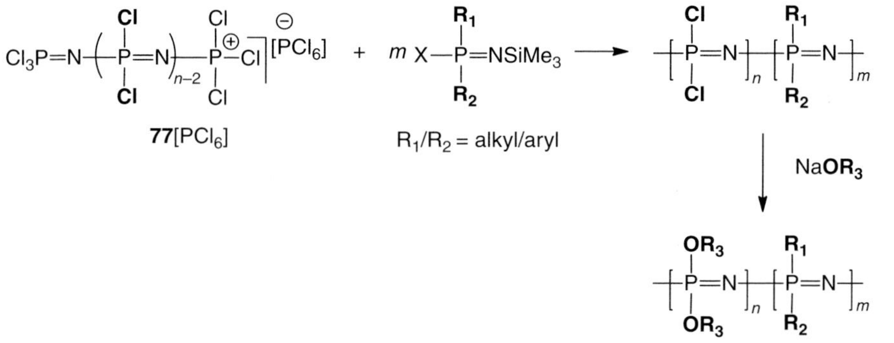

Scheme 4.12 Synthesis of linear polyphosphazene diblock copolymers **84** by sequential living cationic polymerization of phosphoranimines of type $(X)(R_1)(R_2)P{=}N{-}SiMe_3$ over living chains of **76**$[PCl_6]$.

In addition to PP-*b*-PP′ block copolymers, there is another class of diblock copolymers consisting of a PP segment and a non-phosphazenic chain. In fact, since the development of the living cationic polymerization of *N*-trimethylsilyl phosphoranimines, polyphosphazene blocks have been combined with a variety of organic segments, such as poly(dimethylsiloxane) [82], poly(ethylene oxide) [83], polystyrene [84], poly(dimethylaminoethyl methacrylate) [85], poly(propylene glycol) [86], poly(lactic acid) [87], polycaprolactone [88], poly(trimethylene carbonate) [88], and also with the organometallic block poly(ferrocenylsilane) [89]. The hybrid bathocuproines (BCPs) were usually synthesized by macromolecular coupling of two telechelic homopolymers (i.e. polymer chains endowed with end groups reactive towards each other). Linear block copolyphosphazenes, together with other macromolecular architectures (dendritic, comb-like, cyclolinear, star-like, etc.), have been exploited to create a variety of different well-defined nanostructures by self-assembly [90].

Closely related to polyphosphazenes, poly(carbophosphazene)s (**88**) [91], poly(thiophosphazene)s (**89**) [92], and poly(thionylphosphazene)s (**90**) [93], in which the phosphorus has been replaced by an sp^{2-}carbon atom, S(IV) atom or S(VI) atom respectively, have been synthesized by thermal ROP of an appropriate cyclophosphazene derivative (**85–87**, see Scheme 4.13).

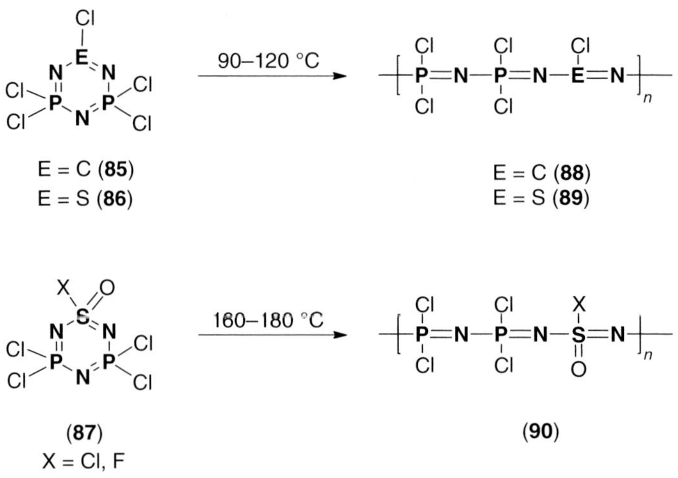

Scheme 4.13 Synthesis of poly(carbophosphazene)s (**88**), poly(thiophosphazene)s (**89**), and poly(thionylphosphazene)s (**90**) by thermal ROP.

4.6 Poly(phosphoester)s

Poly(phosphoester)s, an important class of phosphorus-containing polymers based on the presence of —P(O)—O—R— moieties in the backbone, have received wide attention over the past decade [94]. The phosphorus(V) atom of each repeating unit carries a pendant side chain, which allows for the synthesis of these materials with a high density of functional groups. Moreover, the chemical nature of these side groups leads to a variety of poly(phosphoester)s (i.e. poly (phosphate)s, poly(phosphonate)s, poly(phosphoamidate)s, or poly(phosphite)s, see Figure 4.8). The chemical versatility of poly(phosphoester)s is of pivotal importance for the development of multifunctional polymers having mechanical and chemical properties that can be fine-tuned by altering the backbone (R in Figure 4.8) or the side groups (A in Figure 4.8) [95].

Poly(phosphoester)s are known as very interesting fire-retardant candidates being used as additives to lower the flammability of organic-based materials by acting as an intumescent flame retardant (see also Chapter 8) [96]. Indeed, their thermal decomposition leads to *pyro*-phosphates (via condensation of phosphoric acid) and water vapor, which dilutes the oxidizing and combustible gas phases. Moreover, phosphate anions (i.e. *pyro-* and *poly*-phosphates) participate, with the carbonized residues of organic-based materials, in char formation, which can isolate and protect organic-based materials from the flames and heat, preclude the formation of new free radicals, and hamper the oxygen diffusion to reduce combustion. Beyond flame-retardancy applications, poly(phosphoester)s have also been used for a variety of biomedical applications. In this way, their

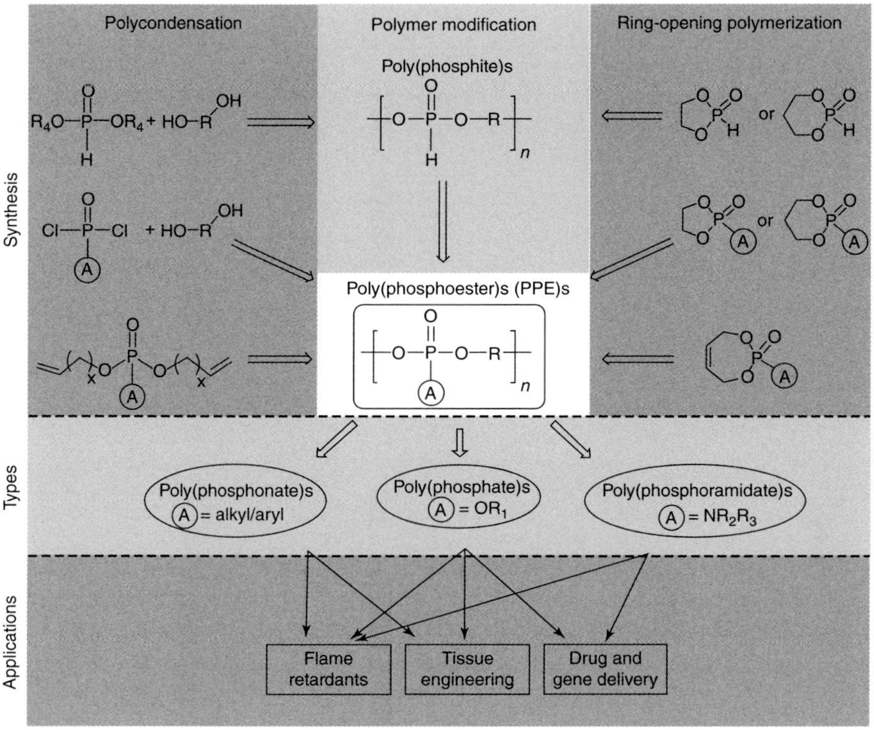

Figure 4.8 Synthetic routes leading to poly(phosphoester)s and fields of application.

high biocompatibility in vivo, controlled degradability, and low cytotoxicity in vitro, have been exploited for the development of nanocarriers for drug and gene delivery (the use of poly(phosphoester)s in biomedical applications has already been reviewed elsewhere) [94, 95, 97].

Synthesis of poly(phosphoester)s can be conducted by polycondensation, polyaddition, and ionic, enzymatic, and metathesis polymerization. *Polycondensation* (step-growth polymerization), is the main route to poly(phosphoester)s. The first reported poly(phosphoester) was prepared by polycondensation of phosphorus oxychloride with 4,4′-(propane-2,2-diyl)diphenol (bisphenol A) [98]. Following this synthetic route, libraries of poly(phosphoester)s have been prepared by using a variety of substituted phosphorus oxychloride and diols (Scheme 4.14) [99]. Polycondensation reactions can be carried out under various experimental conditions (in melt, using Lewis acid catalysis, by high-temperature solution polymerization, or by aqueous interfacial polycondensation with a phase-transfer catalyst) [100].

Polyaddition (step-growth polymerization). Poly(phosphoester)s can be synthesized by the reaction of bis(epoxide)s with phosphorus dichlorides catalyzed by quaternary onium salts [101]. In a recent work, Troev and coworkers prepared functional poly(phosphoester)s (**92**) by the polyaddition reaction of poly(phosphite)s (**91**) and C=C double bonds (Scheme 4.15) [102].

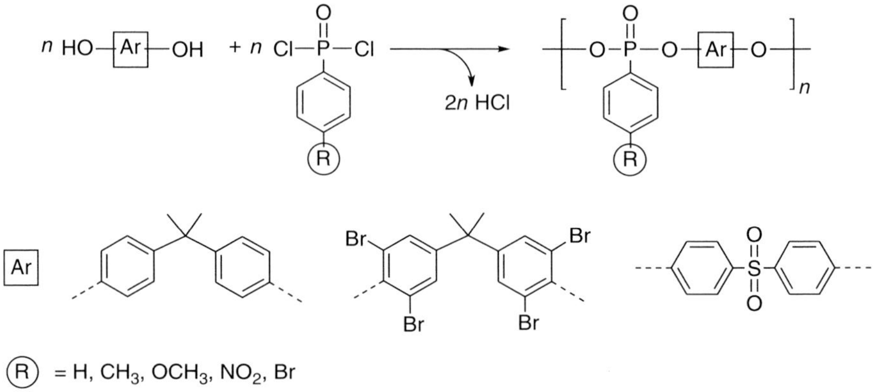

Scheme 4.14 Synthesis of poly(phosphoester)s by polycondensation of bisphenols with aryl phosphodichlorides.

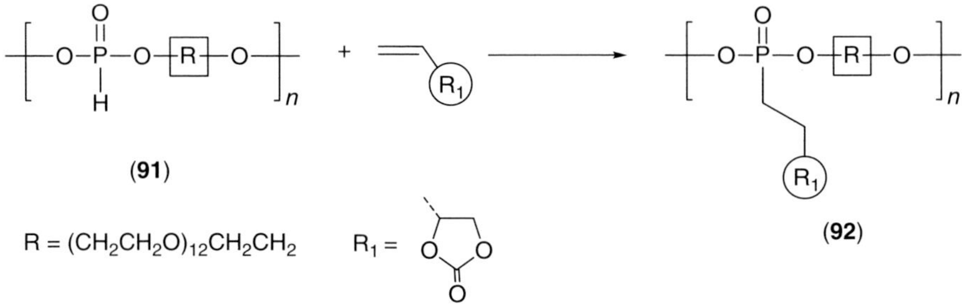

Scheme 4.15 Synthesis of polyphosphoester (**92**) by polyaddition reaction of poly(phosphite) (**91**) with C=C double bonds.

Chain-growth polymerization. Five- and six-membered cyclic phosphates can be polymerized by ROP leading to well-defined (controlled molecular weights and narrow polydispersity indices) poly(phosphoester)s. The ROP can be achieved by cationic, anionic, and insertion mechanisms. Cationic initiators (e.g. protic acids, Lewis acids, or alkylating agents) yield mostly oligomers, and often side products [103], while anionic initiators such as triisobutylaluminum, Grignard reagents, or butyllithium, lead to polymers with high molecular weights [104]. Bases such as trimethylamine [105], superbases [106], and metal carboxylates (tin[II] 2-ethylhexanoate) [107], are also active catalysts to synthesize poly(phosphoester)s by ROP.

Enzymatic polymerization. Jie and Ren-Xi reported the only example of the oligomerization of a cyclic phosphate using porcine pancreas lipase as a catalyst [108]. The as-prepared polyphosphoester exhibited low molecular weights and high polydispersity indexes.

Olefin metathesis polymerization. High molecular weights poly(phosphoester)s are accessible by olefin metathesis polymerization of readily available monomers. Two synthetic pathways based on acyclic diene metathesis (ADMET) [109] and ring-opening metathesis polymerization (ROMP) [110], both initiated by metal carbenes, have been reported.

4.7 Conclusion and Outlook

Over the past decade, the chemistry of phosphorus-containing polymers has made significant advances. Unprecedented building blocks and improved and newly developed polymerization methods have led to a variety of new polymers with fascinating properties stemming from the pnictogen incorporation. In recent years, the heavier analogs, e.g. arsenic and antimony derivatives, have received increasing attention leading to new materials stabilizing the lower oxidation states, facilitating intersystem crossing, etc. For some precursors, advanced polymerization techniques are available and a high maturity in controlling the polymer characteristics has already been achieved, allowing application for example in flame retardants as, for example flame retardants (see Chapter 8). In this last regard, polyphosphazene materials, incorporating P and N in the main chain, are classical examples of inorganic flame retardants. Moreover, their hydrophilic/hydrophobic nature and degradation behavior can be tuned by changing the side groups. Undoubtedly, interesting results have been obtained in the area of conjugated polymers and responsive materials for optoelectronic applications; however, much more work will be needed to tap the full potential of these inorganic–organic hybrid materials. Already now, many polymers containing phosphorus and heavier pnictogen elements play a big role in (industrial) applications. Nonetheless, the rich chemistry never ceases to surprise and thereby fuels new fundamental research questions. The large variability in bonding situations and its implications for optoelectronic and thermo-electric properties, as well as fundamental reactivity questions will continue to inspire researchers to study these materials transitioning from basic research into new applications.

Acknowledgments

The authors have been supported by the Swedish research council – Vetenskapsrådet, Swedish Energy agency – Energimyndigheten (A.O), the Spanish Research Council (MEC, projects; CTQ2014-56345-P, CTQ2017-88357-P, and RYC-2012-09800). The authors thank the COST Action CM1302 (SIPs): "European Network on Smart Inorganic Polymers."

References

1 Becker, G. (1976). Monosubstitutionsreaktionen an substitutierten Disilylphosphinen mit Pivaloylchlorid. *Z. Anorg. Allg. Chem.* 423: 242–254.
2 Wright, V.A. and Gates, D.P. (2002). Poly(*p*-phenylenephosphaalkene): a π-conjugated macromolecule containing P=C bonds in the main chain. *Angew. Chem. Int. Ed.* 41 (13): 2389–2392.
3 Smith, R.C., Chen, X., and Protasiewicz, J.D. (2003). A fluorescent (*E*)-poly(*p*-phenylenephosphaalkene) prepared by a phospha–Wittig reaction. *Inorg. Chem.* 42: 5468–5470.

4 Smith, R.C. and Protasiewicz, J.D. (2004). Conjugated polymers featuring heavier main group element multiple bonds: a diphosphene-PPV. *J. Am. Chem. Soc.* 126: 2268–2269.
5 Moser, C., Orthaber, A., Nieger, M. et al. (2006). The homologous series of 1,1′-ferrocenylenebisdihalophosphanes $(C_5H_4PX_2)_2Fe$ (X = F, Cl, Br, I): precursors for the first metallocene bridged bisphosphaalkene. *Dalton Trans.* 3879–3885.
6 Shah, S. and Protasiewicz, J.D. (2000). 'Phospha-variations' on the themes of Staudinger and Wittig: phosphorus analogs of Wittig reagents. *Coord. Chem. Rev.* 210: 181–201.
7 Esfandiarfard, K., Arkhypchuk, A.I., Orthaber, A., and Ott, S. (2016). Synthesis of the first metal-free phosphanylphosphonate and its use in the "phospha-Wittig-Horner" reaction. *Dalton Trans.* 45: 2201–2207.
8 Morales Salazar, D., Mijangos, E., Pullen, S. et al. (2017). Functional small-molecules & polymers containing P=C and As=C bonds as hybrid π-conjugated materials. *Chem. Commun.* 53: 1120–1123.
9 Fagan, P.J., Nugent, W.A., and Calabrese, J.C. (1994). Metallacycle transfer from zirconium to main group elements: a versatile synthesis of heterocycles. *J. Am. Chem. Soc.* 116: 1880–1889.
10 Hay, C., Fischmeister, C., Hissler, M. et al. (2000). Electropolymerization of π-conjugated oligomers containing phosphole cores and terminal thienyl moieties: optical and electronic properties. *Angew. Chem. Int. Ed.* 39: 1812–1815.
11 Hay, C., Hissler, M., Fischmeister, C. et al. (2001). Phosphole-containing π-conjugated systems: from model molecules to polymer films on electrodes. *Chem. Eur. J.* 7: 4222–4236.
12 Hay, C., Fave, C., Hissler, M. et al. (2003). Synthesis and electronic properties of alternating α,α′-thiophene-phosphole oligomers. *Org. Lett.* 5: 3467–3470.
13 Fave, C., Hissler, M., Kárpáti, T. et al. (2004). Connecting π-chromophores by σ-P–P bonds: new type of assemblies exhibiting σ–π-conjugation. *J. Am. Chem. Soc.* 126: 6058–6063.
14 Matano, Y., Ohkubo, H., Miyata, T. et al. (2014). Phosphole- and benzodithiophene-based copolymers: synthesis and application in organic photovoltaics. *Eur. J. Inorg. Chem.* 2014: 1620–1624.
15 Ashe, A.J. and Drone, F.J. (1985). Synthesis of 1-phenylarsole and 1-phenylstibole. *Organometallics* 4: 1478–1480.
16 Baumgartner, T., Bergmans, W., Kárpáti, T. et al. (2005). From model compounds to extended π-conjugated systems: synthesis and properties of dithieno[3,2-b:2′,3′-d]phospholes. *Chem. Eur. J.* 11: 4687–4699.
17 Baumgartner, T. and Wilk, W. (2006). Synthesis and unexpected reactivity of Si–H functionalized dithieno[3,2-b:2′,3′-d]phospholes. *Org. Lett.* 8: 503–506.
18 Durben, S., Dienes, Y., and Baumgartner, T. (2006). Cationic dithieno[3,2-b:2′,3′-d]phospholes: a new building block for luminescent, conjugated polyelectrolytes. *Org. Lett.* 8: 5893–5896.

19 Reus, C., Stolar, M., Vanderkley, J. et al. (2015). A convenient *N*-arylation route for electron-deficient pyridines: the case of π-extended electrochromic phosphaviologens. *J. Am. Chem. Soc.* 137: 11710–11717.
20 Stolar, M., Reus, C., and Baumgartner, T. (2016). Xylene-bridged phosphaviologen oligomers and polymers as high-performance electrode-modifiers for Li-ion batteries. *Adv. Energy. Mater.* 6: 1600944.
21 Green, J.P., Han, Y., Kilmurray, R. et al. (2016). An air-stable semiconducting polymer containing dithieno[3,2-b:2′,3′-d]arsole. *Angew. Chem. Int. Ed.* 55: 7148–7151.
22 Kato, T., Imoto, H., Tanaka, S. et al. (2016). Facile synthesis and properties of dithieno[3,2-b:2′,3′-d]arsoles. *Dalton Trans.* 45: 11338–11345.
23 Matsumura, Y., Ishidoshiro, M., Irie, Y. et al. (2016). Arsole-containing π-conjugated polymer by the post-element-transformation technique. *Angew. Chem. Int. Ed.* 55: 15040–15043.
24 Ohshita, J., Matsui, S., Yamamoto, R. et al. (2010). Synthesis of dithienobismoles as novel phosphorescence materials. *Organometallics* 29: 3239–3241.
25 Morisaki, Y., Ohashi, K., Na, H.S., and Chujo, Y. (2006). First synthesis of the bismole-containing conjugated polymer. *J. Polym. Sci., Part A: Polym. Chem.* 44: 4857–4863.
26 Zhang, W., Mao, Z., Zheng, N. et al. (2016). Highly planar cross-conjugated alternating polymers with multiple conformational locks: synthesis, characterization and their field-effect properties. *J. Mater. Chem. C* 4: 9266–9275.
27 Schäfer, B., Öberg, E., Kritikos, M., and Ott, S. (2008). Incorporating phosphaalkenes into oligoacetylenes. *Angew. Chem. Int. Ed.* 47: 8228–8231.
28 Geng, X.-L., Hu, Q., Schäfer, B., and Ott, S. (2010). Phosphaalkenes in π-conjugation with acetylenic arenes. *Org. Lett.* 12: 692–695.
29 Geng, X.-L. and Ott, S. (2009). Exploitation of an unprecedented silica-promoted acetylene-allene rearrangement for the preparation of *C,C*-diacetylenic phosphaalkenes. *Chem. Commun.* 7206–7208.
30 Geng, X.-L. and Ott, S. (2011). Acetylene-expanded dendralene segments with exotopic phosphaalkene units. *Chem. Eur. J.* 12153–12162.
31 Ito, S., Shinozaki, T., and Mikami, K. (2017). Stereoselective catalytic synthesis of alkynylated phosphaethenes leading to activation-free gold catalysis. *Eur. J. Org. Chem.* 2017: 6889–6900.
32 Shameem, M.A., Esfandiarfard, K., Öberg, E. et al. (2016). Direct, sequential, and stereoselective alkynylation of *C,C*-dibromophosphaalkenes. *Chem. Eur. J.* 22: 10614–10619.
33 Ghassemi, H. and McGrath, J.E. (1997). Synthesis of poly(arylene phosphine oxide) by nickel-catalyzed coupling polymerization. *Polymer* 38: 3139–3143.
34 Jin, Z. and Lucht, B.L. (2002). Transition metal mediated routes to poly(arylphosphine)s: investigation of novel phosphorus containing conjugated polymers. *J. Organomet. Chem.* 653: 167–176.
35 Jin, Z. and Lucht, B.L. (2005). Poly-*p*-phenylene phosphine/polyaniline alternating copolymers: electronic delocalization through phosphorus. *J. Am. Chem. Soc.* 127: 5586–5595.

36 Lucht, B.L. and St. Onge, N.O. (2000). Synthesis and characterization of poly(*p*-phenylenephosphine)s. *Chem. Commun.* 2097–2098.
37 Rawe, B.W. and Gates, D.P. (2015). Poly(*p*-phenylenediethynylene phosphane): a phosphorus-containing macromolecule that displays blue fluorescence upon oxidation. *Angew. Chem. Int. Ed.* 54: 11438–11442.
38 Patra Sanjib, K., Whittell George, R., Nagiah, S. et al. (2010). Photocontrolled living anionic polymerization of phosphorus-bridged [1]ferrocenophanes: a route to well-defined polyferrocenylphosphine (PFP) homopolymers and block copolymers. *Chem. Eur. J.* 16: 3240–3250.
39 Mizuta, T., Onishi, M., and Miyoshi, K. (2000). Photolytic ring-opening polymerization of phosphorus-bridged [1]ferrocenophane coordinating to an organometallic fragment. *Organometallics* 19: 5005–5009.
40 Resendes, R., Nelson, J.M., Fischer, A. et al. (2001). Tuning the strain and Polymerizability of organometallic rings: the synthesis, structure, and ring-opening polymerization behavior of [2]ferrocenophanes with C–Si, C–P, and C–S bridges. *J. Am. Chem. Soc.* 123: 2116–2126.
41 Vanderark, L.A., Clark, T.J., Rivard, E. et al. (2006). Anionic ring-opening polymerization of a strained phosphirene: a route to polyvinylenephosphines. *Chem. Commun.* 3332–3333.
42 Borucki, S., Kelemen, Z., Maurer, M. et al. (2017). Stereochemical alignment in triphospha[3]ferrocenophanes. *Chem. Eur. J.* 23: 10438–10450.
43 Kargin, D., Kelemen, Z., Krekic, K. et al. (2016). [3]Ferrocenophanes with the bisphosphanotetryl bridge: inorganic rings on the way to tetrylenes. *Dalton Trans.* 45: 2180–2189.
44 Naka, K., Umeyama, T., Nakahashi, A., and Chujo, Y. (2007). Synthesis of poly(vinylene–phosphine)s: ring-collapsed radical alternating copolymerization of methyl-substituted cyclooligophosphine with acetylenic compounds. *Macromolecules* 40: 4854–4858.
45 Naka, K., Nakahashi, A., and Chujo, Y. (2006). Ring-collapsed radical alternating copolymerization of phenyl-substituted cyclooligostibine and acetylenic compounds. *Macromolecules* 39: 8257–8262.
46 Naka, K., Nakahashi, A., and Chujo, Y. (2007). Periodic terpolymerization of cyclooligoarsine, cyclooligostibine, and acetylenic compound. *Macromolecules* 40: 1372–1376.
47 Naka, K., Umeyama, T., and Chujo, Y. (2002). Synthesis of poly(vinylene-arsine)s: alternating radical copolymerization of arsenic atomic biradical equivalent and phenylacetylene. *J. Am. Chem. Soc.* 124: 6600–6603.
48 Tsang, C.-W., Yam, M., and Gates, D.P. (2003). The addition polymerization of a P=C bond: a route to new phosphine polymers. *J. Am. Chem. Soc.* 125: 1480–1481.
49 Noonan, K.J.T. and Gates, D.P. (2006). Ambient-temperature living anionic polymerization of phosphaalkenes: homopolymers and block copolymers with controlled chain lengths. *Angew. Chem. Int. Ed.* 45: 7271–7274.
50 Noonan, K.J.T., Patrick, B.O., and Gates, D.P. (2007). Redox-active iron-containing polymers: synthesis and anionic polymerization of a *C*-ferrocenyl-substituted phosphaalkene. *Chem. Commun.* 3658–3660.

51 Siu, P.W., Serin, S.C., Krummenacher, I. et al. (2013). Isomerization polymerization of the phosphaalkene MesP=CPh_2: an alternative microstructure for poly(methylenephosphine)s. *Angew. Chem. Int. Ed.* 52: 6967–6970.
52 Rawe, B.W., Chun, C.P., and Gates, D.P. (2014). Anionic polymerisation of phosphaalkenes bearing polyaromatic chromophores: phosphine polymers showing "turn-on" emission selectively with peroxide. *Chem. Sci.* 5: 4928–4938.
53 Dück, K., Rawe, B.W., Scott, M.R., and Gates, D.P. (2017). Polymerization of 1-phosphaisoprene: synthesis and characterization of a chemically functional phosphorus version of natural rubber. *Angew. Chem. Int. Ed.* 56: 9507–9511.
54 Romero-Nieto, C., Durben, S., Kormos, I.M., and Baumgartner, T. (2009). Simple and efficient generation of white light emission from organophosphorus building blocks. *Adv. Funct. Mater.* 19: 3625–3631.
55 Leech, M.C. and Crossley, I.R. (2018). Through-conjugation of two phosphaalkyne ('CP') moieties mediated by a bimetallic scaffold. *Dalton Trans.* 47: 4428–4432.
56 Vitvarová, T., Svoboda, J., Hissler, M., and Vohlídal, J. (2017). Conjugated metallo-supramolecular polymers containing a phosphole unit. *Organometallics* 36: 777–786.
57 Allcock, H.R. (2003). *Chemistry and Applications of Polyphosphazenes.* Hoboken, NJ: Wiley.
58 Allcock, H.R. (1972). Recent advances in phosphazene (phosphonitrilic) chemistry. *Chem. Rev.* 72: 315–356.
59 Sun, H. (1997). Molecular structures and conformations of polyphosphazenes – a study based on density functional calculations of oligomers. *J. Am. Chem. Soc.* 119: 3611–3618.
60 Mayer-Gall, T., Knittel, D., Gutmann, J.S., and Opwis, K. (2015). Permanent flame retardant finishing of textiles by allyl-functionalized polyphosphazenes. *ACS Appl. Mater. Interfaces* 7: 9349–9363.
61 Reed, C.S., Taylor, J.P., Guigley, K.S. et al. (2000). Polyurethane/poly[bis(carboxylatophenoxy)phosphazene] blends and their potential as flame-retardant materials. *Polym. Eng. Sci.* 40: 465–472.
62 Gleria, M. and De Jaeger, R. (2005). *Phosphazenes: A Worldwide Insight.* New York: NOVA Science Publisher, Inc.
63 Wilfert, S., Iturmendi, A., Schoefberger, W. et al. (2014). Water-soluble, biocompatible polyphosphazenes with controllable and pH-promoted degradation behavior. *J. Polym. Sci., Part A: Polym. Chem.* 52: 287–294.
64 Andrianov, A.K. (2009). *Polyphosphazenes for Biomedical Applications.* Hoboken, NJ: Wiley.
65 Allcock, H.R. and Kugel, R.L. (1965). Synthesis of high polymeric alkoxy- and aryloxyphosphonitriles. *J. Am. Chem. Soc.* 87: 4216–4217.
66 Carriedo, G.A. (2009). Synthesis and chemical regularity in phosphazene copolymers. In: *Polyphosphazenes for Biomedical Applications* (ed. A.K. Andrianov), 379. Hoboken, NJ: Wiley.
67 Mujumdar, A.N., Young, S.G., Merker, R.L., and Magill, J.H. (1990). A study of solution polymerization of polyphosphazenes. *Macromolecules* 23: 14–21.

68 Zhang, Y., Huynh, K., Manners, I., and Reed, C.A. (2008). Ambient temperature ring-opening polymerisation (ROP) of cyclic chlorophosphazene trimer $[N_3P_3Cl_6]$ catalyzed by silylium ions. *Chem. Commun.* 494–496.
69 D'Halluin, G., De Jaeger, R., Chambrette, J.P., and Potin, P. (1992). Synthesis of poly(dichlorophosphazenes) from $Cl_3P{:}NP(O)Cl_2$. 1. Kinetics and reaction mechanism. *Macromolecules* 25: 1254–1258.
70 Neilson, R.H. and Wisian-Neilson, P. (1988). Poly(alkyl/arylphosphazenes) and their precursors. *Chem. Rev.* 88: 541–562.
71 Taylor, T.J., Presa Soto, A., Huynh, K. et al. (2010). Synthesis of poly(alkyl/arylphosphazenes) via the ambient temperature phosphite-mediated chain-growth polycondensation of (*N*-silyl)bromophosphoranimines. *Macromolecules* 43: 7446–7452.
72 Montague, R.A. and Matyjaszewski, K. (1990). Synthesis of poly[bis(trifluoroethoxy)phosphazene] under mild conditions using a fluoride initiator. *J. Am. Chem. Soc.* 112: 6721–6723.
73 Carriedo, G.A., García Alonso, F.J., Gómez-Elipe, P. et al. (2003). A simplified and convenient laboratory-scale preparation of ^{14}N or ^{15}N high molecular weight poly(dichlorophosphazene) directly from PCl_5. *Chem. Eur. J.* 9: 3833–3836.
74 Carriedo, G.A., Soto, A.P., and Valenzuela, M.L. (2008). The polymerization of $[P(O_2C_{12}H_8)(N_3)]$ and the formation of soluble and insoluble poly(2,2′-dioxy-1,1′-biphenylphosphazene). *Eur. Polym. J.* 44: 1577–1582.
75 Franz, U., Nuyken, O., and Matyjaszewski, K. (1993). Synthesis of poly(phenyl-*o*-tolylphosphazene) via reaction of trimethylsily azide with 2,2,2-trifluoroethyl phenyl-*o*-tolylphosphinite. *Macromolecules* 26: 3723–3725.
76 Honeyman, C.H., Manners, I., Morrissey, C.T., and Allcock, H.R. (1995). Ambient temperature synthesis of poly(dichlorophosphazene) with molecular weight control. *J. Am. Chem. Soc.* 117: 7035–7036.
77 Allcock, H.R., Crane, C.A., Morrissey, C.T. et al. (1996). "Living" cationic polymerization of phosphoranimines as an ambient temperature route to polyphosphazenes with controlled molecular weights. *Macromolecules* 29: 7740–7747.
78 Blackstone, V., Lough, A.J., Murray, M., and Manners, I. (2009). Probing the mechanism of the PCl_5-initiated living cationic polymerization of the phosphoranimine $Cl_3P{=}NSiMe_3$ using model compound chemistry. *J. Am. Chem. Soc.* 131: 3658–3667.
79 Blackstone, V., Pfirrmann, S., Helten, H. et al. (2012). A cooperative role for the counteranion in the PCl_5-initiated living, cationic chain growth polycondensation of the phosphoranimine $Cl_3P{=}NSiMe_3$. *J. Am. Chem. Soc.* 134: 15293–15296.
80 Allcock, H.R., Nelson, J.M., Reeves, S.D. et al. (1997). Ambient-temperature direct synthesis of poly(organophosphazenes) via the "living" cationic polymerization of organo-substituted phosphoranimines. *Macromolecules* 30: 50–56.

81 Allcock, H.R., Reeves, S.D., Nelson, J.M. et al. (1997). Polyphosphazene block copolymers via the controlled cationic, ambient temperature polymerization of phosphoranimines. *Macromolecules* 30: 2213–2215.
82 Prange, R. and Allcock, H.R. (1999). Telechelic syntheses of the first phosphazene siloxane block copolymers. *Macromolecules* 32: 6390–6392.
83 Nelson, J.M., Primrose, A.P., Hartle, T.J. et al. (1998). Synthesis of the first organic polymer/polyphosphazene block copolymers: ambient temperature synthesis of triblock poly(phosphazene–ethylene oxide) copolymers. *Macromolecules* 31: 947–949.
84 Prange, R., Reeves, S.D., and Allcock, H.R. (2000). Polyphosphazene–polystyrene copolymers: block and graft copolymers from polyphosphazene and polystyrene macromonomers. *Macromolecules* 33: 5763–5765.
85 Tian, Z., Liu, X., Chen, C., and Allcock, H.R. (2012). Synthesis and micellar behavior of novel amphiphilic poly[bis(trifluoroethoxy)phosphazene]-co-poly[(dimethylamino)ethyl methacrylate] block copolymers. *Macromolecules* 45: 2502–2508.
86 Allcock, H.R., Cho, S.Y., and Steely, L.B. (2006). New amphiphilic poly[bis(2,2,2-trifluoroethoxy)phosphazene]/poly(propylene glycol) triblock copolymers: synthesis and micellar characteristics. *Macromolecules* 39: 8334–8338.
87 Weikel, A.L., Cho, S.Y., Morozowich, N.L. et al. (2010). Hydrolysable polylactide–polyphosphazene block copolymers for biomedical applications: synthesis, characterization, and composites with poly(lactic-co-glycolic acid). *Polym. Chem.* 1: 1459–1466.
88 Krogman, N.R., Steely, L., Hindenlang, M.D. et al. (2008). Synthesis and characterization of polyphosphazene-block-polyester and polyphosphazene-block-polycarbonate macromolecules. *Macromolecules* 41: 1126–1130.
89 Soto, A.P. and Manners, I. (2009). Poly(ferrocenylsilane-*b*-polyphosphazene) (PFS-*b*-PP): a new class of organometallic–inorganic block copolymers. *Macromolecules* 42: 40–42.
90 Carriedo, G.A., de la Campa, R., and Presa Soto, A. (2018). Polyphosphazenes – synthetically versatile block copolymers ("multi-tool") for self-assembly. *Eur. J. Inorg. Chem.* 2018: 2484–2499. https://doi.org/10.1002/ejic.201800126.
91 Manners, I., Allcock, H.R., Renner, G., and Nuyken, O. (1989). Poly(carbophosphazenes): a new class of inorganic-organic macromolecules. *J. Am. Chem. Soc.* 111: 5478–5480.
92 Dodge, J.A., Manners, I., Allcock, H.R. et al. (1990). Poly(thiophosphazenes): new inorganic macromolecules with backbones composed of phosphorus, nitrogen, and sulfur atoms. *J. Am. Chem. Soc.* 112: 1268–1269.
93 Liang, M. and Manners, I. (1991). Poly(thionylphosphazenes): a new class of inorganic polymers with skeletal phosphorus, nitrogen, and sulfur(VI) atoms. *J. Am. Chem. Soc.* 113: 4044–4045.
94 Yu-Cai, W., You-Yong, Y., Jin-Zhi, D. et al. (2009). Recent progress in polyphosphoesters: from controlled synthesis to biomedical applications. *Macromol. Biosci.* 9: 1154–1164.

95 Steinbach, T. and Wurm, F.R. (2015). Poly(phosphoester)s: a new platform for degradable polymers. *Angew. Chem. Int. Ed.* 54: 6098–6108.

96 Lu, S.-Y. and Hamerton, I. (2002). Recent developments in the chemistry of halogen-free flame retardant polymers. *Prog. Polym. Sci.* 27: 1661–1712.

97 Zhao, Z., Wang, J., Mao, H.-Q., and Leong, K.W. (2003). Polyphosphoesters in drug and gene delivery. *Adv. Drug Delivery Rev.* 55: 483–499.

98 Arvin, J. (1936). Resinous compositions. U.S. Patent 2, 058,394.

99 Cass, W.E. (1952). Organo-phosphorus resinous compositions. U.S. Patent 2, 616,873.

100 Richards, M., Dahiyat, B.I., Arm, D.M. et al. (1991). Interfacial polycondensation and characterization of polyphosphates and polyphosphonates. *J. Polym. Sci., Part A: Polym. Chem.* 29: 1157–1165.

101 Liaw, D.J. (1997). Synthesis of polyphosphates by the polyaddition of bisphenol-S diglycidyl ether and aryl phosphorodichloridates. *J. Polym. Sci., Part A: Polym. Chem.* 35: 2365–2369.

102 Koseva, N., Bogomilova, A., Atkova, K., and Troev, K. (2008). New functional polyphosphoesters: design and characterization. *React. Funct. Polym.* 68: 954–966.

103 Lapienis, G. and Penczek, S. (1977). Cationic polymerization of 2-alkoxy-2-oxo-1,3,2-dioxaphosphorinanes (1,3-propylene alkyl phosphates). *Macromolecules* 10: 1301–1306.

104 Libiszowski, J., Kałużynski, K., and Penczek, S. (1978). Polymerization of cyclic esters of phosphoric acid. VI. Poly(alkyl ethylene phosphates). Polymerization of 2-alkoxy-2-oxo-1,3,2-dioxaphospholans and structure of polymers. *J. Polym. Sci., Part A: Polym. Chem.* 16: 1275–1283.

105 Yasuda, H., Sumitani, M., Lee, K. et al. (1982). High molecular weight poly(2-methoxy-1,3,2-dioxaphospholane 2-oxide) by ring-opening catalysis of tertiary amines. Initiation and stepwise propagation mechanisms as studied by the stoichiometric reaction with triethylamine. *Macromolecules* 15: 1231–1237.

106 Clément, B., Grignard, B., Koole, L. et al. (2012). Metal-free strategies for the synthesis of functional and well-defined polyphosphoesters. *Macromolecules* 45: 4476–4486.

107 Wang, Y.C., Yuan, Y.Y., Wang, F., and Wang, J. (2011). Syntheses and characterization of block copolymers of poly(aliphatic ester) with clickable polyphosphoester. *J. Polym. Sci., Part A: Polym. Chem.* 49: 487–494.

108 Jie, W. and Ren-Xi, Z. (1998). Enzyme-catalyzed ring-opening polymerization of ethylene isopropyl phosphate. *Macromol. Rapid Commun.* 19: 641–642.

109 Steinbach, T., Wahlen, C., and Wurm, F.R. (2015). Poly(phosphonate)-mediated Horner–Wadsworth–Emmons reactions. *Polym. Chem.* 6: 1192–1202.

110 Steinbach, T., Alexandrino, E.M., Wahlen, C. et al. (2014). Poly(phosphonate)s via olefin metathesis: adjusting hydrophobicity and morphology. *Macromolecules* 47: 4884–4893.

5

Synthesis of Inorganic Dendrimers

Anne-Marie Caminade[1, 2]

[1] *Laboratoire de Chimie de Coordination du CNRS, 205 Route de Narbonne, BP 44099, 31077 Toulouse Cedex 4, France*
[2] *Université de Toulouse, LCC-CNRS, CNRS, Toulouse, France*

5.1 Introduction

The world of polymers has a new family discovered 40 years ago [1]: regularly branched polymers, which counted a very low number of members at the beginning, but have largely developed since then. The name of this new family has been uncertain for several years, until the creation of the word "dendrimers" by D.A. Tomalia from two Greek words δέντρο (dendro), which translates to "tree," and μέρος (meros), which translates to "part" [2]. Dendrimers are soft nano-objects, synthesized by the repetition of generally two synthetic steps, creating a new "generation" at the end of each repeating process step, as illustrated in Scheme 5.1.

The first three families of dendrimers were purely organic compounds with nitrogen atoms as branching points, namely poly(propyleneimine) (PPI or DAB for diaminobutane) [1, 3, 4], poly(amidoamine) (PAMAM) [2], and poly(L-lysine) [5, 6]. However, inorganic dendrimers having an inorganic element at each branching point [7], have been proposed also very early, with the synthesis of poly(organosiloxane) dendrimers [8], and that of poly(phosphonium) dendrimers [9]. These two dendrimer families are not the most widely used types of inorganic dendrimers, which are poly(carbosilane) dendrimers [10] for silicon-containing dendrimers, and poly(phosphorhydrazone) dendrimers [11] for phosphorus-containing dendrimers, which will be presented in the first part of their Sections 5.2.1 and 5.3.1 respectively.

5.2 Main Methods of Synthesis of Silicon-Containing Dendrimers

5.2.1 Synthesis of Carbosilane Dendrimers

Carbosilane dendrimers are synthesized by using in alternation alkenylation with Grignard reagents and hydrosilylation with $HSiCl_3$. Tetrachlorosilane

Smart Inorganic Polymers: Synthesis, Properties, and Emerging Applications in Materials and Life Sciences, First Edition. Edited by Evamarie Hey-Hawkins and Muriel Hissler.

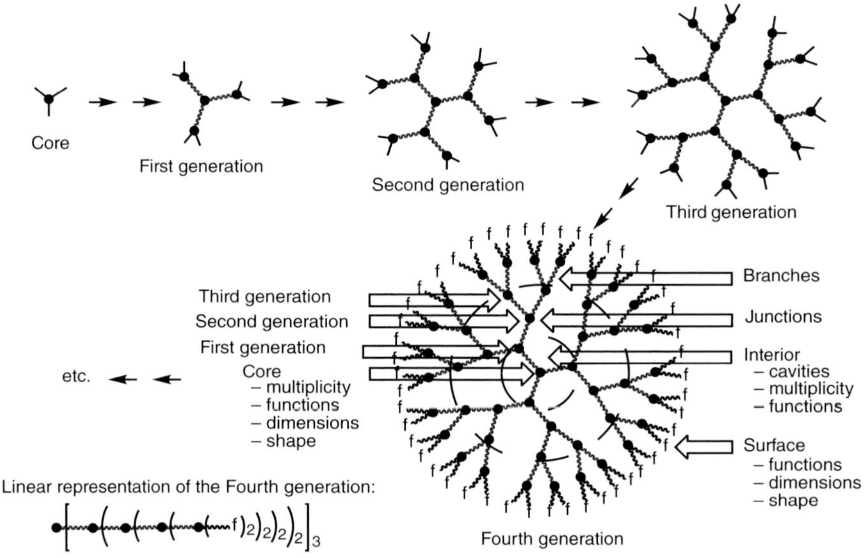

Scheme 5.1 Principle of the step-by-step synthesis of dendrimers, with illustration of the "generations." 2D-full structure and linear representation of generation 4.

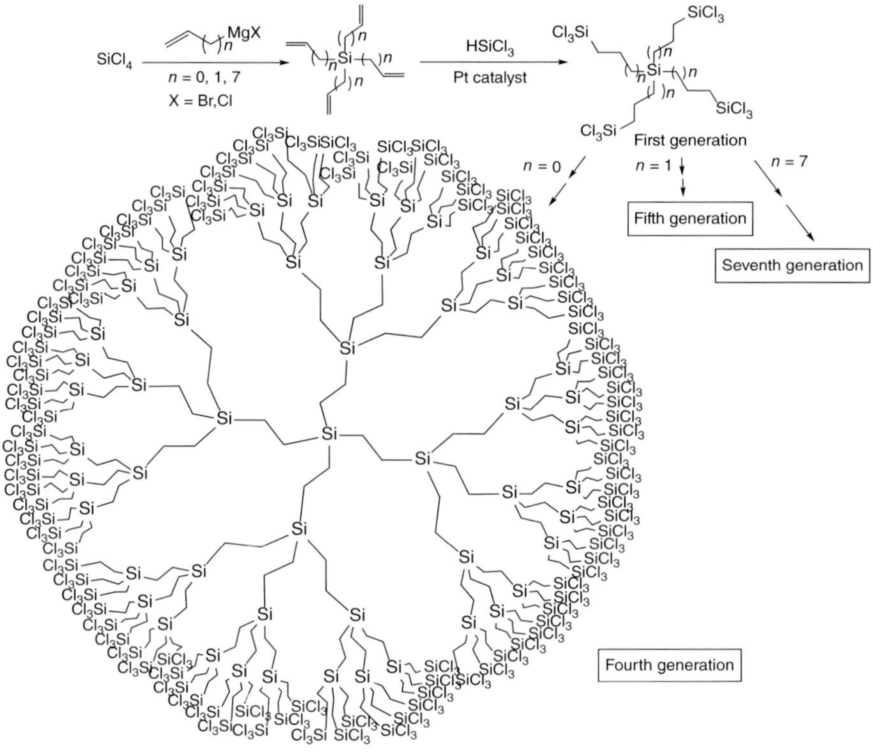

Scheme 5.2 Synthesis of carbosilane dendrimers with a multiplication by 3 at each generation, and the highest generation obtained in each case, depending on *n*.

($SiCl_4$), the most often used precursor of the core unit, is in the first step reacted with a chosen alkenylmagnesium derivative of various lengths. The terminal alkene groups are then hydrosilylated in the presence of a Pt catalyst. The use of $HSiCl_3$ as hydrosilylation agent induces the multiplication by 3 of the number of end-groups at each generation (Scheme 5.2). The highest generation obtainable depends on the length of the alkenyl groups. The vinyl derivative has been used to synthesize dendrimers up to the fourth generation [12], the allyl derivative up to the fifth generation, and the decenyl derivative up to the seventh generation [13].

The branching multiplicity depends on the type of hydrosilylation agent. In particular, replacing $HSiCl_3$ by $HSiCl_2Me$ decreases the branching density as well as the number of functional end-groups. In most cases, these dendrimers are synthesized up to the fifth generation, using vinylmagnesium bromide [14]. The 10th generation, which is the highest generation for all types of silicon-containing dendrimers, has been obtained using allylmagnesium chloride. However, this 10th generation carbosilane dendrimer has been obtained in only 24% purity [15]. The highest generation of all types of dendrimers is the 13th generation, obtained for triazine dendrimers [16]. The third generation of carbosilane dendrimers is displayed in Figure 5.1. This is the precise family that is the most widely used type of carbosilane dendrimers presently, as will be emphasized in Chapter 10.

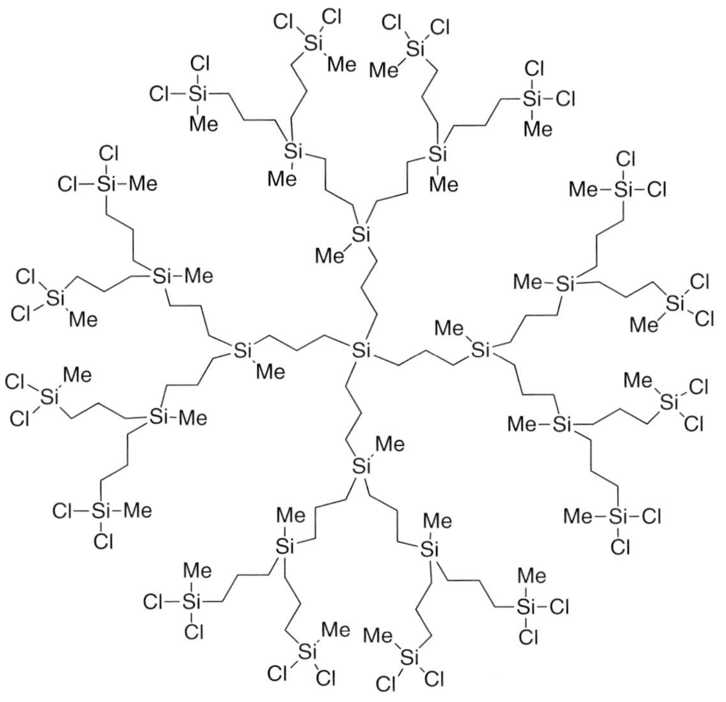

Figure 5.1 The third generation of carbosilane dendrimers with a multiplication by 2 of the terminal functions at each generation.

Another modification of the synthetic procedure toward carbosilane dendrimers consists in the use of alkynyl instead of alkenyl groups, which affords alkene instead of alkane groups in branches of the dendrimers after hydrosilylation. The use of Li—C≡C—Ph and $HSiMeCl_2$ has been carried out successfully up to the third generation [17], which is shown in Figure 5.2.

5.2.2 Synthesis of Other Types of Silicon-Containing Dendrimers

As indicated in Section 5.1, siloxane dendrimers were the very first type of inorganic dendrimers. Various siloxane dendrimers have been synthesized, but they all suffer from the low hydrolytic stability of Si–O–alkyl units compared to carbosilane dendrimers, which explains the wider use of the latter. The first example of synthesis of siloxane dendrimers is shown in Scheme 5.3. It consists in using two types of reactions, starting from the trifunctional $MeSiCl_3$ as the core unit precursor. The first step, nucleophilic substitution by $NaOSi(Me)(OEt)_2$, is followed by reaction with $SOCl_2$, to replace the O-Et groups by Cl [8].

Several types of Si dendrimers are hybrids of carbosilanes and siloxanes. An example is shown in Scheme 5.4. It is a dendron (dendritic wedge), which

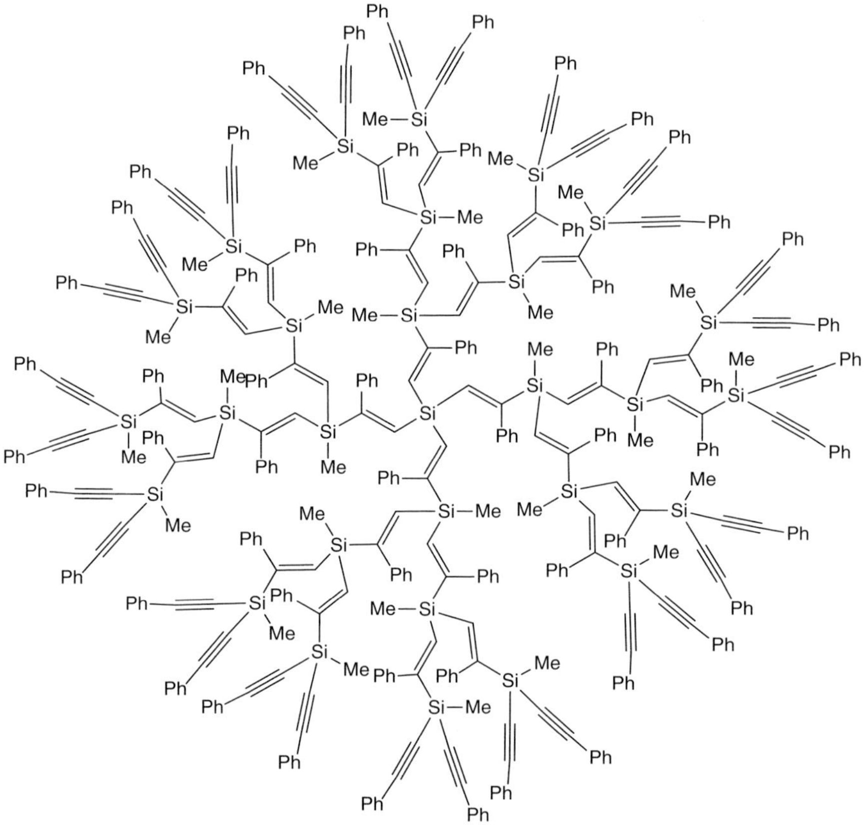

Figure 5.2 Unsaturated carbosilane dendrimer of third generation.

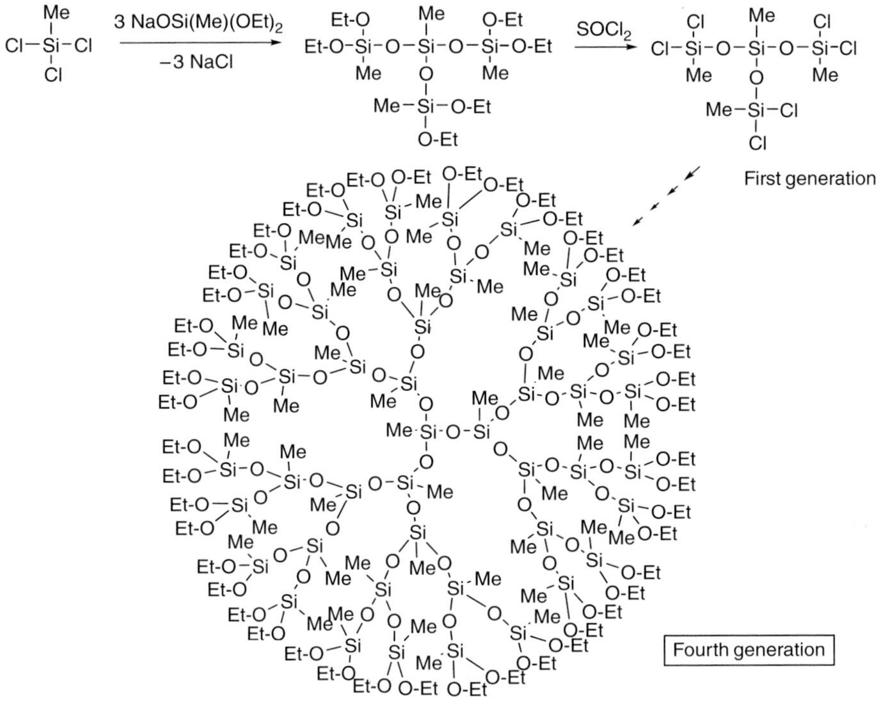

Scheme 5.3 Synthesis of the first silicon-containing dendrimer, of type siloxane.

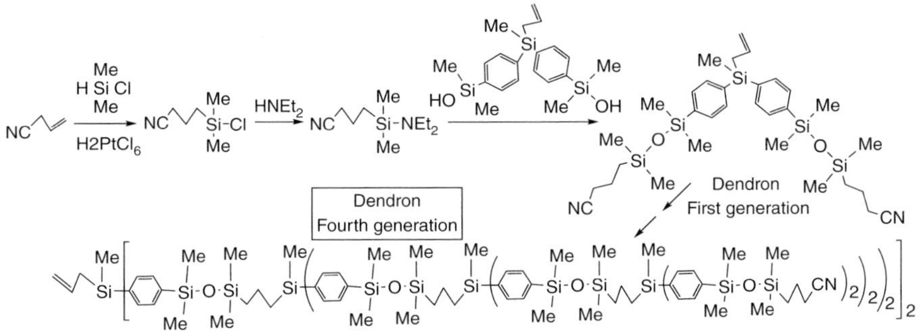

Scheme 5.4 Synthesis of hybrid siloxane carbosilane dendron. The parentheses indicate the level of the branching points.

possesses an allyl group at the core and 16 CN groups as terminal functions. This dendron is synthesized by a convergent process, from the surface toward the core [18].

Other examples of silicon-containing dendrimers are polysilanes, in which all main-chain atoms (not only branching points) are silicon atoms. See example in Scheme 5.5, where each generation is synthesized in two steps: cleavage of Si—Me bonds with triflic acid and formation of Si—Si bonds with lithium derivatives [19].

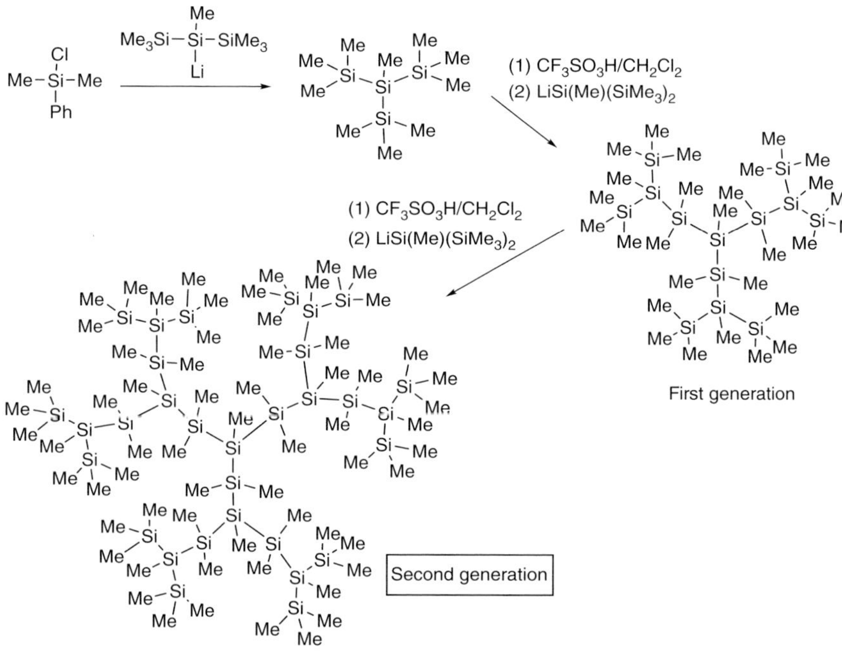

Scheme 5.5 Synthesis of a polysilane dendrimer of generation 2.

5.3 Main Methods of Synthesis of Phosphorus-Containing Dendrimers

5.3.1 Synthesis of Phosphorhydrazone Dendrimers

The synthesis of phosphorhydrazone dendrimers necessitates two steps to multiply by 2 the number of terminal functional units, starting from a core precursor possessing P–Cl functions such as $P(S)Cl_3$ [11] or $N_3P_3Cl_6$ [20]. The first step is a nucleophilic substitution on P–Cl with hydroxybenzaldehyde under basic conditions, and the second step is the condensation of formed carbaldehyde groups with the phosphorhydrazide $H_2NNMeP(S)Cl_2$ (see Scheme 5.6). These reactions have been successfully carried out up to the eighth generation starting from $N_3P_3Cl_6$ core precursor [21] (presumably a higher generation is obtainable) and up to the 12th generation starting from $P(S)Cl_3$ core precursor [22]. The latter is the highest generation dendrimer of all inorganic dendrimers, and the next to the highest of all types of dendrimer (as already indicated, the 13th generation of triazine dendrimers has been reported [16]). The above synthesis procedure is the most widely used method of preparation of phosphorus-containing dendrimers.

Attempts have been made to insert P=O instead of P=S groups by using $H_2NNMeP(O)Cl_2$. The obtained dendrimers are more reactive/less stable [23]. The other approach to the synthesis of phosphorhydrazone dendrimers consists in replacing hydroxybenzaldehyde with hydroxydialdehydes reagents [24]. Using this method, the number of terminal functions increases more rapidly, being

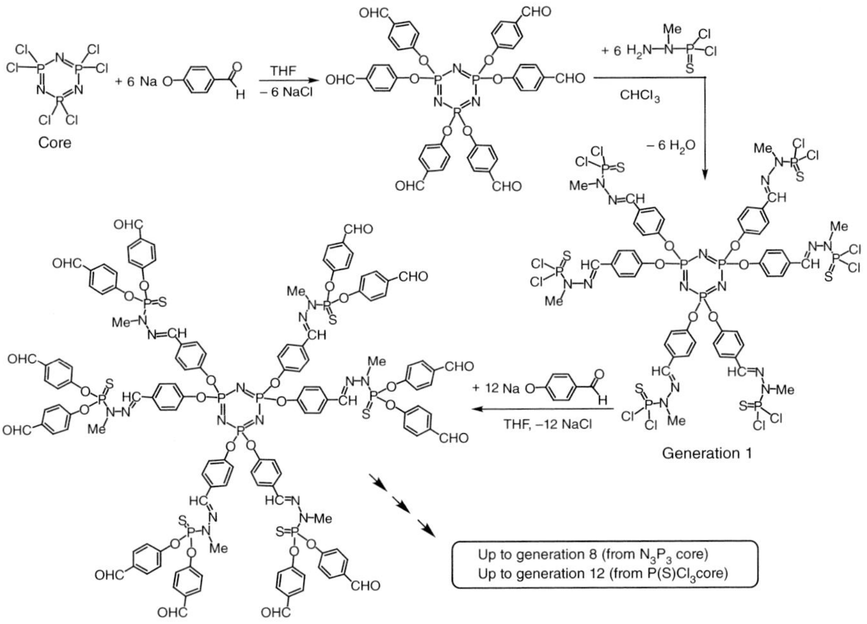

Scheme 5.6 Synthesis of phosphorhydrazone dendrimers from the cyclotriphosphazene (N_3P_3) core.

multiplied by two at each reaction step, not every two steps as in Scheme 5.6. In particular, the hydroxydialdehyde with triazine branching unit has been used in this approach and the corresponding dendrimer was built up to the fourth generation in only four steps (Figure 5.3) [25]. This dendrimer is composed of units coming from two known families of dendrimers: the phosphorhydrazone dendrimers, and the triazine [26] dendrimers. Some of its properties are inherited from its "parents," but it has also its own properties, in particular a color that changes depending on the type of functional terminal units although both parent dendrimers are white, and the possibility of electrochemically monitoring the growth of these dendrimers.

5.3.2 Synthesis of Other Types of Phosphorus-Containing Dendrimers

The very first examples of phosphorus-containing dendrimers are those with phosphonium branching points that provide multiplication by 3 of the number of terminal units at each generation. Up to 40 phosphonium groups have been included in the largest dendrimer of this series. The first step of the synthesis is the selective cleavage of benzyl ether linkages with Me_3SiI and the second step is the alkyl reaction of formed iodo-derivative with tribenzylic ether phosphine (Scheme 5.7) [9]. Different cores have been used, tri-, tetra- or penta-directional [27].

The first method of synthesis of dendrimers possessing phosphate groups as branching points was a convergent-growth process on the surface of porous glass with the use of an automated DNA synthesizer for the chain extension. Adenosine

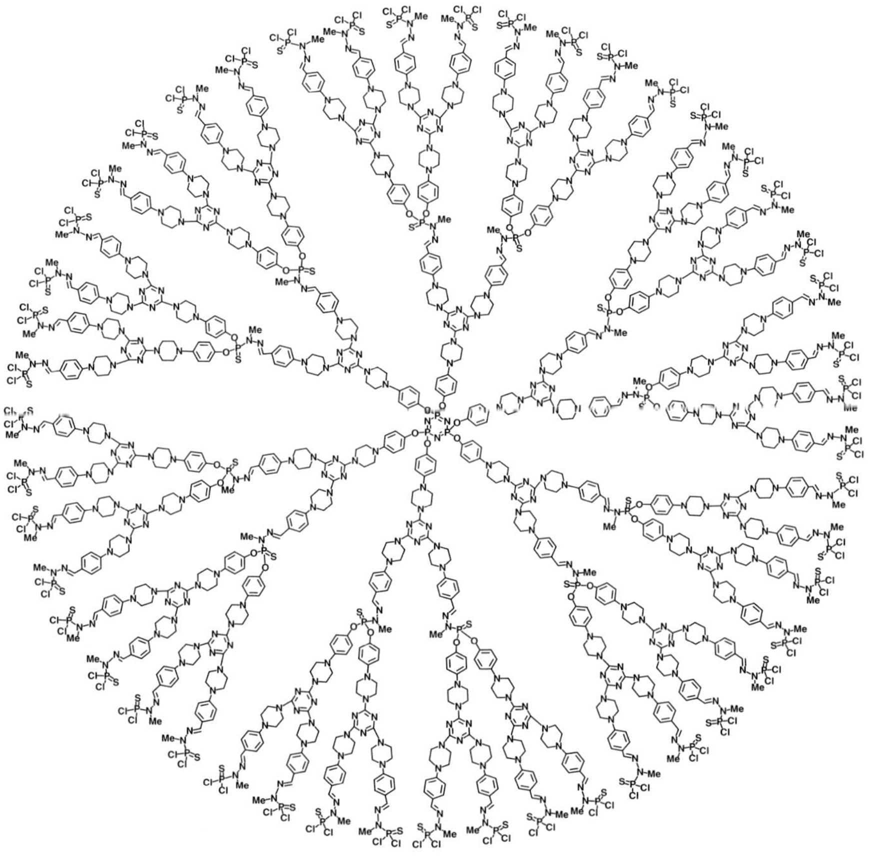

Figure 5.3 Structure of the generation 4 dendrimer built from phosphorhydrazone and triazine units in only four steps.

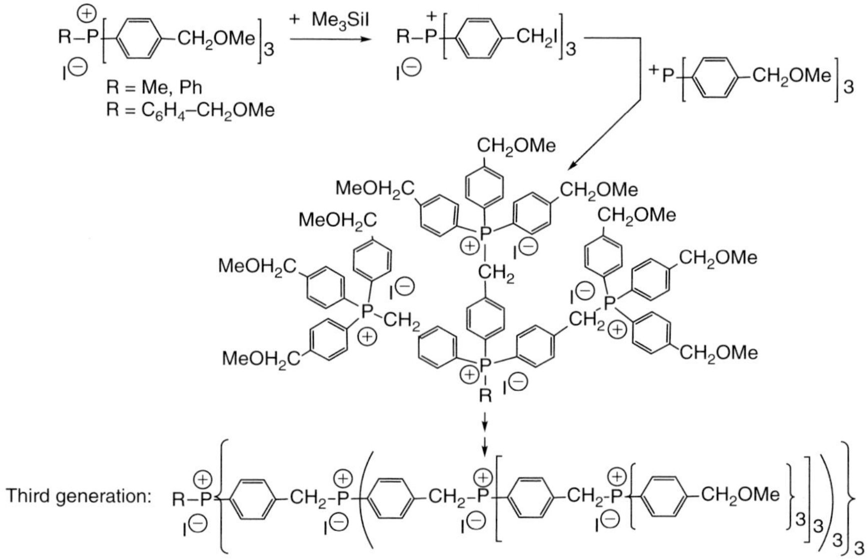

Scheme 5.7 The first example of synthesis of phosphorus-containing dendrimers.

phosphoramidate was used for creating the branching points by coupling two adjacent polymer-bound nucleotide chains. The dendrimers were isolated after cleavage from the solid phase. Various families of dendrimers based on thymidine and adenosine building blocks were synthesized by repetition of chain elongation and branching steps [28].

A family of thiophosphate dendrimers has been synthesized by repetition of two steps: the reaction of a triol with a phosphoramidite possessing two acetate groups, followed by the product oxidation with elemental sulfur. The second step is the deprotection of the acetates. The repetition of both steps was carried out up to the fifth generation [29]. The same type of method is also applicable for the synthesis of dendrimers possessing selenium, oxygen, or boron hydride instead of sulfur linked to phosphorus [30] (Scheme 5.8). It is even possible to build original dendrimers possessing a different type of phosphate at each generation, chosen between P=S, P=Se, and P=O to afford layered dendrimers.

The Staudinger reaction between phosphines and azides, in particular to create P=N—P=S linkages, when using a thiophosphorus azide has been applied either alone, or in combination with the synthesis of phosphorhydrazone dendrimers (Scheme 5.6). This Staudinger reaction is quantitative and generates N_2 as a single by-product. P=N functions are generally easily hydrolyzed, but the presence of the P=S group largely increases its stability. The first series of phosphorus dendrimers including these linkages necessitated three steps to build one generation, starting from aldehyde groups: (i) condensation of methylhydrazine, (ii) reaction with Ph_2PCl [31] under basic conditions, and (iii) Staudinger reaction between the phosphine end groups and an azido dialdehyde, to afford again aldehydes as terminal groups (Scheme 5.9).

Taking advantage of the condensation reaction of hydrazine (**A**) with aldehyde (**D**), and the Staudinger reaction of phosphine (**B**) with azide (**C**), several

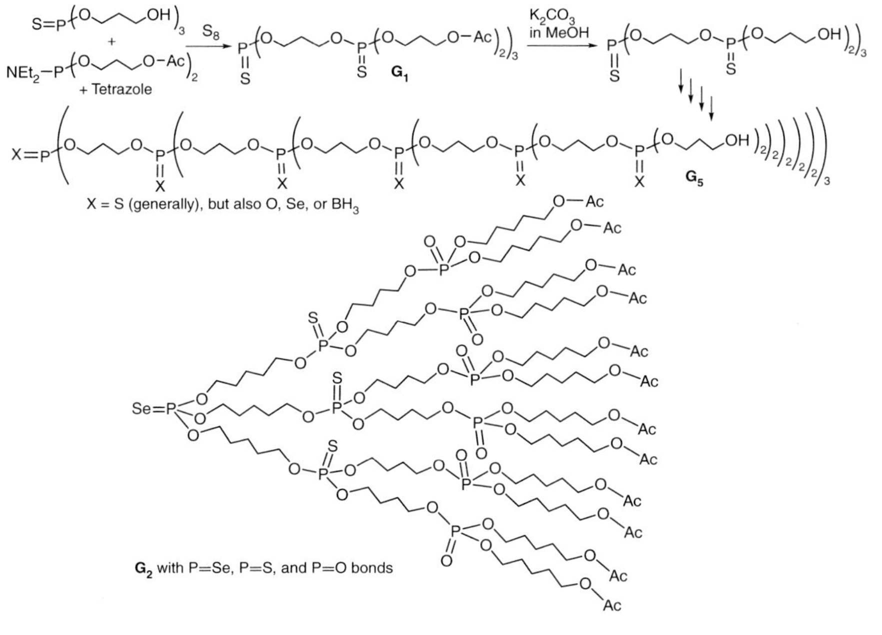

Scheme 5.8 Synthesis of thio-seleno-phosphate layered dendrimers.

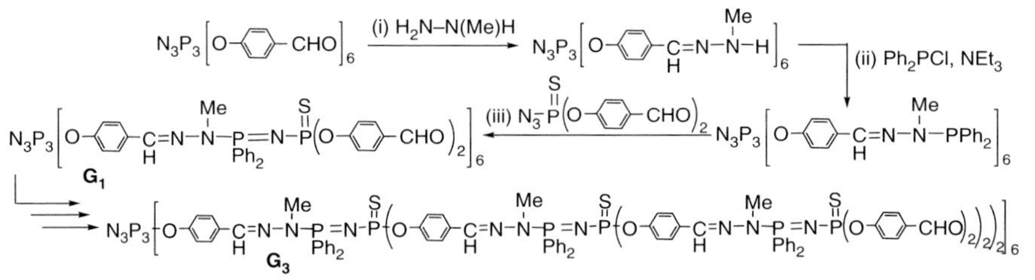

Scheme 5.9 Three-step synthesis of dendrimers based on the Staudinger reaction.

branched monomers have been designed that include two of these functions, to synthesize dendrimers more rapidly: one reaction, one generation. As first, compounds $H_2NNMeP(S)(OC_6H_4PPh_2)_2$ (**AB₂**) and $N_3P(S)(OC_6H_4CHO)_2$ (**CD₂**) have been prepared. The step-wise synthesis of the dendrimers was carried out up to the fourth generation [32] (Scheme 5.10).

The same concept was also extended to monomers having a larger number of functional groups, **AB₅** and **CD₅** monomers to multiply more rapidly the

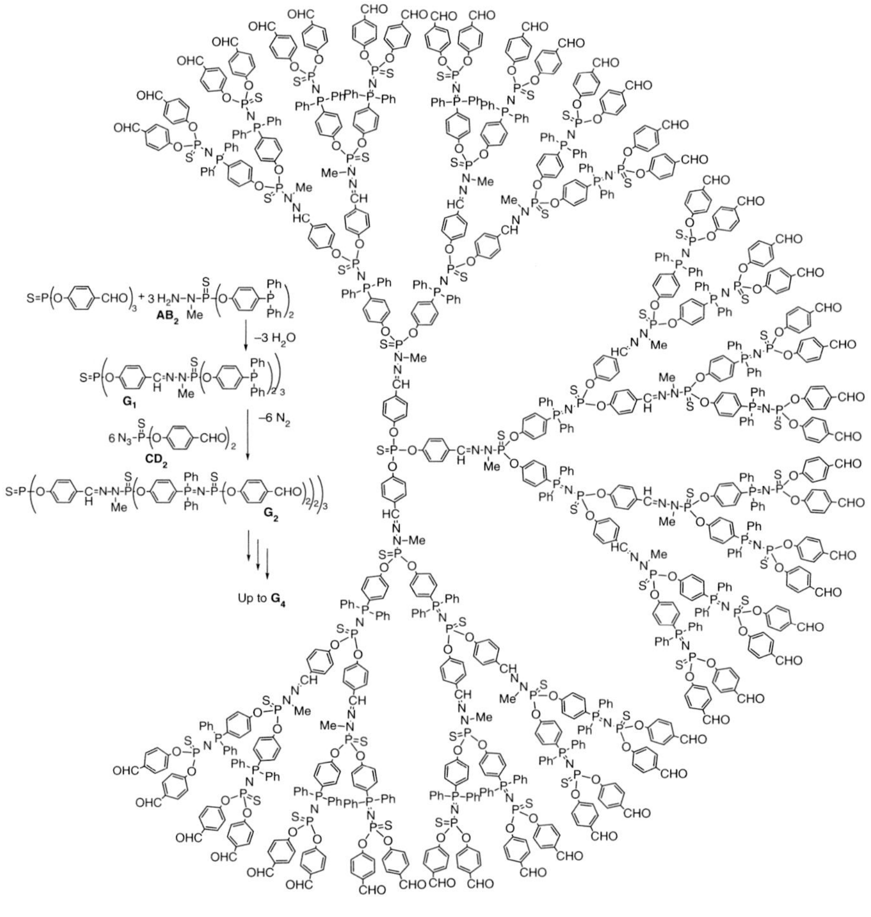

Scheme 5.10 Synthesis of a layered dendrimer based of phosphorhydrazone and P=N—P=S linkages, using two branched monomers, **AB₂** and **CD₂**.

number of terminal groups, based on the specific reactivity of the cyclotriphosphazene. These highly branched monomers may induce problems of steric hindrance rapidly, thus they were first used in combination with the **AB$_2$** and **CD$_2$** monomers. For instance, **AB$_2$** and **CD$_5$** monomers afforded dendrimers up to the fourth generation. Another attempt was made using **AB$_5$** and **CD$_2$**, but a significant steric hindrance was expected and observed at the first generation. Indeed the condensation reaction was abnormally slow and needed for four days at 100 °C to go to completion. However, the next steps occurred without any problem, and the synthesis was also carried out up to the fourth generation. Finally, the synthesis using both **AB$_5$** and **CD$_5$** monomers was attempted. Each step needed several days to go to completion, thus the synthesis was stopped at the third generation, which has 750 phosphine terminal functions after only three reactions (Scheme 5.11). In view of the expected steric hindrance of the third generation outer shell, no attempt was made to try to obtain the next generation [33]. This method is one of the rare examples of a multiplication by five of the number of terminal functions after each step of synthesis. Another example based on N_3P_3 as branching units was previously reported, but in view of the method of synthesis used (reaction of only one side of a diamine used in stoichiometric amounts), the purity of the corresponding dendrimers is highly questionable [34].

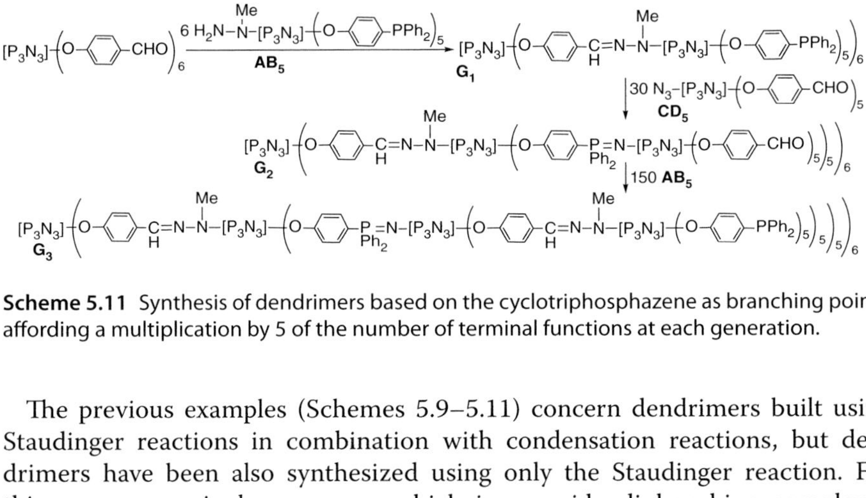

Scheme 5.11 Synthesis of dendrimers based on the cyclotriphosphazene as branching points, affording a multiplication by 5 of the number of terminal functions at each generation.

The previous examples (Schemes 5.9–5.11) concern dendrimers built using Staudinger reactions in combination with condensation reactions, but dendrimers have been also synthesized using only the Staudinger reaction. For this purpose, a single monomer which is an azido diphosphine complexed by BH_3 has been synthesized. The azide was reacted in the first step with a triphosphine core, then the terminal phosphine complexes were deprotected with 1,4-diazabicyclo[2.2.2] octane (DABCO). The next generation was obtained by using the same monomer, followed by its deprotection under the same conditions. These reactions were carried out up to generation 5 (Scheme 5.12), for which the dispersity was very low (1.029). Hyperbranched polymers were also synthesized in only one step by using the deprotection of the monomer, but the measured dispersity was very high for these polymers, from 1.5 to 8 depending on the samples. Furthermore, the properties of the hyperbranched polymers, in particular the intrinsic viscosity, were very different compared to that of the corresponding dendrimers [35].

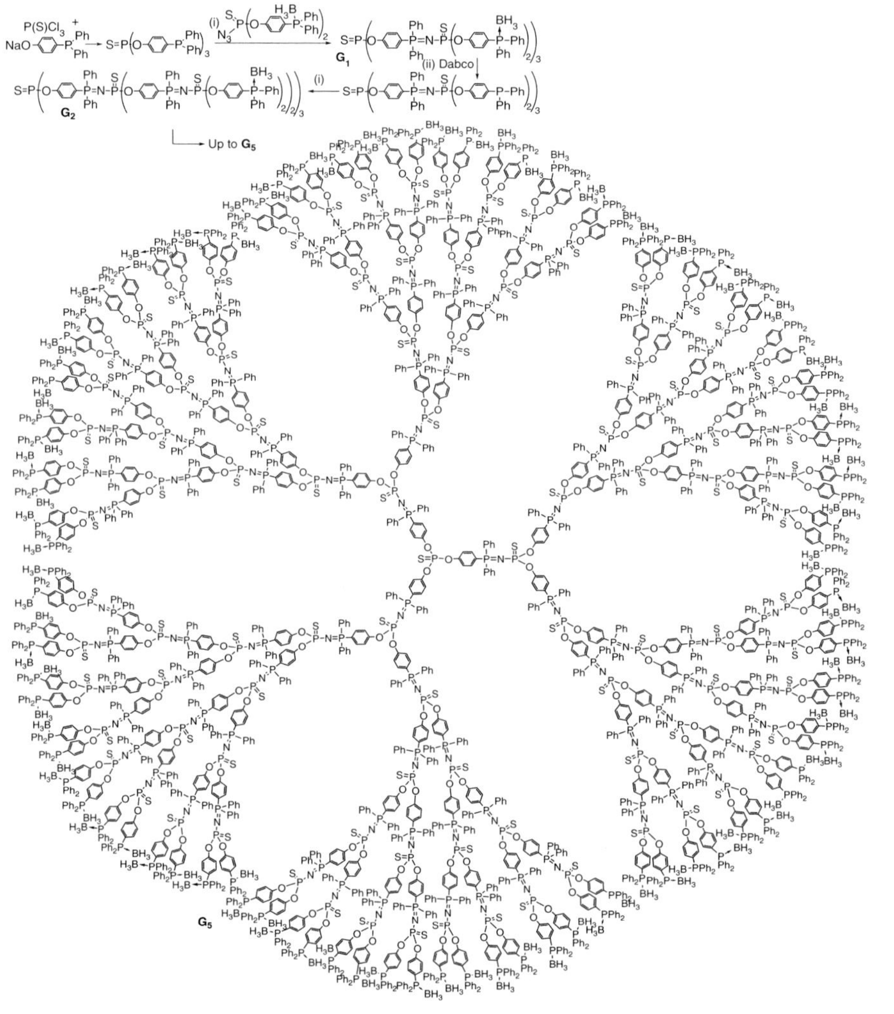

Scheme 5.12 Synthesis of dendrimers based only on Staudinger reactions.

The P=N—P=S linkages have a mesomeric form in which a partial negative charge is located on the sulfur atom ($^{+}$P—N=P—S^{-}), which can be easily and specifically alkylated with alkyl triflates, whereas other P=S groups in the dendrimers (not included in P=N—P=S) are not alkylated [36, 37]. The alkylation on sulfur induces a weakening of the phosphorus sulfur bond, which is easily cleaved using a nucleophilic phosphine such as $P(NMe_2)_3$, to generate tricoordinated phosphorus atoms (P^{III}) inside the structure. These P^{III} have been used to react with various functionalized azides, having primary amines, isothiocyanate [38], but also two aldehydes, affording compound $\mathbf{G_0@G_3}$. Obviously, many reactions were performed from these internal aldehydes [38, 39], but the most interesting is the step-by-step growth of new branches inside the dendrimer, affording finally $\mathbf{G_3@G_3}$ compound possessing 96 carbaldehyde and 48 phenoxy terminal groups [40]. Two synthesis methods have been used for growing the inner branches: (i) the two-step approach displayed in Scheme 5.6 and shown in Scheme 5.13,

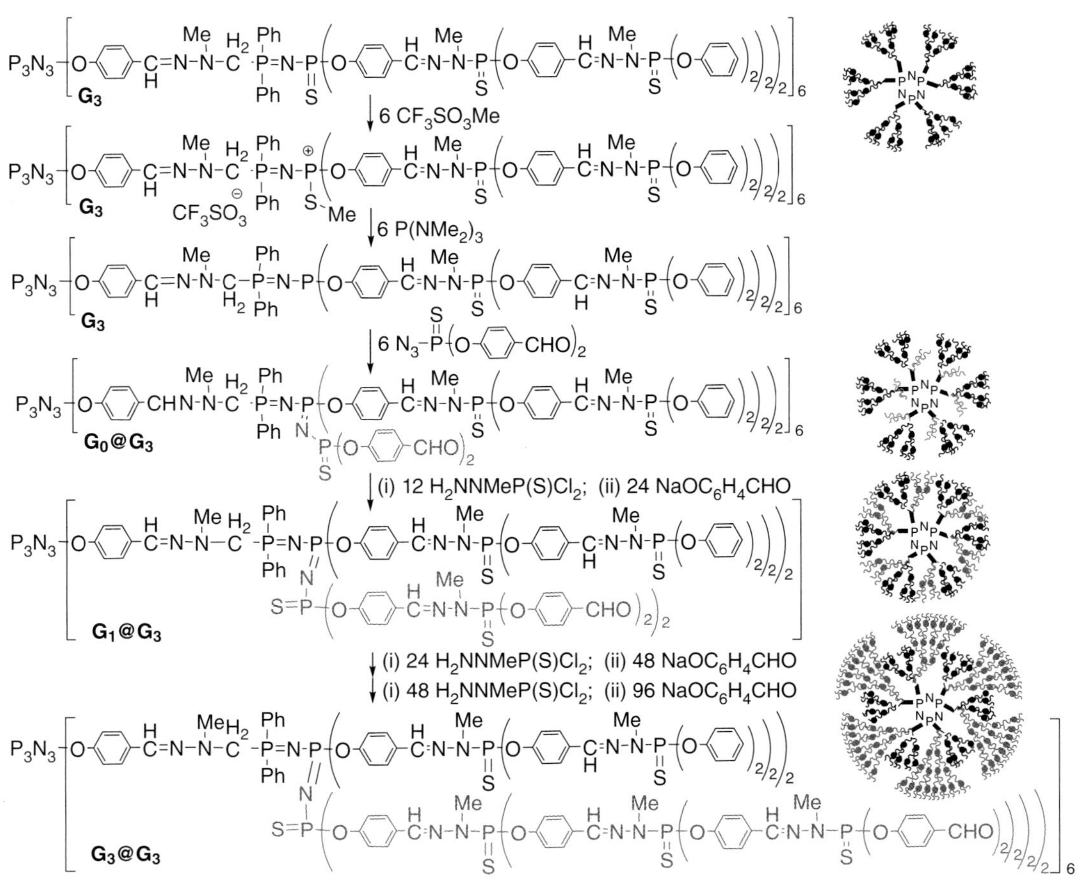

Scheme 5.13 Synthesis of new dendritic branches inside an original dendrimer.

Figure 5.4 Full chemical structure of the dendrimeric structure $\mathbf{G_3@G_3}$, with 6 new branches built inside the main dendrimer.

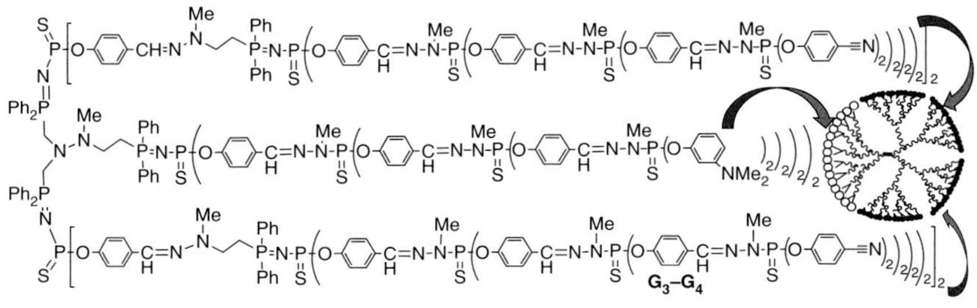

Figure 5.5 A special dendritic structure with 16 tertiary amines and 64 nitriles as terminal groups, as an example of Janus dendrimers.

or (ii) a three-step approach using the Staudinger reaction at one step. This is certainly the most original type of dendrimeric structures ever synthesized. The full chemical structure of this compound is shown in Figure 5.4.

Many different dendritic structures were built from phosphorus-containing dendrons (dendritic wedges), in particular those having a CH_2=CH—$(Ph)_2$ P=N—P=S grouping at the core. Indeed, the presence of the P=N—P=S group activates the vinyl group toward the Michael-type additions with different amines, preferably diamines and hydrazine (both used in large excess), which affords NH_2 group at the core. These functions were then used for the synthesis of a kind of layered segment-block dendritic compound having a nonsymmetrical internal structure, despite the presence of 576 identical dimethylamino end-groups, but also other unsymmetrical dendritic species, for instance those possessing 16 NMe_2 and 64 nitrile terminal groups as shown in Figure 5.5 [41]. This compound pertains to the family of "Janus" dendrimers, i.e. of dendrimers having two faces, with a different type of terminal functions in each side [42].

5.4 Synthesis of Miscellaneous Types of Inorganic Dendrimers

5.4.1 Synthesis of Dendrimers Containing Main Group Elements Other than Si and P

Silicon- and phosphorus-containing dendrimers are the most widely studied types of dendrimers; only a few examples of dendrimers containing other types of main group elements have been prepared. The synthesis of germanium-containing dendrimers is largely inspired by the synthesis of carbosilane dendrimers. The method consists of successive alkenylation and hydrogermylation reactions starting from germanium tetrachloride. However, cleavage of germanium–carbon bonds occurs easily in the vinyl and allyl derivatives, giving by-products that are difficult to remove. The use of longer ω-alkylene chains allows to overcome these problems and to obtain the second generation dendrimer (Scheme 5.14) [43]. Other attempts have been made, in particular by synthesizing dendrons to be grafted to a core, but only the first generation germanium dendrimers were obtained.

A second generation of tin-containing dendrimer has been synthesized by a combination of convergent and divergent processes. The first step is the reaction

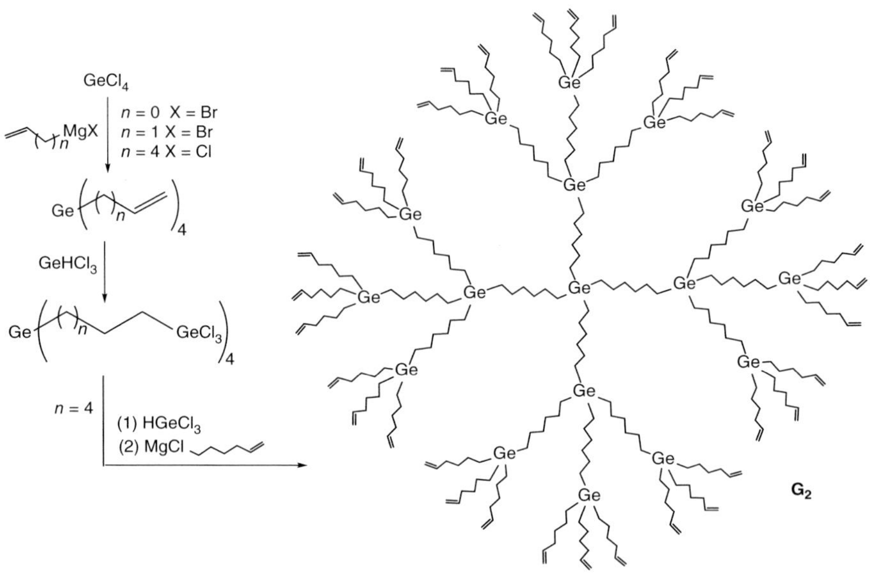

Scheme 5.14 Synthesis of the second generation germanium dendrimer.

of haloalkyltin trihalide with 3 equiv of but-3-enylmagnesium bromide, affording a dendritic unit with Br as core, which is then converted to a Grignard reagent. Treatment of this unit with $SnCl_4$ gave the first generation tin dendrimer. The second generation was obtained by hydrostannation with 12 equiv of Ph_3SnH [44] (Scheme 5.15).

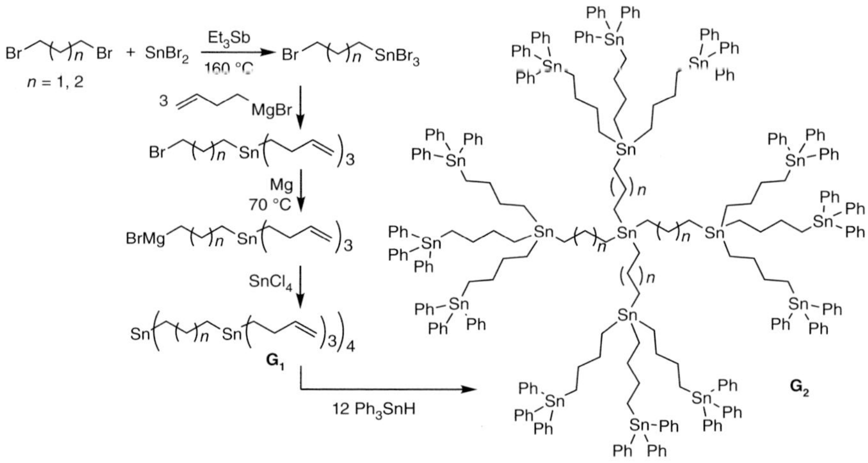

Scheme 5.15 Synthesis of a second generation tin dendrimer.

The bismuth-containing dendrimer of the second generation has been obtained in 20% yield utilizing the direct ortho lithiation of aryl bismuthanes bearing a sulfonamide function at the ortho position of the aromatic ring. The second step is the reaction with bis[2-(diethylaminosulfonyl)phenyl] bismuth iodide, as shown in Scheme 5.16 [45].

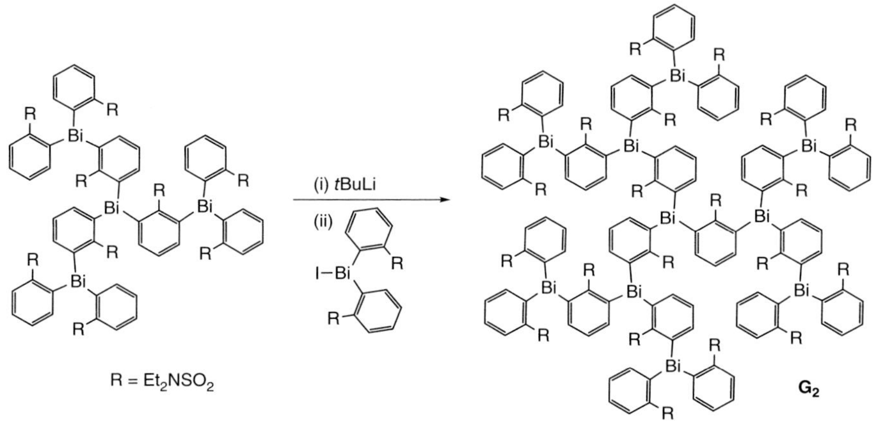

Scheme 5.16 Last step of the synthesis of a second generation of bismuth dendrimers.

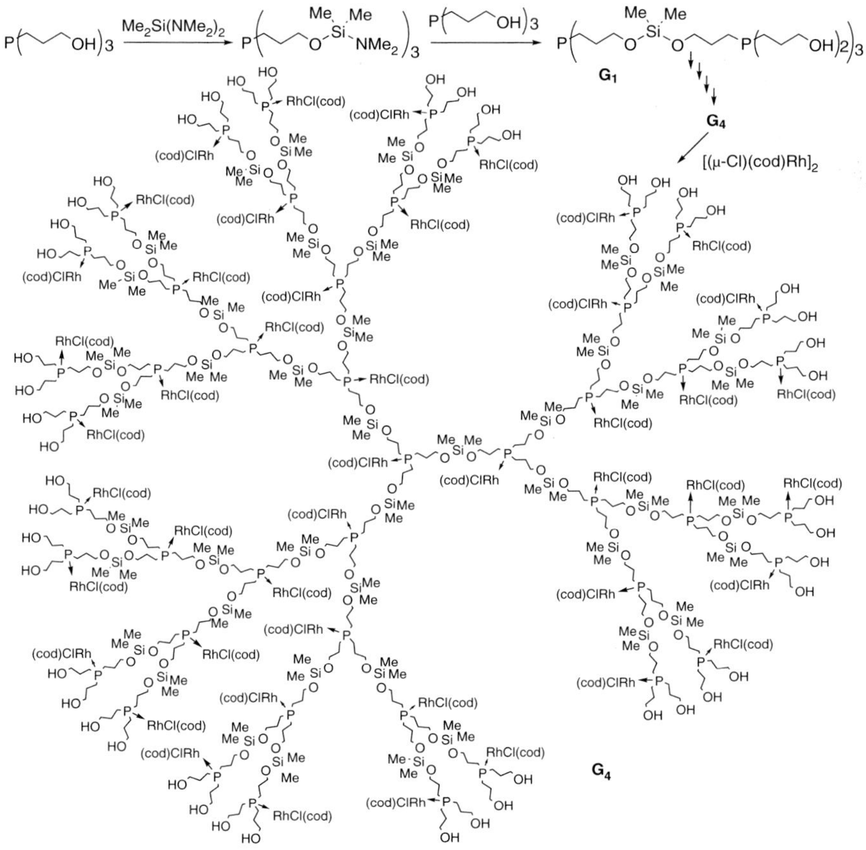

Scheme 5.17 Synthesis of a dendrimer incorporating phosphorus as branching points and siloxanes in the branches.

5.4.2 Synthesis of Hybrid Dendrimers Containing at Least Two Types of Main Group Elements

The largest number of dendrimers incorporating two main group elements concerns the association of silicon and phosphorus. The first example was a dendrimer having phosphorus branching points and siloxane linear linkages. The first step consists of the aminosilanization of alcohol end groups; the second step is a controlled acid–based hydrolysis of the aminosilanes by a single OH of $P[(CH_2)_3OH]_3$. Despite the possibility of multiple reactions, characterization of these compounds indicates dendrimeric rather than hyperbranched growth. Complexation of rhodium affords the corresponding complex, also obtained using the phosphine complex during the synthesis of the dendrimer (Scheme 5.17) [46].

"Onion peel" dendrimers (different entities at each layer) have been prepared associating phosphorus at the core and having carbosilane branching points using combination of convergent and divergent methods, with branches grafted to a core in the final step. The process that gives the first generation dendrimers is shown in Scheme 5.18, but a similar process with branches of the second generation has been also carried out [47].

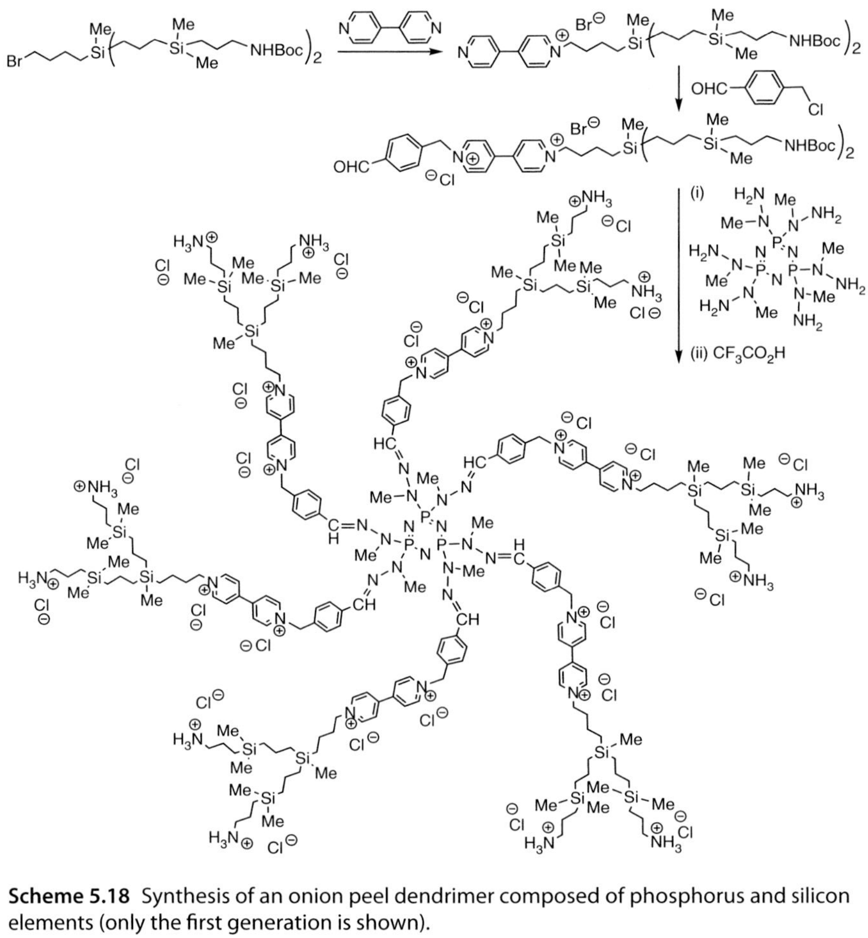

Scheme 5.18 Synthesis of an onion peel dendrimer composed of phosphorus and silicon elements (only the first generation is shown).

Another type of dendrimers incorporating both phosphorus and silicon was obtained by grafting azabisphosphonate derivatives on the surface of carbosilane dendrimers. The deprotection of the phosphonates to phosphonic acid salts was carried out in the last step, to afford water-soluble dendrimers (Scheme 5.19) [48].

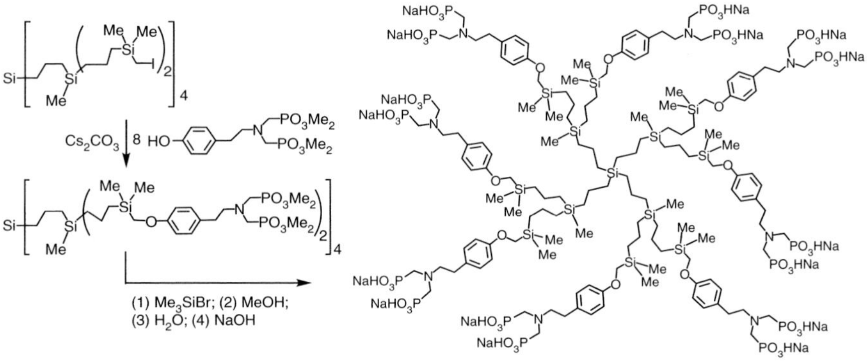

Scheme 5.19 Synthesis of a water-soluble carbosilane dendrimer with phosphonate terminal functions.

Another way to obtain mixed dendrimer structures consists in the coupling of a carbosilane and a phosphorhydrazone dendron by the Staudinger reaction, to afford a small Janus dendrimer, as shown in Scheme 5.20 [49].

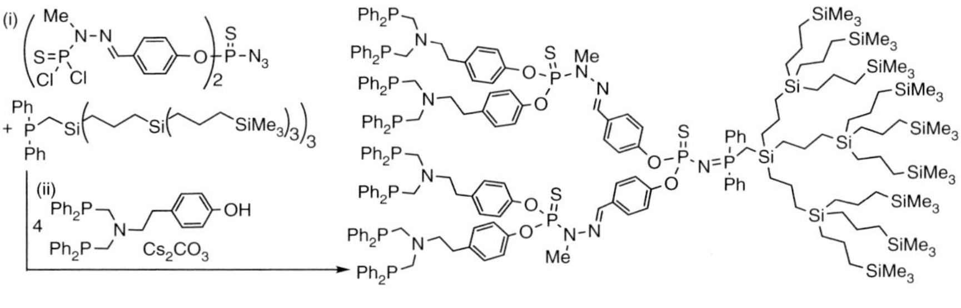

Scheme 5.20 Synthesis of Janus dendrimer based on the association of a phosphorhydrazone dendron with a carbosilane dendron.

Besides P and Si, a few other combinations of main group elements have been used, mainly based on Si, mostly giving small dendrimers. A dendrimer with alternating Si and Ge atoms has been synthesized by a divergent method (Scheme 5.21) [50].

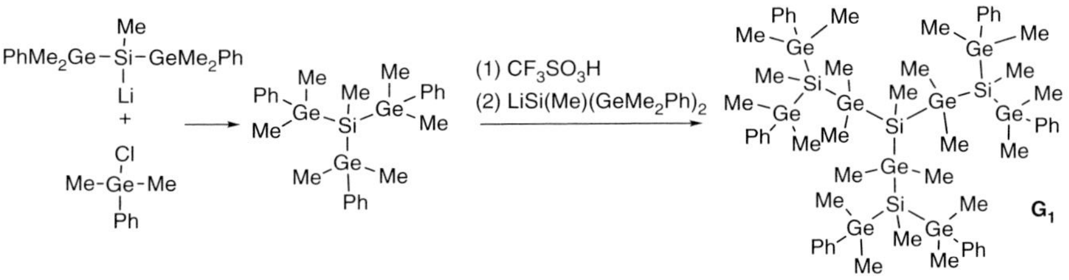

G1

Scheme 5.21 Synthesis of a small dendrimer made by alternating Si and Ge atoms.

Other examples of Ge/Si small dendrimers have been obtained by incorporating C≡C triple bonds between the main group elements. The final step is the grafting of branches to a $GeCl_4$ core (Scheme 5.22) [51].

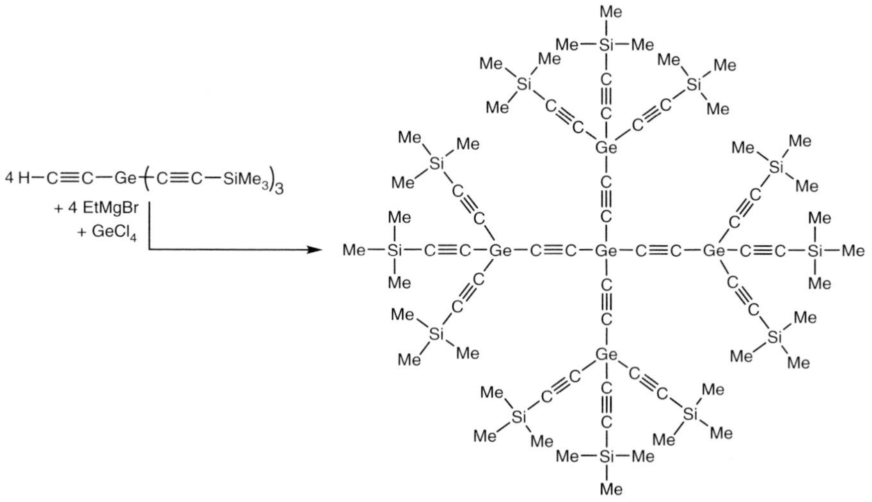

Scheme 5.22 Synthesis of silicon germanium dendrimer incorporating C≡C bonds.

An original dendrimer containing three different types of main groups has been synthesized by the divergent process as shown in Scheme 5.23. This dendrimer contains 1 silicon atom in the core, 4 tin atoms as branching points and 12 germanium atoms in terminal groups [52].

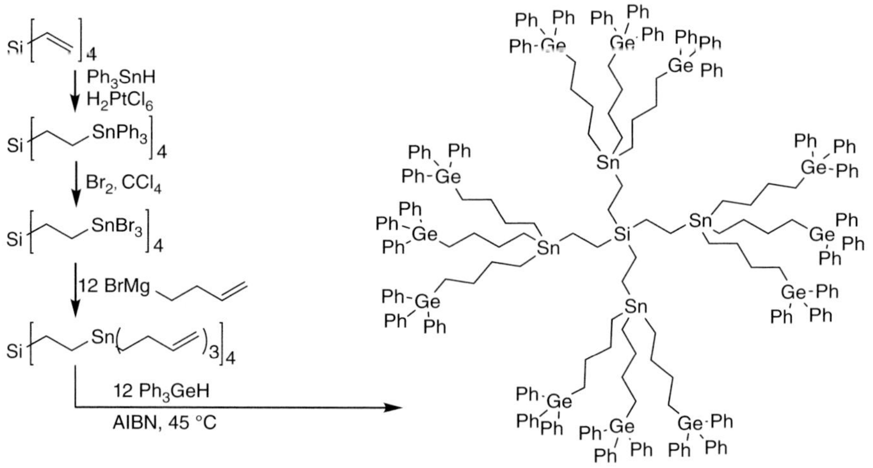

Scheme 5.23 Synthesis of a dendrimer incorporating Si, Sn, and Ge elements.

5.5 Conclusion and Outlook

This chapter has gathered the most important types of inorganic dendrimers, or the most originals, giving their methods of synthesis and their full chemical structures. Among several dozens of types of already synthesized inorganic dendrimers, two remarkable families have emerged for their relatively easy synthesis in high yields at each step, for the possibility to obtain them in very high generations (up to 10 or 12), for their time stability and for the large palette of their properties when modifying their functional terminal units. These two families are the carbosilane dendrimers on one side, as shown in Figure 5.1, and the phosphorhydrazone dendrimers on the other side, as shown in Scheme 5.6. Besides, the synthesis of phosphorhydrazone dendrimers is compatible with different types of modifications of the branches.

In comparison with organic dendrimers, one important difference for most inorganic dendrimers concerns their easy characterization by NMR. Indeed, their branching points are elements that can be detected by NMR, giving an accurate view of the reactions that occur during the growing of the dendritic branches, and offering the possibility to determine if reactions reached completion or not. This is particularly true for ^{31}P (100% natural abundance, spin $I = 1/2$) [53], but also in part for ^{29}Si (4.68%, $I = 1/2$), ^{117}Sn (7.68%, $I = 1/2$), and ^{119}Sn (8.59%, $I = 1/2$), whereas $I = 1$ for ^{14}N used as branching points for most organic dendrimers precludes its practical use in NMR.

Some elements are still missing in the large family of inorganic dendrimers. For instance, carborane has been used many times to decorate the surface of diverse dendrimers, but to the best of our knowledge, there is no example to date of dendrimers based on boron (or carborane) as branching points. Nor does arsenic appear to have ever been used for the synthesis of dendrimers. Elements such as Bi, Sn, or Ge have been used only scarcely for the synthesis of dendrimers, generally giving dendrimers of low generations, and no properties have been so far reported for these compounds.

Besides the synthetic aspects, the main motivation for synthesizing inorganic dendrimers is to observe properties different from those observed with organic dendrimers. Chapter 10 will discuss the different types of properties observed with inorganic dendrimers, and comparisons with organic dendrimers will be done, when possible.

Acknowledgments

The author thanks the CNRS (Centre National de la Recherche Scientifique, France) and the COST Action CM1302 (SIPS): "European Network on Smart Inorganic Polymers."

References

1 Buhleier, E., Wehner, F., and Vögtle, F. (1978). "Cascade-" and "nonskid-chain-like" syntheses of molecular cavity topologies. *Synthesis* 78: 155–158.

2 Tomalia, D.A., Baker, H., Dewald, J. et al. (1986). Dendritic macromolecules – synthesis of starburst dendrimers. *Macromolecules* 19: 2466–2468.

3 de Brabander van den Berg, E.M.M. and Meijer, E.W. (1993). Poly(propylene imine) dendrimers – large-scale synthesis by heterogeneously catalyzed hydrogenations. *Angew. Chem. Int. Ed. Engl.* 32: 1308–1311.

4 Worner, C. and Mulhaupt, R. (1993). Polynitrile-functional and polyamine-functional poly(trimethylene imine) dendrimers. *Angew. Chem. Int. Ed. Engl.* 32: 1306–1308.

5 Denkewalter, R.G., Kolc, J., and Lukasavage, W.J. (1981). Macromolecular highly branched. Homogeneous compound based on lysine units. US Patent 4, 289,872, filed 15 September 1981.

6 Aharoni, S.M., Crosby, C.R., and Walsh, E.K. (1982). Size and solution properties of globular *tert*-butyloxycarbonyl-poly(α,iε-L-lysine). *Macromolecules* 15: 1093–1098.

7 Majoral, J.P. and Caminade, A.M. (1999). Dendrimers containing heteroatoms (Si, P, B, Ge, or Bi). *Chem. Rev.* 99: 845–880.

8 Rebrov, E.A., Muzafarov, A.M., Papkov, V.S., and Zhdanov, A.A. (1989). Three-dimensionally propagating polyorganosiloxanes. *Dokl. Akad. Nauk SSSR* 309: 376–380.

9 Rengan, K. and Engel, R. (1990). Phosphonium cascade molecules. *J. Chem. Soc., Chem. Commun.* 1084–1085.

10 Zhou, L.L., Hadjichristidis, N., Toporowski, P.M., and Roovers, J. (1992). Synthesis and properties of regular star polybutadienes with 32 arms. *Rubber Chem. Technol.* 65: 303–314.

11 Launay, N., Caminade, A.M., Lahana, R., and Majoral, J.P. (1994). A general synthetic strategy for neutral phosphorus-containing dendrimers. *Angew. Chem. Int. Ed. Engl.* 33: 1589–1592.

12 Seyferth, D., Son, D.Y., Rheingold, A.L., and Ostrander, R.L. (1994). Synthesis of an organosilicon dendrimer containing 324 Si—H bonds. *Organometallics* 13: 2682–2690.

13 Van der Made, A.W. and Van Leeuwen, P.W.N.M. (1992). Silane dendrimers. *J. Chem. Soc., Chem. Commun.* 1400–1401.

14 Zhou, L.L. and Roovers, J. (1993). Synthesis of novel carbosilane dendritic macromolecules. *Macromolecules* 26: 963–968.

15 Tatarinova, E.A., Rebrov, E.A., Myakushev, V.D. et al. (2004). Synthesis and study of the properties of the homologous series of polyallylearbosilane dendrimers and their nonfunctional analogs. *Russ. Chem. Bull.* 53: 2591–2600.

16 Lim, J., Kostiainen, M., Maly, J. et al. (2013). Synthesis of large dendrimers with the dimensions of small viruses. *J. Am. Chem. Soc.* 135: 4660–4663.

17 Kim, C. and Kim, M. (1998). Synthesis of carbosilane dendrimers based on tetrakis(phenylethynyl)silane. *J. Organomet. Chem.* 563: 43–51.

18 Morikawa, A., Kakimoto, M.A., and Imai, Y. (1992). Convergent synthesis of siloxane starburst dendrons and dendrimers via hydrosilylation. *Macromolecules* 25: 3247–3253.
19 Sekiguchi, A., Nanjo, M., Kabuto, C., and Sakurai, H. (1995). Chemistry of organosilicon compounds. 323. Polysilane dendrimers. *J. Am. Chem. Soc.* 117: 4195–4196.
20 Slany, M., Bardaji, M., Casanove, M.J. et al. (1995). Dendrimer surface-chemistry – facile route to polyphosphines and their gold complexes. *J. Am. Chem. Soc.* 117: 9764–9765.
21 Launay, N., Caminade, A.M., and Majoral, J.P. (1997). Synthesis of bowl-shaped dendrimers from generation 1 to generation 8. *J. Organomet. Chem.* 529: 51–58.
22 Lartigue, M.L., Donnadieu, B., Galliot, C. et al. (1997). Large dipole moments of phosphorus-containing dendrimers. *Macromolecules* 30: 7335–7337.
23 Lartigue, M.L., Slany, M., Caminade, A.M., and Majoral, J.P. (1996). Phosphorus-containing dendrimers: synthesis of macromolecules with multiple tri- and tetrafunctionalization. *Chem. Eur. J.* 2: 1417–1426.
24 Servin, P., Rebout, C., Laurent, R. et al. (2007). Reduced number of steps for the synthesis of dense and highly functionalized dendrimers. *Tetrahedron Lett.* 48: 579–583.
25 Dib, H., Rebout, C., Laurent, R. et al. (2016). Ordered layered dendrimers constructed from two known dendrimer families: inheritance and emergence of properties. *Chem. Eur. J.* 22: 10736–10742.
26 Chen, H.T., Neerman, M.F., Parrish, A.R., and Simanek, E.E. (2004). Cytotoxicity, hemolysis, and acute in vivo toxicity of dendrimers based on melamine, candidate vehicles for drug delivery. *J. Am. Chem. Soc.* 126: 10044–10048.
27 Rengan, K. and Engel, R. (1991). The synthesis of phosphonium cascade molecules. *J. Chem. Soc., Perkin Trans. 1* 987–990.
28 Hudson, R.H.E. and Damha, M.J. (1993). Nucleic-acid dendrimers – novel biopolymer structures. *J. Am. Chem. Soc.* 115: 2119–2124.
29 Domanski, D.M., Bryszewska, M., and Salamonczyk, G. (2004). Preliminary evaluation of the behavior of fifth-generation thiophosphate dendrimer in biological systems. *Biomacromolecules* 5: 2007–2012.
30 Salamonczyk, G.M., Kuznikowski, M., and Poniatowska, E. (2001). Synthesis and oxygenation of selenophosphate dendrimers. *Chem. Commun.* 2202–2203.
31 Galliot, C., Prevote, D., Caminade, A.M., and Majoral, J.P. (1995). Polyaminophosphines containing dendrimers – syntheses and characterizations. *J. Am. Chem. Soc.* 117: 5470–5476.
32 Brauge, L., Magro, G., Caminade, A.M., and Majoral, J.P. (2001). First divergent strategy using two AB_2 unprotected monomers for the rapid synthesis of dendrimers. *J. Am. Chem. Soc.* 123: 6698–6699.
33 Maraval, V., Caminade, A.M., Majoral, J.P., and Blais, J.C. (2003). Dendrimer design: how to circumvent the dilemma of a reduction of steps or an increase of function multiplicity? *Angew. Chem. Int. Ed.* 42: 1822–1826.
34 Sournies, F., Crasnier, F., Graffeuil, M. et al. (1995). Spherical cyclophosphazene dendrimers to the 5th generation. *Angew. Chem. Int. Ed. Engl.* 34: 578–581.

35 Merino, S., Brauge, L., Caminade, A.M. et al. (2001). Synthesis and characterization of linear, hyperbranched, and dendrimer-like polymers constituted of the same repeating unit. *Chem. Eur. J.* 7: 3095–3105.

36 Larre, C., Caminade, A.M., and Majoral, J.P. (1997). Chemoselective polyalkylations of phosphorus-containing dendrimers. *Angew. Chem. Int. Ed. Engl.* 36: 596–599.

37 Larre, C., Donnadieu, B., Caminade, A.M., and Majoral, J.P. (1998). Phosphorus-containing dendrimers: chemoselective functionalization of internal layers. *J. Am. Chem. Soc.* 120: 4029–4030.

38 Larre, C., Bressolles, D., Turrin, C. et al. (1998). Chemistry within megamolecules: regiospecific functionalization after construction of phosphorus dendrimers. *J. Am. Chem. Soc.* 120: 13070–13082.

39 Larre, C., Donnadieu, B., Caminade, A.M., and Majoral, J.P. (1998). Regioselective gold complexation within the cascade structure of phosphorus-containing dendrimers. *Chem. Eur. J.* 4: 2031–2036.

40 Galliot, C., Larre, C., Caminade, A.M., and Majoral, J.P. (1997). Regioselective stepwise growth of dendrimer units in the internal voids of a main dendrimer. *Science* 277: 1981–1984.

41 Maraval, V., Laurent, R., Donnadieu, B. et al. (2000). Rapid synthesis of phosphorus-containing dendrimers with controlled molecular architectures: first example of surface-block, layer-block, and segment-block dendrimers issued from the same dendron. *J. Am. Chem. Soc.* 122: 2499–2511.

42 Caminade, A.M., Laurent, R., Delavaux-Nicot, B., and Majoral, J.P. (2012). "Janus" dendrimers: syntheses and properties. *New J. Chem.* 36: 217–226.

43 Huc, V., Boussaguet, P., and Mazerolles, P. (1996). Organogermanium dendrimers. *J. Organomet. Chem.* 521: 253–260.

44 Schumann, H., Aksu, Y., and Wassermann, B.C. (2006). Convergent synthesis and characterization of organotin dendrimers $Sn\{(CH_2)_nSn[(CH_2)_4SnPh_3]_3\}_4$ ($n = 3, 4$). *Organometallics* 25: 3428–3434.

45 Suzuki, H., Kurata, H., and Matano, Y. (1997). First synthesis and properties of dendritic bi-*n*-bismuthanes. *Chem. Commun.* 2295–2296.

46 Petrucci-Samija, M., Guillemette, V., Dasgupta, M., and Kakkar, A.K. (1999). A new divergent route to the synthesis of organophosphine and metallodendrimers via simple acid-base hydrolytic chemistry. *J. Am. Chem. Soc.* 121: 1968–1969.

47 Moreno, S., Szwed, A., El Brahmi, N. et al. (2015). Synthesis, characterization and biological properties of new hybrid carbosilane–viologen–phosphorus dendrimers. *RSC Adv.* 5: 25942–25958.

48 Caminade, A.M., Fruchon, S., Turrin, C.O. et al. (2015). The key role of the scaffold on the efficiency of dendrimer nanodrugs. *Nat. Commun.* 6: 7722.

49 Gottis, S., Rodriguez, L.I., Laurent, R. et al. (2013). Janus carbosilane/phosphorhydrazone dendrimers synthesized by the 'click' Staudinger reaction. *Tetrahedron Lett.* 54: 6864–6867.

50 Nanjo, M. and Sekiguchi, A. (1998). Group-14-element-based hybrid dendrimers. Synthesis and characterization of dendrimers with alternating Si and Ge atoms in the chains. *Organometallics* 17: 492–494.

51 Zhilitskaya, L.V., Istomina, E.E., Yarosh, N.O. et al. (2006). Spherical polyunsaturated organosilicon and organogermanium first generation dendrimers of regular structure. *Russ. J. Gen. Chem.* 76: 1864–1869.

52 Schumann, H. and Aksu, Y. (2007). Synthesis and characterization of silicon-based group 14 dendrimers $Si\{(CH_2)_2Sn[(CH_2)_4MPh_3]_3\}_4$ (M = Ge, Sn). *Organometallics* 26: 397–402.

53 Caminade, A.M., Laurent, R., Turrin, C.O. et al. (2010). Phosphorus dendrimers as viewed by P-31 NMR spectroscopy; synthesis and characterization. *C.R. Chim.* 13: 1006–1027.

6

Metallo-Supramolecular Polymers

Jiří Vohlídal[1] *and Muriel Hissler*[2]

[1] *Charles University, Faculty of Science, Department of Physical and Macromolecular Chemistry, Hlavova 8, 12840 Praha 2, Czech Republic*
[2] *Univ Rennes, CNRS, ISCR (Institut des Sciences Chimiques de Rennes) – UMR 6226, 263 avenue du Général Leclerc, Rennes 35000, France*

6.1 Introduction

Advances in supramolecular chemistry [1] have resulted in combining it with the chemistry of covalent polymers, which has given rise to supramolecular polymer chemistry [2–5]. The principles underlying the formation of supramolecular polymers are similar to those applied in the preparation of covalent polymers by stepwise (non-chain) polymerizations for more than 150 years [6]: A reaction known to couple two monofunctional molecules to result in one new molecule is applied to bifunctional (multifunctional) molecules [7, 8]. The principle difference is that no chemical reaction takes place in the formation of a supramolecular polymer. Instead, "monomer molecules" are spontaneously assembled to form supramolecular chains by attractive intermolecular interactions between their mutually compatible terminal units. This process is based on the molecular recognition (the lock–key principle) [1, 5], and since it generally has a low activation energy, it is controlled by equilibrium thermodynamics. Therefore, this process is referred to as assembling and not as polymerization, and it proceeds spontaneously upon mixing suitable, mutually compatible molecules.

In a typical supramolecular polymer, building units are assembled by means of multiple hydrogen bonding or by host–guest interactions (see Figure 6.1), and less frequently either by aromatic donor–acceptor interactions or by two or more types of the above interactions [7–9]. Apart from the mentioned modes, the coordination binding is also classified as supramolecular binding due to its molecular recognition capability that allows a defined self-organization of chemical systems either on the basis of design or by selection of their components. This binding mode is the basis of metallo-supramolecular polymers (MSPs).

Smart Inorganic Polymers: Synthesis, Properties, and Emerging Applications in Materials and Life Sciences, First Edition. Edited by Evamarie Hey-Hawkins and Muriel Hissler.

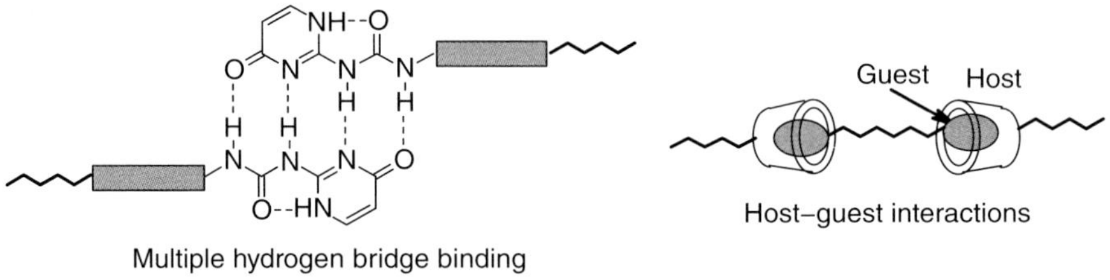

Figure 6.1 Examples of linkages in purely supramolecular polymers.

6.2 Constitutional Dynamic Polymers: Dynamers

Thermodynamic control generally limits the extent of the assembling process that therefore proceeds only up to the stage where the monomer(s)–polymer equilibrium is achieved. This provides supramolecular polymers the capability to disassemble or reassemble in response to the changes in external conditions (temperature, light) as well as to the changes within the system composition (concentration, pH, solvent, ionic strength, etc.). This property is called constitutional dynamics and it makes supramolecular polymers highly responsive and adaptable [9]. Similar properties are also exhibited by covalent polymers wherein monomeric units are linked to chains by significantly reversible bonds, such as polyazomethines (Schiff base polymers), polyhydrazones, or polymers with linkages formed by Diels–Alder type additions [10–14]. As a result, a new class of polymers, namely constitutional dynamic polymers shortly referred as dynamers [8] have been established.

Dynamers can be divided into: (i) *supramolecular dynamers,* based on the supramolecular interactions (Figure 6.1) such as multiple hydrogen bonding, host–guest interactions (e.g. encapsulation of coumarine into β-cyclodextrin end-groups), aromatic donor–acceptor interactions, or mixed non-covalent interactions; (ii) *covalent dynamers,* based on the reversible formation of covalent bonds (Figure 6.2); and (iii) *metallo-supramolecular dynamers,* based on the coordination of chelate end-groups of building units (mostly organic) to metal ions (Figure 6.3). The MSPs represent a transition between supramolecular and covalent dynamers.

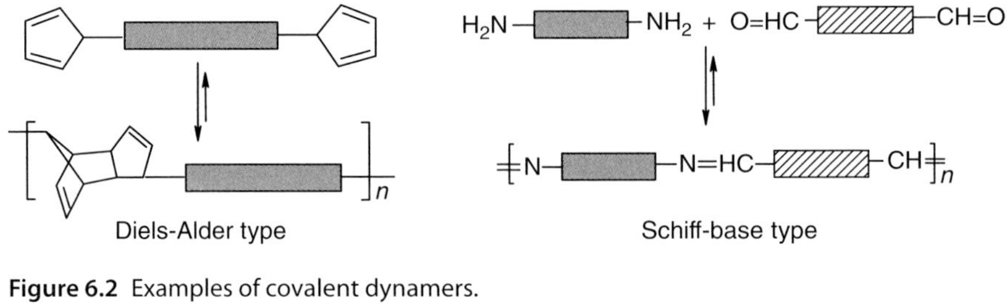

Figure 6.2 Examples of covalent dynamers.

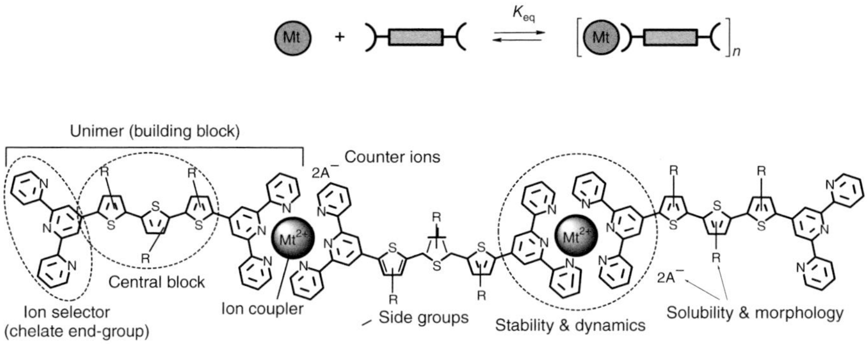

Figure 6.3 Schematic structure of a metallo-supramolecular polymer (dynamer).

6.3 Main Types of Metallo-Supramolecular Polymers (MSPs) and Terminologies Related to Them

Typical MSPs are constituted of precisely defined chelate building units (defined molecules) connected into chains by coordination to metal ions. Metal ions connecting chelating building units of MSPs are therefore termed *ion couplers*. The building units mostly can be classified as oligomers capped with chelate end-groups often referred to as *ion selectors*. The oligomeric nature of unimers brings about ambiguity in the terms monomer and oligomer in the field of MSPs because: (i) a monomer related to the unimer is not the "monomer" related to the MSP; (ii) under the dynamics promoting conditions (solutions or melt), the dynamer chains comprise only a small number of enchained repeating units – they are oligomeric chains. Such a dynamer is thus a "superior oligomer" of hierarchically lower oligomer(s). Ambiguity in the terms monomer and oligomer in the field of MSPs is obvious. Therefore, Ciferri [7] proposed the term *unimer* for the building units of MSPs.

Similarly to organic polymers, MSPs can have a linear, branched, or network-chain architecture. Chain branching is achieved either by using metal ions capable of binding more than two, typically three, chelate end-groups, such as lanthanoide ions, or by using unimers containing three or more chelate end-groups. The related organic/inorganic compounds, metal-organic frameworks (MOFs), are not considered as MSPs because they lack longer flexible subchains.

The MSP chains usually contain main-chain cations, so they are in general ionic polymers of the ionene type [15, 16]. Despite the partial shielding by counter anions, the electrostatic repulsions between main-chain cations promote extended conformations of MSP chains. This supports inter-macromolecular interactions of MSP chains, which promote their aggregation in solutions. An increase in the solubility of MSPs in nonpolar and medium polar solvents is achieved by introducing pendant alkyl groups to unimeric units. However, this approach is often insufficient and thus it is not easy to find a solvent providing good MSP films. Moreover, future technologies require materials that are soluble in "green solvents" such as alcohols and aqueous solutions. The desired solubility of MSPs can be achieved by introducing ionic side groups onto their unimeric units. So modified MSPs are then ionic polymers of the mixed (hybrid) ionene-polyelectrolyte type, since their chains contain both charged main-chain atoms (structure feature of ionenes) as well as pendent ionic groups (structure feature of polyelectrolytes).

Besides the above described "typical" MSPs, there are two other types of compounds that can be regarded as MSPs described in the literature: (i) materials derived from telechelic polymers capped with chelate end-groups, which are linked to reversible linear chains through coordination of the end-groups to metal ions and (ii) polymers with pendent chelate groups, which are linked to reversible network polymers also by coordination of the chelate groups to metal ions [17]. To illustrate basic concepts, only MSPs formed by self-assembly of unimers with defined structure with metal ions, and their properties and applications are discussed in this chapter, with special reference to conjugated MSPs. These MSPs are derived from unimers with both conjugated central blocks and ion selectors and they are a constitutional dynamic counterpart of covalent conjugated polymers. Therefore, they are of interest as materials with potential applications based on the light/electricity inter-conversion and nonlinear optical phenomena, such as materials for light-emitting devices, photovoltaic cells, electrochromic devices, ion conductors, and biosensors [7–9, 18–21].

6.4 MSP Dynamers Derived from Unimers with Defined Structure

Typical unimers for preparation of regular MSPs are composed of a central block and chelating end-groups that act as ion selectors. To achieve a reasonable stability of an MSP in solution, the ion selectors should be at least bidentate ligands such as, 2,2′-biphenyl or 1,10-phenanthroline. Nevertheless, tridentate ligands are typically used for these purposes [9, 18–21]. As such, ligands are mostly quite large: they represent a significant proportion of the unimers as well as of the resulting MSP. Coordination linkages thus currently form more than one half of the MSP structure and, therefore, significantly contribute to the functional properties of all MSPs.

6.4.1 Synthesis of Unimers

Perhaps the most popular ion selector of conjugated MSPs is 2,2′:6′,2″-terpyridin-4′-yl (*tpy*). It is usually connected to a conjugated central block or its parts by various coupling reactions (Suzuki–Miyaura, Negishi, Sonogashira, Stille, Kumada, etc.) of commercially available 4′-bromo-2,2′:6′,2″-terpyridine (Br-*tpy*) with a relevant derivative of the central block or its precursor (Figure 6.4).

Alternatively, 4′-(hetero)aryl terpyridines are prepared by the two-step Krohnke method [22] based on the ring-closing reaction of 1,5-bis

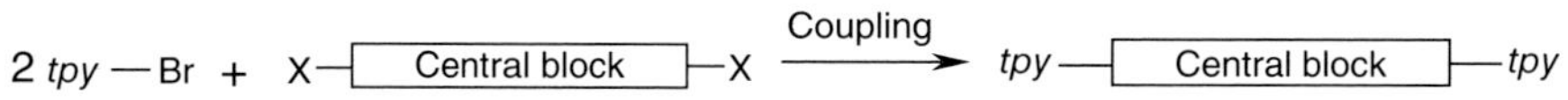

Figure 6.4 The most often used pathway to unimers.

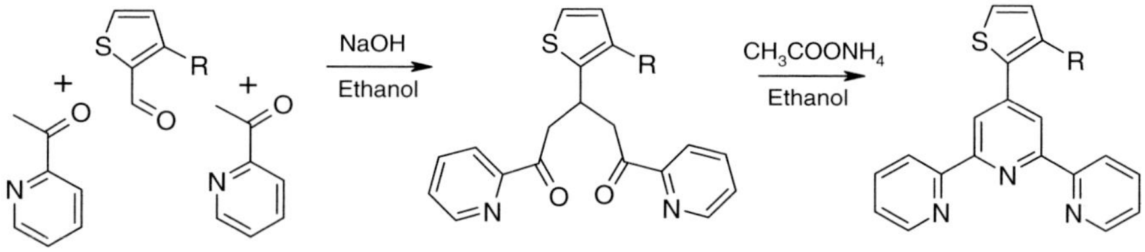

Figure 6.5 Scheme of Krohnke type synthesis of terpyridine derivatives.

(pyridin-2-yl)(3-aryl)penta-1,5-diones with ammonium acetate, as illustrated in Figure 6.5. The 4′-substituted terpyridines are obtained upon appropriate functionalization at the 4′-(hetero)aryl ring, which depends on the chosen coupling type, used for the preparation of desired unimers by coupling reactions [23].

It is worth mentioning here that 2,2′:6′,2″-terpyridine and its bromoderivatives are quite expensive materials. Other ion selectors for conjugated MSP dynamers can be obtained from cheaper chelidamic acid (4-hydroxypyridine-2,6-dicarboxylic acid) by its condensation with suitable amines such as benzene-1,2-diamine giving 2,6-bis(benzimidazol-2-yl)pyridine (*bzimpy*, Figure 6.6) [24].

The obtained bromoderivative is then coupled with a central block derivative to obtain the desired unimer. Nevertheless, bis(*bzimpy*) unimers usually suffer from too low solubility, which can be increased by the replacement of the hydrogen of the imidazole ring with a methyl group giving 2,6-bis(1-methylbenzimidazol-2-yl)pyridine. Another ion selector group derived from chelidamic acid, 2,6-bis(oxazolin-2-yl)pyridine-4-yl, usually referred to as *pybox* (Figure 6.7), gives significantly more soluble unimers [25–27]. This group can be prepared from the bromoderivative of chelidamic acid as described in Figure 6.7.

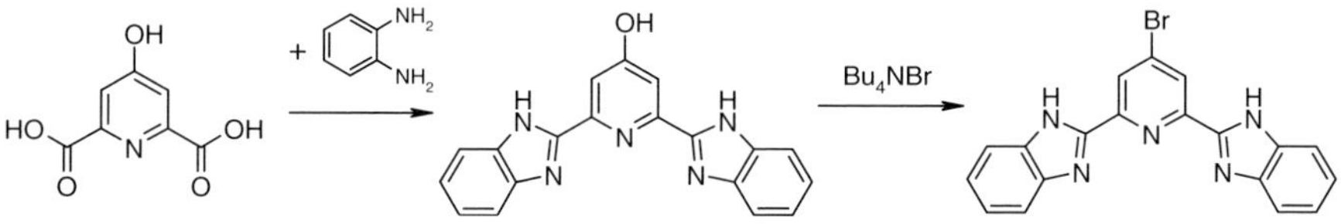

Figure 6.6 Reaction path from chelidamic acid to activated *bzimpy* ligand.

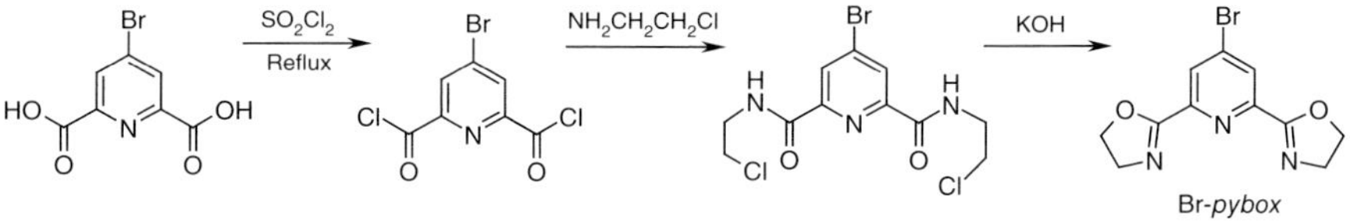

Figure 6.7 Preparation of activated *pybox* ligand.

6.4.2 Central Blocks of Conjugated Unimers

A huge variety of unimers with oligo(arylene), oligo(arylene-vinylene), oligo (arylene-ethynylene) central blocks, and corresponding dynamers are described in the literature [9, 19–21]. The unimers with unsubstituted central blocks as well as their MSPs are mostly sparingly soluble, which limits their processing advantages. Therefore, more soluble unimers with functional groups attached to their central blocks are widely used. Substitution at ion selector groups is less frequent since the *tpy*-ring substitution also undesirably affects the binding properties of *tpy* groups.

Introduction of functional groups brings about not only better solubility but also a significant distortion of the coplanarity of conjugated π-bonds, which significantly affects the spectral properties of the unimers as well as those of related dynamers. The substituent effect is clearly seen on α,ω-bis(*tpy*)bithiophene unimers (Figure 6.8). Introduction of hexyl groups inclined to the *tpy* units gives a blueshift (hypsochromic shift) of 30 nm and their introduction to near neighboring central positions gives a blueshift of 60 nm, which makes this structure spectrally identical with its half [28].

The substituent effects should be taken into account while designing a unimer. Different strategies of the synthesis of unimers with and without substituents attached to the central blocks are mentioned in the following text.

(i) Synthesis of a whole central block, direct functionalization of its end units for the chosen type of coupling reaction, and final coupling of the functionalized block with a precursor of the ion selector group (for example Br-pybox or Br-*tpy*) (Figure 6.9).
(ii) Preparation of two halves of the desired symmetric unimer (half of the central block with attached ion selector group), functionalization of the block

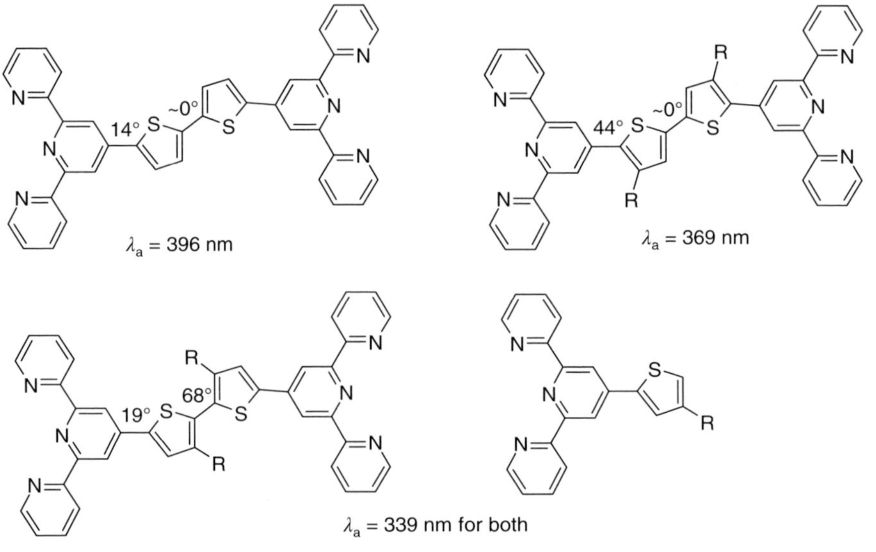

Figure 6.8 The effect of substituents on absorption maximum, λ_a, of bis(*tpy*)bithiophenes and their chain distortion calculated by the DFT method. Source: Data from Svoboda et al. 2011 [23] and Bláhová et al. 2014 [28].

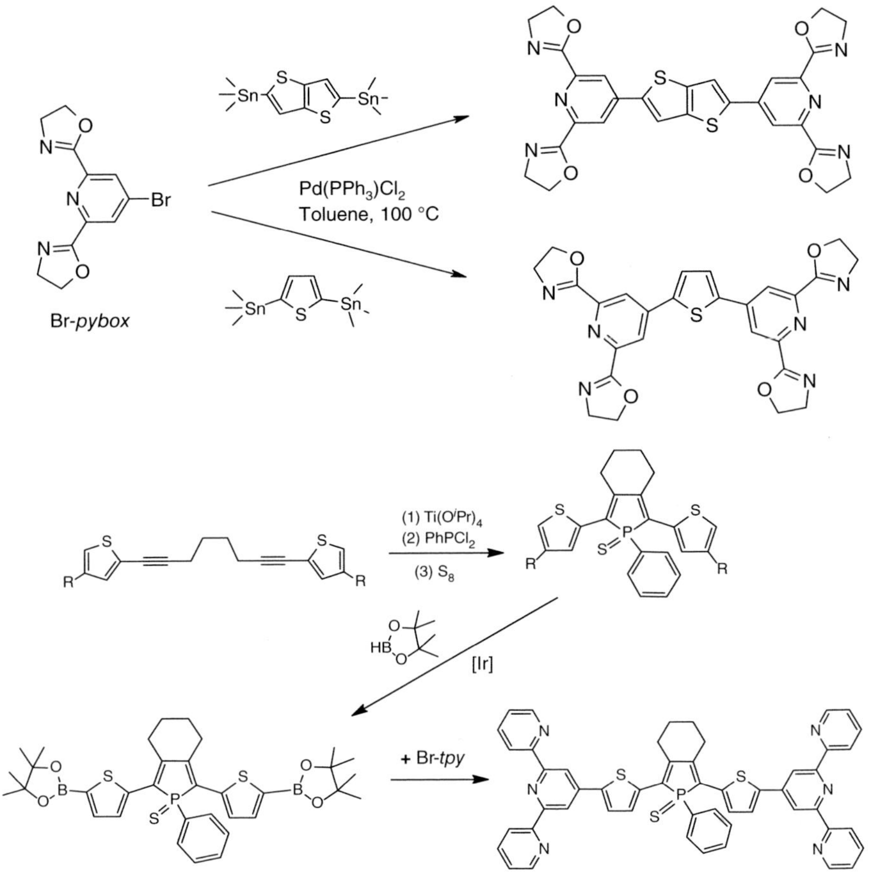

Figure 6.9 Synthesis of unimers according to the approach (i). Source: Data from Hladysh et al. 2017 [27] and Vitvarova et al. 2017 [29].

end unit (unit at the end opposite to the ion-selector group), and a homo-coupling of functionalized half-unimer molecules giving the final unimer (Figure 6.10).

(iii) Preparation of functionalized end fragments of a designed unimer, such as for example *tpy*-(hetero)arylene-Br fragment, and their subsequent co-coupling with appropriately functionalized central part of the designed unimer (Figure 6.11).

(iv) Unimers carrying ionic side groups are usually prepared by modifications of end-functionalized nonionic side groups, e.g. by ionization of 6-bromohexyl groups (Figure 6.12). The ionization of 6-bromohexyl side groups attached to α,ω-bis(*tpy*)oligothiophenes can be more or less inhibited by the terminal *tpy* groups as it was observed when attempting to replace bromine atoms by imidazolium or ammonium groups. It is worth mentioning that the ionization reactions smoothly proceed on polythiophenes but they were found to either proceed poorly or not at all on *tpy*-unimers. This has been ascribed to the inhibitive effect of *tpy* end-groups [30, 32]. Ionization of side groups only slightly affects the absorption spectra of unimers as well as dynamers but dramatically changes their solubility.

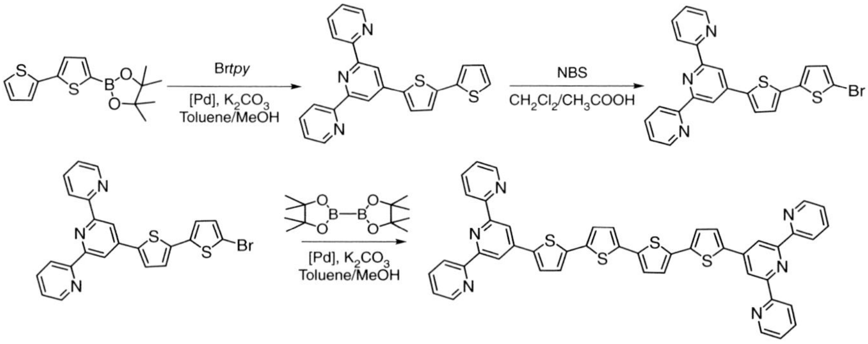

Figure 6.10 Synthesis of bis(*tpy*)terthiophene unimer by the approach (ii) (according to Ref. [30]).

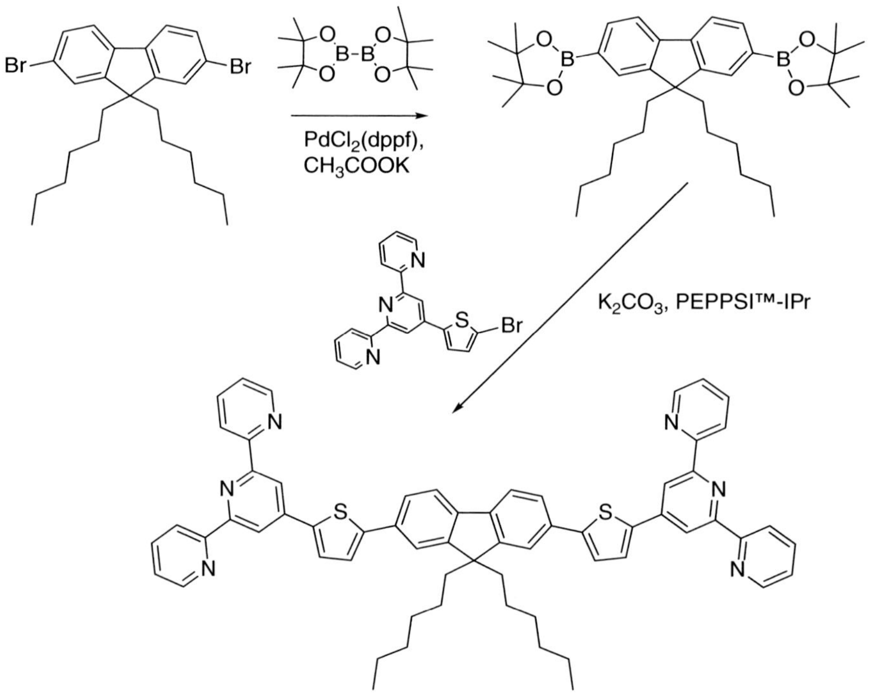

Figure 6.11 Synthesis of unimers according to the approach (iii). Source: After Hrma et al. 2017 [31].

6.4.3 Assembly and Characterization of MSP Dynamers

The solution assembly of MSP dynamers from defined unimers and compatible ion couplers usually proceeds spontaneously. MSPs are thus mostly simply prepared by mixing solutions of equimolar amounts of corresponding metal salt and unimer and their films are prepared by casting a solution on a chosen substrate and subsequent evaporation of solvent. Assembly of MSPs is mostly easy to follow through the changes in the UV/vis and luminescence spectra due to complexation

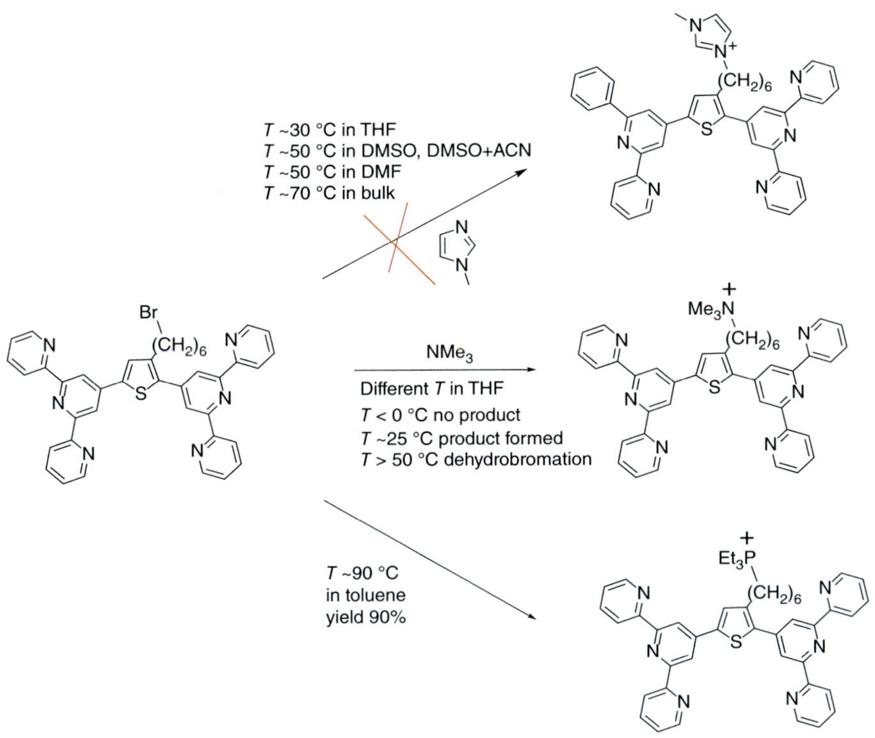

Figure 6.12 Preparation of ionic unimers; approach (iv). Source: After Štenclová et al. 2016 [30] and Štenclová et al. 2015 [32].

Figure 6.13 Course of R-*tpy* assembling with Mt^{2+} ions when adding the ion into the ligand solution.

and, in the case of conjugated MSPs, also due to the extended delocalization of electrons. In the case of unimers with *tpy* ion selectors, the assembling is accompanied with the anti-to-syn conformation of nitrogen atoms of the *tpy* groups that is effectively monitored by NMR spectroscopy through the changes in NMR shifts of hydrogenes H^6 of *tpy* units (Figure 6.13) in particular (other H-atoms show smaller shifts).

In an uncomplexed *tpy* group that has anticonformation of N atoms, the signals of H^6 atoms are found at around 8.7 ppm: at 8.70 ppm for R = H; 8.71 ppm for R = ethynyl; and 8.74 ppm for R = thiophen-2-yl. In octahedrally coordinated $[Zn(tpy)_2]^{2+}$ species, *tpy* groups acquire syn conformation where H^6 atoms are close to each other thus shielding each other, which results in a high-field shift of their signal about 1 ppm for Zn^{2+} complexes (to 7.73, 7.77, and 7.84 ppm for R=H, —C≡CH and thiophen-2-yl, respectively). The shift is still much higher for $[Fe(tpy)_2]^{2+}$ species (7.06–7.08 ppm). In mono-*tpy* $[Zn(tpy)]^{2+}$ species the signal shifts back to c. 8.85–8.90 ppm [33].

$[Fe(tpy)_2]^{2+}$ species almost do not dissociate in the presence of excess Fe^{2+} ions, and exhibit a metal-to-ligand charge transfer (MLCT) band in visible spectral region. This band is strong and broad ranging from c. 400 to 700 nm. The band maximum occurs at c. 550 nm for terpyridine complex and at c. 590–640 nm for various Fe-MSPs. Fe-MSPs do not show luminescence. The MLCT bands were also observed for Ru^{2+}-MSPs derived from *tpy* unimers [34], which unlike the Fe-MSPs show a weak luminescence at low temperatures and bis(*tpy*) complexes of Ru(II) and Os(II) [35, 36]. The R-*tpy* ligands and terpyridine were also found to form MLCT complexes with Ag^0 and Au^0 surface species occurring on the surfaces of Ag and Au nanoparticles [37–39].

Assembly of MSPs in solution can be potentially monitored by viscometry, but only qualitatively, since the degree of polymerization (average number of monomeric units built in a polymer molecule) of MSPs in solution usually depends on the solution concentration. Nevertheless, the dependence of the solution viscosity on the metal-to-unimer mole ratio (Mt/U) at a constant unimer concentration usually shows a maximum at the ratio equal to 1 for MSPs derived from ditopic unimers [31, 40].

Analysis of MSPs by the size exclusion chromatography (SEC) method can provide qualitative information on the dynamics of Mt–U systems, i.e. the rate of establishing complexation equilibrium but hardly any information on the actual molar mass of MSPs in a given solution. For example, SEC records of Zn-MSPs derived from bis(*tpy*) unimers show only a peak of the free unimer, which indicates a complete dissociation of injected MSP chains (typically, 20 μl of a solution) upon gradual high dilution in SEC columns. This demonstrates fast establishment of the dissociation equilibrium, i.e. fast constitutional dynamics of these Zn-MSPs [29, 32]. However, SEC records of Fe-MSPs derived from bis(*tpy*) unimers, which have equilibrium constants comparable to that of Zn-MSPs [33, 41, 42], mostly show a series of more or less good resolved peaks of dimers to higher oligomers [21, 29, 31, 32]. This proves that Fe-MSP molecules are time stable enough to at least partly survive a 1000-fold or higher dilution in SEC columns within c. 30 minutes of their analysis: they exhibit a slow constitutional dynamics. This corresponds to the differences in the rates of ligand exchanges for $[Fe(tpy)_2]^{2+}$ and $[Zn(tpy)_2]^{2+}$ species observed much earlier for isotopically labeled ligands (a minute for Zn– but a month for Fe– species) [41]. Slow constitutional dynamics show also MSPs with ion couplers such as Ru^{2+}, Ni^{2+}, Co^{2+}, and Cu^{2+} ions [43].

Recently, the molar mass (or DP) averages of Zn-dynamers derived from α,ω-bis(*tpy*)terthiophenes dissolved in solution were determined by the time-resolved transient absorption spectroscopy [44]. At these measurements, the ground-state chromophores are first excited to the S_1 state by about an intense 50 fs pump pulse and, immediately afterwards, the time-resolved absorption spectra of the excited chromophores are collected on the femtosecond scale time resolution. The measurements were done on solutions of free unimers and their Zn-dynamers. Knowledge of the spectroscopic characteristics of the free unimers allowed determining their unreacted fractions in equilibrium dynamer solutions (0.09 for unsubstituted terthiophene unimer and of 0.21 for its counterpart bearing hexyl groups on the thiophene ring carbons near-neighboring to the *tpy* end-groups). A reversible assembly of unimer species to dynamer chains should give the Schulz–Flory distribution of the degrees of polymerization, X, of dynamer chains [45]. Hence, the number-average, $\langle X \rangle_n$, and weight-average, $\langle X \rangle_w$ degrees of polymerization of the dynamers could be calculated using the equations: $\langle X \rangle_n = 1/(1-p)$ and $\langle X \rangle_w = (1+p)/(1-p)$, where p is the degree of unimer conversion to oligomer chains. Thus, the obtained values $\langle X \rangle_n \approx 11$; $\langle X \rangle_w \approx 21$ for unsubstituted and $\langle X \rangle_n \approx 5$; $\langle X \rangle_w \approx 11$ for hexyl-substituted dynamers in dimethylsulfoxyde (DMSO) solutions (1.10^{-4} M) indicate a substantial effect of the attached hexyl groups on the extent of assembling of a unimer into a dynamer.

Interesting MSPs have been recently prepared by the reaction of *tpy* with $AgNO_3$ [46]. The obtained product formed needle-shaped elastic crystals of complicated twinned modulated structures, which were light sensitive and exhibited a strong X-ray absorption. The crystal structure model based on the supercell approach provided evidence of an infinite chain of Ag atoms with the Ag–Ag distance of 2.985(2) Å surrounded by a helical arrangement of the *tpy* ligands. In this structure, each *tpy* ligand is bidentatelly coordinated to one Ag(I) atom and monodentatelly to the neighboring Ag(I) atom thus giving rise to a helical chain. The Ag–N distances ranging from 2.228(4) to 2.609(4) Å are comparable with the distances presented in Ref. [37]. The chain conformation and the unit cell packing are shown in Figure 6.14.

6.4.4 Properties of Conjugated MSPs

The main attention in this field is focused on tuning the bandgap and light absorption of these materials, which are a combination of the properties of unimer(s) and metal-coordination linkages formed in assembling MSPs. Electronic spectroscopic properties of conjugated unimers can be, to a great extent, anticipated from the properties of related covalent conjugated polymers. Regarding the stability of MSPs in air, central blocks without longer sequences of ethynylene or vinylene units, such as oligo(hetero)arylene blocks and the blocks of alternating (hetero)arylene and ethynylene, or vinylene units are preferred. The structure dependence of the UV/vis absorption maxima of α,ω-bis(*tpy*) unimers is illustrated in the examples shown in Figure 6.15.

The bandgap of a conjugated sequence is a function of the number of conjugated multiple bonds in the sequence, the capability of the units constituting this sequence to share their valence electrons with neighboring units, and, eventually, the presence of intramolecular charge transfer within this sequence. Unimers with oligo(phenylene) central blocks, which have been studied since the start of investigation of MSPs due to their high thermal as well as photochemical stability, are not too favorable owing to the high aromaticity of benzene rings, which limits the delocalization of electrons. The absorption maximum λ_a

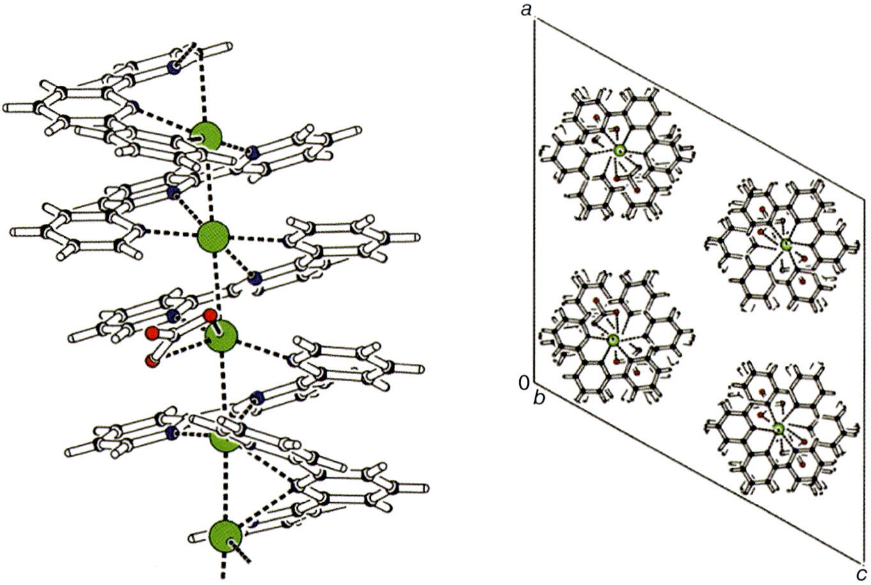

Figure 6.14 Structure motif (left) and crystal packing (right) of $[Ag(tpy)NO_3]_n$ chains [46].

of oligo(1,4-phenylenes) changes from 280 nm for a dimer to only 323 nm for a heptamer [47].

As a rule, the higher the aromaticity, the lower the extent of the delocalization of electrons. Aromaticity of the conjugated rings discussed here decreases in the sequence: benzene > thiophene > phosphole [48]. Thus in contrast to oligo(1,4-phenylene) unimers, λ_a of oligo(thiophene) unimers increases about 100 nm when going from mono- to quaterthiophene central block [28] (Figure 6.15, first row) and, for quaterthiophene unimer, reaches the λ_a value of 441 nm that is in the range observed for poly(hexylthiophenes) [49]. The difference in the effects of thiophene and benzene rings on the λ_a value of unimers with fluorene central block (Figure 6.15, second row) is also documented. The replacement of 1,4-phenylene units with thiophene-2,5-diyl units results in the bathochromic shift of 50 nm [31].

The first example on the third row indicates almost no difference in λ_a upon replacing a central fluorene unit by a carbazole unit [50], although the latter has a nitrogen atom with nonbonding electrons that can be delocalized. This can be ascribed to the polyphenylene nature of the unimer main chain. Unlike the previous case, the interconnection of a thiophene ring of a bithiophene unit through a nitrogen atom bridge gives remarkable redshift of λ_a (c. 60 nm) compared to the unimer with a dialkylsilanediyl bridge [50]. Still higher redshifts can be achieved by introducing phospole rings. A replacement of the central thiophene unit in bis(*tpy*)terthiophene by a less aromatic phosphole unit resulted in the redshift of 60 nm and the λ_a of the obtained unimer, 482 nm, is among the highest ones observed for bis(*tpy*) unimers [29]. In addition, low-lying lowest unoccupied molecular orbital (LUMO) energy level [51] and

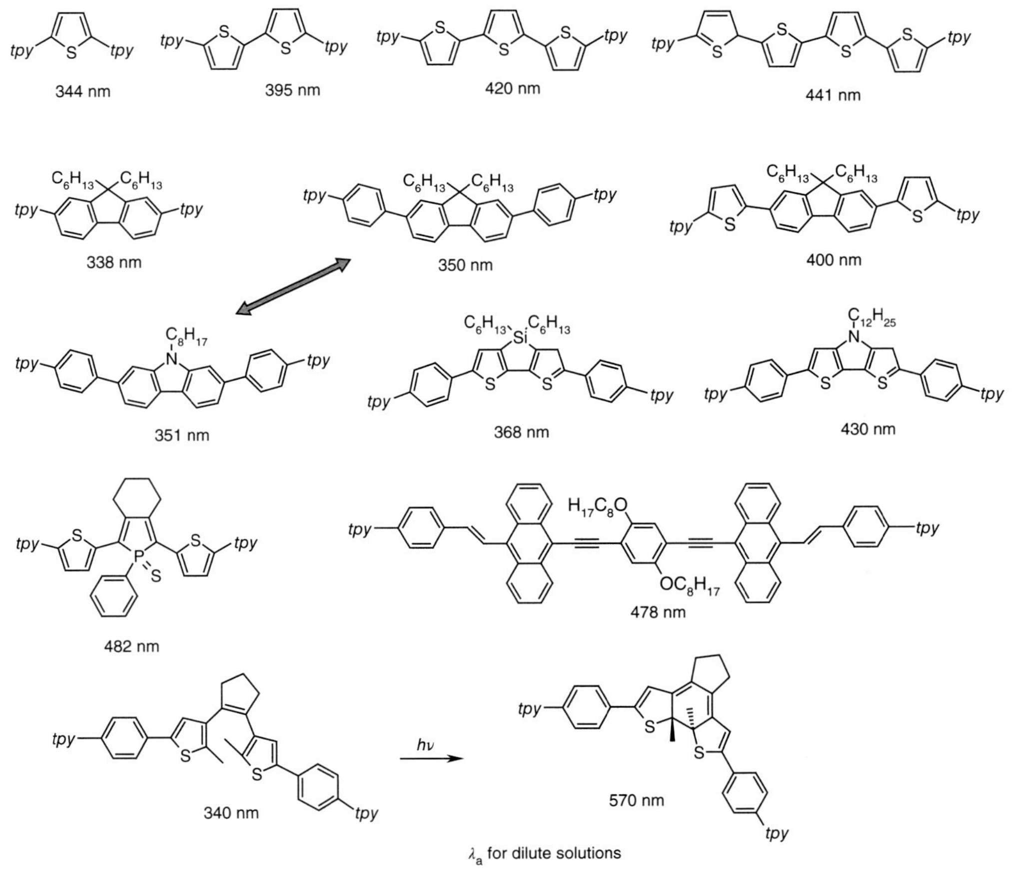

Figure 6.15 Illustration of the structure effects on the absorption maximum wavelength of α,ω-bis(*tpy*) unimers.

the reactive P-atom offer the possibility for effective tuning of the properties of the phosphole by means of chemical modification [52–57]. A much longer bis(*tpy*)oligo(arylene-ethynylene/vinylene) unimer with anthracene and benzene rings (Figure 6.15) showed a comparable value $\lambda_a = 478$ nm [58]. Zhong et al. [59] prepared an interesting unimer with 1,2-bis(thiophene-2,5-diyl)cyclopentene central moiety that upon photoinduced cyclization gives a new unimer with $\lambda_a = 570$ nm (Figure 6.15, the last row).

Changes in the photophysical properties of a conjugated unimer when transiting to its original MSP form are (i) extension of the chain segments where electrons can delocalize and (ii) contributions and effects of coordination linkages. The chain extension due to polymerization typically results in the redshift of λ_a about 50 ± 30 nm if the MLCT band is absent. When present, the MLCT band occurs at a longer wavelength than the MLCT band of the complex parent chelate molecule (terpyridine in the case of *tpy*-unimers, see Section 6.4.3), because it is contributed by the transitions in the central blocks of the enchained unimeric units [28, 32, 39, 46]. Hence, the Fe^{2+}-MSPs derived from bis(*tpy*) unimers have a violet to green color that secondarily depends on the central block structure, while the colors of MSPs without MLCT, such as Zn^{2+} or Co^{2+} MSPs strongly depends on the unimer structure. An electrode oxidation of an Fe^{2+}-MSP cancels the MLCT band and thus also its spectral contribution originating from its blue-region absorption. This is the principle of electrochromism of such MSPs [19–21, 60].

Coordination linkages with transition metals having an open shell electronic structure (not fully occupied d-orbitals) usually weaken or quench the fluorescence of MSPs, which can be used for monitoring the kinetics of assembling [40]. The closed shell metal (such as Zn or Cd) linkages provide MSPs with luminescence that can be tuned by mixing their components in various ratios (even white emission is accessible in this way [31]) or by preparing MSP copolymers from a few different unimers and/or a few ion couplers. A special case is presented by MSPs with lanthanoide ion couplers, Ln^{3+}-MSPs, that exhibit a high luminescence yield and emission spectra with resolved emission bands attributable to the emission from ions [26, 61] (see Figure 6.16, sharp bands correspond to the bands of $^5D_0 \rightarrow {}^7F_J$ electronic transitions within Eu^{3+} ions: $J = 1, 2, 3$, and 4 for the bands at 594, 617, 650, and 670 nm, respectively). In luminescent Ln^{3+}-MSPs derived from bis(*tpy*) and bis(pybox) unimers, the unimeric units act as sensitizers from which the excitation energy is transferred to the Ln^{3+} ions that emit light (antenna effect).

Recently, Rais et al. studied the photophysical processes in thin solid films of α,ω-bis(*tpy*)terthiophene and its MSP with Zn^{2+} ions by the femtosecond transient absorption spectroscopy and observed an ultrafast singlet fission process with the time constant of 160 fs after photoexcitation of the MSP into its second electronic excited state [62]. This seems to be the first reported MSP material that exhibits a singlet fission process. In contrast, the free unimer thin films showed no sign of the singlet fission but the exciton relaxation through excimers. The formation of excimers in a free unimer has been also shown in Ref. [29] and is supported by the wide-angle X-ray scattering (WAXS) diffractograms that showed a high structural ordering in the unimer but low ordering in the MSP, which may be

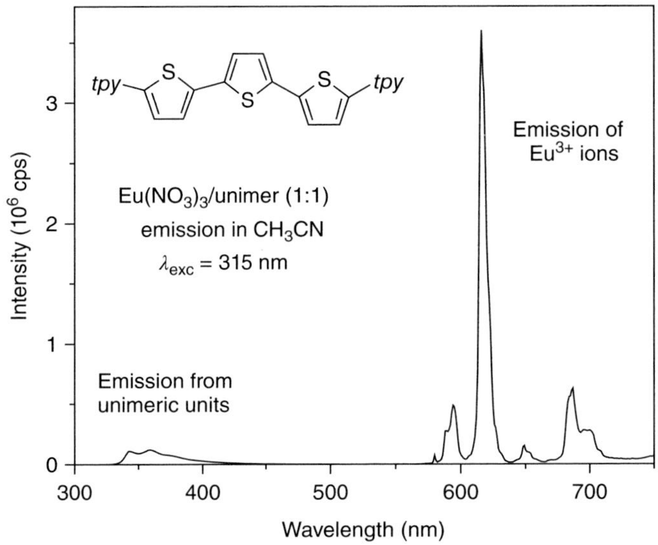

Figure 6.16 Luminescence spectrum of Eu^{3+}-MSPs derived from bis(*tpy*) unimer (unpublished data of author J.V.).

the probable reason for the differences in the energy relaxation pathways of these two materials. Therefore, the octahedral geometry of Zn^{2+} linkages that prevents the formation of excimers appears to be one of the key factors for the observed singlet fission.

6.5 Potential Applications and Outlook

There is no doubt that dynamic MSPs fulfill the demanding criteria that are imposed on the processable and recyclable functional polymeric materials [63] to comply with future technologies. The MSP dynamers derived from unimers with well-defined structures can be used as recyclable, degradable, stimulus-responsive, and self-healing materials. The use of various metal ions and unimers allows for fine-tuning of the physical properties for the engineering of functional materials for different devices and purposes. Indeed, based on the properties of these dynamic MSPs, a number of their potential applications on functional devices or samples are described and suggested in literature and have also been documented (see Section 6.4.4).

6.5.1 Electrochromic Devices Based on the Modification of the Absorption Properties

Electrochromism based on absorption properties show the MSPs with the intense MLCT band that occurs far from the absorption band of enchained unimeric units, which disappears upon the change in the oxidation state of metal ions, such as Fe-MSPs. This offers the applications in electrochromic displays that change their optical properties in response to the applied external electric potential. For example, Kurth and coworkers prepared a series of unimers that can be coordinated to Fe(II), Ru(II), Cu(II), and Co(II) ions leading to MSPs having blue-violet, red, green, and orange colors [60, 64].

Another type of electrochromic devices based on heterometallo (Fe/Ru) MSPs were developed by Higuchi and coworkers [65]. The color of their polymers stems from the different contributions of the Fe (blue) and Ru (red) MLCT complex moieties. Furthermore, these two moieties present different redox properties, which the authors used in an electrochromic device that presented multicolor electrochromism from purple to orange and to colorless with increasing applied potential from 0 to 1.1 V vs. Ag/Ag^+. A long-term device stability (5000–10 000 cycles) with a very high optical contrast ΔT (around 68%) between the colored and bleached states and a coloration efficiency of around 188.2 and 242.1 cm^2/C for the switching between 0 and 0.9 V and 0 and 1.1 V, respectively has been developed.

6.5.2 Electrochromic Devices Based on the Modification of the Emission Properties

Higuchi et al. [66] synthesized an MSP containing alternating Eu(III) and Fe(II) ions (Figure 6.17). Its emission spectrum in ethane-1,2-diol showed the characteristic emission of Eu(III) ions attributed to the $^5D_0 \rightarrow {}^7F_{1-4}$ transitions, but the emission was weak (quantum yield of 0.07) owing to the presence of the Fe(II) complex with an MLCT band (λ_a of 450–650 nm). However, a thin film of the polymer on indium tin oxide (ITO) showed reversible electrochromic behavior on applying 2.0 V vs. Ag/AgCl for the bleaching and 0 V vs. Ag/AgCl for coloring indicating that the color change is triggered by the electrochemical redox between the Fe(II) and Fe(III) states. The emission properties are also affected by the applied potential. A significant increase in photoluminescence was observed at 2.0 V vs. Ag/AgCl. In contrast, the red emission was re-quenched at 0 V vs. Ag/AgCl. The ON–OFF switching of the emission by the electrochemical redox could be repeated for at least 10 times.

Higuchi et al. [67] have also developed an amperometric glucose biosensor based on the glucose oxidase immobilized on Co(II)-MSP derived from simple 1,4-bis(*tpy*)benzene prepared by a simple one-step process. The combined redox properties of the enzyme and Co-MSP have been used for the precise detection of glucose. This modified electrode showed good amperometric sensing performance, with the linear range, sensitivity, response time, and

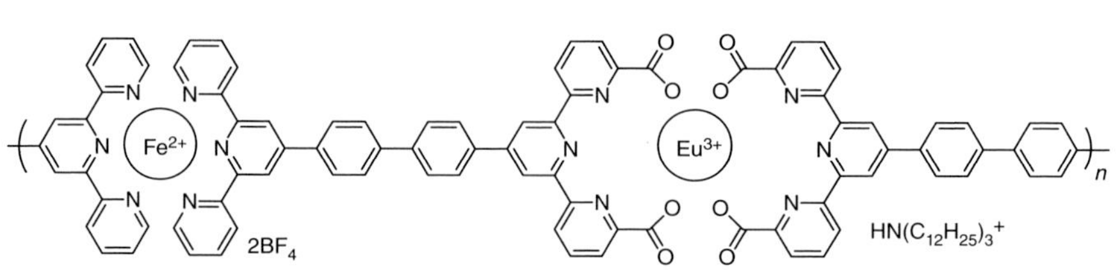

Figure 6.17 Structure of MSP showing electrochromism by modification of its emission.

limit of detection (LOD) of 0.1–1.4 mM (correlation coefficient $R^2 = 0.999$), 47.1 μAmM^{-1}/cm^2, 5–20 seconds, and 82.8 μM, respectively. Furthermore, they also showed that use of Co-MSP solves the problem of enzyme denaturation in biosensors indicating a strong stability of the biosensors. The characteristics proved applicability of the Co-MSP electrode material for glucose monitoring.

6.5.3 Ion Conductivity

Ion-conducting materials received strong attention since they can be used for energy-related applications such as fuel cells and secondary batteries. In this research field, the MSPs can be interesting candidates since their structure involves hydrophilic metal ions and hydrophobic rigid ligands, properties that are important for achieving both high ionic conductivity and durability of the material. Higuchi and coworkers [68] developed material based on gold nanoparticles, Au-NPs, onto which *tpy* groups have been anchored by a sulfide bridge by reaction with HS-*tpy*. Finally, the decorated Au-NPs were co-self-assembled with bis(*tpy*)unimers and Fe^{2+} ions to give a material showing ionic conductivity of 10^{-2} S/cm. The idea lying behind this is that the ionic coordination linkages spontaneously assemble close to each other thus making ionic channels (domains) that enable effective transport of ions. It is very probable that also thiophene-2-yl-*tpy* can be used for these purposes since it preferably anchors on Au-NPs by a thiophene ring leaving the *tpy* group free [39].

6.5.4 Actuators

Mauro and coworkers [69] have synthetized Zn-MSPs from a bis(*tpy*)unimer with photo-switchable diazophenylene central blocks and other central blocks with side groups thereby increasing the solubility of MSPs (Figure 6.18). Irradiation of a gel of the MSP system with a low-power LED lamp induced trans-to-cis isomerization of azo units resulting in a mechanical response of the gel.

6.5.5 Outlooks

The synthesis and characterization of supramolecular polymers have made significant progress in recent years and, as a result, the MSPs have been used in different application fields. Nevertheless, the structure–properties relationships need to be investigated in more detail to widen the scope of applications and

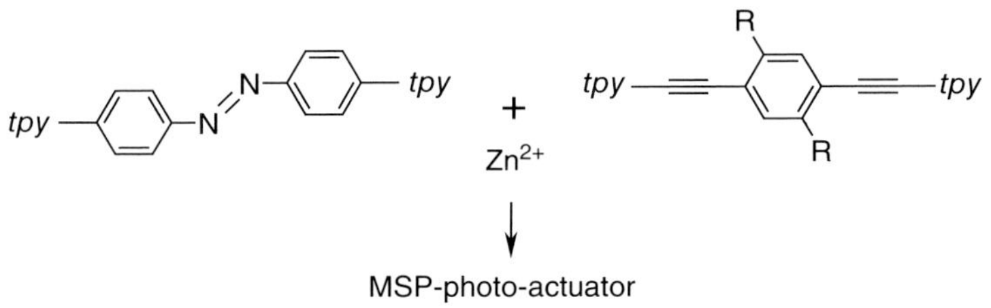

Figure 6.18 Design of the MSP photo-actuator. Source: After Borr et al. 2016 [69].

explore the hidden potential of these materials. This depends not only on the development of further functional systems but also on the development of efficient techniques and protocols for preparing functional devices based on MSPs.

Acknowledgments

The authors thank the COST Action CM1302 (SIPs): "European Network on Smart Inorganic Polymers."

References

1 Lehn, J.-M. (1988). Supramolecular chemistry – scope and perspectives molecules, supermolecules, and molecular devices. *Angew. Chem. Int. Ed. Engl.* 27: 89–112.
2 Lehn, J.-M. (1993). Supramolecular chemistry – molecular information and the design of supramolecular materials. *Makromol. Chem. Makromol. Symp.* 69: 1–17.
3 Kato, T., Kihara, H., Kumar, U. et al. (1994). A liquid-crystalline polymer network built by molecular self-assembly through intermolecular hydrogen-bonding. *Angew. Chem. Int. Ed. Engl.* 33: 1644–1645.
4 Stpourcain, C.B. and Griffin, A.C. (1995). Thermoreversible supramolecular networks with polymeric properties. *Macromolecules* 28: 4116–4121.
5 Lehn, J.-M. (1995). *Supramolecular Chemistry – Concepts and Perspectives.* Weinheim: Wiley-VCH.
6 Mark, H. (1973). Early days of polymer science. *J. Chem. Educ.* 50: 757–760.
7 Cifferi, A. (2005). *Supramolecular Polymers*, 2e. New York: Taylor & Francis.
8 Lehn, J.-M. (2005). Dynamers: dynamic molecular and supramolecular polymers. *Prog. Polym. Sci.* 30: 814–831.
9 Yang, L., Tan, X., Wang, Z., and Zhang, X. (2015). Supramolecular polymers: historical development, preparation, characterization, and functions. *Chem. Rev.* 115: 7196–7239.
10 Reutenauer, P., Buhler, E., Boul, P.J. et al. (2009). Room temperature dynamic polymers based on Diels–Alder chemistry. *Chem. Eur. J.* 15: 1893–1900.
11 Kolomiets, E. and Lehn, J.-M. (2005). Double dynamers: molecular and supramolecular double dynamic polymers record contains structures. *Chem. Commun.* 12: 1519–1521.
12 Kuhl, N., Bode, S., Bose, R.K. et al. (2015). Acylhydrazones as reversible covalent crosslinkers for self-healing polymers. *Adv. Funct. Mater.* 25: 3295–3301.
13 Folmer-Andersen, J.F. and Lehn, J.-M. (2011). Thermoresponsive dynamers: thermally induced, reversible chain elongation of amphiphilic poly(acylhydrazones). *J. Am. Chem. Soc.* 133: 10966–10973.
14 Shunsuke Fujii, S. and Lehn, J.-M. (2011). Structural and functional evolution of a library of constitutional dynamic polymers driven by alkali metal ion recognition. *Angew. Chem. Int. Ed.* 48: 7635–7638.

15 Barón, M., Hellwich, K.-H., Hess, M. et al. (2009). Glossary of class names of polymers based on chemical structure and molecular architecture (IUPAC recommendations 2009). *Pure Appl. Chem.* 81: 1131–1186.
16 Hess, M., Jones, R.G., Kahovec, J. et al. (2006). Terminology of polymers containing ionizable and ionic groups and of polymers containing ions. *Pure Appl. Chem.* 78: 2067–2074.
17 Appel, E.A., del Bario, J., Loh, X.J., and Scherma, O.A. (2012). Supramolecular polymeric hydrogels. *Chem. Soc. Rev.* 41: 6195–6214.
18 Chiper, M., Hoogenboom, R., and Schubert, U.S. (2009). Toward main chain metallo-terpyridyl supramolecular polymers: the metal does the trick. *Macromol. Rapid. Commun.* 33: 565–578.
19 Wild, A., Winter, A., Schlutter, F., and Schubert, U.S. (2011). Advances in the field of π-conjugated 2,2′:6′,2″-terpyridines. *Chem. Soc. Rev.* 40: 1459–1511.
20 Husson, J. and Knor, M. (2012). 2,2′:6′,2″-Terpyridines functionalized with thienyl substituents: synthesis and applications. *J. Heterocycl. Chem.* 49: 453–478.
21 Winter, A. and Schubert, U.S. (2016). Synthesis and characterization of metallo-supramolecular polymers. *Chem. Soc. Rev.* 45: 5311–5357.
22 Krohnke, F. (1976). The specific synthesis of pyridines and oligopyridines. *Synthesis* 1976 (1): 1–24.
23 Svoboda, J., Štenclová, P., Uhlík, F. et al. (2011). Synthesis and photophysical properties of α,ω-bis(terpyridine)oligothiophenes. *Tetrahedron* 67: 75–79.
24 Addison, A.W. and Burke, P.J. (1981). Synthesis of some imidazole- and pyrazole-derived chelating agents. *J. Heterocycl. Chem.* 18: 803–805.
25 Zhang, W.H., Chien, S.W., and Hor, T.S.A. (2011). Recent advances in metal catalysts with hybrid ligands. *Coord. Chem. Rev.* 225: 1991–2024.
26 Duerrbeck, A., Gorelik, S., Hobley, J. et al. (2015). Highly emissive, solution-processable and dynamic Eu(III)-containing coordination polymers. *Chem. Commun.* 8656–8659.
27 Hladysh, S., Václavková, D., Vrbata, D. et al. (2017). Synthesis and characterization of metallo-supramolecular polymers from thiophene-based unimers bearing pybox ligands. *RSC Adv.* 7: 10718–10728.
28 Bláhová, P., Zedník, J., Šloufová, I. et al. (2014). Synthesis and photophysical properties of new α,ω-bis(*tpy*)oligothiophenes and their metallo-supramolecular polymers with Zn^{2+} ion couplers. *Soft Mater.* 12: 214–229.
29 Vitvarova, T., Svoboda, J., Hissler, M., and Vohlidal, J. (2017). Conjugated metallo-supramolecular polymers containing a phosphole unit. *Organometallics* 36: 777–786.
30 Štenclová, P., Šichová, K., Šloufová, I. et al. (2016). Alcohol- and water-soluble bis(*tpy*)quaterthiophenes with phosphonium side groups: new conjugated units for metallo-supramolecular polymers. *Dalton Trans.* 45: 1208–1224.
31 Hrma, M., Sichova, K., Svoboda, J., and Vohlídal, J. (2017). Assembling of bis(*tpy*)fluorenes with Zn^{2+} and Fe^{2+} ions into metallo-supramolecular polymers with highly efficient white-light emission. *Polymer* 122: 22–33.

32 Štenclová-Bláhová, P., Svoboda, J., Šloufová, I., and Vohlídal, J. (2015). Alcohol-soluble bis(*tpy*)thiophenes: new building units for constitutional dynamic conjugated polyelectrolytes. *Phys. Chem. Chem. Phys.* 17: 13743–13756.

33 Vitvarová, T., Zedník, J., Bláha, M. et al. (2012). Effect of ethynyl and thiophene-2-yl substituents on complexation of 4′-substituted 2,2′:6′,2″-terpyridines with Zn^{2+} and Fe^{2+} ions and spectroscopic properties of the ligands and formed complex species. *Eur. J. Inorg. Chem.* 2012: 3866–3874.

34 Duprez, V., Biancardo, M.O., Spanggaard, H., and Krebs, F.C. (2005). Synthesis of conjugated polymers containing terpyridine-ruthenium complexes: photovoltaic applications. *Macromolecules* 38: 10436–10448.

35 Constable, E.C., Thompson, A.M.W.C., Tocher, D.A., and Daniels, A.M. (1992). Synthesis, characterization and spectroscopic properties of ruthenium(II)-2,2′-6′,2″-terpyridine coordination triades X-ray structures of 4′-(*N,N*-dimethylamino)-2,2′-6′,2″-terpyridine and bis(4′-(*N,N*-dimethylamino)-2,2′-6′,2″-terpyridine)ruthenium(II) hexafluorophosphate acetonitrile solvate. *New J. Chem.* 16: 855–867.

36 Sauvage, J.-P., Collin, J.-P., Chambron, J.-C. et al. (1994). Ruthenium(II) and osmium(II) bis(terpyridine) complexes in covalently-linked multicomponent systems: synthesis, electrochemical behavior, absorption spectra, and photochemical and photophysical properties. *Chem. Rev.* 94: 993–1019.

37 Šloufová, I., Procházka, M., and Vlčková, B. (2015). Identification of two Ag-2,2′:6,2″-terpyridine surface species on Ag nanoparticle surfaces by excitation wavelength dependence of SERS spectra and factor analysis: evidence for chemical mechanism contribution to SERS of Ag(0)-*tpy*. *J. Raman Spectrosc.* 46: 39–46.

38 Šloufová, I., Vlčková, B., Procházka, M. et al. (2014). Comparison of SERRS and RRS excitation profiles of $[Fe(tpy)_2]^{2+}$ (*tpy*-2,2′:6′,2″-terpyridine) supported by DFT calculations: effect of the electrostatic bonding to chloride-modified Ag nanoparticles on its vibrational and electronic structure. *J. Raman Spectrosc.* 45: 338–348.

39 Pruskova, M., Sutrova, V., Slouf, M. et al. (2017). Arrays of Ag and Au nanoparticles with terpyridine- and thiophene-based ligands: morphology and optical responses. *Langmuir* 33: 4146–4156.

40 Munzert, S.M., Schwarz, G., and Kurth, D.G. (2016). Tailoring length and viscosity of dynamic metallosupramolecular polymers in solution. *RSC Adv.* 6: 15441–15450.

41 Hogg, R. and Wilkins, R.G. (1962). Exchange studies of certain chelate compounds of the transitional metals. Part VIII. 2,2′,2″-terpyridine complexes. *J. Chem. Soc.* 1962: 341–350.

42 Irving, H. and Williams, R.J.P. (1953). The stability of transition-metal complexes. *J. Chem. Soc.* 1953: 3192–3210.

43 Munzert, S.M., Schwarz, G., and Kurth, D.G. (2016). Kinetic studies of the coordination of mono- and ditopic ligands with first row transition metal ions. *Inorg. Chem.* 55: 2565–2573.

44 Rais, D., Menšík, M., Štenclová-Bláhová, P. et al. (2015). Time-resolved transient optical absorption study of bis (terpyridyl) oligothiophenes and their metallo-supramolecular polymers with Zn (II) ion couplers. *J. Phys. Chem. A* 119: 6203–6214.

45 Stepto, R., Chang, T., Kratochvil, P. et al. (2015). Definitions of terms relating to individual macromolecules, macromolecular assemblies, polymer solutions, and amorphous bulk polymers (IUPAC recommendations 2014). *Pure Appl. Chem.* 87: 71–120.

46 Šloufová, I., Vlčková, B., Mojzeš, P. et al. (2018). Probing the formation, structure, and reactivity of Zn(II), Ag(I), and Fe(II) complexes with 2,2′:6′,2″-terpyridine on Ag nanoparticles surfaces by time evolution of SERS spectra, factor analysis, and DFT calculations. *J. Phys. Chem. C* 122: 6066–6077.

47 Read, W. and Freitag, D. (1968). Oligophenyls, oligophenylenes, and polyphenyls, a class of thermally very stable compounds. *Angew. Chem. Int. Ed. Engl.* 7: 835–902.

48 Salzner, U., Lagowski, J.B., Pickup, P.G., and Poirier, R.a. (1998). Comparison of geometries and electronic structures of polyacetylene, polyborole, polycyclopentadiene, polypyrrole, polyfuran, polysilole, polyphosphole, polycyclopentadiene, polypyrrole, polyfuran, polysilole, polyphosphole, polythiophene, polyselenophene and polytellurophene. *Synth. Met.* 1998 (96): 177–189.

49 Ko, S., Hoke, E.T., Pandey, L. et al. (2012). Controlled conjugated backbone twisting for an increased open-circuit voltage while having a high short-circuit current in poly(hexylthiophene) derivatives. *J. Am. Chem. Soc.* 134: 5222–5232.

50 Li, F., Hu, Z., Qiao, H. et al. (2016). Terpyridine-based donor–acceptor metallo-supramolecular polymers with tunable band gaps: synthesis and characterization. *Dyes Pigm.* 132: 142–150.

51 Krygowski, T.M. and Cyranski, M.K. (2009). Aromatic phosphorus heterocycles. In: *Aromaticity in Heterocyclic Compounds*, vol. 93 (ed. L. Nyulaszi and Z. Benko), 95–99. Springer Science & Business Media.

52 Fave, C., Hissler, M., Kárpati, T. et al. (2004). *J. Am. Chem. Soc.* 126: 6058–6063.

53 Hissler, M., Lescop, C., and Réau, R. (2005). *J. Organomet. Chem.* 690: 2482–2487.

54 Matano, Y. and Imahori, H. (2009). *Org. Biomol. Chem.* 7: 1258–1271.

55 Su, H.-C., Fadhel, O., Yang, C.-J. et al. (2006). *J. Am. Chem. Soc.* 128: 983–995.

56 Fadhel, O., Gras, M., Lemaitre, N. et al. (2009). *Adv. Mater.* 21: 1261–1265. Hobbs, M.G. and Baumgartner, T. (2007). *Eur. J. Inorg. Chem.* 3611–3628.

57 Casado, J., Réau, R., and López Navarrete, J.T. (2006). *Chem. Eur. J.* 12: 3759–3767.

58 Wild, C.F., Winter, A., Hager, M.D. et al. (2010). π-conjugated 2,2′:6′,2″-bis(terpyridines): systematical tuning of the optical properties by variation of the linkage between the terpyridines and the π-conjugated systém. *Eur. J. Org. Chem.* 2010: 1859–1868.

59 Zhong, Y.W., Vila, N., Henderson, J.C., and Abruna, H.D. (2007). Dithienylcyclopentenes-containing transition metal bisterpyridine complexes directed toward molecular electronic applications. *Inorg. Chem.* 48: 991–999.

60 Pai, S., Moos, M., Schreck, M.H. et al. (2017). Green to red electrochromic Fe(II) metallo-supramolecular polyelectrolytes self-assembled from fluorescent 2,6-bis(terpyridyl)pyrimidine bithiophene. *Inorg. Chem.* 56: 1418–1432.

61 Mauro, M., Bellemin-Laponaz, S., and Cebrián, C. (2017). Metal-containing polymers as light-responsive materials and beyond. *Chem. Eur. J.* 23: 17626–17636.

62 Rais, D., Pfleger, J., Mensik, M. et al. (2017). Singlet fission in thin films of metallo-supramolecular polymers with ditopic thiophene-bridged terpyridine ligands. *J. Mater. Chem. C* 5: 8041–8051.

63 Horie, K., Baron, M., Fox, R.B. et al. (2004). Definitions of terms relating to reactions of polymers and to functional polymeric materials – (IUPAC recommendations 2003). *Pure Appl. Chem.* 76: 889–906.

64 Han, F., S., Higuchi, M., and Kurth, D.G. (2007). Metallo-supramolecular polymers based on functionalized bis-terpyridines as novel electrochromic materials. *Adv. Mater.* 2007 (19): 3928–3931.

65 Hu, C.-W., Sato, T., Zhang, J. et al. (2013). Multi-colour electrochromic properties of Fe/Ru-based bimetallo-supramolecular polymers. *J. Mater. Chem. C* 1: 3408–3413.

66 Sato, T. and Higuchi, M. (2013). An alternately introduced heterometallo-supramolecular polymer: synthesis and solid-state emission switching by electrochemical redox. *Chem. Commun.* 2013 (49): 5256.

67 Hsu, C.-Y., Sato, T., Moriyama, S., and Higuchi, M. (2016). A Co(II)-based metallo-supramolecular polymer as a novel enzyme immobilization matrix for electrochemical glucose biosensing. *Eur. Polym. J.* 83: 499.

68 Pandey, R.K., Chakraborty, C., Rana, U. et al. (2016). An insight into ion-conduction phenomenon of gold nanocluster ligand based metallo-supramolecular polymers. *J. Mater. Chem. A* 4: 4398–4401.

69 Borr, E., Stumb, J.-F., Bellemin-Laponnaz, S., and Mauro, M. (2016). Light-powered self-healable metallosupramolecular soft actuators. *Angew. Chem. Int. Ed.* 55: 1313–1317.

7

Applications of Heteroatom-Based Oligomers and Polymers in Optoelectronics

Matthew P. Duffy, Pierre-Antoine Bouit, and Muriel Hissler

Univ Rennes, CNRS, ISCR (Institut des Sciences Chimiques de Rennes) – UMR 6226, 263 avenue du Général Leclerc, Rennes 35000, France

7.1 Introduction

Organic semiconductors (i.e. small molecules, oligomers, and polymers) are by far the most promising functional materials for applications in less expensive and flexible electronic devices such as light-emitting diodes (OLEDs), organic field effect transistors (OFETs), and organic photovoltaic solar cells (OPVSc) [1]. This interest is mainly due to the possibility of tailoring their physical properties and supramolecular organization via molecular structural variations [1]. For example, the optical and electronic properties of organic semiconductors depend on the chemical structure of the conjugated oligomer/polymer carbon backbone (highest occupied molecular orbital–lowest unoccupied molecular orbital [HOMO–LUMO] gap, electronic density, etc.) and the interaction between the individual molecules (supramolecular arrangement, morphology). Predictive structure–property relationships can be established and the optimization of physical properties to suit a desired function is possible via chemical engineering at the molecular level. This process is well illustrated by the development of advanced electroluminescent organic materials following the seminal reports of efficient OLEDs based on small molecules and conjugated polymers [2]. Till date, new materials with specific properties are under investigation for the development of efficient and easily manufacturing organic devices. The control of their properties can be achieved by varying the chemical composition of conjugated systems, and due to the high flexibility of organic synthesis, many strategies can be considered to diversify these structures. A fruitful approach involves the incorporation of heteroatoms into the backbone of conjugated frameworks [3], with two main approaches being possible for the development of polymers. The first involves the use of heterocyclopentadiene subunits since their electronic properties depend on the nature of the heteroatoms (Figure 7.1). Metalloles of Group 14 elements (siloles, germoles, stannoles) exhibit high electron affinity [4], whereas those of Groups 15 (pyrrole) and 16 (furan, thiophene) behave as electron-rich aromatic pi-systems [5] and the phosphole can be regarded as a chameleon-like structure combining the two properties [6].

Smart Inorganic Polymers: Synthesis, Properties, and Emerging Applications in Materials and Life Sciences, First Edition. Edited by Evamarie Hey-Hawkins and Muriel Hissler.

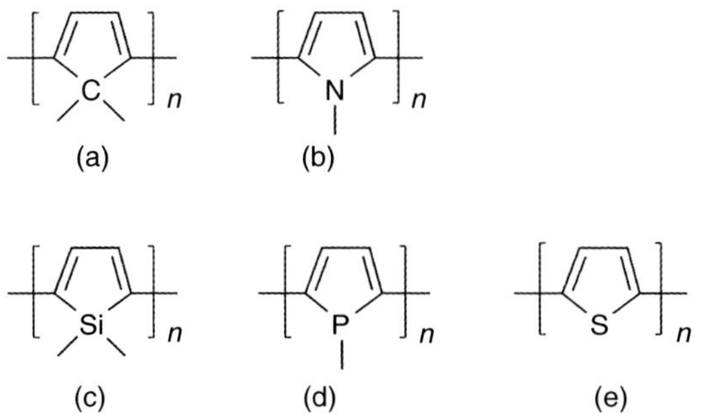

Figure 7.1 Poly(heterocyclopentadienes).

Based on these metalloles, a large number of oligomers and polymers have been developed offering a combination of novel electronic properties, excellent stabilities, and structural diversity. It can be highlighted that polymers including the heavier elements have much narrower bandgaps than their corresponding second row analogs (Figure 7.1: silole vs. cyclopentadiene; phosphole vs. pyrrole). This effect is particularly prominent for the Group 15 derivatives with the estimated bandgap for poly(phosphole) being half of that of poly(pyrrole). The second approach for the development of polymers is to replace the vinylene bridges of poly(*p*-phenylene vinylene), PPV, by a heteroatom possessing a lone pair that can participate in the π-conjugation [7]. Representative examples of this class of derivatives are the polyanilines and poly(*p*-phenylenesulfides). In résumé, the introduction of heteroatoms and/or heterocycles allows the emergence of new electronic and geometric properties directly related to the nature of the item [6]. Furthermore, the incorporation of heavier heteroatoms-based polymers into devices has been achieved only since the past 10–15 years, despite decades of research on such species. In this chapter, we will highlight the different kinds of heteroatom-based materials that have been used in electronic devices, such as OLEDs, organic photovoltaic (OPV) cells, dye-sensitized solar cells (DSSCs), OFETs, and electrochromic cells.

7.2 Organic Light-Emitting Diodes (OLEDs)

OLED technology is attracting considerable attention due to its use in flat-panel displays and solid-state lighting [1, 2]. An OLED technology is a flat light-emitting technology based on small molecules or polymers that are able to generate electrically stimulated emission of light. The multilayer structure of an OLED consists of several organic layers (hole-transport layer [HTL], emissive layer [EML], and electron-transport layer [ETL]) sandwiched between two electrodes (transparent conducting oxide anode and a metallic low work function cathode), as depicted in Figure 7.2 [1].[1] When a voltage is applied between the electrodes, holes are injected into the HTL from the anode and electrons are injected into the ETL

1 Some additional layers (hole blocking layer, electron blocking layers) can also be inserted to increase the performances.

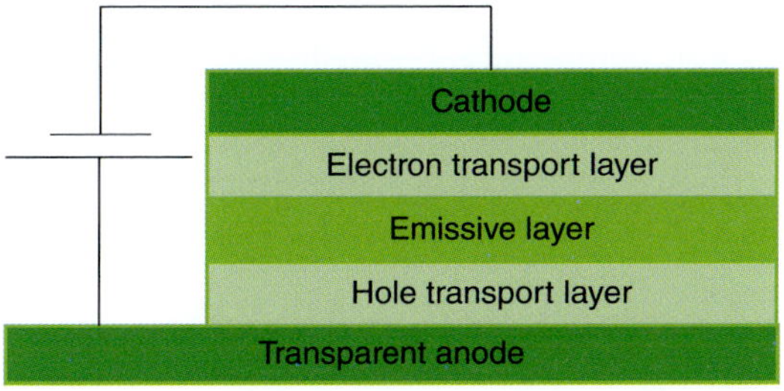

Figure 7.2 OLED structure.

layer from the cathode. Then, these charge carriers migrate under the applied electric field, usually by a hopping process and recombine to form singlet and triplet excitons (bound excited-state electron–hole pairs). The location of the recombination zone in the diode is a function of the charge mobility of the organic material as well as that of the electric field distribution. After diffusion, the exciton recombines and a photon is emitted.

In order to evaluate the performance characteristics of an OLED, a set of electrical and optical characterizations can be made. The electrical parameters can be determined from I–V–L (current–voltage–luminance) characteristics [8].

(1) The internal quantum efficiency η_{int} is the ratio of the total number of photons generated within the device to the number of injected charges.

$$\eta_{int} = \frac{N_{\text{internal emitted photons}}}{N_{\text{injected charges}}} = \eta_r\ \chi \cdot \phi_{PL}$$

(2) The external quantum efficiency η_{ext} is the ratio of the total number of photons emitted from the device into the viewing direction to the number of injected charges.

$$\eta_{ext} = \eta_{int} \cdot \eta_{out} = \eta_r \cdot \chi \cdot \phi_{PL} \cdot \eta_{out}$$

η_r denotes the probability that holes and electrons recombine to form excitons; χ the probability to produce emitting species; ϕ_{PL} the fluorescence or phosphorescent quantum yield; and η_{out} the fraction of generated photons emitted by the device.

(3) The luminous efficiency η_L in candelas per ampere (cd/A) is the ratio of the luminous intensity in the forward direction and the current through the device.

$$\eta_L = \frac{S \cdot L}{I}$$

L denotes the luminance in cd/m^2; S the surface of the diode in m^2; and I the current in A.

(4) The luminous power efficiency or luminous efficacy η_P in lumen per watt (lm/W) represents the output light power from a device per electrical power input (measured in watts). The efficacy of a light source takes into account the sensitivity of human vision to the different wavelengths of the visible spectrum.

$$\eta_P = \frac{\eta_L \cdot \pi}{V}$$

where V is the working voltage.

Since high-quality light is important for illumination, it is important to qualitatively define the color quality of OLEDs. Three different light output characteristics can be used.

(i) The Commission Internationale d'Eclairage (CIE) chromaticity coordinates (*x*,*y*) (chromaticity). Locate the emission color in a chromaticity diagram allowing an optical comparison of all different sources. The CIE chromaticity coordinates of pure white light are (0.33, 0.33).
(ii) The Correlated Color Temperature (CCT) is the color appearance of a light source, measured by the source's chromaticity in reference to the blackbody locus.
(iii) The Color Rendering Index (CRI) is a number ranging from 0 to 100 measuring how accurate the colors of objects can be rendered under a given illumination condition.

As already mentioned, the multilayer structure of an OLED consists of several organic layers (HTL, EML, and ETL) which play specific roles in the production of light via electroluminescence (EL). Therefore, each organic material must fulfill certain requirements. Over time, scientists have begun to understand how the manipulation of the chemical structure of these heteroatomic oligomers/polymers allows the flexibility of fine-tuning the chemical, optical, and electrochemical properties of these systems, and have begun incorporating them into OLED structures.

7.2.1 Application as Charge-Transport Layer

A charge-transport layer must facilitate the charge transport and injection from the device's electrodes to the active layer of the OLED (see footnote 1) [8]. In order to fulfill these requirements, different heteroatom-based materials have been synthesized and studied. For example, phosphine oxide (PO)-based materials present interesting electron-transport and injection capabilities and hole and exciton blocking properties as well as good morphological stability. At the beginning, arylphosphine **1** (Figure 7.3) [9] was the first phosphine-based material to be used as a HTL and exhibited better performances than its nitrogen counterpart in an OLED. Later, it was discovered that π-extended phosphine oxides (Figure 7.3) possessed good electron-transporting abilities due to the highly polar PO group [10].

For example, triphenylphosphine oxide was used to replace tris-(8-hydroxyquinoline)aluminum (Alq_3) allowing to reduce the driving voltage for luminance output. Then, chemical engineering was performed around the phosphine oxide unit leading to a more efficient ETL. For example, the phosphine oxide unit was attached to pyrene (**2**, Figure 7.3) [11] or spirobifluorene (**3–6**, Figure 7.3) π-systems [12]. The nature of the substituents plays an important role for improving the OLED performance. The pyrene substituents allow an enhancement of the conductivity due to its packing ability and the spirobifluorene in SPPO2 **5** and SPPO21 **6** (Figure 7.3) facilitates the electron injection, allowing direct injection from the Al cathode [13]. SPPO21 **6** has also shown to be an effective ETL in an OPV cell using a device configuration

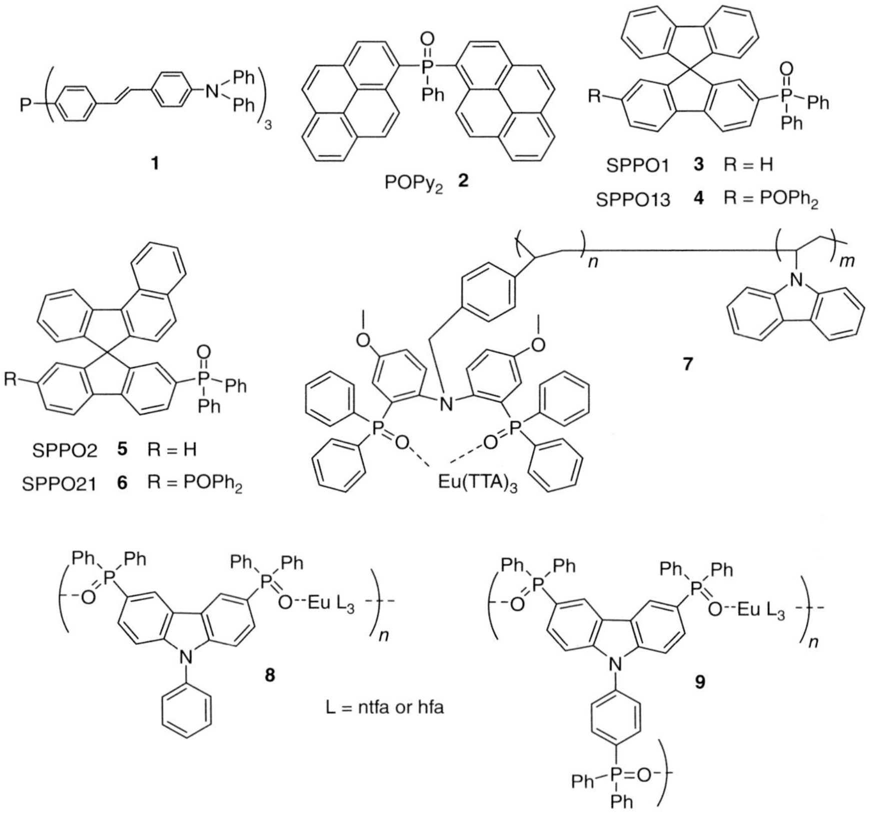

Figure 7.3 Phosphine materials as charge-transport and injection layers.

[ITO/PEDOT:PSS/P3HT:PCBM/SPPO21 6/Al] [14]. Compound **6** protects the active polymer from Al diffusion into the active layer leading to higher performances, and a lowering of the open circuit voltage (V_{oc}) is observed since the coordination ability of the phosphine oxide onto the Al surface leads to band bending. It is a more effective interlayer than LiF in terms of open-circuit voltage and efficiency.

Since phosphine oxide units have low vibrational frequency leading to the suppression of nonradiative transitions via vibrational relaxation, this unit is used as a ligand for the formation of lanthanide polymers having high quantum yields. For example, Europium-complexes-based polymers **7**, **8**, and **9** including bidentate and tridentate phosphine oxides with carbazole units, respectively, have been prepared and studied. The carbazole unit plays the role of a hole transporter in the emitting layer and the electron-injection ability of phosphine oxide compounds is preserved leading to an effective EL device reaching a luminance of 149 and 188 cd/m^2 with compounds **7** and **8**, respectively [15].

Another efficient class of organophosphorus electron transporters are π-conjugated phospholes. Tsuji/Sato/Nakamura et al. synthesized benzo[*b*] phosphole oxides **10a** and sulfides **10b** (Figure 7.4) and tested them as

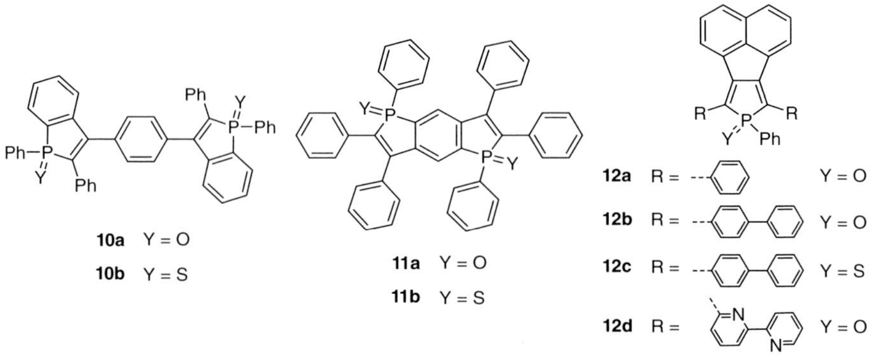

Figure 7.4 ETL phosphole-based molecules used as charge-transport layers.

electron-transporting layers in OLEDs [16]. They showed that **11b** had a higher electron mobility than **10a** (**10a**: 5×10^{-6}/(V s), **10b**: 2×10^{-3}/(V s)) indicating that the decrease in polarity from P=O to P=S improved the electron mobility by preventing electron trapping with the highly polarized P=O sections. They co-deposited **10a** or **10b** with cesium (Cs) to form an electron-transport material (ETM), using the device configuration [ITO/PEDOT:PSS/α-NPD/Alq_3/ETM:Cs/Al]. The device using **10b** as the ETM had the best performance with a luminance of 1000 cd/m^2, driving voltage of 5.0 V, luminance efficiency of 1.8 lm/W, and a current efficiency of 2.8 cd/A.

Compounds **10a,b**, **11a,b** and α,α′-diarylacenaphtho[1,2-*c*]phospholes **12a–d** (Figure 7.4) have been used as ETLs in OPV devices [17]. The benzophosphole sulfides **10b** and **11b** presenting high glass-transition temperatures and resistance toward crystallization, have performed the best in OPV devices featuring a porphyrin donor and fullerene acceptor with a device configuration: [ITO/PEDOT:PSS/BP/BP:SIMEF/SIMEF/ETL/Al]. (V_{oc} = 0.71 V, J_{sc} = 10.8 mA/cm^2, FF = 0.61, and a η = 4.6%, with **11b**: V_{oc} = 0.72 V, J_{sc} = 10.4 mA/cm^2, FF = 0.61, and a η = 4.6%) (see Section 7.3).

The molecular engineering performed around the acenaphtho[1,2-*c*] phospholes by tuning the α-aryl substituent modified the degree of π-conjugation, electron-accepting ability, solid-state packing, and thermal stability of the phosphole materials. The best performing device [ITO/PEDOT:PSS/P3HT:IC_{70}BA/ETL/Al] contained **12b** as the ETL (V_{oc} = 0.76 V, J_{sc} = 8.8 mA/cm^2, FF = 0.62, and a η = 4.2%). These examples emphasize the potential of π-conjugated phosphole oxides and sulfides as charge-transport materials for organic electronics.

Besides phosphole rings, silole rings can also be used for the development of ETMs [18]. The silole ring presents high electron affinity and low LUMO energy since this ring presents a σ*–π* conjugation between the σ* orbitals of the exocyclic bonds on silicon and the π*orbitals of the butadiene moiety (negative hyperconjugation). Compounds **13–15** (Figure 7.5) present a higher electron affinity compared to Alq_3 leading to higher electron mobility and improved EL efficiencies [19]. For example, compound **14** presents an electron mobility higher than that of Alq_3 (2×10^{-4} cm^2/(V s)) and OLEDs including compound **14** as the ETM present half-life time of the luminance nearly three times longer than for a device

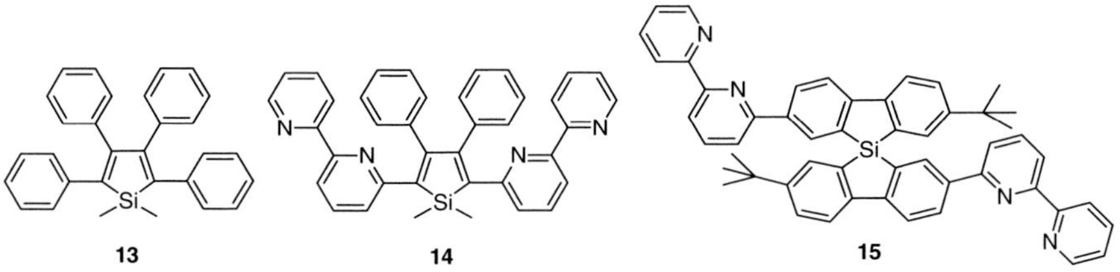

13 14 15

Figure 7.5 Silole-based ETLs used as charge-transport layers.

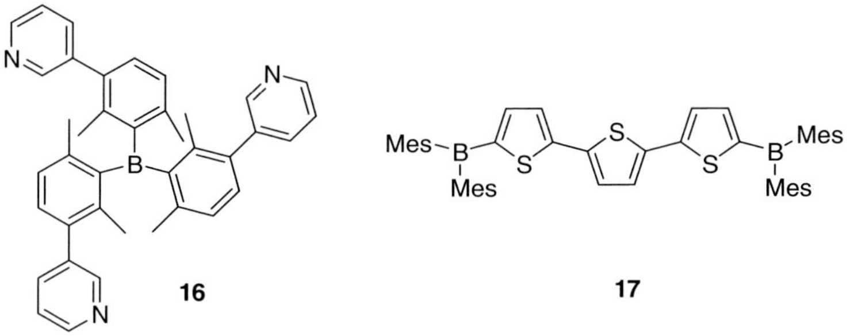

16 17

Figure 7.6 Boron-based ETMs.

without it. It is likely that the use of compound **14** as an ETM improves the charge transporting and thus the longevity.

Different boron-based ETMs have been developed. For example, tris-[3-(3-pyridyl)mesityl]borane **16** and 5,5″-bis-(dimesitylboryl)-2,2′:5′,2″-terthiophene **17** (Figure 7.6) have been used as ETMs in OLEDs. Compared with OLEDs built with a conventional ETM (Alq_3), higher luminance and larger current density were achieved with the two compounds as the ETM [20].

7.2.2 Application as Host for Phosphorescent Complexes

To be an efficient host material for a triplet emitter in an OLED, the substance must satisfy several requirements: (i) should be both physically and chemically stable, (ii) should be efficient in both, hole- and electron-transporting, (iii) must contain an appropriate triplet energy level in order to transfer energy to the triplet emitter. The first report of a phosphine oxide-based material used as a host for a triplet emitter was in 2006 [21]; since then a plethora of PO materials connected to different cores have been studied. Typically, these hosts contain electron-withdrawing and electron-donating groups. The phosphine oxide is electron-withdrawing and facilitates the electron transport. Usually, for the electron-donating groups, a tertiary amine is used, as it aides in the hole-transport. Some of the most efficient PO hosts are connected to carbazole groups, which are well known to have good hole-transporting abilities. When compound **18** (Figure 7.7) was used as host for the dopant FIrpic in a sky blue phosphorescent organic light-emitting diode (PhOLED), it achieved a high external quantum efficiency of 23.9%, with a low driving voltage [22]. In the same article, Lee et al. also showed that the position of the phosphine oxides on

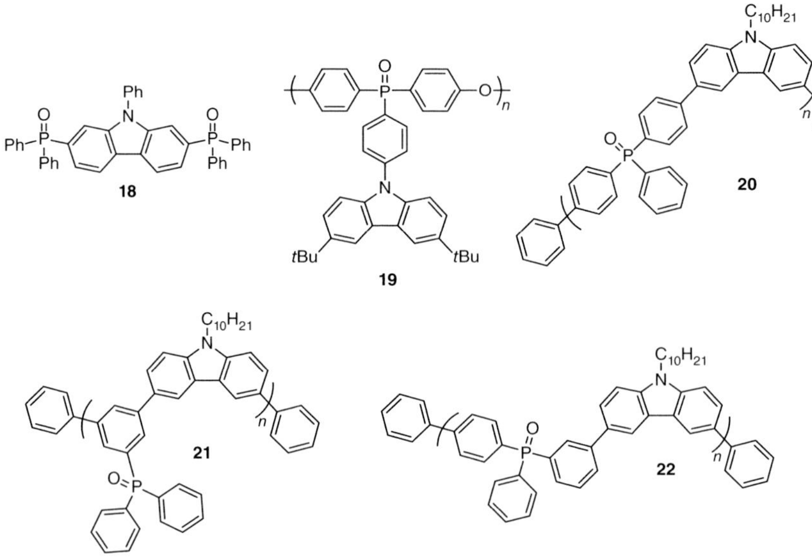

Figure 7.7 Selected hosts based on phosphine oxides and carbazole.

the carbazole also has a significant effect on the performance of the host. When the phosphine oxides were placed in the 2,7 positions, i.e. **18**, the quantum efficiency increased by more than 50%, and the power efficiency was significantly improved, compared to that of carbazole with phosphine oxide substituents in the 3,6 positions.

Different bipolar polymeric hosts **19–22** (Figure 7.7) based on phosphine oxide and carbazole scaffolds have been synthesized and studied. For example, Ding, Wang, and coworkers showed that polymer **19** presents a high thermal stability and a high E_T of 2.96 eV and an efficiency of 10.8% can be reached in an OLED device using FIrpic as the emitter [23]. H.-J. Jiang et al. showed that triphenyl phosphine oxide and carbazole-based polymers **20–22** can be used as host materials for green phosphorescent dopants of PhOLEDs [24]. They fabricated PhOLEDs with the configuration ITO/PEDOT:PSS (40 nm)/host (**20–22**): 5% $Ir(mppy)_3$ (40 nm)/1,3,5-tris(1-phenyl-1*H*-benzimidazol-2-yl)benzene (35 nm)/LiF (0.8 nm)/Al (100 nm). The PhOLED including polymer **20** as the host material shows electroluminescent performance with maximum current efficiency of 2.16 cd/A, maximum external quantum efficiency of 0.7%, maximum brightness of 1475 cd/m^2, and reduced efficiency roll-off of 7.14% at 600 cd/m^2. The best performance has been obtained with polymer **20**. This data indicates that the introduction of triphenyl phosphine oxide into the polycarbazole main chain through different linking modes has been an effective way to adjust the thermal stability, photoelectric properties, and device performance of polymer host materials.

Phosphazenes are compounds where P and N are linked by alternating single and double bonds. However, this compound should be considered as a ylide structure (—P^+—N^-—) and not as a π-conjugated framework (see Chapter 4) [25].

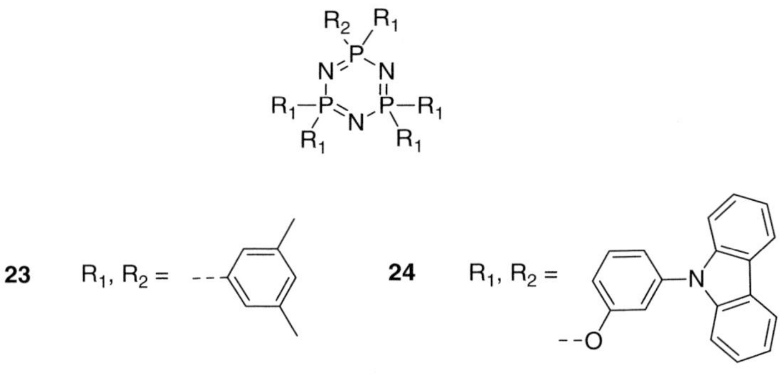

Figure 7.8 Phosphazene derivatives used in OLEDs.

This backbone is easily functionalized to afford linear polyphosphazenes or cyclic phosphazenes. These polyphosphazenes and cyclic phosphazenes are interesting candidates as hosts for triplet emitters since they possess a relatively high HOMO–LUMO gap. Simple aryl-substituted cyclic phosphazenes such as **23** (Figure 7.8) were tested as a matrix for Ir(dbfmi) emitters. Low efficiencies were obtained with these hosts despite balanced charge transport (power efficiency max = 7.6 lm/W) [26]. Carbazole-functionalized derivatives (for example **24**, Figure 7.8) were then designed [27]. The triplet energy level of this amorphous matrix was evaluated at 3 eV, which is compatible with FIrpic and $Ir(mppy)_3$. Multilayered OLEDs [ITO/PEDOT:PSS/poly-TPD/EML/BCP/BCP:Cs_2CO_3/Al] with FIrpic as the emitter showed twice the luminance/PCE (power conversion efficiency) of a PVK (poly *N*-vinylcarbazole) reference device that was prepared with this host. These two examples show that chemical diversity can be easily introduced on the phosphazene core to obtain hosts for PhOLEDs. It is particularly interesting to note that in both cases the matrices are amorphous and thermally stable.

7.2.3 Application as Emitting Materials

A large number of heteroatom-based oligomers and polymers presenting very interesting emissive or charge-transport properties [28], have been synthesized but only a few of them have been incorporated in devices. Most of the emitters used in OLEDs contain mainly phosphorus and silicon.

Phosphine oxide-based materials are interesting emitters because they are thermally stable and have good morphological stability. Furthermore, they incorporate a pyramidal phosphorus, which prevents π-stacking in the solid state and therefore promotes solid-state emission. A highly efficient light blue emitter **25** (Figure 7.9) was synthesized and used to construct a double-layer device utilizing **25** as the emitting, electron-transporting, and electron-injecting material. An η_{ext} of 4.3% was achieved with chromaticity CIE coordinates (0.15, 0.07) [29]. Ma, Yang, and coworkers synthesized a series of solution-processable linear oligomer emitters consisting of a phosphine oxide center connected to oligo-phenylene/fluorene cores and end-capped with *N*-phenylnaphthalen-1-amine groups [30]. They constructed solution-processed OLEDs utilizing

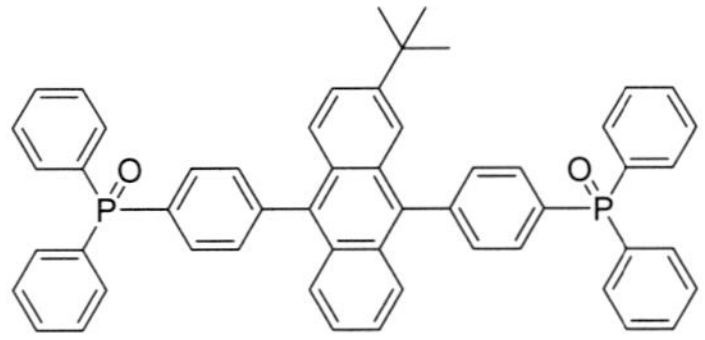

25

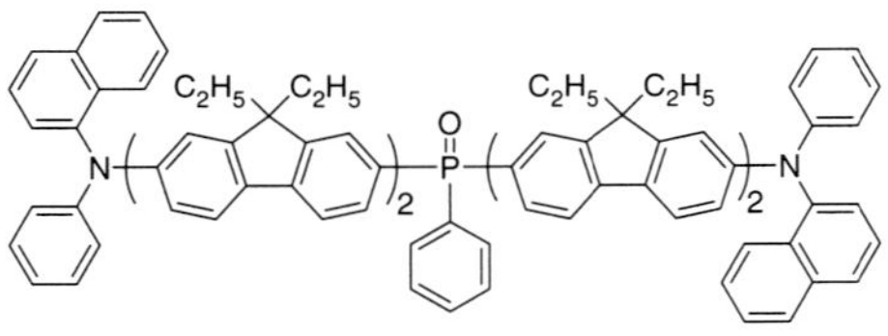

26

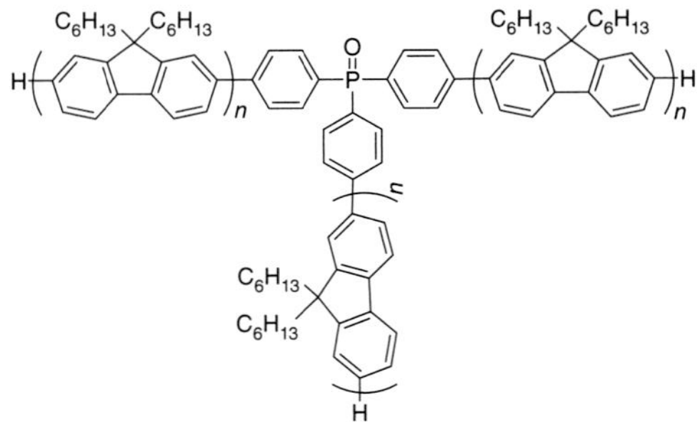

27 $n = 1$
28 $n = 2$
29 $n = 3$

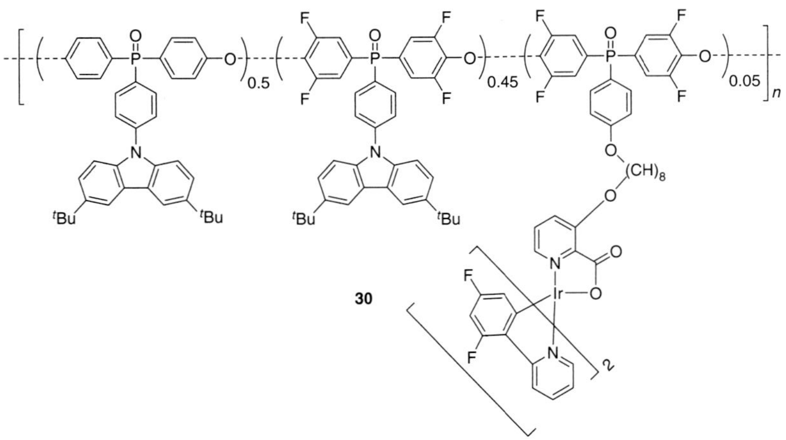

30

Figure 7.9 Examples of phosphine oxide-based emitters.

26 (Figure 7.9) as the emitter. A maximum current efficiency of 2.36 cd/A, a maximum power efficiency of 1.86 lm/W, and a maximum external quantum efficiency of 2.06% with CIE coordinates (0.15, 0.11) were achieved for the deep blue OLED. Later, Yang, Wu, and coworkers developed two-dimensional oligofluorenes connected to a central triphenyl phosphine oxide **27–29** (Figure 7.9) [31]. The device utilizing **29** as the emitter had the best performance, with a maximum current efficiency of 1.88 cd/A, and η_{ext} of 3.39% with CIE coordinates (0.16, 0.09). This example highlights the increase in efficiency based on 2D frameworks compared to their 1D counterparts. Ding, Wang, and coworkers generated highly efficient blue electrophosphorescent polymers composed of a fluorinated poly(arylene ether phosphine oxide) backbone with grafted carbazole and FIrpic units [32]. The best performing device utilized **30** (Figure 7.9) as the emitter and achieved a luminous efficiency of 19.4 cd/A, an η_{ext} of 9.0%, and CIE coordinates (0.18, 0.33). Although the majority of the phosphorescence comes from the FIrpic moieties, the poly(arylene ether phosphine oxide) backbone plays a definite role in the charge transport of the emissive polymer. In a further study, the authors introduced an additional yellow phosphorescent emitter [(fbi)$_2$Ir(acac)] into the polymer backbone and achieved a white organic light-emitting diode (WOLED) [33]. The simultaneous blue and yellow triplet emission allowed the generation of white EL, and the device had a luminous efficiency of 18.4 cd/A, an η_{ext} of 7.1%, and CIE coordinates (0.31, 0.43).

Several phosphole-based polymers have been synthesized and studied [28f–h]. They are efficient emitters and/or charge transporters but only a few of them have been used as emitters in OLEDs. Phosphole-based oligomers had been used for the first time, as EML in OLEDs in 2003 by Réau, Hissler, and coworkers [34]. For example, they demonstrated that compound **31** (Figure 7.10) can be used as an OLED emitter [35]. The devices employed in this study had the

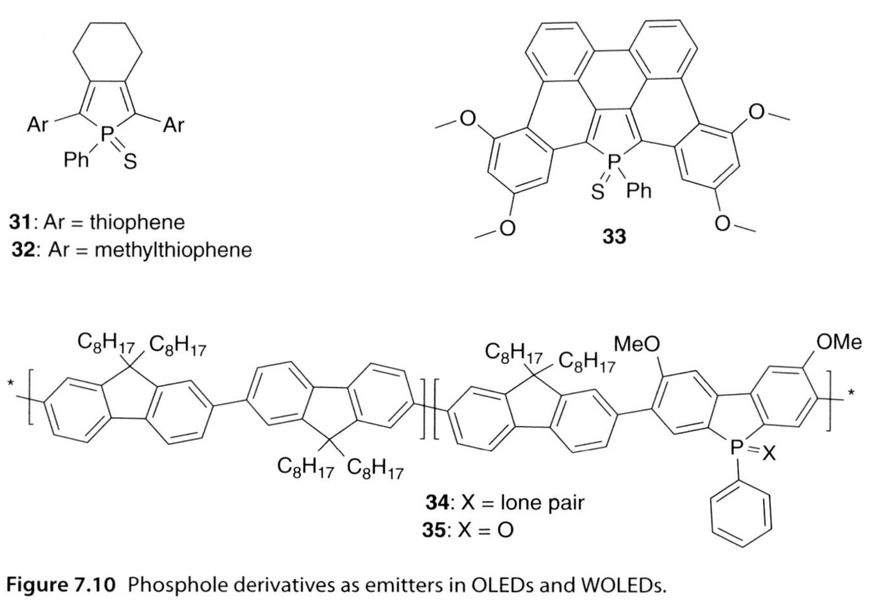

Figure 7.10 Phosphole derivatives as emitters in OLEDs and WOLEDs.

following configuration: [ITO/PEDOT:PSS/α-NPD/EML: **31**/Alq_3/Mg:Ag/Ag] using Alq_3 as an ETL and α-NPD as a HTL. The different organic layers were deposited by thermal evaporation under high vacuum. The thermal stability of the thioxo derivative **31** (Figure 7.9) allowed to get homogeneous thin films having an emission wavelength centered at 550 nm and good device performance (MB = 38 000 cd/m^2, η_{max} = 0.80). Since phosphole derivative **31** presents interesting hole-transport properties and an emission that overlaps with the absorption spectrum of the highly red fluorescent dye 4-(dicyanomethylene)-2-tert-butyl-6-(1,1,7,7-tetramethyl-julolidin-4-yl-vinyl)-4H-pyran (DCJTB), it was used as host for a red emitter [34, 35]. Although the dopant concentration (1.4 wt%) was not optimized, this approach led to an important enhancement of the η_{ext} (1.83%) and a maximum brightness of 37 000 cd/m^2 (λ_{max} = 617 nm).

Phosphole emitters also appeared as original materials for the construction of WOLEDs. Since organophosphorus compounds are redox and thermally stable, and present an orange emission, Réau, Hissler, et al. co-evaporate **32** (Figure 7.10) with a blue emitter, 4,4′-bis(2,2′-diphenylvinyl)biphenyl (DPVBi) for the construction of the active layer in WOLEDs. Effectively, white emission could potentially be obtained by combining orange and blue emissions, if the ratio of the two colors is correct [36]. It is important to note that, the absorption spectra of phosphole-thiophene derivatives span the 320–524 nm range, and overlap with the emission spectrum of DPVBi to a large extent, potentially allowing energy transfer from the DPVBi host to the phosphole dopants. First, Réau, Hissler, and coworkers developed a multilayer WOLED on a glass substrate with a configuration of [ITO/CuPc (10 nm)/α-NPB (50 nm)/EML: doped-DPVBi (50 nm)/Alq_3 (10 nm)/LiF (1.2 nm)/Al (100 nm)] [37]. Doping of the blue matrix with 0.2% of compound **32** leads to an OLED exhibiting a small turn-on voltage (5.2 V) with current and power efficiencies of 7.0 cd/A and 2.3 lm/W, respectively. Its electroluminescent spectrum presents well-balanced emission characteristics of the DPVBi (452 nm, full width at half maximum (FWHM) = 70 nm) and of the dopant **32** (548 nm, FWHM = 115 nm). The resulting CIE coordinates (0.31, 0.39) are close to those of pure white light (0.33, 0.33) and are independent of the driving current.

Fully planarized phosphorus-containing polycyclic aromatic hydrocarbon [38] (**33**, Figure 7.10) presents high luminescence in the orange-red region in dilute solutions and this fluorescence remains high in a solid matrix. This compound has been used as orange dopant in a blue emitting matrix for the development of WOLEDs [38]. The WOLEDs having the configuration [ITO/CuPc/α-NPD/EML/DPVBi/BCP/Alq_3/LiF/Al] exhibit a white emission as evidenced by the CIE coordinates when the EML is generated by co-subliming **33** (0.8 wt%) with the α-NPD. The device utilizing emitter **33** in the EML (EL = α-NPD: **33** [1.1 wt%]) had a turn-on voltage of 5.55 V, brightness of 1122 cd/m^2, η_{ext} of 1.67%, power efficiency of 0.96 lm/W, current efficiency of 3.69 cd/A, and the CIE coordinates (0.32, 0.37).

The first investigation of the potential application of low molecular weight dibenzophospholes in OLEDs by vacuum sublimation was unsuccessful, as the device characteristics showed instability with increasing driving current [37]. In 2008, Huang and coworkers reported the synthesis of phosphafluorene (that is, dibenzophosphole) copolymers (**34** and **35**, Figure 7.10) [39]. Since, both polymers were emissive and had high decomposition temperatures (400 °C), they were employed in a polymeric light-emitting diode (PLED) devices [configuration: ITO/PEDOT:PSS/polymer/Ba/Al]. The device with **34** as the EML, presented blue EL (CIE coordinates: (0.21, 0.24)) with brightness up to 1423 cd/m^2 whereas **35** exhibited white emission (CIE coordinates: (0.34, 0.36)), but with lower brightness (142 cd/m^2). The nature of the substituents on the P-atom of the phosphafluorene derivative has a strong impact on the properties of the polymer, changing the blue EL to a white one. All the described examples of phosphole-based emitters clearly highlight the potential of these materials for the development of OLEDs and WOLEDs.

The cyclic phosphazenes core is an interesting platform, which can be easily functionalized by substituents having specific optical properties. For example, pyrene (pyrenic emitters **36** and **37**, Figure 7.11) and phosphorescent complexes (Ir-complex **38**, Figure 7.11) have been used to functionalize the cyclic phosphazene core [40]. They all form amorphous thin films with a blue-green emission. Single-layer OLEDs displayed low performances due to the lack of electron-transporting ability of the layers. The use of an ETL (2,2′,2"-(1,3,5-Benzinetriyl)-tris(1-phenyl-1-H-benzimidazole)) (TPBi), improved the properties with the configuration [ITO/PEDOT:PSS/**36**–**38**/TPBI/Ba:Ag]. The best performing compound was **36** (11 000 cd/m^2 at 12 V, η_{ext} of 0.72% at 8 V). Phosphorescent **38** has also been incorporated in a PVK matrix, to afford a device [ITO/PEDOT/PSS/PVK:PBD:**38**/TPBI/Ba/Si], with an η_{ext} of 7%. A series of phosphazene-containing dendrimers decorated in the periphery by polyaromatic moieties (e.g. anthracene, pyrene) have been synthesized by Caminade, Majoral, and coworkers [41]. Films prepared from the pyrene-decorated dendrimers showed a characteristic excimeric emission. The fourth generation dendrimer **39** (Figure 7.11) diluted in a PVK matrix was explored in OLED devices. However, the luminance appeared to be low and the threshold voltage very high (>20 V).

Another interesting class of compounds concerns the functionalization of π-systems with silicon atoms, allowing a control of the bandgap and the emission color of the polymers.

Several silyl-substituted poly(*p*-phenylene vinylenes) have been developed and possess high photoluminescence quantum yields (PLQYs). For example, Kim et al. [42] synthesized Si-containing polymers **40** (Figure 7.12) having an emission centered at 515 and 550 nm with a PLQY of 60%. In a single-layer OLED (ITO/**40**/Al), the EL emission spectrum is similar to that recorded in solution and η_{ext} reaches 0.05%. The η_{ext} can be improved to 0.2% by using a hole-blocking and electron-injecting layer of 2-(4-biphenylyl)-5-(4-tert-butyl-phenyl)-1,3,4-oxadiazole (PBD) between the light-emitting layer and the cathode. The introduction of a bulky trialkylsilyl group on PPV (polymer **41**, Figure 7.12) increases the torsion angle between the phenylene and the vinylene systems leading to a blue shift of the emission [43]. In the device, polymer **41** exhibits light green EL emission with an η_{ext} of 0.3% [44]. Further chemical modifications have been made on the PPV core in order to establish structure–properties relationship. For example, silyl-disubstituted PPVs having different side chain lengths ranging from C1 to C18 have been synthesized and studied. The silyl-disubstituted

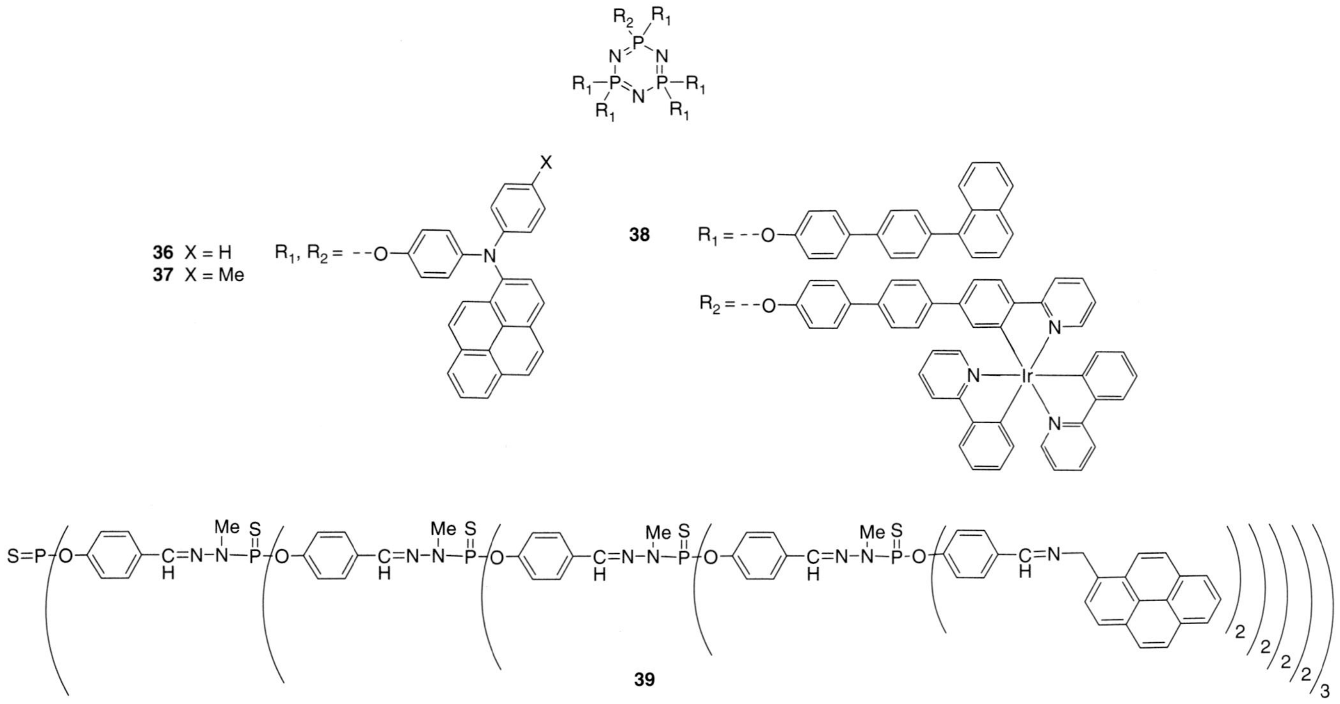

Figure 7.11 Phosphazene derivatives used as emitter in OLEDs.

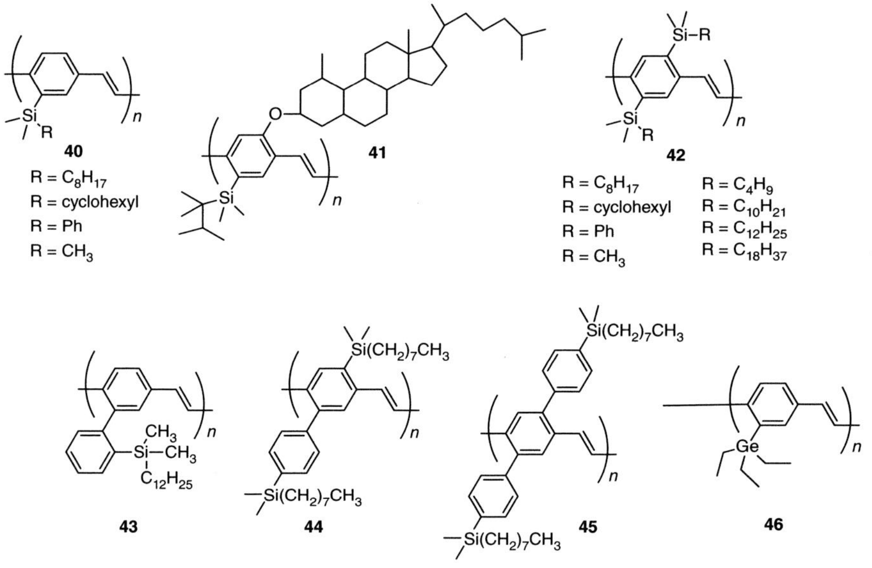

Figure 7.12 Silyl- and germyl-substituted PPVs.

PPV **42** having long chains presents better processability and film-forming properties. In a single-layer OLED (ITO/**42**/Al), η_{ext} reaches only a value of 0.05% [45]. The introduction of more sterically demanding groups like cyclohexyl or phenyl also leads to higher glass temperature and processability but PLEDs with these polymers such as ITO/PVK/**42**/Al emits green-yellow light with a low $\eta_{ext} = 0.08\%$ [46]. PPV derivatives **43–45** (Figure 7.12) possessing pendant phenyl group substituted with trialkylsilyl substituents in different positions have also been synthesized and studied. The polymer including trialkylsilyl substituent in meta position of the pendant phenyl **43** [47] has been used as an emitter in the OLED with a configuration: ITO/PEDOT/**43**/Al:Li. Moderate performance ($\eta_{ext} = 0.08\%$, maximum of brightness = 570 cd/m^2 at 43 V) have been recorded due to a high-energy barrier between the HOMO band of **43** and the ITO/PEDOT work function. The introduction of trialkylsilyl substituents in the ortho position of pendant phenyl rings does not allow to clearly establish structure/properties relationships since polymers **44** and **45** present a maximum of luminance of 3098 and 383 cd/m^2, respectively, with similar OLED configuration [48].

The synthesis of PPV including a heavier atom like germanium has been undertaken. A germylated-PPV **46** (Figure 7.12) has been synthesized by the thermoconversion route since classical reactions such as Gilch polymerization usually used for the formation of PPVs give only insoluble material. The device having the following configuration: ITO/**46**/LiF/Al emits a green light with a moderate efficiency of 0.015 lm/W and a maximum brightness of 600 cd/m^2 [49].

A better control of the HOMO–LUMO gap and the EL properties of PPVs can be achieved with the formation of copolymers. The synthesis of polymers **47** (Figure 7.13) has been achieved by Jin et al. [47]. The HOMO and LUMO energy

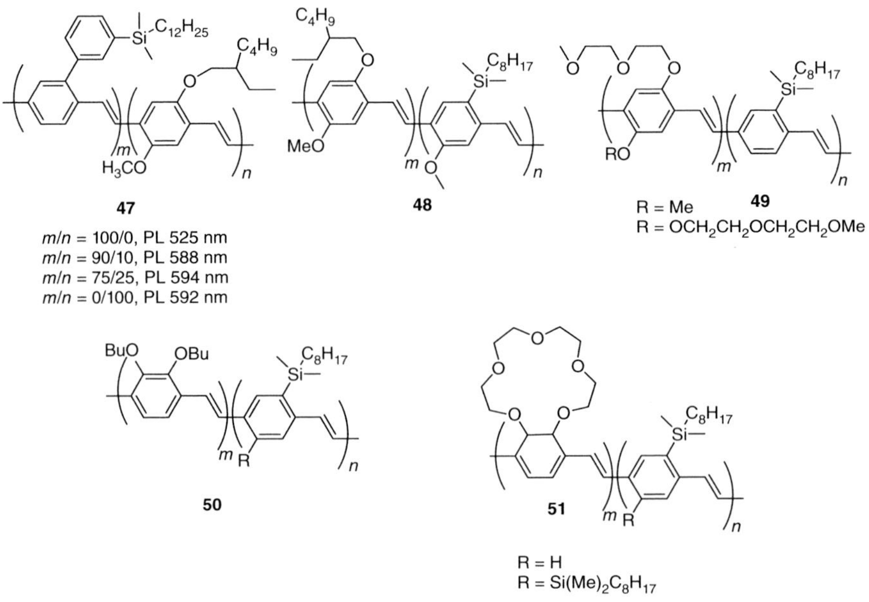

Figure 7.13 Chemical structures of PPV co-polymers alkoxy- and trialkylsilyl-substituted phenylenes.

level and the emission color was tuned by adjusting the co-monomer ratio. It was demonstrated that the performance, namely the turn-on voltage of the OLED is directly related to the gap between the HOMO level of the polymer and the work function of the ITO electrode. The OLED (ITO/**47**/Al:Li) incorporating polymer **47** as the emitter presents a turn-on voltage of 2.3 V, a maximum of brightness of >19 000 cd/m^2, η_L of 2.9 lm/W: performance which are higher than the OLED performance built with homo-polymers.

Another way to tune the HOMO level of these polymers is to introduce electron donor substituents. Holmes, Friend, and coworkers synthesized PPV copolymers containing alkoxy- and trialkylsilyl-substituted phenylene rings [50] in a random distribution **48–51** (Figure 7.13). The best performance was observed for devices made with **50**. EL efficiencies were up to 0.72 cd/A with a maximum luminance of 1384 cd/m^2 at 12 V and turn-on voltages of 4.0 V. In polymer **49**, the presence of oligo(ethylene oxide) pendant groups allowed to create light-emitting electrochemical cells (LECs: ITO/**49**/LiOTf/Al) with an efficiency of 0.5 lm/W.

There are several ways to control the bandgap of a polymer. One approach is to introduce bulky substituents to distort the polymer conjugated backbone (vide supra), another way is by the precise control of the conjugation lengths. For example, Kim et al. introduced a silicon atom in PPV block polymers to confine the conjugation length allowing to obtain blue EL materials **52–55** (Figure 7.14) [51]. Depending on the structure of the polymer **52–55**, emission can be tuned between 410 and 520 nm. Single-layer OLEDs (ITO/polymer/Al) have been constructed with polymers **52** and **54**, which had a turn-on voltage of around 7 V. The same approach has been used with polythiophenes.

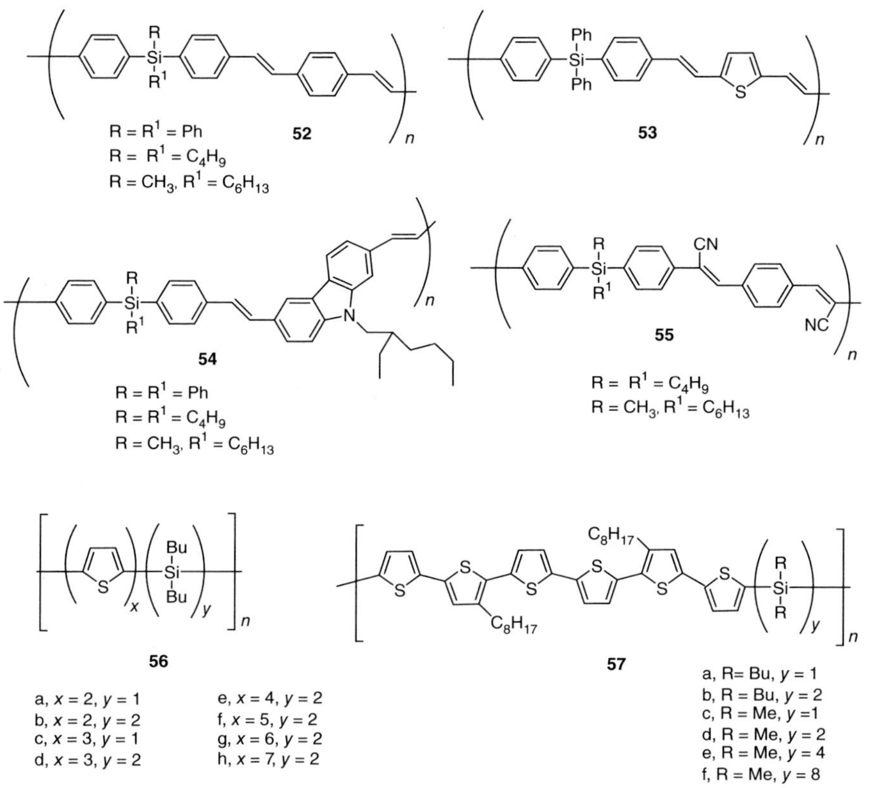

Figure 7.14 Chemical structures of PPV and polythiophene including silicon atoms.

Hadziioannou and coworkers controlled the conjugation length of oligothiophene blocks by using silanylene units (polymers **56a–d** and **57a–f**, Figure 7.14) as conjugation interrupters allowing a tuning of the emission from blue to orange [52]. Later, Yoshino et al. were able to change the EL color from green to red by increasing the oligothiophene block length (polymer **56d–h**) [53].

Besides phosphole rings, silole rings can also be used as building blocks for the development of emissive materials. Polyphenylsiloles are weakly emissive in diluted solutions but show intense luminescence in the aggregated state. This exalted emission results from stiffening of their molecular structure due to restriction of the intramolecular rotations (RIRs) of the peripheral rotors that promote the radiative decay of the excited state [54]. This observation prompted Chen et al. to develop silole-substituted polyacetylenes (PAs) **58–60** (Figure 7.15) [55]. As expected, the fluorescence of these polymers **58–60** is very low ($\phi_{PL} < 0.5\%$) in solution but the PLQY is increased by 20–50 times in solid state for polymers **59** and **60**. In the case of polymer **58**, an energy transfer is observed from the silole to the PA generating a red emission (λ_{El} = 664 nm). The introduction of a longer link between the silole unit and the polymer allows to observe the specific emission of the silole at 512 nm. The development of single-layer devices with these polymers gives only

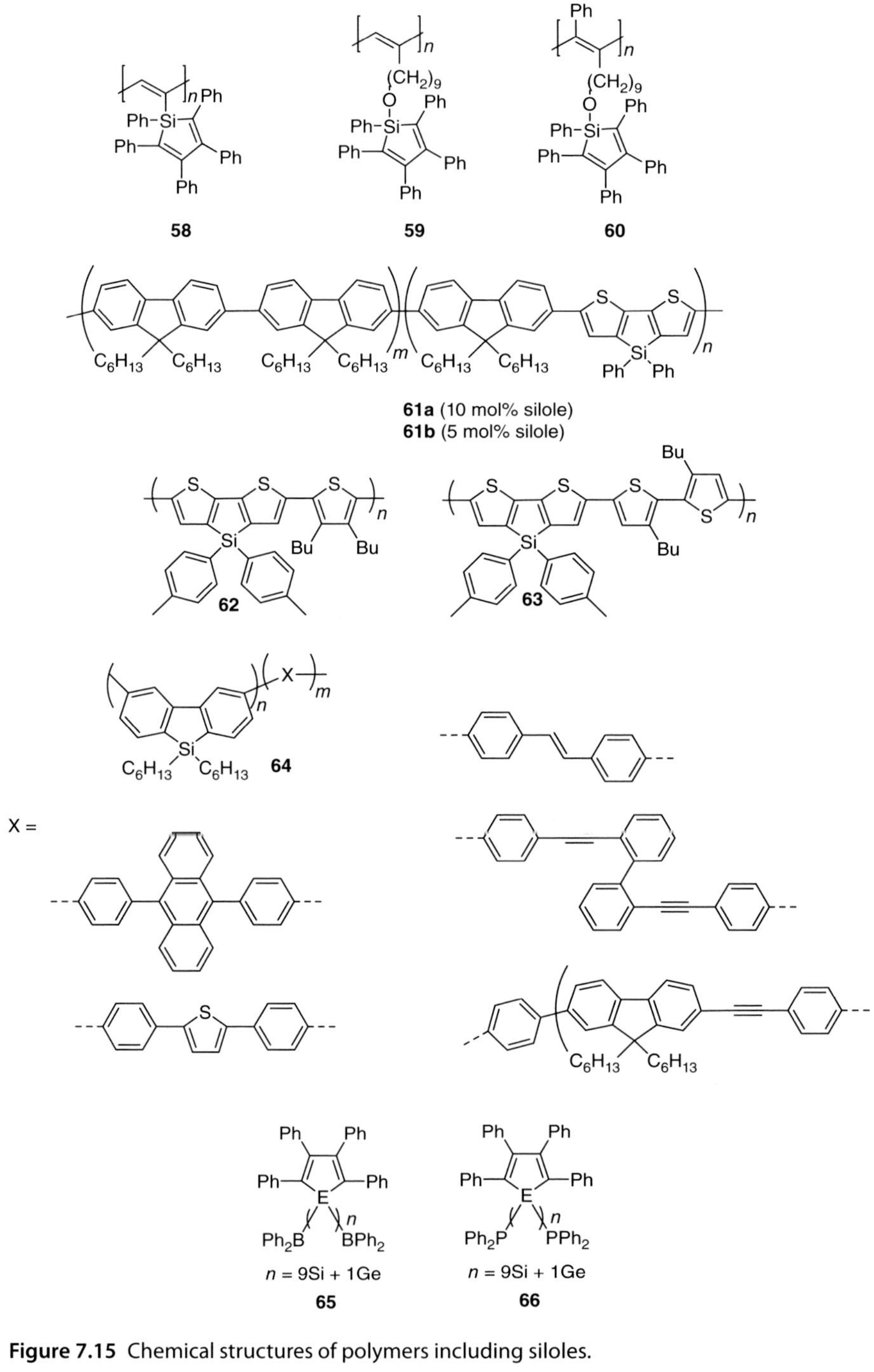

Figure 7.15 Chemical structures of polymers including siloles.

moderate EL efficiency (0.013%) due to limited charge-transport processes. The performance can be increased with the development of multi-layer OLEDs (ITO/60-PVK/BCP/Alq_3/Al). A current efficiency of 1.45 cd/A, a η_{ext} of 0.55%, and maximum of brightness of 1118 cd/m^2 can be recorded.

Siloles can also be included in the polymeric backbone **61–64** (Figure 7.15) [56]. For example, Liu et al. [56a] used two random silole-containing copolymers as emitting layer in a double-layer device (ITO/BTPD-PFCB/**61a,b**/Ca). In the case of polymer **61a**, a maximum brightness of 25 900 cd/m^2 and a η_{ext} of 1.64% are recorded. When the 4,4-diphenyldithienosilole unit is incorporated in polythiophene (polymers **62** and **63**), only weak performances are observed for single-layer devices (ITO/polymer/Mg–Ag) [56b]. A series of 3,6-(dialkyl)silafluorene-based polymers **64** [56c] (Figure 7.15) presenting a wide bandgap and ultraviolet light emission have been synthesized and were used as emitters in OLEDs. The monomer containing vinylene, anthracene, and tri-arylamine moieties incorporated into the poly(3,6-silafluorene) backbone can form efficient deep-blue emitting copolymers with EL efficiency of 1.1–1.9%. Furthermore, electroluminescent poly(silole-co-germole)s-containing borane/phosphine-ends **65** and **66** (Figure 7.15) have been synthesized in high yield and have similar emission properties (around 520 nm) [57]. The emission color of **65/66** mixture is green and the maximum brightness of the device is 2760 cd/m^2 with a luminous efficiency of 0.67 lm/W. In this case, the boranyl/phosphine end groups did not show an appreciable B-P-dative effect on the luminescent properties of poly(silole-co-germole) chain.

7.3 Photovoltaic Cells (Organic Solar Cells [OSCs] and Dye-Sensitized Solar Cells [DSSCs])

Fossil fuel alternatives, such as harvesting energy directly from sunlight using photovoltaic technology, are one of the ways to address growing global energy needs [58]. Since the discovery of silicon solar cells in the 1950s, tremendous advances in solar cell research and development have been made. DSSC and organic solar cell (OSC) devices are promising alternatives for producing clean and renewable energy, as there is the potential to fabricate them onto large areas of lightweight flexible substrates by solution processing at a low cost.

Actual organic/hybrid solar cells are mainly divided into two categories: DSSCs in which organic dyes are used for absorption of light (the charge transport is insured by an inorganic semiconductor such as TiO_2) followed by OSCs in which two organic materials (a donor and an acceptor) are used for both absorption of light and charge transport in a p–n junction.

Generally, a *n*-DSSC [1f, g] device consists of four main components: a photoanode of transparent fluorine-doped tin oxide (FTO) glass covered with a wide bandgap nanocrystalline n-type semiconductor (TiO_2, ZnO, SnO_2, Nb_2O_5, etc.); a sensitizer (dye molecules, usually a push–pull organic dye or coordination complex) anchored on the surface of the semiconductor; a redox-coupled electrolyte (cobalt complexes, I^-/I_3^-,...); and a counter electrode of FTO coated

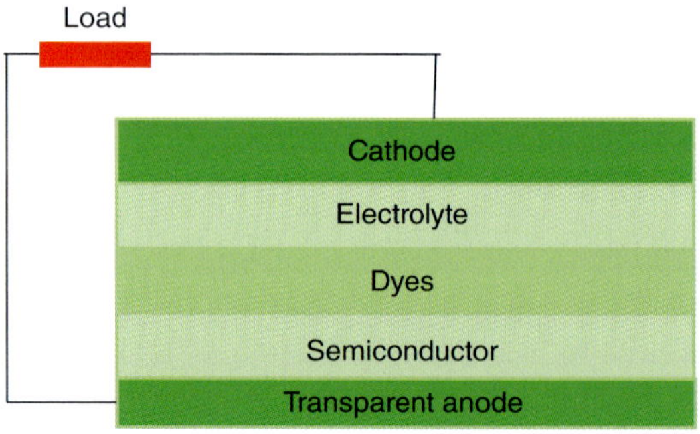

Figure 7.16 DSSC structure.

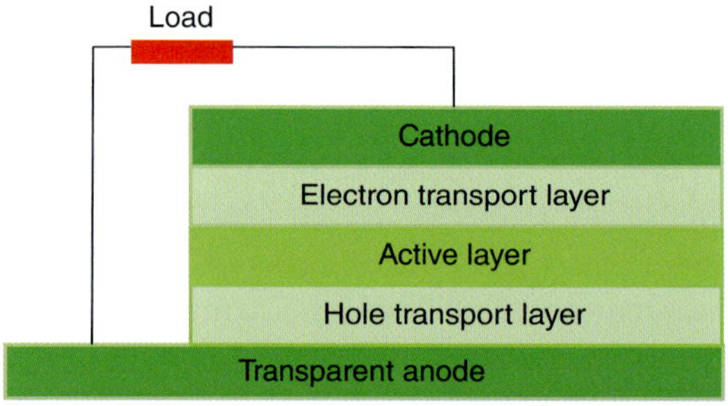

Figure 7.17 Structure of a bulk heterojunction OPV device.

with platinum nanoparticles (Figure 7.16). Under irradiation, the chemisorbed dye is photo-excited to create exciton pairs, which are rapidly split at the nanocrystalline surface. The electrons are injected into the conduction band of the semiconductor and the holes, staying on the dyes, are recovered by the redox mediator in the liquid electrolyte. The oxidation state of the redox mediator is further recovered by electrons transferred from the counter electrode. A closed circuit is thereby established to continuously convert the solar light to electricity.

If the two semiconductors (n- and p-type) are fully organic, the device is referred to as an OSC. A bulk heterojunction (BHJ) OSC device consists of an active layer composed of a nanoscale blend of donor and acceptor materials sandwiched between two electrodes having different work functions (Figure 7.17) [1h, 59]. The general working principle first involves the photoexcitation of the donor material by the absorption of light to generate excitons. After the exciton dissociation/charge separation process at the donor–acceptor interface, the separated electrons and holes move toward their respective electrodes, driven by either the built-in electric field or the charge carrier concentration gradient, leading to a photocurrent or photovoltage.

For an organic material to be suitable for OSC applications, the compound must fulfill certain requirements. It must simultaneously possess strong absorption ability in the visible–near infrared spectral range, have suitable HOMO–LUMO energy levels to transfer electrons to the n-type semiconductor (or p-type in the case of an acceptor), have high hole or electron mobility and good film-forming properties in the case of OSC. As in OLEDs, additional charge-transport layers (such as an ETL or a cathode buffer layer) can be inserted into the devices.

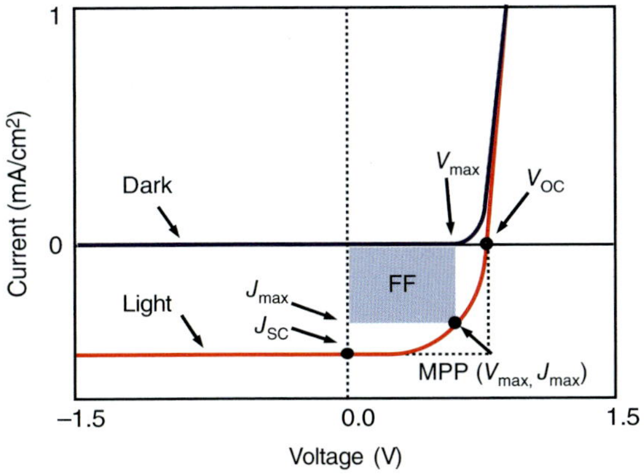

$$\eta = \frac{P_{out}}{P_{in}} = FF\,\frac{V_{OC} \times J_{sc}}{P_{in}}$$

$$FF = \frac{V_{max} \times J_{max}}{V_{OC} \times J_{sc}}$$

Figure 7.18 *I–V* curve of a PV device and associated characteristic values.

To evaluate the performance of a photovoltaic (PV) device, some characteristic values are compared. They are all measured from the I/V curve as described in Figure 7.18. Those values are the PCE η (in %), the short-circuit current density J_{sc} (mA/cm^2), the open circuit voltage V_{oc} (in V) and the fill-factor, FF (without unity, maximal value = 1 or 100%).

7.3.1 Dyes for Dye-Sensitized Solar Cells (DSSCs)

In 2010, Matano, Imahori, and coworkers used a phosphole-based oligomer in a DSSC [60]. Their 1-hydroxy-1-oxodithienophosphole (**67**, Figure 7.19) was used as an accepting moiety, as well as a novel anchoring group. A PCE (η) of 1.8% was measured, together with a short-circuit current density (J_{sc}) of 7.4 mA/cm^2, an open-circuit voltage (V_{oc}) of 0.46 V, and a FF of 0.54. In 2014, the same authors developed a new push–pull based phosphole, featuring a triarylamine donor, a terthiophene π-system, and a phosphole sulfide acceptor, bound to the semiconductor through a carboxylic acid **68** (Figure 7.19) [61]. The best PCE η was 5.6%, (J_{sc} of 12.5 mA/cm^2, a V_{oc} of 0.63 V, and an FF of 0.70). Even if the efficiencies remain moderate, these examples highlight the versatility of using phospholes oligomers for future dyes in DSSCs.

Following the same strategy, silole-based oligomers were also prepared [62]. The best efficiency achieved with this strategy used compound **69** (Figure 7.20) featuring a triarylamine donor group, a π-backbone composed of EDOT-

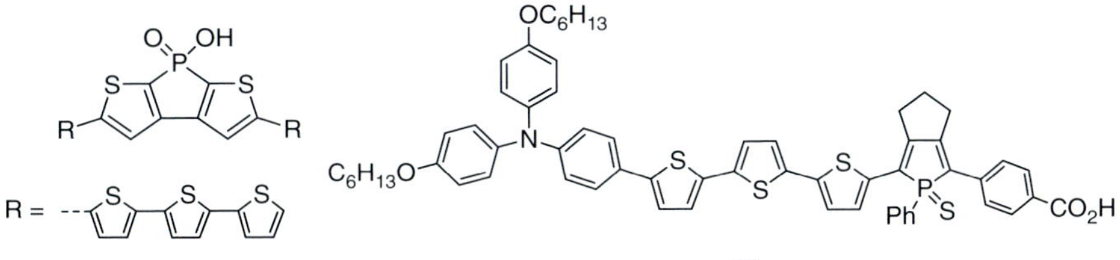

67 **68**

Figure 7.19 Phosphole-based dyes utilized in DSSCs.

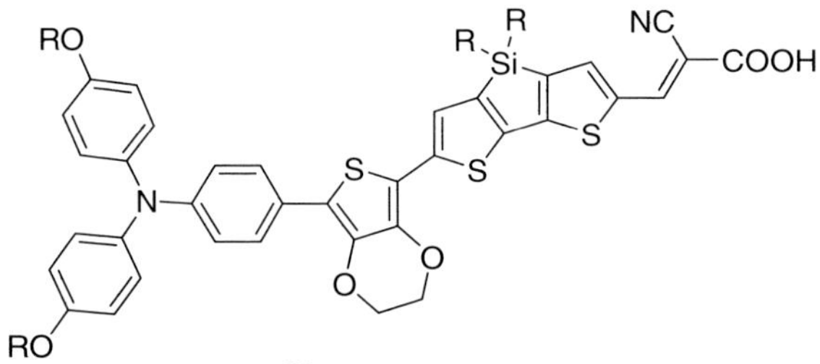

69

Figure 7.20 Silole-based dyes utilized in DSSCs.

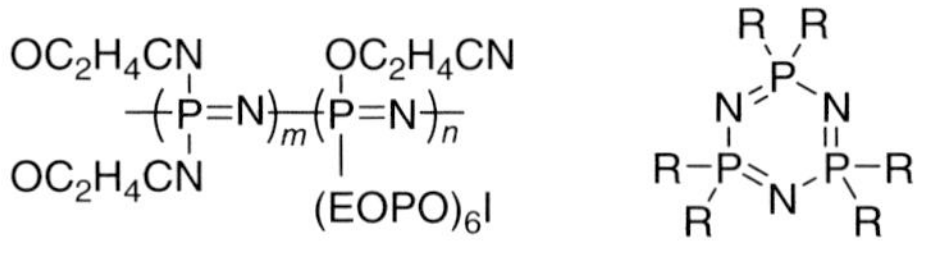

70

71: R = O(CH$_2$CH$_2$O)$_2$CH$_3$

Figure 7.21 Structures of phosphazenes. EOPO, ethylene oxide-co-propylene oxide.

dithienosilole (3,4-Ethylenedioxythiophene) and a cyanoacrylic anchoring group [63]. A PCE of 10% could be obtained with a volatile electrolyte J_{sc}: 17.94 mA/cm^2, V_{oc}: 0.73 V; FF: 0.73).

Some P-containing materials were also tested as the electrolyte in DSSCs. To this end, phosphazenes were selected. Polyphosphazenes were first tested as electrolytes in Li ion batteries [64]. Then, they were used as electrolytes in solid-state DSSCs. Electrolytes based on **70** (Figure 7.21) displayed good penetration into TiO$_2$ and a PCE of 2.7% was obtained [65]. Pore penetration of phosphazenes was also studied by Allcock and coworkers [66]. They showed that "small" polyphosphazenes such as **71** (Figure 7.21) help the penetration into the pores. However, a low PCE was obtained.

7.3.2 Donors in Organic Solar Cells (OSCs)

The first example of a phosphole-based copolymer used as a donor toward fullerene as the acceptor in an OSC was published by Matano et al. [67] Polymers **72–73**, (Figure 7.22) differing in the substituent on the P-atom (P=NSO$_2$C$_8$H$_{17}$ or P=O) have been synthesized and used as absorber and electron donor in solar cells. Devices [ITO/PEDOT:PSS/**73**/PC$_{71}$BM/Al] including polymer **73**, showed low efficiency (PCE: 0.65%). However, the PCE of the device with **73** was an order of magnitude higher than that with **72**, showing that the P-substituents play a key role. In 2015, it was demonstrated that the use of dithienophosphole

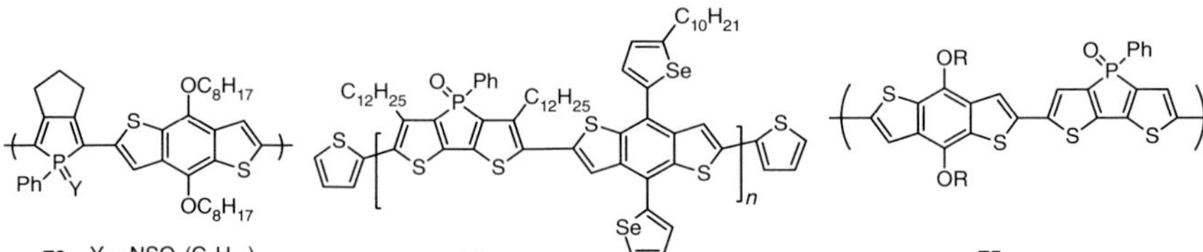

72 Y = NSO$_2$(C$_8$H$_{17}$)

73 Y = O

74

75

Figure 7.22 Phosphole-containing polymers used for OSC devices.

oxide-containing copolymers could generate highly efficient materials for OPV [68]. A BHJ-incorporating polymer **74** (Figure 7.22) as the donor and $PC_{71}BM$ as the acceptor were prepared [ITO/PEDOT:PSS/polymer:$PC_{71}BM$ (1 : 4, weight ratio)/Ca/Al]. The best device performance had a PCE of 7.08% (V_{oc} = 0.85 V, J_{sc} = 14.8 mA/cm^2, FF = 56.3%). It was highlighted that the polarizability of the dithienophosphole oxide moiety contributes to the high-excited state dipole moment, and thus facilitates the charge transfer to the acceptor. D–A copolymers with similar structure were also prepared (**75**, Figure 7.22) but with lower PCE in the devices [69]. As shown in the whole Section 7.3.2, it must be kept in mind that the PCE in OSC depends on many factors (electronic/structural properties of the compounds, manufacturing processes of the cells, use of additives, etc.) and it is difficult to make clear structure–properties relationships between the different polymers.

Silole-based oligomers and polymers were also used as donors in BHJ solar cells together with fullerene derivatives. In particular, the use of donor–acceptor (D–A)-based copolymers, where dibenzosiloles/dithienosiloles are used as the donor and various acceptors were used as the acceptor, were found to be efficient. Hence, this D–A structure allows for efficient absorption of the UV–vis-NIR (near infrared) radiations, insuring good light harvesting properties [70]. Cao and coworkers prepared low-bandgap polymer **76** (Figure 7.23) based on a dibenzosilole-thiadiazole structure. The BHJ with PCBM ([6,6]-phenyl C_{61} butyric acid methyl ester) as the acceptor led to a PCE of 5.4% (V_{oc} = 0.90 V, J_{sc} = 9.5 mA/cm^2, FF = 51%) [71]. The mobility of the polymer (10^{-3} cm^2/(V s)) appeared to be three times higher than that of its analog when the SiR_2 part is replaced by CR_2. The replacement of the biphenyl unit of the dibenzosilole by a bithiophene gives a redshift of the absorption. The D–A copolymer **77**, where the acceptor is a benzothiadiazole was prepared by Yang and coworkers (Figure 7.23) [72]. This low-bandgap polymer possesses good hole mobility (3×10^{-3} cm^2/(V s)). When $PC_{71}BM$ was used as the n-semiconductor, a PCE of 5.1% was obtained in the BHJ (V_{oc} = 0.68 V, J_{sc} = 12.7 mA/cm^2, FF = 55%). Photocurrent is generated over the whole visible range showing that both donor and acceptor absorption allows to fully cover the visible range. The modification of the acceptor in polymer **78** (Figure 7.23), where the thiadiazole is replaced by a thienopyrroledione, slightly increased the bandgap and the BHJ with $PC_{71}BM$ reached 7–8% PCE [73, 74]. Several polymers based on this backbone were independently prepared and the PCE appeared to be sensitive to both aliphatic chains on the silicon atom, as well as to the experimental conditions (solvent of evaporation, concentration, annealing temperature, additives, ...). However, this is not specific to siloles as BHJ efficiencies are known to depend on many interdependent factors linked with the preparation of the cells. When naphtha[2,3-c]thiophene-4,9-dione or polybithiopheneimide were used as the acceptor (polymers **79–80**, Figure 7.23), the efficiencies were still very high (PCE: 5.2%/6.4%, respectively) [75, 76].

The structural modifications are not limited to the accepting part. The donor can also be modified as in the silicon-bridged dithienocarbazole-based polymers (**81**, Figure 7.23) [77]. Here again, good electron mobilities (up to 0.1 cm^2/(V s)) and high PCE in BHJ devices with $PC_{71}BM$ as the acceptor (η = 5.2%; V_{oc} = 0.82 V, J_{sc} = 11.1 mA/cm^2, FF = 57%) were measured.

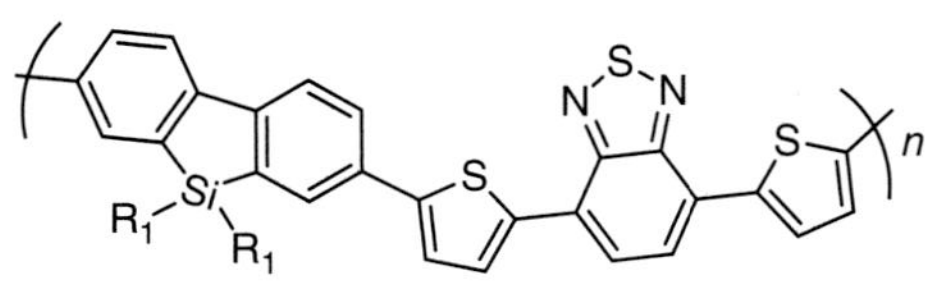

76

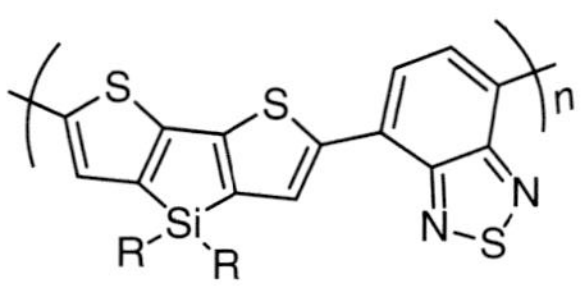

77

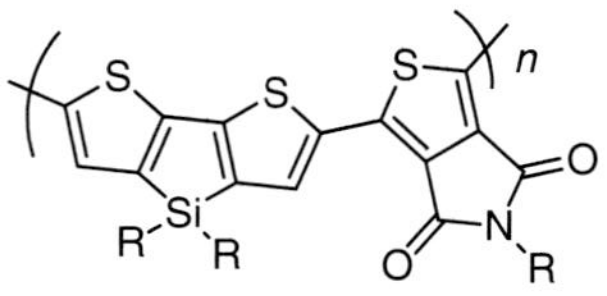

78

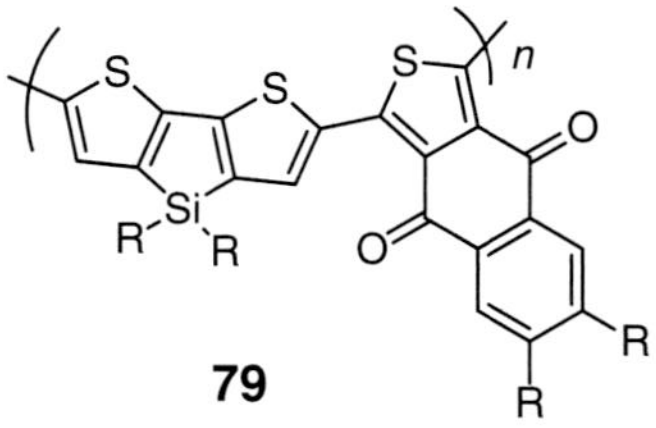

79

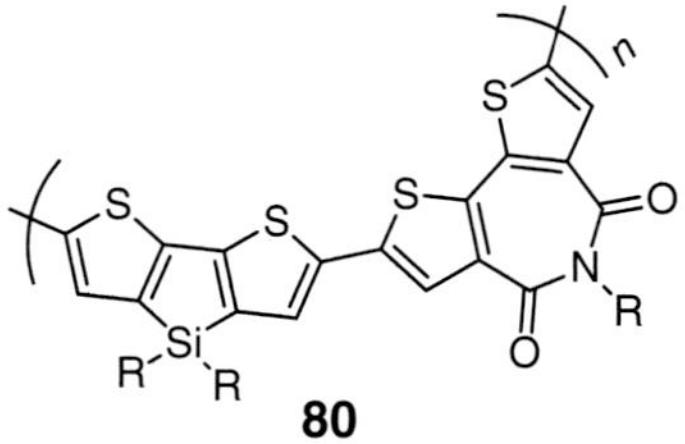

80

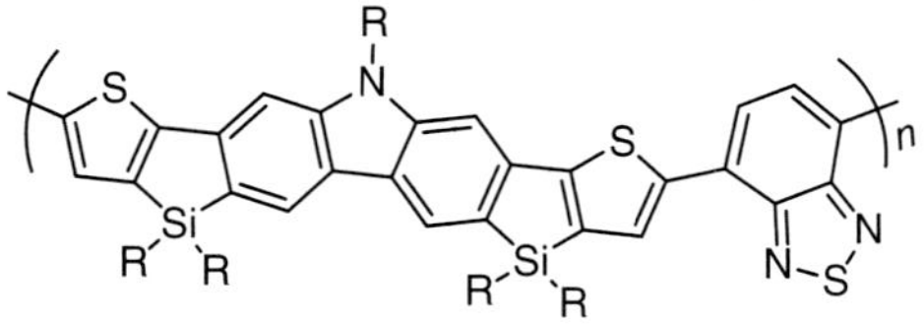

81

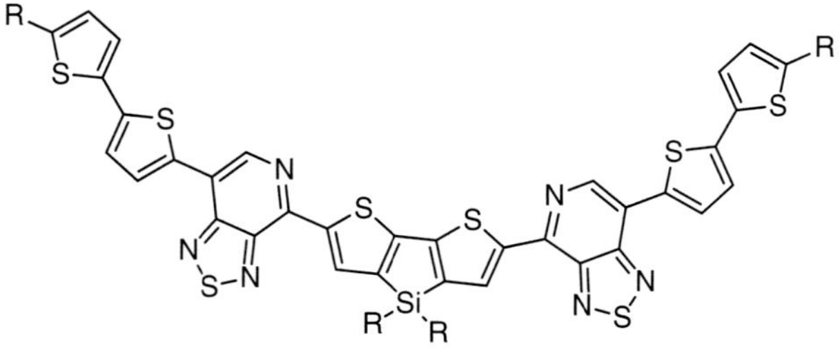

82

Figure 7.23 Si-based polymers used in OSCs.

These examples (this selection does not intend to be exhaustive) clearly show that introducing low-bandgap polymers with dithienosiloles into the backbone is an efficient strategy for preparing OPV materials.

Beside polymers, silole-containing oligomers were also prepared for OPV applications. The groups of G.C. Bazan and A.J. Heeger have been particularly active and successful in this area. In 2011, they reported a solution processable dithienosilole-based oligomer (**82**, Figure 7.23). In a BHJ device with $PC_{71}BM$, a PCE of 6.7% was obtained using a very low amount of additives (V_{oc} = 0.78 V, J_{sc} = 14.4 mA/cm^2, FF = 59%). Following this study, they reported several other examples of dithienosilole oligomers with efficiencies in the same order of magnitude [78, 79].

Similarly to siloles, the first examples of Ge-based conjugated polymers used in optoelectronic devices were based on the dibenzogermole scaffold (**83**, Figure 7.24) [80]. The compounds displayed decent hole mobility measured in OFET devices (0.04 cm^2/(V s)) and good on/off ratio (10^6). Their efficiency in OSC with $PC_{71}BM$ was found to be 1.5% (V_{oc} = 0.76 V, J_{sc} = 4.1 mA/cm^2, FF = 62%). Here also, the use of dithienogermole structures allowed to increase the efficiency of the devices. Homopolymers based on the dithienogermole monomer were first prepared [81]. They showed broad UV–vis absorption 500 nm < λ_{max} < 600 nm) and HOMO–LUMO level compatibles with applications in optoelectronics. Hole mobility in an OFET device was observed to be moderate (μ_{max} = 2×10^{-3} cm^2/(V s)). The corresponding BHJ with $PC_{71}BM$ were also moderate (PCE = 1.7%). The rather low charge transport can be a reason for this efficiency. The first example of D–A copolymer based on dithienogermole was prepared by So, Reynolds, and coworkers [82]. The alternating copolymer **84** absorbs until 735 nm which is similar to its Si-analog, but possesses a slightly higher HOMO. In an inverted BHJ with $PC_{71}BM$ as the acceptor, **84** displayed a PCE of 7.3% (V_{oc} = 0.85 V, J_{sc} = 12.6 mA/cm^2, FF = 68%) while its Si analog possessed a PCE of 6.6% under the same conditions. A PCE up to 8.5% could be obtained with the same polymer in slightly different conditions (V_{oc} = 0.86 V, J_{sc} = 15.9 mA/cm^2, FF = 63%) [83]. The fact that Ge-polymers display better performances than their Si counterparts is sometimes observed but is not a general trend. As in many studies about OPV, so many interdependent parameters are involved that it is complicated to draw direct conclusions between the PCE and molecular structure. However, it is clear that both Si and Ge polymers allow to obtain high efficiencies in OPV devices. Most of the acceptors (thiadiazole, thienopyrroledione, etc.) tested in the D–A structures for the dithienosiloles were also tested for the Ge-containing polymers and high efficiencies were also reported [75, 84].

Silole-based polymers allow to obtain efficient OPV materials. This is partly due to the good ability of these materials to transport holes. This property can

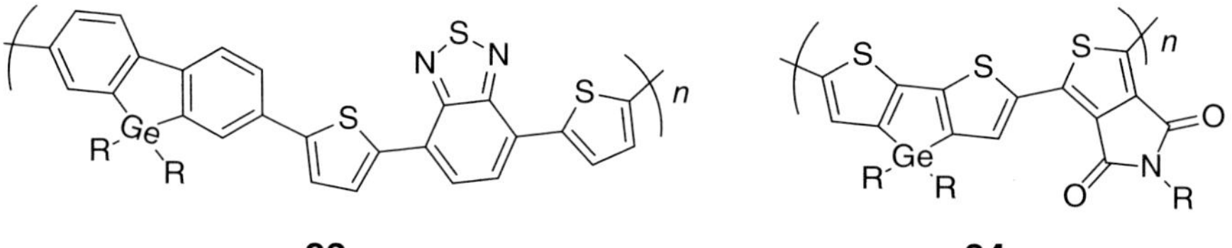

83 **84**

Figure 7.24 Ge-based polymers used in OSCs.

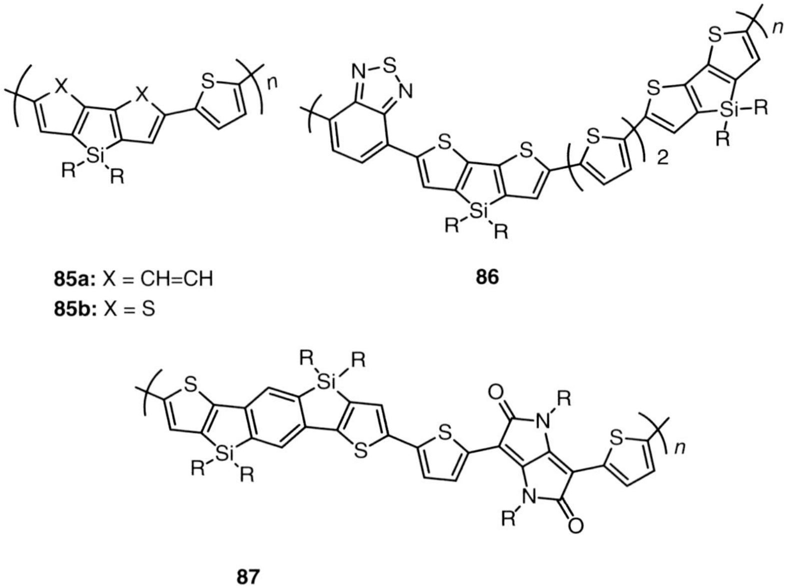

Figure 7.25 Si-based polymers used in OFETs.

also be evaluated in OFETs. In 2006, Fachetti, Marks, and coworkers developed dibenzosiloles/dithienosiloles-based polymers **85** (Figure 7.25) [85]. OFETs devices with a Si/SiO_2 wafer were prepared and showed good hole mobilities (0.05–0.08 $cm^2/(V\ s)$) as well as on/off ratio of around $10^4/10^5$. Similarly for the OPV part, many variations of the polymer backbone were performed (for example polymer **86**, Figure 7.25) [86]. In this study, various polymers with different monomeric units were prepared and compound **86**, with its complex monomer based on two dithienosilole units showed a moderate hole mobility (0.02 $cm^2/(V\ s)$) with moderate on/off ratio (10^3). Mobility up to 0.1 $cm^2/(V\ s)$ could be obtained with polymer **87** [87]. Many examples previously showed in the OPV Section 7.3.2 (Figure 7.25) were also studied in transistors, mainly to understand the charge transport in the solar cells.

This section demonstrates that phosphole/silole/germole-based conjugated oligomers/polymers possess strong UV–vis absorption together with good hole-transporting properties. They are thus highly interesting for applications in OPV and OFET devices. The fact that their efficiencies reached, particularly in OPV devices (~8%), corroborates this assumption. These novel conjugated polymers thus possess a bright future in this field.

7.4 Application in Electrochromic Cells

Electrochromism is a phenomenon exhibited by some materials whose optical properties change when a potential is applied. In 2011, Baumgartner, Rodríguez-López, and coworkers synthesized a dendritic material around

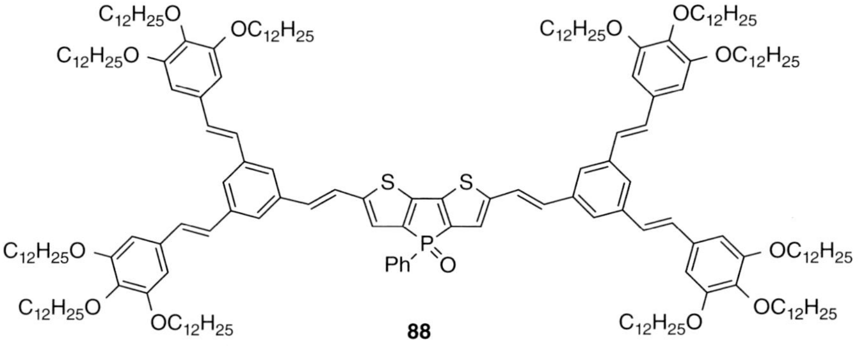

88

Figure 7.26 Phosphole-containing material used for electrochromic devices.

a dithienophosphole **88** (Figure 7.26) [88]. They constructed a device with the following configuration: [FTO/**88**/electrolyte (Bu_4NClO_4/THF (tetrahydrofuran) 0.1 M)/FTO]. When they applied a potential greater than 1.6 V, under UV-light, the photoluminescence of the film changed from orange to a light-green emission. This was the first time electrochromism was studied with a phosphorus heterocycle.

7.5 Conclusion

This chapter surveys the main classes of heteroatom-containing polymers used in electronic devices. Only few polymers have made the cut to date and have been incorporated into electronic devices including OLEDs, OPV cells, DSSCs and electrochromic cells. The chemistry of these π-conjugated systems with embedded heteroatom-moieties has been widely developed leading to a plethora of different structures. Structure–property relationships have still to be established in order to fully exploit the potential of heteroatom-moieties for the construction of conjugated frameworks. Nevertheless, it has already been clearly established that heavier atom derivatives offer specific advantages in comparison with their widely used carbon, sulfur, or nitrogen analogs. There are many other heteroatom-scaffolds described in the literature (containing Bi, S, Te, or As) [89–92], which could potentially be used for the development of new efficient π-conjugated materials for optoelectronic applications. The next challenge is clearly to exploit these novel organic materials for the fabrication of optoelectronic devices.

Acknowledgments

This research is supported by the Ministère de la Recherche et de l'Enseignement Supérieure, Région Bretagne, the University of Rennes 1, CNRS, China-French associated international laboratory in "Functional Organophosphorus Materials."

The authors thank the COST Action CM1302 (SIPs): "European Network on Smart Inorganic Polymers."

Abbreviations

1D	one-dimensional
2D	two-dimensional
α-NPB	*N,N′*-di(1-naphthyl)-*N,N′*-diphenyl-(1,1′-biphenyl)-4,4′-diamine
α-NPD	*N,N′*-di(1-naphthyl)-*N,N′*-diphenyl-(1,1′-biphenyl)-4,4′-diamine
η	power conversion efficiency
A	ampere
Alq_3	tris-(8-hydroxyquinoline)aluminum
BCP	bathocuproine
BP	benzoporphyrin
BPhen	bathophenanthroline
BTPD-PFCB	bis-tetraphenylenebiphenyldiamine-perfluorocyclobutane
CBP	4,4′-*N,N′*-dicarbazolebiphenyl
CCT	correlated color temperature
cd	candela
CIE	commission internationale de l'éclairage
cm	centimeter
CRI	Color Rendering Index
CuPc	copper(II) phthalocyanine
CPB	4,4′-*N,N′*-dicarbazole-biphenyl
DPVBi	4,4′-bis(2,2′-diphenylvinyl)biphenyl
DSSC	dye-sensitized solar cell
E_{red}	electrode potential (reduction)
E_T	triplet energy level
EL	electroluminescence
EML	emitting layer
ETL	electron-transporting layer
ETM	electron-transport material
eV	electron volt
$[(fbi)_2Ir(acac)]$	bis[2-(9,9-diethyl-9*H*-fluoren-2-yl)-1-phenyl-1*H*-benzimidazolate-κN,κC](acetylacetonato)-iridium(III)
FCNIrpic	bis[2-(5-cyano-4,6-difluorophenyl)pyridinato-C^2,*N*](picolinato)iridium(III)
FET	field-effect transistor
FF	fill-factor
FIrpic	bis[2-(4,6-difluorophenyl)pyridinato-C^2,*N*](picolinato)iridium(III)
FCNIrpic	bis[2-(5-cyano-4,6-difluorophenyl)pyridinato C^2,*N*](picolinato) iridium(III)
FTO	fluorine-doped tin oxide

HOMO	highest occupied molecular orbital
Hfa	hexafluoroacetylacetonate
HTL	hole-transporting layer
Ir(dbfmi)	mer-tris(*N*-dibenzofuranyl-*N*′-methylimidazole)iridium(III)
$Ir(mppy)_3$	(tris[2-(4-methyl-phenyl)-pyridine]iridium (III))
$Ir(ppy)_3$	tris[2-phenylpyridinato-C^2,*N*]iridium(III)
I	current
ITO	indium tin oxide
J_{sc}	short-circuit current
L	luminance
lm	lumens
LUMO	lowest unoccupied molecular orbital
m	meter
MB	maximum brightness
mCP	*N*,*N*′-dicarbazolyl-3,5-benzene
Me	methyl
Mes	2,4,6-trimethylphenyl-(=Mesityl–)
M_n	number-average molecular weight
η_L	luminous efficiency
η_{int}	internal quantum efficiency
η_{ext}	external quantum efficiency
η_P	luminous power efficiency or efficacy
NPB	*N*,*N*′-bis (naphthalen-1-yl)-*N*,*N*′-bis(phenyl)benzidine
ntfa	3-(2-naphthoyl)-1,1,1-trifluoroacetonate
OFET	organic field-effect transistor
OLED	organic light-emitting diode
OPV	organic photovoltaic cell
OSC	organic solar cell
P3HT	poly(3-hexylthiophene-2,5-diyl)
PBD	2-(4-biphenylyl)-5-(4-tert-butyl-phenyl)-1,3,4-oxadiazole
PCBM	[6,6]-phenyl C_{61} butyric acid methyl ester
$PC_{71}BM$	[6,6]-phenyl C_{71} butyric acid methyl ester
PCE	power conversion efficiency
PEDOT:PSS	poly(3,4-ethylenedioxythiophene)-poly(styrenesulfonate)
PhOLED	phosphorescent organic light-emitting diode
PLED	polymeric light-emitting diode
poly-TPD	poly(4-butylphenyl-diphenyl-amine)
PVK	poly *N*-vinylcarbazole
S	surface
s	second
SIMEF	bis(triorganosilymethyl)[60]fullerene
TADF	thermally activated delayed fluorescence
TTA	2-thenoyltrifluoroacetonate
V	volt
V_{oc}	open circuit voltage
W	watt
WOLED	white organic light-emitting diode

References

1 (a) Müllen, K. and Wegner, G. (1998). *Electronic Materials: The Oligomer Approach*. Weinheim: Wiley-VCH. (b) Skotheim, T.A. and Reynolds, J.R. (2007). *Handbook of Conducting Polymers*, 3e. Boca Raton, FL: CRC. (c) Müllen, K. and Scherf, U. (2006). *Organic Light Emitting Devices: Synthesis, Properties and Applications*. Weinheim: Wiley-VCH. (d) Scholz, S., Kondakov, D., Lussem, B., and Leo, K. (2015). *Chem. Rev.* 115: 8449. (e) Grimsdale, A.C., Chan, K.L., Martin, R.E. et al. (2009). *Chem. Rev.* 109: 897. (f) Gratzel, M. (2005). *Inorg. Chem.* 44: 6841. (g) Hagfeldt, A., Boschloo, G., Sun, L. et al. (2010). *Chem. Rev.* 110: 6595. (h) Cheng, Y.J., Yang, S.H., and Hsu, C.S. (2009). *Chem. Rev.* 109: 5868. (i) Sirringhaus, H. (2014). *Adv. Mater.* 26: 1319.

2 (a) Kraft, A., Grimsdale, A., and Holmes, A.B. (1998). *Angew. Chem. Int. Ed. Engl.* 37: 403. (b) Martin, R.E. and Diederich, F. (1999). *Angew. Chem. Int. Ed.* 38: 1350. (c) Burroughes, J.H., Bradley, D.D.C., Brown, A.R. et al. (1990). *Nature* 347: 539. (d) Garnier, F., Haijaloui, R., Yassar, A., and Srivastava, P. (1994). *Science* 265: 1684.

3 (a) McCullough, R.D. (1998). *Adv. Mater.* 10: 93. (b) Mitschke, U. and Bäuerle, P. (2000). *J. Mater. Chem.* 10: 1471. (c) Toth, A., Schaub, T.A., Meinhardt, U. et al. (2017). *Chem. Sci.* 8: 3494. (d) Larranaga, O., Romero-Nieto, C., and de Cozar, A. (2017). *Chem. Eur. J.* 23: 17487. (e) Duffy, M.D., Delaunay, W., and Bouit, P.-A. et al. (2016). *Chem. Soc. Rev.* 45: 5296.

4 (a) Yamagushi, S., Itami, Y., and Tamao, K. (1998). *Organometallics* 17: 4910. (b) Yamagushi, S., Endo, T., Uchida, M. et al. (2000). *Chem. Eur. J.* 6: 1683. (c) Tamao, K., Yamagushi, S., Shiozaki, M. et al. (1992). *J. Am. Chem. Soc.* 114: 5867.

5 Albert, I., Marks, T., and Ratner, M. (1997). *J. Am. Chem. Soc.* 119: 6575.

6 Hissler, M., Dyer, P.W., W, P., and Réau, R. (2003). *Coord. Chem. Rev.* 244: 1.

7 (a) Matsumi, N., Naka, K., and Chujo, Y. (1998). *J. Am. Chem. Soc.* 120: 10776. (b) Matsumi, N., Naka, K., and Chujo, Y. (1998). *J. Am. Chem. Soc.* 120: 5112.

8 (a) Farinola, G. and Ragni, R. (2011). *Chem. Soc. Rev.* 40: 3467. (b) Forrest, S.R., Bradley, D.D.C., and Thompson, M.E. (2003). *Adv. Mater.* 15: 1043.

9 Padmaperuma, A.B., Schmett, G., Fogarty, D. et al. (2000). *MRS Proc.* 621: Q3.9.1.

10 Ha, M.Y. and Moon, D.G. (2008). *Synth. Met.* 158: 617.

11 Oyamada, T., Sasabe, H., Adachi, C. et al. (2005). *Appl. Phys. Lett.* 86: 033503.

12 Avilov, I., Marsal, P., Brédas, J.L., and Beljonne, D. (2004). *Adv. Mater.* 16: 1624.

13 (a) Jeon, S.O., Yook, K.S., Joo, C.W., and Lee, J.Y. (2009). *J. Mater. Chem.* 19: 5940. (b) Jeon, S.O., Yook, K.S., Joo, C.W., and Lee, J.Y. (2009). *J. Phys. D: Appl. Phys.* 42: 225103. (c) Jeon, S.O., Yook, K.S., Lee, J.Y. et al. (2011). *Appl. Phys. Lett.* 98: 073306.

14 Jeon, S.O., Yook, K.S., Chin, B.D. et al. (2010). *Sol. Energy Mater Sol. Cells* 94: 1389.

15 (a) Hasegawa, Y., Natori, S., Fukudome, J. et al. (2018). *J. Phys. Chem. C* 122: 9599. (b) Xu, H., Zhu, R., Zhao, P., and Huang, W. (2011). *J. Phys. Chem. C* 115: 15627.
16 Tsuji, H., Sato, K., Sato, Y., and Nakamura, E. (2009). *J. Mater. Chem.* 19: 3364.
17 Matano, Y., Saito, A., Suzuki, Y. et al. (2012). *Chem. Asian J.* 7: 2305.
18 (a) Tamao, K., Uchida, M., Izumizawa, T. et al. (1996). *J. Am. Chem. Soc.* 118: 11974. (b) Yu, G., Yin, S., Liu, Y. et al. (2005). *J. Am. Chem. Soc.* 127: 6335. (c) Zhan, X., Risko, C., Amy, F. et al. (2005). *J. Am. Chem. Soc.* 127: 9021. (d) Lee, .H., Jang, B.B., and Kafafi, Z.H. (2005). *J. Am. Chem. Soc.* 127: 9071. (e) Murata, H., Malliaras, G.G., Uchida, M. et al. (2001). *Chem. Phys. Lett.* 339: 161.
19 Uchida, M., Izumizawa, T., Nakano, T. et al. (2001). *Chem. Mater.* 13: 2680.
20 (a) Noda, T. and Shirota, Y. (1998). *J. Am. Chem. Soc.* 120: 9714. (b) Tanaka, D., Tadeka, T., Chiba, T. et al. (2007). *Chem. Lett.* 36: 262.
21 Burrows, P.E., Padmaperuma, A.B., Sapochak, L.S. et al. (2006). *Appl. Phys. Lett.* 88: 183503.
22 Son, H.S., Seo, C.W., and Lee, J.Y. (2011). *J. Mater. Chem.* 21: 5638.
23 Shao, S., Ding, J., Ye, T. et al. (2011). *Adv. Mater.* 23: 3570.
24 Jiang, H.-J., Zhang, Q.-W., He, X. et al. (2017). *Chin. J. Polym. Sci.* 35: 611.
25 Rothemund, S. and Teasdale, I. (2016). *Chem. Soc. Rev.* 45: 5200.
26 Schrögel, P., Hoping, M., Kowalsky, W. et al. (2011). *Chem. Mater.* 23: 4947.
27 Soh, M.S., Santamaria, S.A.G., Williams, E.L. et al. (2011). *J. Polym. Sci., Part B: Polym. Phys.* 49: 531.
28 (a) Matsumi, M., Miyata, M., and Chujo, Y. (1998). *J. Am. Chem. Soc.* 120: 5112. (b) Matsumi, M., Miyata, M., and Chujo, Y. (1999). *Macromolecules* 32: 4467. (c) Matsumi, M., Miyata, M., and Chujo, Y. (2000). *Macromolecules* 33: 3956. (d) Matsumi, M., Miyata, M., and Chujo, Y. (1998). *J. Am. Chem. Soc.* 120: 10776. (e) Makioka, Y., Hayashi, T., and Tanaka, M. (2004). *Chem. Lett.* 33: 44. (f) Mao, S.S.H. and Tilley, T.D. (1997). *Macromolecules* 30: 5566. (g) Baumgartner, T., Neumann, T., and Wirges, B. (2004). *Angew. Chem. Int. Ed.* 43: 6197. (h) Baumgartner, T. and Wilk, W. (2006). *Org. Lett.* 8: 503. (i) Xu, C., Yamada, H., Wakamiya, A. et al. (2004). *Macromolecules* 37: 8978. (j) Yamaguchi, S. and Tamao, K. (2005). *Chem. Lett.* 34: 2. (k) Romero-Nieto, C., Durben, S., and Kormos, I.M. (2009). *Adv. Funct. Mater.* 19: 3625.
29 Chien, C.-H., Chen, C.-K., Hsu, F.-M. et al. (2009). *Adv. Funct. Mater.* 19: 560.
30 Liu, C., Gu, Y., Fu, Q. et al. (2012). *Chem. Eur. J.* 18: 13828.
31 Liu, C., Li, Y., Li, Y. et al. (2013). *Chem. Mater.* 25: 3320.
32 Shao, S., Ding, J., Wang, L. et al. (2012). *J. Am. Chem. Soc.* 134: 15189.
33 Shao, S., Ding, J., Wang, L. et al. (2012). *J. Am. Chem. Soc.* 134: 20290.
34 Fave, C., Cho, T.Y., Hissler, M. et al. (2003). *J. Am. Chem. Soc.* 125: 9254.
35 Su, H.C., Fadhel, O., Yang, C.J. et al. (2006). *J. Am. Chem. Soc.* 128: 983.
36 Geffroy, B., Lemaitre, N., Lavigne, J. et al. (2007). *Nonlinear Optics and Quantum Optics*, vol. 37, 9.

37 (a) Fadhel, O., Gras, M., Lemaitre, N. et al. (2009). *Adv. Mater.* 21: 1261. (b) Joly, D., Tondelier, D., Deborde, V. et al. (2012). *Adv. Funct. Mater.* 22: 567.
38 (a) Bouit, P.-A., Escande, A., Szűcs, R. et al. (2012). *J. Am. Chem. Soc.* 134: 6524. (b) Riobé, F., Szűcs, R., Bouit, P.A. et al. (2015). *Chem. Eur. J.* 21: 6547.
39 Chen, R.F., Zhu, R., Fan, Q.L., and Huang, W. (2008). *Org. Lett.* 10: 2913.
40 (a) Bolink, H.J., Santamaria, S.G., Sudhakar, S. et al. (2008). *Chem. Commun.* 618. (b) Bolink, H.J., Barea, E., Costa, R.D. et al. (2008). *Org. Electron.* 9: 155.
41 Brauge, L., Vériot, G., Franc, G. et al. (2006). *Tetrahedron* 62: 11891.
42 Kim, S.T., Hwang, D.-H., Li, X.C. et al. (1996). *Adv. Mater.* 8: 979.
43 Hoeger, S., McNamara, J., Schricker, S., and Wudl, F. (1994). *Chem. Mater.* 6: 171.
44 Zhang, C., Hoeger, S., Pakbaz, K. et al. (1994). *J. Electron. Mater.* 23: 453.
45 Chen, Z., Huang, W., Wang, L. et al. (2000). *Macromolecules* 33: 9015.
46 Ahn, T., Ko, S., Lee, J., and Shim, H. (2002). *Macromolecules* 35: 3495.
47 Jin, S., Jang, M., and Suh, H. (2002). *Chem. Mater.* 14: 643.
48 Jin, Y., Song, S., Park, S.H. et al. (2008). *Polymer* 49: 4559.
49 Hwang, D.-H., Lee, J.-I., Cho, N.-S., and Shim, H.-K. (2004). *J. Mater. Chem.* 14: 1026.
50 (a) Garay, R.O., Mayer, B., Karasz, F.E., and Lenz, R.W. (1995). *J. Polym. Sci., Part A: Polym. Chem.* 33: 525. (b) Martin, R.E., Geneste, F., Chuah, B.S. et al. (2001). *Synth. Met.* 122: 1. (c) Chuah, B.S., Geneste, F., Holmes, A.B. et al. (2000). *Macromol. Symp.* 154: 177.
51 (a) Kim, H.-K., Ryu, M.-K., Kim, K.-D. et al. (1998). *Macromolecules* 31: 1114; (b) Kim, K.-D., Park, J.-S., Kim, H.-K. et al. (1998). *Macromolecules* 31: 7267.
52 (a) Malliaras, G.G., Herrema, J.K., Wildeman, J. et al. (1993). *Adv. Mater.* 5: 721. (b)Herrema, J.K., van Hutten, P.F., Gill, R.E. et al. (1995). *Macromolecules* 28: 8102.
53 Yoshino, K., Hirohata, M., Sonoda, T. et al. (1999). *Synth. Met.* 102: 1158.
54 (a) Tang, B.Z. and Qin, A. (eds.) (2013). *Aggregation-Induced Emission: Fundamentals.* Wiley. (b) Hong, Y., Lam, J.W.Y., and Tang, B.Z. (2011). *Chem. Soc. Rev.* 40: 5361–5388. (c) Mei, J., Leung, N.L.C., Kwok, R.T.K. et al. (2015). *Chem. Rev.* 115: 11718–11940.
55 Chen, J., Xie, Z., Chen, J.W.Y. et al. (2003). *Macromolecules* 36: 1108.
56 (a) Liu, M.S., Luo, J., and Len, A.K.Y. (2003). *Chem. Mater.* 15: 3500. (b) Ohshita, J., Kimura, K., Lee, K.-H. et al. (2007). *J. Polym. Sci., Part A: Polym. Chem.* 45: 4588. (c) Mo, Y.Q., Deng, X.Y., Jiang, X., and Cui, Q.H. (2009). *J. Polym. Sci., Part A: Polym. Chem.* 47: 3286. (d) Liao, L., Cirpan, A., Chu, Q. et al. (2007). *J. Polym. Sci., Part A: Polym. Chem.* 45: 2048.
57 Roh, S.-H., Cheong, H., Park, J. et al. (2014). *J. Nanosci. Nanotechnol.* 14: 6438.
58 (a) Brabec, C.J., Sariciftci, N.S., and Hummelen, J.C. (2001). *Adv. Funct. Mater.* 11: 15. (b) Günes, S., Neugebauer, H., and Sariciftci, N.S. (2007). *Chem. Rev.* 107: 1324. (c) Cao, W. and Xue, J. (2014). *Energy Environ. Sci.* 7: 2123. (d) Chen, H.-Y., Kuang, D.-B., and Su, C.-Y. (2012). *J. Mater. Chem.* 22: 15472.
59 Delgado, J.L., Bouit, P.-A., Fillipone, S. et al. (2010). *Chem. Commun.* 46: 4853.

60 Kira, A., Shibano, Y., Kang, S. et al. (2010). *Chem. Lett.* 39: 448.
61 Matano, Y., Hayashi, Y., Nakano, H., and Imahori, H. (2014). *Heteroat. Chem.* 25: 533.
62 Lin, L.-Y., Tsai, C.-H., Wong, K.-T. et al. (2010). *J. Org. Chem.* 75: 4778–4785.
63 Zeng, W., Cao, Y., Bai, Y. et al. (2010). *Chem. Mater.* 22: 1915–1925.
64 Blonsky, P.M., Shriver, D.F., Austin, P., and Allcock, H.R. (1984). *J. Am. Chem. Soc.* 106: 6854.
65 Xiang, W., Zhou, Y., Yin, X. et al. (2009). *Electrochim. Acta* 54: 4186.
66 Fei, S.-T., Lee, S.-H.A., Pursel, S.M. et al. (2011). *J. Power Sources* 196: 5223.
67 Matano, Y., Ohkubo, H., Miyata, T. et al. (2014). *Eur. J. Inorg. Chem.* 2014: 1620.
68 Park, K.H., Kim, Y.J., Lee, G.B. et al. (2015). *Adv. Funct. Mater.* 25: 3991.
69 Bui, T.T.T., Park, S., Hoang, Q.V. et al. (2016). *Synth. Met.* 215: 235.
70 Boudreault, P.L.T., Michaud, A., and Leclerc, M. (2007). *Macromol. Rapid Commun.* 28: 2176.
71 Wang, E., Wang, L., Lan, L. et al. (2008). *Appl. Phys. Lett.* 92: 033307–033303.
72 Hou, J., Chen, H.-Y., Zhang, S. et al. (2008). *J. Am. Chem. Soc.* 130: 16144.
73 Chu, T.-Y., Lu, J., Beaupré, S. et al. (2011). *J. Am. Chem. Soc.* 133: 4250.
74 Chu, T.-Y., Lu, J., Beaupré, S. et al. (2012). *Adv. Funct. Mater.* 22: 2345.
75 Guo, X., Zhou, N., Lou, S.J. et al. (2012). *J. Am. Chem. Soc.* 134: 18427–18439.
76 Cui, C., Fan, X., Zhang, M. et al. (2011). *Chem. Commun.* 47: 11345.
77 Wu, J.-S., Cheng, Y.-J., Lin, T.-Y. et al. (2012). *Adv. Funct. Mater.* 22: 1711.
78 van der Poll, T.S., Love, J.A., Nguyen, T.Q., and Bazan, G.C. (2012). *Adv. Mater.* 24: 3646.
79 Kyaw, A.K.K., Wang, D.H., Gupta, V. et al. (2013). *Adv. Mater.* 25: 2397.
80 Allard, N., Aïch, R.B., Gendron, D. et al. (2010). *Macromolecules* 43: 2328.
81 Ohshita, J., Miyazaki, M., Tanaka, D. et al. (2013). *Polym. Chem.* 4: 3116.
82 Amb, C.M., Chen, S., Graham, K.R. et al. (2011). *J. Am. Chem. Soc.* 133: 10062.
83 Constantinou, I., Lai, T.-H., Zhao, D. et al. (2015). *ACS Appl. Mater. Interfaces* 7: 4826.
84 Wang, Q., Zhang, S., Ye, L. et al. (2014). *Macromolecules* 47: 5558.
85 Usta, H., Lu, G., Facchetti, A., and Marks, T.J. (2006). *J. Am. Chem. Soc.* 128: 9034.
86 Beaujuge, P.M., Pisula, W., Tsao, H.N. et al. (2009). *J. Am. Chem. Soc.* 131: 7514.
87 Ashraf, R.S., Chen, Z., Leem, D.S. et al. (2011). *Chem. Mater.* 23: 768–770.
88 Romero-Nieto, C., Marcos, M., Merino, S. et al. (2011). *Adv. Funct. Mater.* 21: 4088.
89 Linshoeft, J., Baum, E.J., Hussain, A. et al. (2014). *Angew. Chem. Int. Ed.* 53: 12916.
90 He, G., Kang, L., Torres Delgado, W. et al. (2013). *J. Am. Chem. Soc.* 135: 5360.
91 Parke, S.M., Narreto, M.A.B., Hupf, E. et al. (2018). *Inorg. Chem.* https://doi.org/10.1021/acs.inorgchem.8b00149.
92 Green, J.P., Han, Y., Kilmurray, R. et al. (2016). *Angew. Chem. Int. Ed.* 55: 7148–7151.

8

Inorganic Polymers as Flame-Retardant Materials

Raghvendra Kumar Mishra[1], Tarik Eren[2], and De-Yi Wang[1]

[1] *IMDEA Materials Institute, C/Eric Kandel, 2, 28906 Getafe, Madrid, Spain*
[2] *Department of Chemistry, Yildiz Technical University, 34220, Istanbul, Turkey*

8.1 Introduction

The hazardous nature of flames promotes significant damage to both lives as well as property. It may also lead to intense human suffering in addition to capital losses. As a result, notable studies have been focused on the development of useful flame-resistant materials in an effort to avoid death and injury by minimizing the effects of fire to a tolerable level. Nonflammable materials for example metals and concrete do not burn and in addition help in combustion or discharge the flammable vapors when subject to flame. It is also important to consider materials that are able to resist a flame for a definite time period. The facets of a material play a predominant role before flashover that can be governed by flame-retardant materials. Post-flashover flames are usually considered perilous for systems. Flame-retardant materials are normally associated with reaction-to-flame to decelerate or slow down the combustion of flammable materials [1]. Several factors, for example, ignition time, oxygen index, heat intensity, smoke generation, flame spread rate as well as the discharge of toxic gases, have to be taken into consideration for evaluating both flammable and flame-retardant materials. The applications of flame-retardant materials are seen in several areas, which include adhesives, lubricants, structural elements for cars as well as aircrafts, electronic devices, structure, motor and space sectors, and so on [2], as demonstrated in Figure 8.1.

Besides metal and ceramic materials, polymers are nowadays one of the primary facets of molecular science and also materials science and consequently perform an influential role in the environment around us. In this manner, flame-retardant polymeric materials have received substantial demand from both scientific as well as commercial sectors. Previous researches outline the synthesis as well as the characterization of inorganic polymers having excellent thermal durability and intrinsic flame retardancy, which is completely achieved by their molecular structure, with or without the integration of any kind of additives or even fillers [3, 4]. Because of the notable changeability of the backbone, these inorganic polymers show better performance as compared to

Smart Inorganic Polymers: Synthesis, Properties, and Emerging Applications in Materials and Life Sciences,
First Edition. Edited by Evamarie Hey-Hawkins and Muriel Hissler.

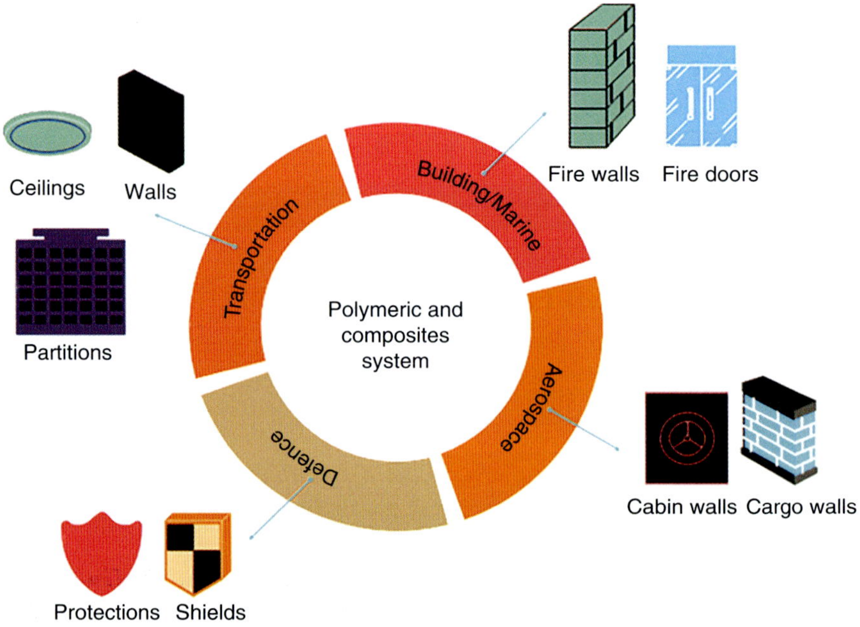

Figure 8.1 Application of flame-retardant materials in daily life.

their organic alternatives. Likewise, certain inorganic polymers manifest low to lowest flammability, which enables them to be considered as flame-retardant materials in thermoplastics as well as in thermosetting matrix.

Furthermore, in addition to characteristics such as environmental adaptability, intrinsically flame-retardant features, solid polymer electrolytes with low glass-transition temperature, and so on, inorganic polymers assume better user-safety roles due to their excellent properties. Accordingly, this chapter deals with a discussion of the flame-retardant behavior of inorganic polymer materials that have been illustrated in scientific literature. Substantial deliberation will be focused on morphology, structural, and flame-retardant interrelationships. This chapter will likewise present the state-of-the-art and polymer structure with cutting edge breakthroughs in inorganic polymers and their flame-retardant functions [5–7].

8.2 Importance of Flame-Retardant Materials

Flames have always remained a threat to society, for most part of the reported historical past. However, in the current age, fires risks have reduced considerably since open fires in houses have been occasional and also because they are effectively controlled. Taking a look at the current fire statistics, fire as a concern is contracting, and therefore it is justified to a have a thorough understanding of fire hazards, fire risks, along with the application of flame retardants in materials, globally. Therefore, before delving into the flame-retardant behavior of inorganic polymeric materials, this chapter begins with a description of the fundamentals of what exactly is implied by fire hazard, fire risk, and also the reason why flame-retardancy strategies are employed presently [8]. To start with, all carbon-based materials, from wood to plastics, are combustible if heat and oxygen are available. Therefore, since oxygen is plentifully accessible, combustion is a consistent force on earth. Without doubt, the few materials that are quicker to burn as compared to others are, for example, any carbon-based material such as graphite and diamond whenever sufficient oxygen and heat are available. Another example, namely, virgin organic polymeric materials are categorized into a wide range of carbon-based flammable materials and, in some instances, may be more flammable compared to natural materials such as wood, cotton, or cellulosic mass [9]. In the case of plastics, specifically petroleum or bio-based materials use of which is escalating in most areas of everyday life, it must be noted that these types of materials promote fires: when they hold fire, they may be prone to burn and set other items on fire as well. The perils of fire in a home and the possibility of it being burnt down to the floor may be fairly small, but this is not the case when the house is made of dry timber clapboard, in which case the home environment is highly conducive for potential fires. Similarly, electronic products also have the propensity for fire risk. In particular, when a battery short-circuits, it leads to arcing on the circuit board, and this event can burn the electronic device and contribute to injuries [10]. Figure 8.2 provides data on burnt areas in hectares by these types of fires in European countries and in the EU Mediterranean region. Hence, it is very important to apply modern flame-retardant materials that help to minimize fire hazards and secure devices [11]. Therefore, the main objective of this particular chapter is to present a discussion about the role of inorganic polymers as a flame-retardant materials.

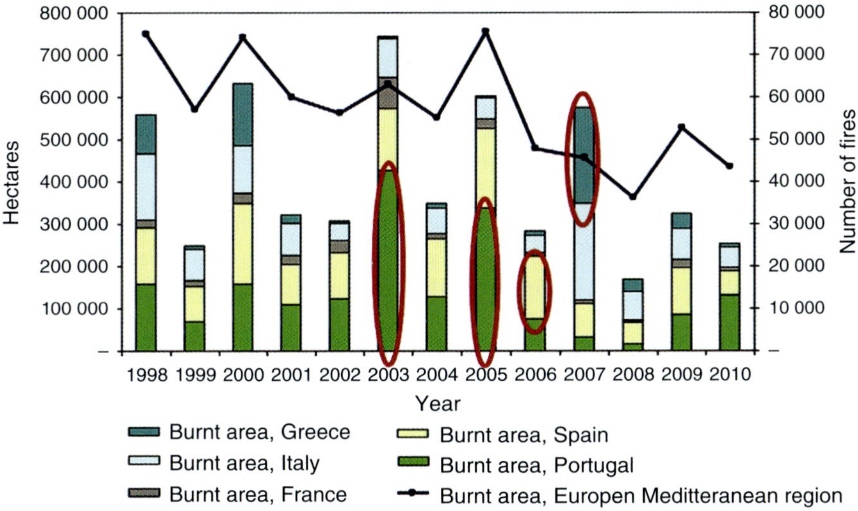

Figure 8.2 The numbers of fire and total hectares burned by the fire with respect to time. Source: Adapted from San-Miguel-Ayanz et al. 2013 [11] with permission.

8.3 Application of Inorganic Polymer as a Flame-Retardant Material

The principal objective of this section is to review the exciting developments in the vast field of inorganic polymers in the last decade with particular reference to their flame-retardant properties [12]. As organic polymers consist of carbon and hydrogen in their chemical arrangement, they are significantly flammable. Combustion of polymers comprises four successive stages: (i) preheating of the polymer, (ii) decomposition of the polymer, (iii) ignition of the polymer, and consequently (iv) combustion as well as flame propagation. The first step when the polymer encounters a heating source is referred to as preheating. The preheating stage is when the materials absorb sufficient heat energy to degrade [13]. It is dependent upon thermal characteristics such as the heat capacity, thermal conductivity, and the latent heat of fusion of the polymer. After adequate heat is actually taken in by the polymer it begins to degrade. Initially, the weaker bonds split and the combustion output, both gaseous and flammable, is discharged into the atmosphere. These types of gases spark in the presence of the ignition source (oxygen as well as flammable gases), and start burning. The ignition of the material renders the heat for combustion and also the propagation of fire throughout decomposition [14]. This action continues as far as any one of the components of the fire triangle such as heat, fuel, or oxygen is available. Combustion is retained with the occurrence of free radicals that are produced during the course of the fire in the gas phase. If the material and sufficient heat energy are collectively available, the combustion continues until finally an explosion occurs. When the energy required for the fire is consistently available, there can be an immobile flame. In case the available energy is lower than the required energy, the combustion rate declines and finally, the flame extinguishes. Flame retardants interfere with the course of action of combustion at any of these steps including heating, ignition, flame spread, decomposition, and therefore, the flame-retardant features of a polymer have significant impact on the safeguarding of material from fire and in protecting lives [15].

Irrespective of the common nature of organic polymers, the inorganic macromolecules have inorganic constituents within their principal chain. These types of polymers have drawn substantial attention, because organic polymers possess several inadequacies: for example, low softening temperature ranges or even low degradation temperature ranges; susceptibility to degradation from oxygen, ozone, or high-energy emission; and dissolution or swelling when involved with solvents or even fluids [16]. The attributes of inorganic polymers are low-temperature overall flexibility, high thermal as well as oxidative constancy, flame retardancy, unique kind of chemical reactivity, and so on. These types of inorganic polymers are the connecting link between polymer science and ceramics science. While exploring the application of inorganic polymers as flame retardants, it should be noted that all these inorganic polymers turn out to be ceramics under the application of heat, and so they have been employed as flame retardants because they can simply change into ceramic char on burning [17]. This char is additionally stable and avoids the movement of flame. Phosphorus-containing fire retardants are the most widespread materials because they have a synergistic effect, and in addition, they are eco-friendly materials. Phosphorus-based flame retardants create a char coating on the polymer, and therefore they protect the unburned section of the polymer from the flame, and as a result, the polymer is secured from both oxygen and heat damage. The nitrogen-dependent flame retardants display positive effects with phosphorus-containing materials. Nitrogen-based flame retardants do not discharge the harmful gases; in fact, they generate inflammable gases throughout the course of combustion [18]. Silicon-based flame retardants are likewise considered eco-friendly because they never generate hazardous gases to the environment throughout burning, and they produce silica carbon in the course of combustion, which is thermally durable and also shields the material from additional decomposition. The phosphorus-, nitrogen-, and silicon-based flame retardants are sustainable materials because they will not generate dangerous gaseous products while they burn [19, 20]. Each component such as phosphorus and nitrogen possesses a unique functionality, for example, phosphorus contributes to developing the char, nitrogen assists the swelling of char, and consequently silicon helps to form a thermally stable char. Finally, there are several kinds of inorganic polymers, all of which have great potential to be flame retardants [21]; in spite of this, only a limited number of studies on combining flame-retardant materials with inorganic polymers have been presently reported. An overview of inorganic polymers and their application in flame retardancy as well as in high-temperature application are reported in Table 8.1.

8.3.1 Polysiloxanes

Polysiloxanes are a unique category of organosilicon polymers; they have several desirable features, for example, outstanding heat, radiation or chemical stability, effective solubility in well-known organic solvents, and relatively nontoxic characteristics in addition to reliable mechanical properties. Attributes such as thermal stability, elasticity at low-temperature ranges, and thermo-oxidative durability at higher temperatures are the vital characteristics of polysiloxanes. Authors analyzed the thermal degradation behavior of polydimethylsiloxane (PDMS) end-blocked with trimethylsiloxy groups at various heating rates; it was observed that the heating rate has an important effect on the type of volatile yield produced throughout the thermal degradation of PDMS in helium or even in air atmosphere. In addition, a variety of polysiloxanes have been employed to produce high-performance and flame-retardant thermosetting resins. The results prove that polysiloxane with an appropriate structure can enhance the overall performance of the resin. A common illustration is the completely end-capped hyperbranched polysiloxane with significant branching degree and amine-groups, which has been used to modify the bismaleimide resin; the modified resin displayed excellent flame retardancy, resilience, and strength as well as thermal endurance [29].

In this regard, a phosphorus-based hyperbranched polysiloxane with plenty of 9,10-dihydro-9-oxa-10-phosphaphenanthrene-10-oxide (DOPO) as well as silanol groups has been prepared by hydrolysis of phosphorus-containing

Table 8.1 An overview of inorganic polymer and their application in flame retardancy as well as high-temperature application operations.

Types of inorganic polymers	Structure	Remarks	References
Polysiloxane	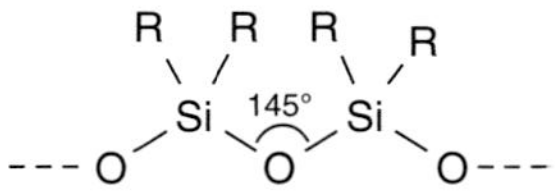	• Low glass transition temperature, flexibility, permeability to gases, low surface energy, transparent toward UV radiation; thermal stability due to Si—O bonds. • An effective flame retardant without affecting the environment.	[22]
Polyphosphazenes	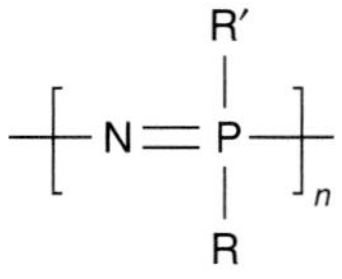	• Backbone has alternating phosphorus and nitrogen elements and with inorganic or organic side groups, which are accompanied by phosphorus. • High flexibility due to the backbone, radiation resistance, high refractive index, ultraviolet and visible transparency. • It is effective in flame retardation and the flammability is tuned by composition without altering the physical and mechanical properties.	[23]
Polysilanes	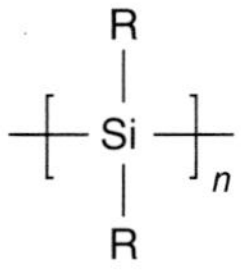	• The exclusive Si—Si bond allows sigma delocalization, and solubility can be tuned with varying attached group. • It has strong electronics absorption, conductivity, photoconductivity, and photosensitivity. • It shows the thermal resistant up to 573 K and it has been used as environment-friendly flame-retardant materials as well as high-performance application.	[24]
Polysilynes	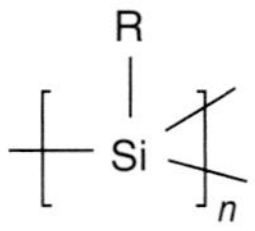	• It is network polysilanes, organic group R attached to silicon has at least four carbon atoms, soluble in organic solvents, amorphous. • It is converted to silicon carbide on pyrolysis with a high yield. • It is thermally stable up to 300 °C, it can act as an excellent flame retardant without emitting toxic and corrosive products.	[25]

(Continued)

Table 8.1 (Continued)

Types of inorganic polymers	Structure	Remarks	References
Ferrocene-based polymers	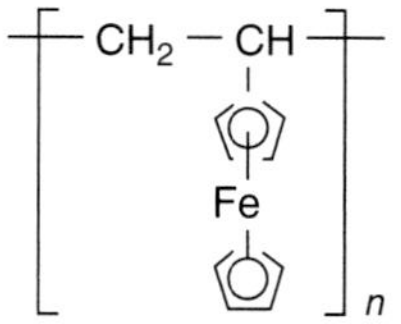	• It offers interesting electronic as well as magnetic properties. • It has electron donating, super-fast electrochemical response ability. • It shows high-temperature thermal stability, and it can be used in a various forms for flame-retardant applications.	[26]
Poly(carboranylenesiloxanes)	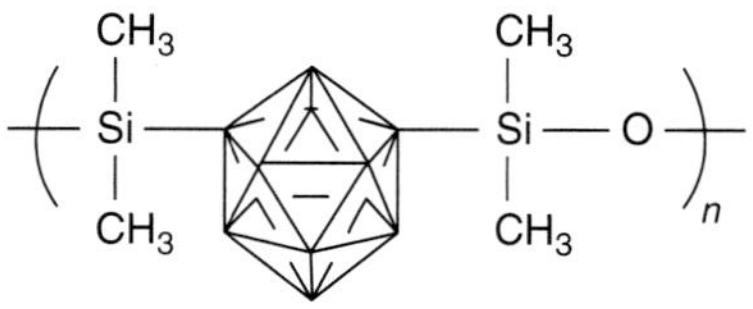	• It exhibits outstanding thermal and oxidative properties, and it has high-temperature stability up to 1000 °C. • It is a potential candidate for high temperature and flame retardancy applications.	[27]
Polyaryletherketone-*p*-carboranes	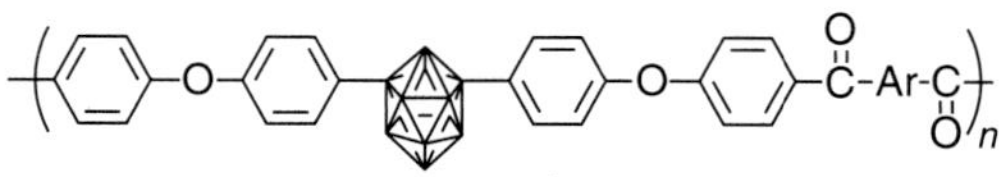	• It is a carborane-based thermoplastic, and it offers excellent thermomechanical properties and the ability to withstand solvent attack. • It has high-temperature stability in both air as well as inert atmosphere. • This type of thermoplastics can be used as flame-retardant materials due to high thermo-oxidative stability.	[28]

Table 8.2 Characteristic data from TG analyses of cured CE and P–HSi/CE resins.

Samples	T_{di} (°C)	T_{max} (°C)		Y_c at 800 °C (wt%)
		T_{max1}	T_{max2}	
Cyanate ester (CE)	373	443	515	33.8
Silanol groups (P–Hsi)(5 wt%)/ cyanate ester(P–HSi5/CE)	412	448	516	36.8
Silanol groups (P–Hsi)(10 wt%)/ cyanate ester (P–HSi10/CE)	425	446	518	39.6
Silanol groups (P–Hsi)(15 wt%)/ cyanate ester (P–HSi15/CE)	431	444	515	43.6
Silanol groups (P–Hsi)(20 wt%)/ cyanate ester (P–HSi20/CE)	414	445	516	41.4
Silanol groups (P–Hsi)(25 wt%)/ cyanate ester (P–HSi25/CE)	407	449	518	41.8
Silanol groups (P–Hsi)(30 wt%)/ cyanate ester (P–HSi30/CE)	393	446	516	40.6
Silanol groups (P–Hsi)(35 wt%)/ cyanate ester (P–HSi35/CE)	381	449	518	39.2

Source: Ye et al. 2013 [30]. Reproduced with permission of Elsevier.

triethoxysilane; silanol groups are employed to establish high-class flame-retarding cyanate ester resin. Cyanate ester resin has been selected as the basic resin because of its excellent potency in several areas such as electric as well as electronic sectors, aerospace, automobile, and so on. The thermal degradation of the original and modified cyanate ester resins is generally found to be distinct, as observed from the ultimate degradation rate temperature (T_{max}), which was around 445–515 °C, respectively; this shows that the original and modified cyanate ester resin possess comparable degradation processes. The related T_{di} at which the weight reduction of the sample attains 5 wt%, the T_{max}, and also char produce (Y_c) at 800 °C are outlined in Table 8.2. It is observed that all resins possess identical T_{max}, which reveals that the inclusion of silanol groups does not alter the chemistry of most of the molecular chains in each and every cross-linked network [30].

Limiting oxygen index (LOI) value has been employed to analyze the flame-retardant behavior of cured cyanate ester (CE) as well as P–HSi/CE resins. As presented in Figure 8.3, modified cyanate ester resin possesses much higher LOI as compared to pristine resin; furthermore, a minute inclusion of P–HSi is able to enhance the flame-retardant behavior of the resin; the LOI value of P–HSi5 (5 wt%)/cyanate ester system is 37%, which is nearly 1.3 times the cyanate ester resin. If the quantity of P–HSi increases, the LOI value enhances and attains the utmost value (41%) at 20 wt% P–HSi, which is close to 1.5 times of the LOI value of the cyanate ester resin. This is related to the structural configuration of the cured resin with respect to the amount of P–HSi along with the intrinsic flammability of phosphorus. Related elemental studies of P–HSi15

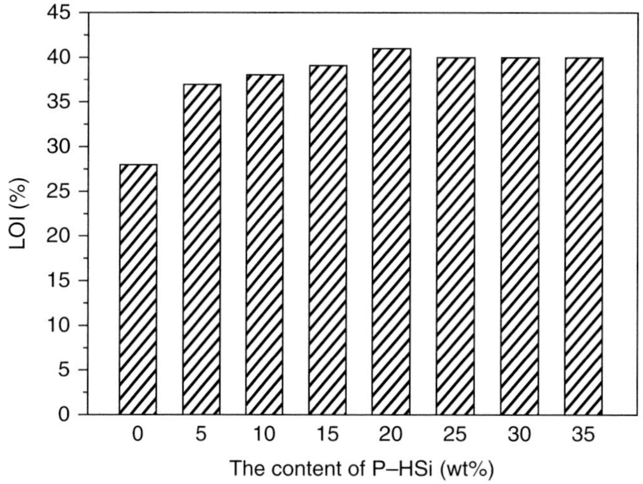

Figure 8.3 Limiting oxygen index values of cured cyanate ester and silanol groups/cyanate ester resins. Source: Adapted from Ye et al. 2013 [30] with permission.

Table 8.3 Elemental compositions of the residues for cured CE and P–HSi15/CE resins.

Sample		Elemental composition (wt%)				
		C	N	O	P	Si
Cyanate ester (CE)	Interior	93.08	4.89	2.01	0.00	0.00
	Exterior	89.84	6.49	3.66	0.00	0.00
Residue of P–HSi15(15 wt%)/ Cyanate ester (CE)	Interior	86.10	6.54	4.26	0.97	2.11
	Exterior	78.81	5.32	8.26	2.13	5.47

Source: Ye et al. 2013 [30]. Reproduced with permission of Elsevier.

(15 wt%)/CE resin indicate that the percentage of both P and Si components in the outside layer is more prominent as compared to that in the inner layer, which indicates that both P and Si components of P–HSi are able to move in the direction of the surface of the resin, and create a tiny layer, which has the ability to inhibit the movement of the flame by keeping back the flammable gases and insulating the heat from the unburned resins, as shown in Table 8.3. With the optimized content of P–HSi, the modified cyanate ester resins have enhanced flame retardancy and thermal stability properties [30].

The silicon–nitrogen based flame retardants are a reactive kind of flame retardants, which helps to improve the flame-retardant behavior of polymers. In this way, branched benzoxazine-based polysiloxanes (PBz–PSO) have already been produced by the 3 + 2 condensation polymerization reaction. Benzoxazine-containing triethoxysilane complex (Bz–TES) is achieved with aminopropyltriethoxysilane, paraformaldehyde, and naphthol (in the form of precursors for the production of benzoxazine compounds). After the reaction, purified Bz–TES is received without additional filtration. The fire-retardant behavior of the cross-linked PBz–PSO samples has been examined with their LOI.

Table 8.4 The molecular weight data of PBz-PSO samples and the thermal stability and flame-retardant property of cross-linked PBz-PSO samples.

Sample	Thermal properties from TGA		LOI
	Temperatures at 10 wt% loss, T_{d10} (°C)	Char yield at 800 °C (wt%)	
Cross-linked PBz-PSO-23 (molar amounts of Bz-TES are 2/3)	350	50	>45
Cross-linked PBz-PSO-24 (molar amounts of Bz-TES are 2/4)	385	44	>45
Cross-linked PBz-PSO-22 (molar amounts of Bz-TES are 2/2)	410	55	>45

Source: Hsieh et al. 2013 [31]. Reproduced with permission of Elsevier.

The test results are mentioned in Table 8.4. All samples display LOI values above 45. The higher LOI values illustrate the higher fire-retardant properties of the cross-linked PBz–PSO samples. These results reveal that the produced PBz–PSO samples are a reactive kind of fire-retardant materials for polymers. For thermosetting polymers, PBz–PSO can be integrated straightaway with the resin and in situ turn out to be cross-linked during the curing of the resins. The cured products are found to be an interpenetrating network of the thermosetting matrix along with cross-linked PBz–PSO flame retardants. As a result, PBz–PSO acts as a fire retardant as well as an efficient reinforcing agent for thermosetting resins [31].

Cotton is a prominent material in civilian as well as military applications because of its excellent mechanical, biodegradable, as well as air permeable properties. Nevertheless, cotton experiences specific disadvantages, for example, inadequate water repellency and swift ignition. Several attempts have been formulated to establish fire-retardant cotton fabric. In this manner, iodine butyl-*N*-sulfonate amino polysiloxane ((IB-*N*-SA) PDMS) has been prepared with dual functional properties (e.g. water repellency as well as heat retardancy) by using poly(4-iodobutoxy)methyl siloxane and guanidine sulfamate. It has been employed to modify the cotton fabrics; the burning characteristics of cotton fabrics modified with (IB-*N*-SA) PDMS have been examined using a cone calorimeter: the test results are demonstrated in Figure 8.4 and the test results linked to the combustibility are listed in Table 8.5.

The cotton fabric modified with (IB-*N*-SA) PDMS achieved a reduction in its heat release rate (HRR), peak heat release rate (PHRR), and total heat release rate (THR), as demonstrated in Figure 8.4a,b and Table 8.5. It shows that (IB-*N*-SA) PDMS can inhibit the combustion of cotton fabrics, which is confirmed by the increase of time to ignition and the enhancement of the fire performance index (FPI) value as mentioned in Table 8.5. A higher FPI shows that the material has much better fire retardancy. EHC (effective heat combustion), as well as mass loss criteria, are likewise vital, which can reveal the combustibility of cotton fabrics, as demonstrated in Figure 8.4c,d. It is observed that the effective heat combustion value of the modified cotton fabric reduced as compared to that of the unmodified

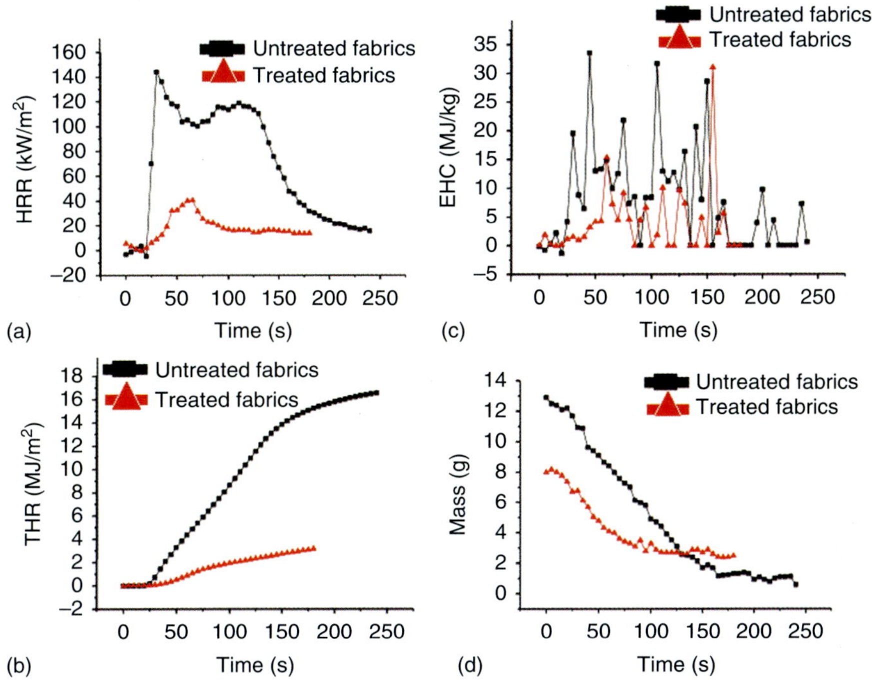

Figure 8.4 Combustion curves of treated and untreated cotton fabrics with (IB-*N*-SA) PDMS: (a) HRR, (b) THR, (c) EHC, and (d) mass curves. Source: Adapted from Dong et al. 2015 [32] with permission.

Table 8.5 Combustion data by a cone calorimeter.

Sample	TTI (s)	FPI (s m²/kW)	PHRR (kW/m²)	Average EHC (MJ/kg)	CO (kg/kg)	CO_2 (kg/kg)	CO_2/CO
Unmodified cotton fabrics	2	0.01	144.23	12.57	0.10	2.33	23.30
Modified cotton fabrics	15	0.37	40.70	10.17	0.34	2.44	7.18

Source: Dong et al. 2015 [32]. Reproduced with permission of Elsevier.

cotton fabric. The mass loss value of the modified cotton fabric is significantly less than that of the unmodified cotton fabric and it confirms the limited combustion of cotton fabric. Lower CO_2/CO ratio indicated incompetent combustion of materials. From Table 8.5, it can be observed that the CO_2/CO ratio of modified cotton fabric is reduced considerably. These values prove that the fire retardancy of cotton fabric is improved because of the application of (IB-*N*-SA) PDMS [32].

A polysiloxane (known as APID) having phosphorus, nitrogen, and also benzene rings have been prepared in the form of a fire retardant. APID has hydrophobicity as well as outstanding thermo-oxidative stability. APID is employed to enhance the fire retardancy and the mechanical performance of ammonium polyphosphate (APP)/PP composites. The fire-retardant efficiency

Table 8.6 LOI and UL-94 test results of the PP composites.

Samples	Components of the composites (wt%)			LOI (vol%)	UL-94 (1.6 mm)	
	APID	APP	PP		Rating	Dripping
PP	0	0	100	17.6	NR	Yes
PP1	0	25	75	23.2	NR	Yes
PP2	25	0	75	24.1	NR	Yes
PP3	12.5	12.5	75	28.9	V-0	No
PP4	8.33	16.67	75	29.1	V-0	No
PP5	16.67	8.33	75	26.8	V-2	Yes
PP6	6.25	18.75	75	28.4	V-0	No
PP7	18.75	6.25	75	26.4	NR	Yes
PP8	15	10	75	27.6	V-1	No
PP9	10	15	75	29.8	V-0	No
PP10	9.2	13.8	77	28.2	V-0	No
PP11	8.8	13.2	78	27.4	V-1	No

Source: Zhao et al. 2018 [33]. Reproduced with permission of Elsevier.

of the polypropylene composites with different amounts of flame retardants have been examined, and the test results are listed in Table 8.6. Virgin PP is an extremely combustible thermoplastic polymer, with a low LOI value and no score at the UL-94 test accompanied with severe dripping [33].

Figure 8.5 shows the (a) HRR, (b) THR, (c) mass loss, (d) carbon monoxide production (COP), (e) carbon dioxide production (CO_2P), and (f) total smoke production (TSP) results of virgin polypropylene (PP) as well as PP composites, the HRR result of virgin PP confirmed a sharp peak, along with the HRR attained its peak value of 980 kW/m^2 at 160 seconds. The PP9 exhibited the lowest possible PHRR of 168 kW/m^2 at 335 seconds. In contrast to the HRR curve of PP1 (APID and APP 0 wt%), the PP4 (APID 8.33 wt% and APP 16.67 wt%), and PP6 (APID 6.25 wt% and APP 18.75 wt%) manifested a pair of maxima, that has been related to two aspects: firstly, the composites degraded and an intumescent coating was created; secondly, these types of intumescent char degraded and subsequently the underlying material degraded and new intumescent shielding char was formed [33].

An efficient approach to enhance the thermal insulation along with the stability of polysiloxane foam (SIF) by manipulating the chain length of hydroxyl-terminated PDMS (OH-PDMS) has recently been discussed. SIF displayed a better level of fire-retardant behavior as compared to traditional polymer systems. This feature is because the combustion outputs of silicone-based polymers are silica; however, the combustion yield of carbon-based polymers is carbon dioxide. Figure 8.6 illustrates the burning operation of SIF under air. The entire combustion operation lasted for about 100 seconds. Toward the end of combustion, an off-white char layer is noticed on the surface of the samples. This layer works as a shielding layer to avoid the spreading of flame;

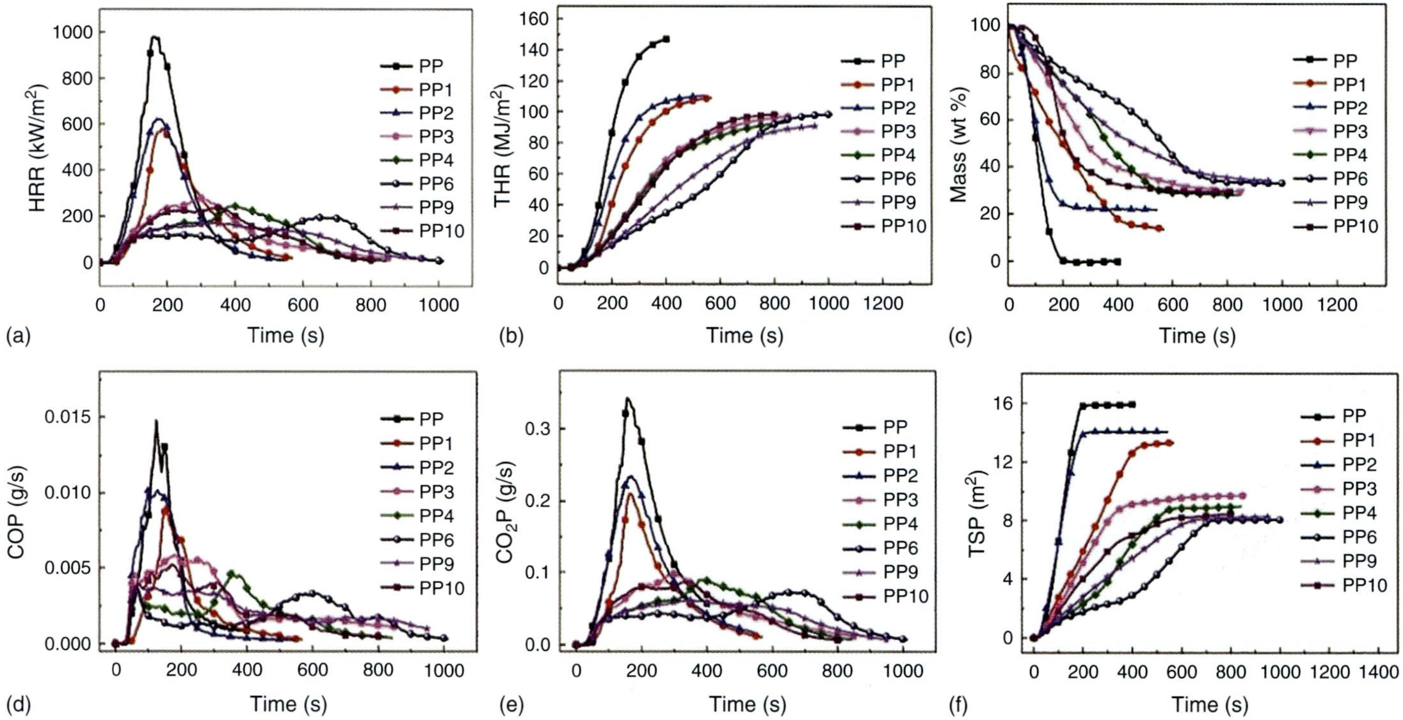

Figure 8.5 (a) HRR, (b) THR, (c) mass loss, (d) COP, (e) CO_2P, and (f) TSP curves of PP/flame-retardant composites. Source: Zhao et al. 2018 [33]. Reproduced with permission of Elsevier.

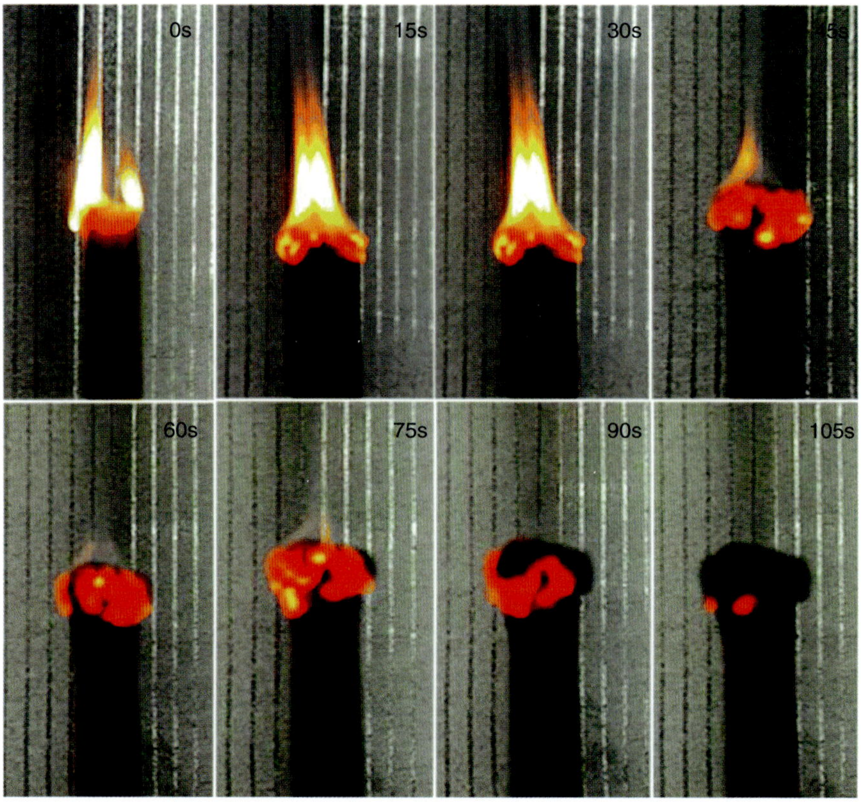

Figure 8.6 Combustion process of polysiloxane foams sample under air. Source: Zhang et al. 2018 [34]. Adapted with permission of Royal Society of Chemistry.

however, the char is found to be brittle in nature. Figure 8.7 demonstrates the LOI of SIF. The LOI calculates the minimum oxygen concentration of the flowing gas (e.g. nitrogen as well as oxygen) that is needed for promoting flame combustion. This index is used to examine the fire-retardant behavior of the materials. The LOI of SIF changes with the chain length of the polysiloxane: LOI of polysiloxane increases with increase of chain length. In conclusion, there are many scientific studies, that have already proven that polysiloxane is excellent for flame-retardant application [34], as mentioned in Table 8.7.

8.3.2 Polyphosphazenes

Polyphosphazenes are a group of functional organic–inorganic hybrid materials, which have several remarkable features. It is widely known that polyphosphazenes are being employed as biomaterials, optical materials, electrical materials, hybrid materials, high-temperature materials, and so on, because of their excellent thermal stability as well as structural configuration. Exhibiting equivalent structures to carbon nanotubes, polyphosphazenes possess interesting capability to act as flame retardants. These types of materials display

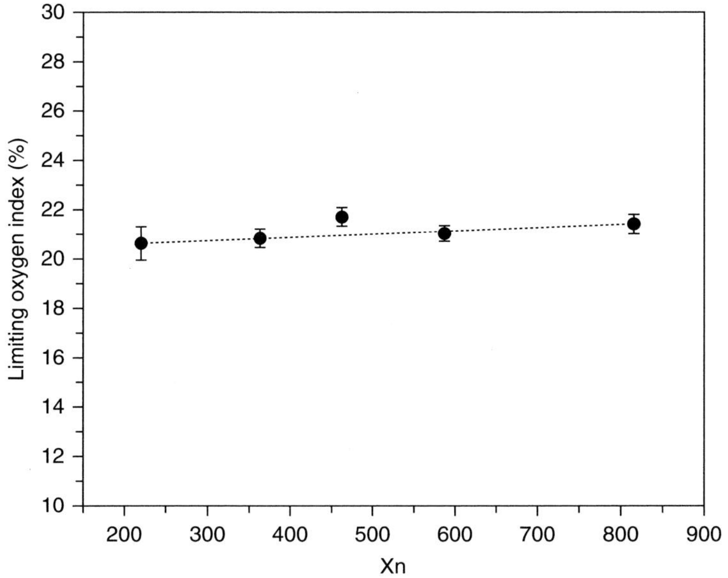

Figure 8.7 Limiting oxygen index of SIFs with different chain lengths of OH-PDMS. Source: Zhang et al. 2018 [34]. Adapted with permission of Royal Society of Chemistry.

Table 8.7 Various types of a flame-retardant system based on polysiloxane.

Type of polysiloxane	Remarks	References
Polysiloxane grafting multiwalled carbon nanotubes/polycarbonate nanocomposites	• Improvement in flame retardancy, tensile strength, flexural strength, a flexural modulus of the polycarbonate was achieved.	[35]
Epoxy-functional polysiloxane(as flame retardant)/poly(butylene terephthalate)/aluminum diethylphosphinate composites	• The epoxy-functional polysiloxane greatly encouraged the formation of cross-linking char residue. • The condensed-phase action was proposed to be the dominant flame-retardant role of the composite.	[36]
Multi-maleimide terminated branched polysiloxane(as flame retardant)/bismaleimide/diallyl bisphenol resins	• Reduces the risk of catching fire and combustion intensity, and decreases the smoke production of the resin.	[37]
Carboxyl-containing polysiloxane(as flame retardant)/polycarbonate composite	• Increases the flame-retardant effectiveness and reduces the degradation speed of the polycarbonate.	[38]
Polysiloxane-containing nitrogen(as flame retardant)/epoxy resin	• Improvement in flame retardancy, thermal stability, a limited oxygen index of the epoxy resin.	[39]
Ladder-like multifunctional polysiloxane (PN-PSQ), with a large number of amine groups and a phosphaphenanthrene(as a flame retardant)/bismaleimide resin	• Improvement in the curing process, flame retardant thermal and dimensional stability properties of the bismaleimide resin.	[40]

(Continued)

Table 8.7 (Continued)

Type of polysiloxane	Remarks	References
Guanidyl- and phosphorus-containing polysiloxane(flame retardant)/ cotton fabrics	• Improvement in fire performance index, flame-retardant effect, reduction in combustion heat, CO/CO_2 ratio of the cotton fabric.	[41]
Nitrogen- and phosphorus-containing polysiloxane/cotton fabrics	• Enhancement in thermal and thermo-oxidative stability and flame-retardant properties of the cotton fabric; longer time to ignition.	[42]
Polysiloxane–polypropylene system	• Enhancement of the thermal stability at high temperature and higher char residue, flame-retardant properties of the polypropylene.	[43]
Polysiloxane solid powder combined with organophosphate in a polycarbonate matrix	• Improvement in flame retardancy, reduction and inhibition of CO production rate as well as combustion heat release of the polycarbonate composites.	[44]

high LOI value, and are classified into three types. The first type is the linear polyphosphazenes or phosphazene polymers with cyclophosphazene units in the principal chain or side chain [45]. Although there are several positive aspects with respect to linear polyphosphazenes, the primary drawbacks include poor production rate as well as their high costs, which limit their application. The second type of polyphosphazenes, which has drawn substantial attention is cyclotriphosphazene. These types of materials have a principal chain of —P=N— units and display self-extinction in the flame study. Cyclotriphosphazenes can be applied to synthesize micro- or nanoscale polymeric materials by means of condensation polymerization. All these micro- or nanoscale polymeric materials comprise polyphosphazene nanotubes, microspheres, nanofibers, and also nanochains [46]. Poly(cyclotriphosphazene-*co*-4,4′-sulfonyldiphenol) (PZS) nanotubes have already been created through one-pot strategy with desired microstructures. These kinds of nanomaterials can show excellent thermal stability and fire retardancy as well as radiation resistance. In such a way, the impact of poly(dianilinephosphazene) (PDAP) on the processability, thermal tendency, crystallinity, morphology, and mechanical attributes as well as the flammability characteristics of low-density polyethylene (LDPE) has been evaluated. It has been confirmed that LDPE modified with PDAP is able to offer remarkable enhancement in the flame-retardant characteristics of the LDPE, as presented in Table 8.8. There is no doubt that the LOI value enhanced with respect to the PDAP content, both phosphorus and nitrogen components in PDAP, have specific roles as fire retardants. It is considered that the coexistence of phosphorus along with nitrogen in the same molecule can offer a synergistic result in the properties of PDAP/LDPE blends [47].

Table 8.8 Mechanical and limited oxygen index of the low-density polyethylene composites (LDPE).

Sample	Tensile strength (MPa)	Limiting oxygen index
LDPE	22.84	17.4
LDPE (90%)/PDAP (10%)	21.33	24.6
LDPE (85%)/PDAP (15%)	18.41	26.7
LDPE (80%)/PDAP (20%)	16.74	28.4

Source: Chen et al. 2002 [47]. Reproduced with permission of John Wiley and Sons.

During the synthesis of polyurethane/poly(bis(carboxylatophenoxy) phosphazene) foamed blends, a few carboxylic acid groups of the poly(organophosphazene) are able to react with the isocyanate groups throughout the polymerization as well as the curing process of the polyurethane foam. This can result in the inclusion of the poly(organophosphazene) into the polyurethane network. The burning tendencies of these kinds of foams have been examined by a horizontal flame test as well as by LOI analysis. The data obtained from these two procedures suggested an improvement in fire resistance at loadings of 20 wt% or above. The thermal stability of the prepared foams has been evaluated by thermogravimetric analysis (TGA). The char output at both 400 and 600 °C improved in relation to the virgin polyurethane via increasing the quantity of polyphosphazene from 5 to 20 wt% [48]. The functionalized polyphosphazene (FR@PZS) nanotubes covered with a cross-linked DOPO-containing fire retardant properties have been prepared by means of one-step approach and were also analyzed for their fire-retardant properties. The stages of preparation of fire-retardant covered PZS nanotubes (FR@PZS) is demonstrated in Figure 8.8. This fire retardant has been introduced into the epoxy resin (EP) to create samples for the study of their fire-retardant properties [49].

Cone calorimeter is a very popular instrument for analyzing the combustion performance of a wide variety of materials subjected to flames. The HRR together with THR vs. the time graph of EP composites is shown in Figure 8.9. The pristine EP is remarkably flammable, with a PHRR value of 1820.7 kW/m^2. A clear decrease in the PHRR is noticed for the EP/PZS3.0 (3 wt%), a 36.7% decrement as compared to pristine EP. Including FR@PZS into EP additionally reduces the PHRR values. Because of this, the PHRR of EP/FR@PZS3.0 (3 wt%) is reduced by 46.0% as compared to EP, suggesting the optimum flames safety efficiency in association with all these samples. The enhancement is related to catalytic carbonization of DOPO-based flame retardants as well as the barrier effect of PZS. Figure 8.9b implies that the THR of EP composites shows a reducing pattern of PHRR. EP/FR@PZS3.0 (3 wt%) exhibits lesser THR value as compared to EP/PZS3.0 (3 wt%) by increasing the FR@PZS content. After introducing 3.0 wt% content of FR@PZS in EP, the THR of EP/FR@PZS3.0 is decreased to 72.4 MJ/m^2, 27.1% less than that of pristine EP. The significant decline in the fire hazards of EP/FR@PZS is related to the gas as well as condensed phase action: the catalytic

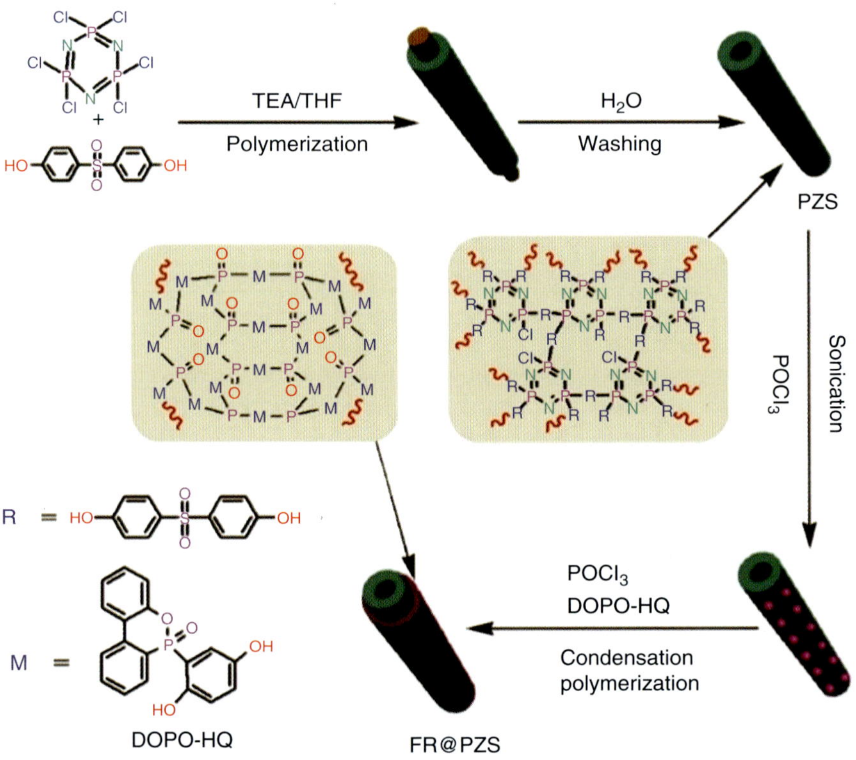

Figure 8.8 The fabrication method of functionalized polyphosphazene nanotubes (FR@PZS) wrapped with a cross-linked DOPO-based flame retardant. Source: Qiu et al. 2017 [49]. Adapted with permission of Elsevier.

carbonization has an impact on of the wrapped FR, which reduces the discharge of degradation products; the physical barrier effect of randomly dispersed PZS network structure slows down the heat as well as mass transfer [49].

The smoke production rate (SPR), and the total smoke release (TSR) graph of EP nanocomposites, are shown in Figure 8.10. Owing to its particular multi-aromatic architectural structures, pristine EP displays high production rate of harmful smoke with more PSPR and TSR values. EP/PZS exhibits a small decrease in the PSPR and TSR. In addition, the inclusion of FR@PZS to EP considerably decreases the PSPR value. It is reduced from 0.47 m^2/s for pristine EP to 0.21 m^2/s for EP/FR@PZS3.0 (3 wt%) with a deduction of 55.3%. An EP/FR@PZS3.0 (3 wt%) sample additionally reflects the minimum TSR with a minimization of 44.1%, as compared to that of pristine EP. The above data prove that the EP/FR@PZS3.0 (3 wt%) composite displays the most beneficial flame retardancy among all the samples, which implies the thick shielding char layers in the form of barriers by the assistance of polyphosphazene and consequently DOPO-based FR in EP/FR@PZS composites are found more efficient as compared to the EP/PZS.

The preparation of cross-linked organic–inorganic polyphosphazene nanoshells (PZM) with amino-rich active groups has been discussed through

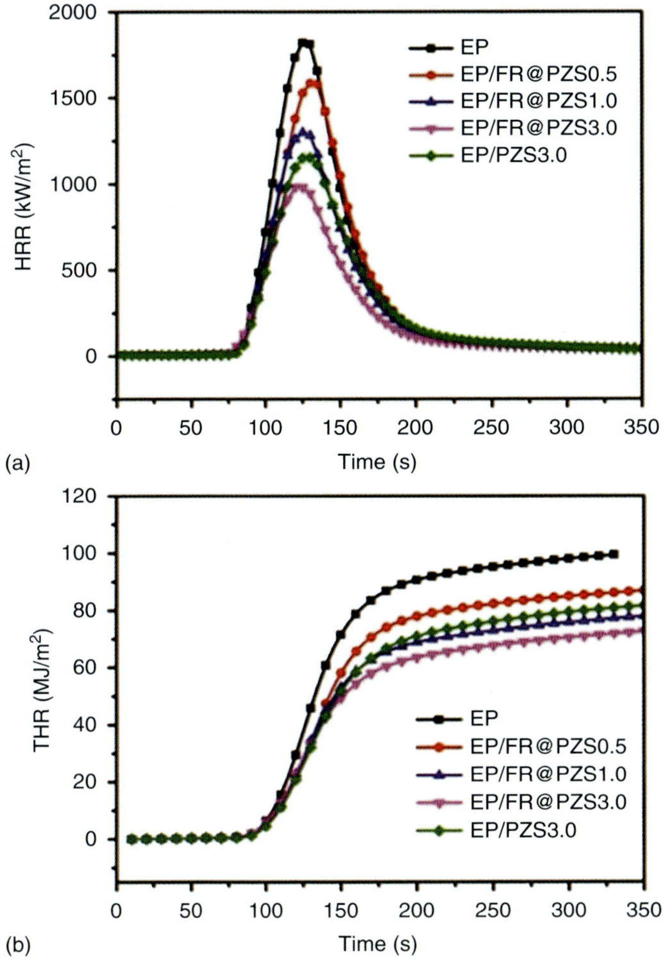

Figure 8.9 EP curves of (a) HRR vs. time, (b) THR vs. time and its nanocomposites obtained using cone calorimeter. Source: Qiu et al. 2017 [49]. Adapted with permission of Elsevier.

a condensation polymerization process of hexachlorocyclotriphosphazene (HCCP) along with 4,4′-diamino diphenyl ether (4,4′-ODA) on silica (SiO_2) nanospheres in the form of templates. Subsequently, cuprous oxide nanoparticles (Cu_2O NPs) have been prepared via synchronized chemical reduction on the surface of PZM. Consequently, the resulting SiO_2@PZM@Cu spheres have been introduced into the epoxy resin to produce samples for analysis of their flame-retardant properties. This core–shell nanostructure is generally employed to enhance the fire retardancy as well as harmful effluents removal of epoxy resins (EP) through synergistic catalytic effect. The fabrication path of SiO_2@PZM@Cu spheres is presented in Figure 8.11. Various crucial parameters received using the cone calorimeter are time to peak heat release rate (TPHRR), peak heat release rate (PHRR), THR, the peak smoke production rate (PSPR), the TSR, and maximum average heat rate emission (MAHRE) values, which are listed

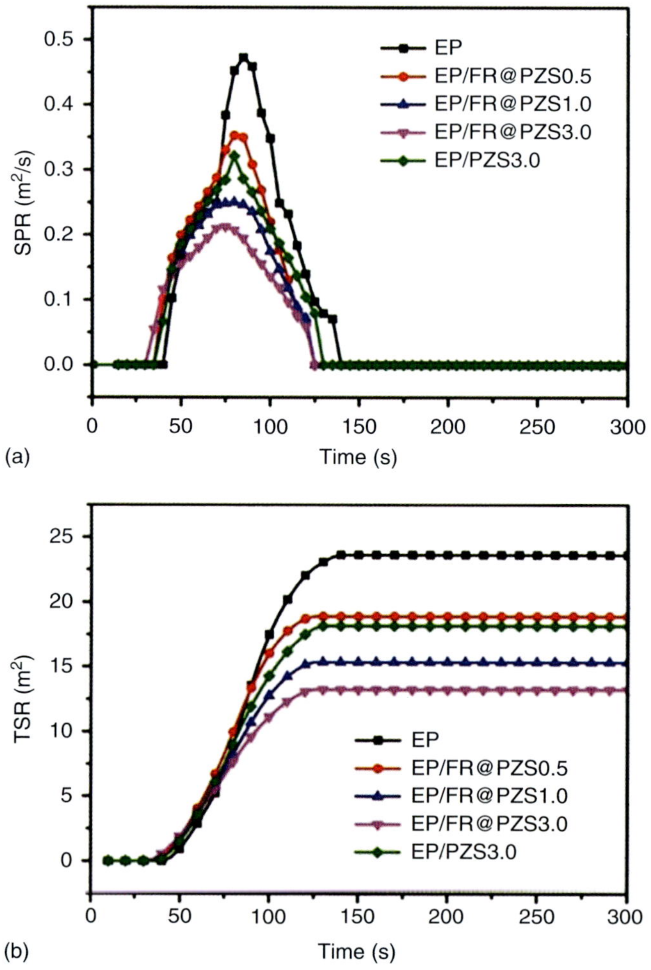

Figure 8.10 EP curves of (a) SPR vs. time and (b) TSR vs. time and its nanocomposites obtained using cone calorimeter. Source: Qiu et al. 2017 [49]. Adapted with permission of Elsevier.

in Table 8.9. The considerable improvement of flame retardancy as well as toxin effluents removal is due to the synergistic action between the catalytic effect of SiO_2 in conjunction with Cu_2O NPs and also the intumescent effect of PZM [50].

In another research, melamine-based polyphosphazene-wrapped ammonium polyphosphate (PZMA@APP) with abundant amino active groups has been produced and consequently applied in the role of an effective fire retardant. PZMA@APP showed superior fire-retardant performance and also the smoke reduction to the EP composites. The polyphosphazenes-encapsulated APP (PZMA@APP) with the amino active groups on the surface is anticipated to considerably reduce the interfacial tension between PZMA@APP along with EP matrix. The developed arrangement of the polyphosphazenes-encapsulated APP and cross-linked structure of EP/PZMA@APP is demonstrated in Figure 8.12 [51].

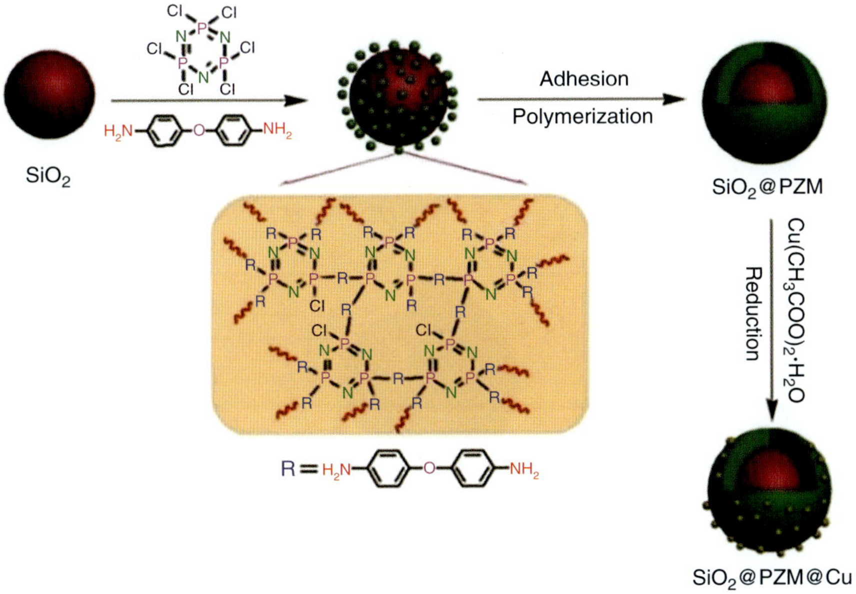

Figure 8.11 Synthetic route of SiO_2@PZM@Cu. Source: Qiu et al. 2017 [50]. Adapted with permission of Elsevier.

Table 8.9 Cone calorimeter data of EP and its nanocomposites.

Sample	TPHRR (s)	PHRR (kW/m^2)	THR (MJ/m^2)	PSPR (m^2/s)	TSR (m^2)	MARHE (kW/m^2)
EP	125	1915.3	107.6	0.48	23.6	667.4
EP/SiO_2	133	1776.9	95.6	0.37	21.6	614.8
EP/SiO_2@PZM	135	1363.4	86.8	0.35	17.7	564.2
EP/SiO_2@PZM@Cu1 (1 wt%)	125	1289.3	78.0	0.27	16.6	494.3
EP/SiO_2@PZM@Cu2 (2 wt%)	130	1188.8	73.9	0.25	13.7	479.9

Source: Qiu et al. 2017 [50]. Reproduced with permission of Elsevier.

The flame retardancy performances of polymer composites are measured by cone calorimeter test assessment, which shows the combustion scenario of a variety of materials. Figure 8.13a,b provide the HRR and THR relative to the time behavior of the EP composites. The PHRR and THR values for the EP having 5 wt% of PZMA@APP have been notably reduced by 58.7% and 52.7%, respectively. However, introducing 7.5 wt% of PZMA@APP into EP results in the additional decrease of PHRR as well as THR values. At the same time, the PHRR and the THR values of EP/APP10.0, have been decreased by 66.4% and 59.4%, respectively, as compared to pristine EP. Specifically, the inclusion of 10 wt% PZMA@APP leads to a lowering of PHRR around 75.6%, and an optimum 65.9% decrement of THR for combustion. The PHRR and the THR values of an EP/PZMA@APP10.0 sample are found to be lower as compared

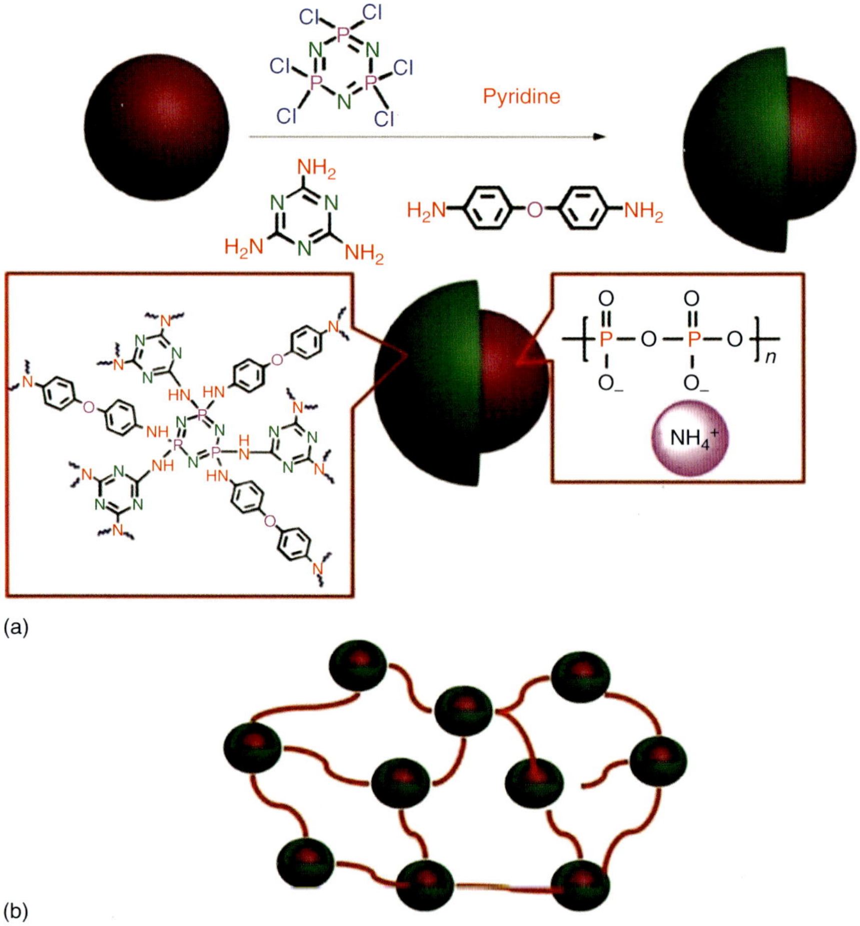

Figure 8.12 (a) The structure of PZMA@APP, (b) the cross-linked structure of EP/PZMA@APP. Source: Qiu et al. 2018 [51]. Reproduced with permission of Elsevier.

to those of an EP/APP10.0 sample with similar content. This suggests that better flame safety capabilities of EP/PZMA@APP10.0 composites in contrast to EP/APP10.0 composites. The smoke production rates (SPR) vs. time curves of EP and its composites have been provided in Figure 8.13c. Pristine EP with a multi-aromatic structure discharges huge amount of smoke with excessive SPR values throughout combustion; the SPR values of the EP/PZMA@APP composites diminishes when increasing the content of PZMA@APP in EP matrix, and consequently EP/PZMA@APP10.0 exhibits the lowest SPR value, as compared to that of EP/APP10.0 with similar content. The remarkable reduction in the flames hazards of EP/PZMA@APP are attributed to the catalytic carbonization ripostes between the APP and PZMA phase, which slows down the pyrolysis and provides safety to materials during a fire [51].

Polyphosphazane, with an array of fascinating features and applications, proves that it is one of the important class of inorganic polymers. Polyphosphazenes

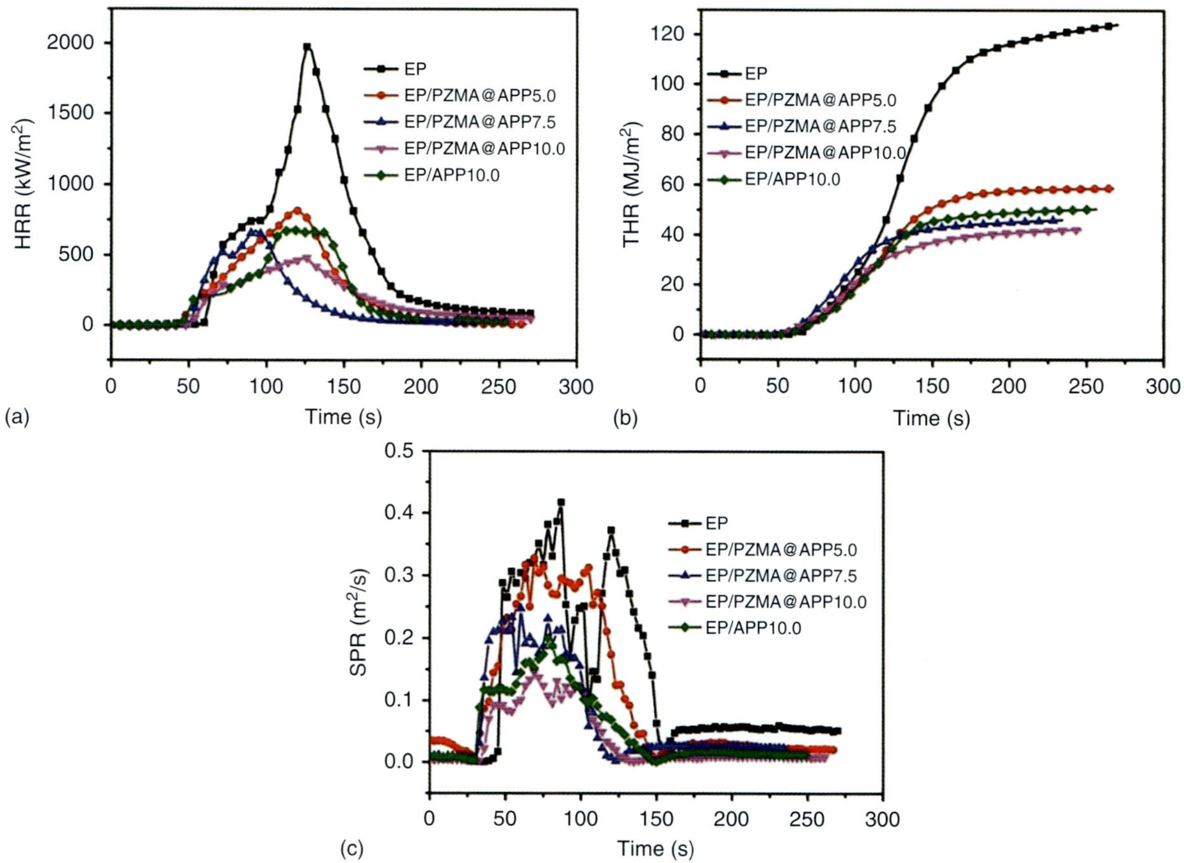

Figure 8.13 EP curves of (a) HRR vs. time, (b) THR vs. time, and (c) SPR vs. time and its composites. Source: Qiu et al. 2018 [51]. Reproduced with permission of Elsevier.

are employed in the role of flame retardants, additives, functional polymers, and in other specific applications. Moreover, the general applications of these kinds of polymers are O-rings, fluoroelastomer seals, gaskets, insulating foams, optic devices, polymer conductors, coatings, fiber, and rubbers. The fascinating fire-retardant behavior, as well as high-temperature applications of these polymers, is outlined in Table 8.10.

8.3.3 Polysilane and Polysilynes

Materials having silicon atoms in their molecular structure have drawn significant attention because they are frequently used in the form of precursors for ceramic materials. Also, polysilanes have been used as a thermally stable inorganic polymer with a backbone composed completely of silicon–silicon bonds with two substituents connected to every single silicon atom. There are several kinds of polysilanes that can be used in a variety of applications. For example, poly(dimethylsilane) is a type of linear preceramic polymer having silicon atoms in its main chain [62]. The existence of Si atoms in its structure enhances the fire retardancy of thermoplastic and mechanical properties of the thermoplastic matrix. Poly(dimethylsilane) has additionally been employed in the preparation of polycarbosilane, which is a key ceramic precursor for producing silicon carbide. Poly(diorganosilane)s, are reliable as well as stable in oxidative and hydrolytic conditions, and are therefore used conveniently without inert gas methods. Thermal characteristics of varieties of polysilanes have been investigated by differential scanning calorimetry (DSC) as well as in TGA methods [63]. Thermal properties of poly(methyl-*n*-propylsilane) (PMPrS) have been analyzed through DSC; the melting temperature is dependent on the thermal history as well as on the molecular weight of the PMPrS. The dual melting of PMPrS has been noticed during isothermal crystallization. This tendency is an indication of the double-melting characteristic recognized for semi-crystalline aromatic polymers, for example, PEEK, PET, or PBT and has also been noticed for another type of polysilanes [64]. The double melting is generally attributed to a melting-recrystallization that occurs throughout the heating, as observed by DSC. The thermogravimetry analysis of polydimethyl–methylsilane demonstrates the solitary-step decomposition under a nitrogen environment. The decomposition of the material begins at 263 °C. The TGA of polydimethyl–methylsilane under N_2 demonstrated a two-step decomposition which begins at 212 °C with a 10% weight reduction of the polymer; this can be related to the vaporization of the accumulated solvent in the polymer, decomposition of low molecular weight polymer as well as to the presence of other contaminants. Later on, at 345 °C, reduction in 27% weight has been observed, which can be related to the evolvement of methane as well as other hydrocarbon gases because of cleavage of the Si—CH_3 bond. This weight loss remains upto 550 °C and around 50% residue is found up to 700 °C. The TGA of the copolymer of polydimethyl–methylphenylsilane revealed a loss of weight (10%) at 219 °C, which can be caused by the loss of accumulated solvents and decomposition of low molecular weight polymers along with the presence of additional impurities. It has been found that at 299 °C a weight loss (16%) of the polymer takes place, which can

Table 8.10 An overview of polyphosphazene-based materials for flame-retardant and high-temperature application.

Type of system	Remarks	References
Phosphazene cyclomatrix network polymers	• Excellent thermal stability and flame-retardant properties.	[52]
Hybrid polyphosphazene nanotubes	• Good thermal stability and excellent char performance (70% even at a high temperature of 800 °C).	[53]
Hybrid polyphosphazene nanotubes(HPPN)/PP system	• High-performance flame- retardant behaviors of PP/HPPN blend at 30 wt% of HPPN. HPPN acted as flame retardant as well as reinforcing agent for PP.	[53]
Epoxy curing system containing linear polyphosphazene moieties	• Good flame-retardant properties, extraordinary nonflammability due to phosphorus–nitrogen combination in linear polyphosphazene-based epoxy resin segments.	[54]
Flame-retardant textiles based on allyl-functionalized polyphosphazenes	• Functionalized polyphosphazenes functioned as a flame retardant in cotton/polyester blend. • Improvement in limiting oxygen index and flame-retardant behavior of cotton/polyester blended fabrics.	[55]
Octa-aminophenyl polyhedral oligomeric silsesquioxanes – polyphosphazene/epoxy resins system	• Reduction in heat release rate, increment in char formation, released nonflammable gases such as CO_2, NH_3, and N_2, the high-performance flame-retardant behavior of the epoxy resin.	[56]
Thermoplastic polyurethane/ polyphosphazene blends	• Increase in the onset of degradation temperature and char yields of TPU, improved limiting oxygen index, increase in flame retardancy of TPU/PPZ blend.	[57]
Hexamethoxycyclotriphosphazene/ polyurethane foams	• Improvement in thermal, and flame-retardant properties of PU foams.	[58]
Polyphosphazene–polybenzoxazine networks/polybenzoxazine phenolic thermosets	• Improvement in flame retardancy, limiting oxygen index, smoke density rating, vertical burning rating.	[59]
Polyphosphazene microspheres (PZS)@molybdenum disulfide nanoflower (MoS_2) hierarchical hybrid architecture/epoxy resin	• Reduction in peak heat release rate and decrease 30.3% maximum in total heat release, improved flame-retardant performance and char formation of the epoxy resin composites.	[60]
Low-density polyethylene/ ethylene-vinyl acetate/ polyphosphazene	• Improvement in limiting oxygen index, char formation, flame retardancy of the composites.	[61]

Table 8.11 Flammability properties of polysilynes.

Sample	Poly(methylsilyne)	Poly(phenylsilyne)
Heat release capacity (J/g K)	92	32
Total heat (kJ/g)	15	13
Char yield (%)	64	77
Maximum mass loss rate ($\times 10^3$/s)	0.34	0.12
T_{max} (°C)	447	425
T_{onset} (°C)	126	140

Source: Ma et al. 2007 [67]. Reproduced with permission of Elsevier.

be linked to the cleavage of Si—Me and Si—Si bonds as well as to the expansion of hydrocarbon gases [65].

Polysilynes, which are basically network polysilanes, are a comparatively new category of preceramic polymers. They were initially prepared by Bianconi along with Weidman, in 1988 [66]. They show a wide UV band that falls into the visible region in contrast to the polysilanes, and as a result, they possess the visible absorption as well as emission features. Flammability in addition to thermal characteristics of poly(methylsilyne) and poly(phenylsilyne) have been investigated in various studies. The TGA results of poly(methylsilyne) and poly(phenylsilyne) indicates that, after 600 °C, there is no additional mass loss. Subsequently, these polymers transfer greater than 60% char because of their thermal decomposition. It is also observed that poly(phenylsilyne) possesses better thermal stability as compared to poly(methylsilyne) because its decomposition is sluggish and creates much more char. The thermal decomposition yield of poly(methylsilyne) is siloxanes due to the oxidation of the polymer.

Both poly(methylsilyne) and poly(phenylsilyne) possess poor flammability. The flammability of polysilynes originates from the methyl or phenyl group. The flammability as well as thermal decomposition characteristics of polysilynes, is shown in Table 8.11 [67].

8.3.4 Ferrocene-Based Polymers

Metal-based polymers are the key group of polymeric materials. The growth and development of these types of materials are attributed to the metal they consist of and are likely to have considerably better features from those of traditional organic polymers. In this fashion, ferrocene-containing polymers hold excellent attributes such as air, heat, and also photochemical stability. Considering the earlier studies of ferrocene in the role of flame-retardant materials, the aim of this section is to provide a general idea of the ferrocene-based polymer for flame-retardant applications [68]. The introduction of ferrocene into the polymer backbone enhances the thermal stability as well as fire retardancy of polymeric materials. It is applied as a filler, without having to be bonded into the polymer structure, or it is can be bonded specifically into the main chains through functionalization. Ferrocene is proficient in generating much more char if bonded into the backbone of the polymer. The properties of ferrocene-based flame-retardant materials have additionally been demonstrated in the analysis of diamine monomer [69]. This monomer comprises a variety of functional groups, such as amides, ethers, heterocyclic pyridines, ferrocene, and a different R-group. LOI values of these polymers have been analyzed and an enhancement of 5–5.5% has been observed for the ferrocene-containing polymer. Based on the research on the synergistic effects of ferrocene (Fc) as a flame retardant with molybdenum disulfide (MoS_2), it was perceived that both are appropriate to produce nanosheets and contribute to improving the flame-retardant behavior in polystyrene. Also, it has been discussed that ferrocenyl compounds are employed to enhance the flame retardancy as well as to minimize the smoke generation for polymer materials simultaneously, due to their ability to develop chars. Authors have prepared certain ferrocenyl derivatives and consequently employed them in polyvinyl chloride (PVC), to provide decent smoke suppression as well as flame retardancy [70]. Non-phosphorus polymer poly((3,3′-diphenyl diacetylethylenediamino)-1,1′-ferrocene) (PDPFDE), based on ferrocene, benzene, and amine groups, has been prepared through an aza-Michael addition reaction, and introduced into an epoxy resin to enhance flame retardancy. The thermal stability, the gaseous pyrolysis yield, burning characteristic of the PDPFDE, and a variety of EP/PDPFDE composites have been reported, and the curing process of EP/PDPFDE composites is mentioned in Figure 8.14 [71].

LOI and UL-94 vertical burning are reliable measurements for estimating the flame retardancy of materials. The related composition and also the data on the EP composites are outlined in Table 8.12. Pristine epoxy resin (EP-0) is quickly

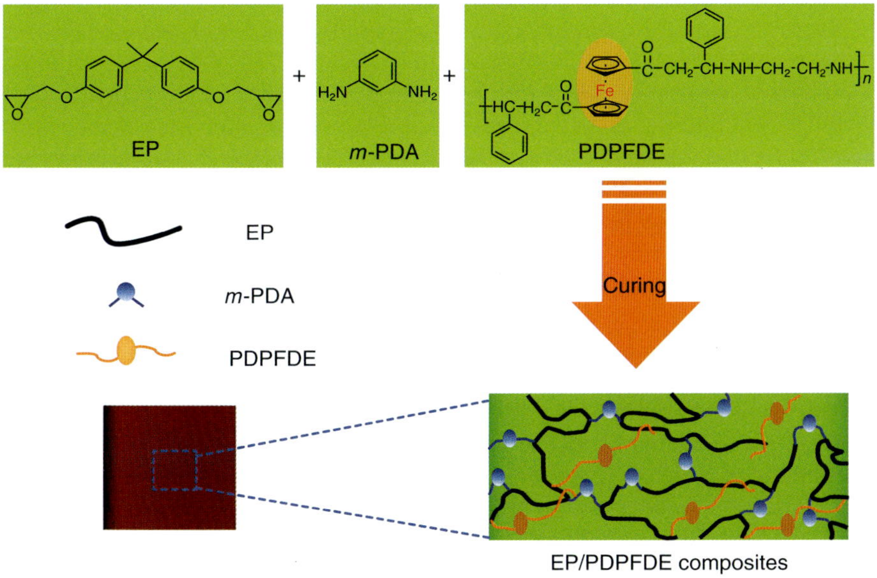

Figure 8.14 A schematic representation of the curing process of EP/PDPFDE composites. Source: Liao et al. 2017 [71]. Adapted with permission of American Chemical Society.

Table 8.12 Detailed formulation and flame retardancy of the composites.

Sample	EP-0	EP-3	EP-4	EP-5	EP-6	EP-7
PDPFDE (wt%)	0	3	4	5	6	7
LOI (%)	24.0	27.7	28.5	29.1	29.6	28.2
Dripping	NO	NO	NO	NO	NO	NO
Grade	NR	V-2	V-1	V-1	V-1	V-2

Source: Liao et al. 2017 [71]. Reproduced with permission of American Chemical Society.

flammable with an LOI value of 24.0% and has not been categorized under the UL-94 rating. A 3.0 wt% of PDPFDE composite with EP, enhanced the LOI value to 27.7% and the UL-94 rating attained was V-2. The LOI value reached an optimum value of 29.6% and the UL-94 rating achieved was V-1 for EP composites with 6.0 wt% PDPFDE. Nevertheless, a 7.0 wt% loading of PDPFDE dropped the flame retardancy of the EP composites; the LOI value reduced to 28.2% and this formulation failed to achieve the V-1 rating. This result may be due to the slower charring rate as compared to the decomposition rate. This trend showed that there is always an optimum stability point between the catalyzing decomposition as well as the char generation of the PDPFDE [71].

A number of ferrocene-based polyamides have been produced through a polycondensation reaction of a ferrocenyl diamine with a variety of diacid chlorides in *N*-methyl-2-pyrrolidone (NMP) using trimethylchlorosilane (TMSCl) in the role of the activating agent. Ferrocenyl diamine (FDA) material with cutting-edge characteristics has been produced by means of the condensation reaction of 1,1′-ferrocenedicarbonyl chloride with 2 M of 2,6-bis(4-aminophenoxy)pyridine. The intrinsic viscosities, solubility, thermal stability, and fire-retardant behavior of the resultant polymers have been investigated. The polymers exhibited effective heat resistance and flame retardancy [72]. The smoke generation of the epoxy film and the effective smoke suppression characteristics of the epoxy cross-linked structure together with ferrocene (FC) on epoxy-based intumescent flame-retardant (IFR–EP) coating have already been examined. The test results confirm that the cross-linked arrangement of epoxy/polyamide resin efficiently improve thermal stability and consequently decrease smoke in the initial stages of combustion and, in addition, decrease the generation of toxin and combustible gases and boost the generation of nonflammable gases. Alternatively, the FC is able to control smoke by radical reactions at the beginning of the combustion phase and enhance the making of denser residue char. Furthermore, the FC is also able to efficiently minimize the total smoke rate in the entire combustion process [73]. The dynamic smoke generation tendencies have been investigated via cone calorimeter test method, and the results have been mentioned in Table 8.13. As is listed in Table 8.13, the PHRR gives the risk of materials with relevance to catching fire; the peak smoke production rate (PSPR); and the total smoke rate (TSR), which is the term denoting the strength of smoke generation under test situations. In addition, it is essential to examine the smoke factor

Table 8.13 Typical data for EP and EP/PA from cone calorimeter tests.

Resin	PHRR	PSPR	TSR	SF	SP
EP	2658.40	0.46	2237.70	4567.40	787.70
EP/PA	802.70	0.13	1969.60	685.00	344.20

Source: Chen et al. 2016 [73]. Reproduced with permission of John Wiley and Sons.

(SF) as well as the smoke parameter (SP), which reveals the smoke generation behaviors throughout combustion; low SF, as well as SP values suggests that the smoke generation is significantly reduced. Table 8.13 further demonstrates that all the cone calorimeter test data of EP/PA are greatly reduced as compared to that of EP, showing that the epoxy cross-linked structure is able to efficiently minimize the smoke generation [73].

8.3.5 Carborane-Containing Polymers

Boron-containing compounds such as boric acid, borax, zinc borates, and ammonium pentaborate are well-known fire retardants used in paints, coatings, and commodity plastics. This section will review one of the organoboron compounds, carboranes, its attachment to the polymer backbone, end-use applications, and fire test performance. There are many carborane clusters such as *close-*, *nido-*, and *arachno-*carboranes that have been synthesized and developed; however, closed-cage carboranes are over-rated due to easier modification of the carborane cluster to obtain polymerizable groups on the cage [74, 75]. Figure 8.15 shows the example of the carborane clusters mainly used in polymer synthesis. 1,2-$C_2B_{10}H_{12}$ (*o*-carborane) is a white solid and is quantitatively converted to 1,7-$C_2B_{10}H_{12}$ (*m*-carborane) between 400 and 500 °C. The isomer, *p*-carboranes, can be obtained after thermal treatment at 700 °C. The stability of the icosahedral cage during degradation in the presence of strong acids and oxidant allows extensive chemistry to be applied without disrupting the carborane clusters [76]. There are several examples of the numerous types of carborane polymers that have been prepared and have served in numerous applications [77, 78]. Here, the primary emphasis is placed on the carborane-type polymers that exhibit unique thermal stability.

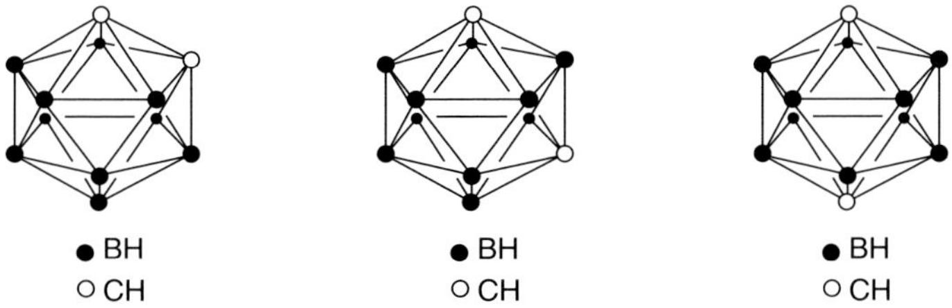

Figure 8.15 The three isomers: *o*-, *m*-, *p*-carboranes. Source: Adapted from Lu and Hamerton 2002 [75] with permission.

8.3.5.1 Poly(carboranylenesiloxanes)

The first commercially available carborane products were based on siloxy-*m*-carborane polymers marketed as DEXSIL and UCARCIL used for high-temperature gas chromatography [76]. In the presence of $FeCl_3$ as catalyst, silyl group functionalized carboranes and dichlorosilanes were given to the polymeric species in a simple condensation reaction having a backbone structure, as shown in Figure 8.16.

Bulk condensation of (dimethylmethoxy)-*m*-carborane and (dichlorodimethyl)-silane occurs in the presence of $FeCl_3$ and the rate-determining step is determined by the ^1H NMR [79]. The large entropy of the reaction showed that the transition step is highly ordered. However, further work on other Lewis acid catalysts is desirable due to the cross-linking problem encountered with the use of $FeCl_3$. Carborane cage caused a dramatic effect in the thermal and oxidative stability of the elastomer derived from poly(carborane–siloxane) and provided exceptional stability in the 350–450 °C range. Several publications have been released from that time. Keller and coworkers reported the development of new derivatives of poly(carborane–siloxane)-type polymers exhibiting outstanding thermal and oxidative properties [80–82]. Novel linear polymers containing alternating diacetylene and siloxane units were prepared in high yields. The syntheses are one-pot procedures that involve the reaction of 1,4-dilithiobutadiyne with the appropriate dichloro functional siloxane–carborane compound. For example, Henderson and Keller studied the effects of incorporating acetylenic units into the poly(carborane–siloxane) elastomers for cross-linking purposes [80]. Pyrolysis of poly(carborane–siloxane–acetylene) to 1000 °C gave a black solid residue in 85% yield. The oxidative stabilities of silane, siloxane, and/or carborane-bearing polymers were compared after heat treatment. It was observed that ceramic yields for the silane (13%) and siloxane (30%) were much lower than that for the carborane (64%) monomer. After the initial chars are formed under nitrogen at 1000 °C, the ceramic residues were further tested by heating to sample at 1000 °C for their oxidative stability under air at high temperatures. The char from the carborane-bearing polymers showed the most oxidative weight retention as 100% and enhanced the thermo-oxidative stability performance. Carborane-bearing polymers showed no weight loss and actually exhibited a slight weight gain at elevated temperatures under air atmosphere. This slight weight gain might be associated with the oxidation of boron to B_2O_3 and of silicon to SiO_2 present in the polymer backbone [81]. Moreover, carborane units present in the backbone demonstrated superior thermal and oxidative stability, relative to nonbearing carborane cage in the siloxane or silane-based thermoset, indicating the importance of boron in the design of polymeric

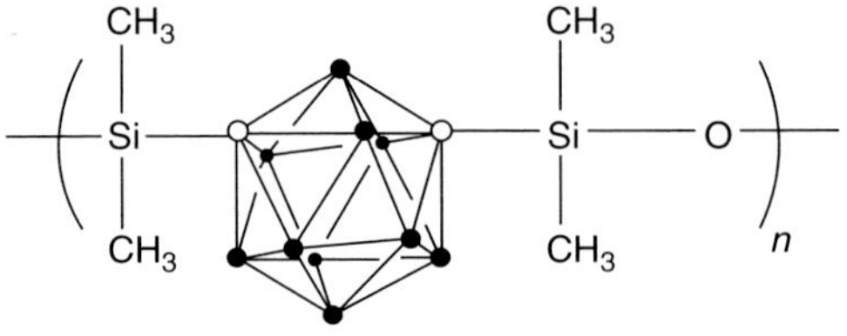

Figure 8.16 The general structure for the poly(carborane–siloxane). Source: Adapted from Apedaile et al. 2012 [79] with permission.

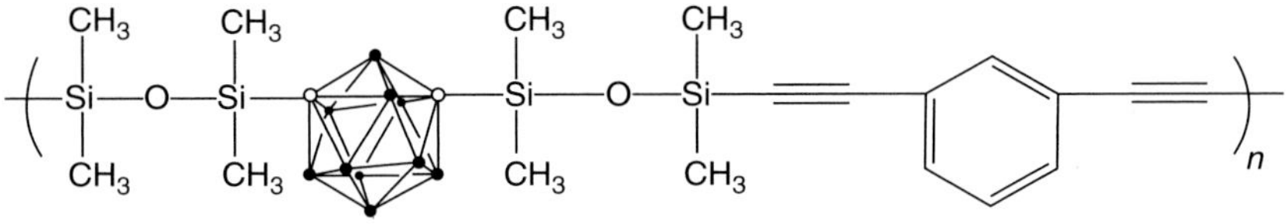

m-PCSAA

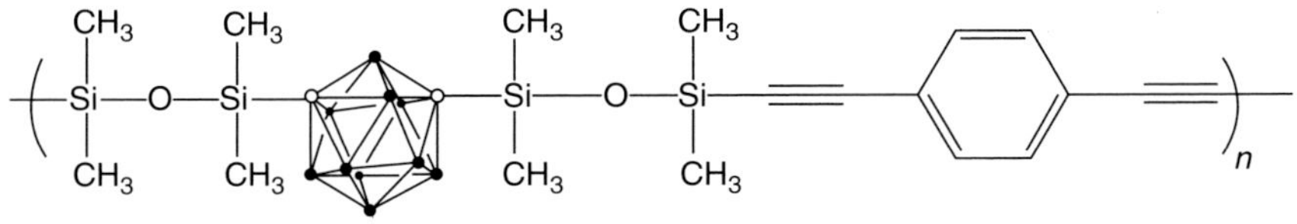

p-PCSAA

Figure 8.17 Backbone aromatic group containing carboranylenesiloxanes *m*-PCSAA and *p*-PCSAA. Source: Adapted from Kolel-Veetil et al. 2013 [83] with permission.

materials used under high thermal conditions. Kolel-Veetil et al. synthesized the structural isomers of thermo-oxidatively stable poly(carborane–siloxane–aryl acetylene) (PCSAA), namely, *m*-PCSAA and *p*-PCSAA, via the reaction of the dimagnesium salts of *m*-diethynylbenzene or *p*-diethynylbenzene with 1,7-bis(chlorotetramethyldisiloxyl)-*m*-carborane [83]. Figure 8.17 shows the structures of synthesized polymers. The developed polymers had exceptional thermo-oxidative properties and *m*-PCSAA showed 5% weight loss at 349 °C and 84% under argon and 93% in air, while *p*-PCSAA gave a 5% weight loss at 400 °C and 84% char yield under argon and 93% in air atmosphere at 1000 °C.

Wang and coworkers synthesized 1,2-bis(4-trimethylsilylethynylphenyl)-carborane (CBTMS) to improve the thermo-oxidative stability of acetylenic aromatic compounds [84]. The thermoset exhibited extreme thermo-oxidatively stable properties and showed a weight gain in air at elevated temperature and a char yield of 98.8% at 1000 °C in air. Moreover, the ceramic residue was stable up to 1000 °C in air and there was only 0.33% weight loss. It was observed that there was a weight gain under air atmosphere. This is related to the oxygen capturing ability of boron and silicon atoms and to the subsequent formation of B_2O_3 and SiO_2 layers at elevated temperatures. *o*-Carborane-containing poly(siloxane–aryl acetylene)s were prepared by the coupling reaction between poly(siloxane–aryl acetylene) and decaborane in the presence of CH_3CN [85]. Thermosets obtained after heat treatment showed excellent thermo-oxidative stability with over 85% residue yield at 1000 °C in air. Scanning electron microscopy (SEM) analysis illustrates the surface layer formation [86]. It is concluded that the boron and silicon atoms present on the surface are oxidized to form a borosilicate layer during the burning process. Increasing the layer thickness and density of the char further prevents the penetration of oxygen into the sample and then improves the thermal stability of the material. A new and convenient synthesis of poly(carborane–siloxane)s has been synthesized by the polycondensation of 1,7-bis(dimethylhydroxylsilyl)-*m*-carborane and cyclotrisilazanes in the presence of $(NH_4)_2SO_4$ [87]. TGA showed the superior thermal stability of the polymers and thermosets, with their char yields of over 83% both in the air (800 °C) and nitrogen (1000 °C). Introduction of pendant vinyl groups into the polymer backbone increased the cross-linking as well as the thermal stability under pyrolysis condition. These results indicated that they were potential candidates of ceramic precursors and fire-retardant coatings. A series of carborane-containing solvent-free liquid silicone resins were synthesized by a hydrosilylation reaction with silicone resins containing different vinyl contents and 1,7-bis(3-hydridotertramethyldisiloxanyl)-*m*-carborane evaluation for their potential application as adhesives with high temperature resistance and anti-oxidative coatings [88]. The silicone–carborane based resin showed better thermal stability and thermo-oxidative stability than the silicon-based resin. It is also observed that the weight retention in the air was 93.5% at 800 °C and predominantly higher than the carborane free counterpart. Furthermore, adhesive properties of carborane-silicone resin were evaluated by lap shear strength (LSS) test. It was observed that the carborane unit bearing adhesive had higher LSS as compared to the carborane-free adhesive at an applied temperature and adhered to stainless steel more efficiently.

Preceramic carborane-containing polycarbosilanes with excellent thermo-oxidative stability were studied after the pyrolysis of an *o*-carborane-containing poly(silylene–aryl acetylene) (CB-PSA) thermoset at 1000, 1200, 1300, and 1450 °C [89]. SEM and X-ray photoelectron spectroscopy (XPS) studies showed that the Si–B–C ceramic prepared at 1450 °C exhibited better oxidation resistance than those prepared at 1000, 1200, and 1300 °C. Formation of borosilicate film on the surface of the ceramics increased oxidation resistance; however, increasing the amount of the carbon and hydrogen contents of the ceramics resulted in lower oxidation resistance. Huang and coworkers analyzed the oxidation behavior of carborane-incorporated poly(dimethylsilylene-ethynylenephenyleneethynylene) at 800, 1000, or 1200 °C for one hour under static air [90]. Oxidation of carborane-containing polymers at 800 °C for one hour gave a residue yield of 86.2%, which is close to that obtained after pyrolysis in argon (84.9%). The long-term isothermal oxidative stability of the oxidation products was also analyzed for 12 hours at the same condition and the residue yield was 85.4%, which is only a slight decrease as compared to the value received for one hour oxidation. The formation of protective borosilicate layer was analyzed by elemental composition, XPS, and SEM. The results showed that the resulting borosilicate layer was the main mechanism that provided remarkable oxidation at 1000 °C in air and enhance the resistance of the carbon materials. Degradation mechanism of polyorganocarboranesiloxane-based compounds showed that the framework of main chain (Si–O–Si) survived after high-temperature pyrolysis after analysis by the solid-state ^{29}Si NMR experiments [91]. A similar analysis was performed on the XPS. The silicon atom (Si 2p) binding energy value in the outside of layer deviates from 102.51 eV (Si–O–Si) to 103.24 eV (SiO_2) completely and manifests the development of the SiO_2 layer. Litt and coworkers synthesized variable poly(carborane siloxane)-type polymer coatings for low earth orbit space structures [92]. Carborane siloxane polymers can take upto 10 oxygen atoms on oxidation with a considerable weight increase (~50%) due to the presence of 10 boron atoms in carborane. Hence, increase in density was also observed from 1 to 1.92 g/cm^3 for $B_{10}O_{15}x(SiO_2)_3$. Linear carborane–carbosilane–phenylacetylene copolymer has been synthesized as a precursor for thermosets and ceramics for the protection of carbon fibers from oxidation in an oxidizing environment. The novel linear copolymers can be processed conveniently and converted into thermoset or ceramics since they are either liquids or low melting solids at room temperature and are soluble in most organic solvents. Surface treatment of carbon fibers with poly(carborane–carbosilane–phenylacetylene) can provide a ceramic coating at elevated temperatures [93]. The carbon fibers were impregnated into a different concentration of poly(carborane–carbosilane–phenylacetylene) solution prepared in dichloromethane. After air drying, the fibers were thermally treated at 250/350/450/550 °C and cured for one hour to obtain a ceramic film on the surface of the carbon fibers to enhance oxidation protection. Oxidation aging test at 600, 700, 800, and 900 °C in a flow of air (20 cm^3/min) for 120 minutes was performed on ceramic-coated carbon fibers to determine the durability of the protection stage. The weight loss of the uncoated carbon fiber was higher than that of the coated fiber that had a mass residue of up to 74.46% heating at 800 °C. The tensile strength was also tested and it was revealed that coated fiber caused a decrease in the strength in 18.61% with a 3988 MPa value as against uncoated fiber.

8.3.5.2 Carborane-Containing High-Performance Thermoplastics

Polyetherketones is one of the typical examples for the high engineering plastics. Semicrystalline aromatic polyetherketones are commercially available under the trade name as Ultrapek®, PEEK. Incorporation of carborane into the polyetherketone's backbone might result in better thermomechanical properties and the ability to withstand solvent attack [94]. The properties of carborane units and the effects of incorporating them into the chains of the aromatic poly(ether ketone)s have been reported [95]. Polycondensations of 1,12-bis(4-phenoxypheny1)-*p*-carborane monomer with 4,4′-oxydibenzoic acid and with biphenyl-4,4′-dicarboxylic acid proceeded rapidly at room temperature in trifluoromethanesulfonic acid catalyst to deliver viscous red solutions of poly(ether ketone–carborane)s (Figure 8.18) [96].

TGA of biphenyl-bearing functional polymer showed a 90% char yield at 850 °C under argon atmosphere, with the onset of decomposition at c. 440 °C. Thermal analysis under air indicated even less weight loss under the same conditions due to the formation of a boron oxide layer. Extended pyrolysis of the polymer at 1600 °C under argon produced a final ceramic yield of 79% [95]. Polybenzoxazines are a class of novel phenolic resins and are promising alternative materials for applications in electronics and aerospace industries. However, the stability of polybenzoxazines under oxidizing atmosphere is rather poor and limits their uses under special conditions. Jia and coworkers synthesized polybenzoxazine precursors containing *m*-carborane units in the main chain through click reaction of diazidomethyl *m*-carborane and diacetylene bisbenzoxazine to obtain carborane-based polybenzoxazines with high thermo-oxidative stability [97]. The char yield of cured carborane-bearing

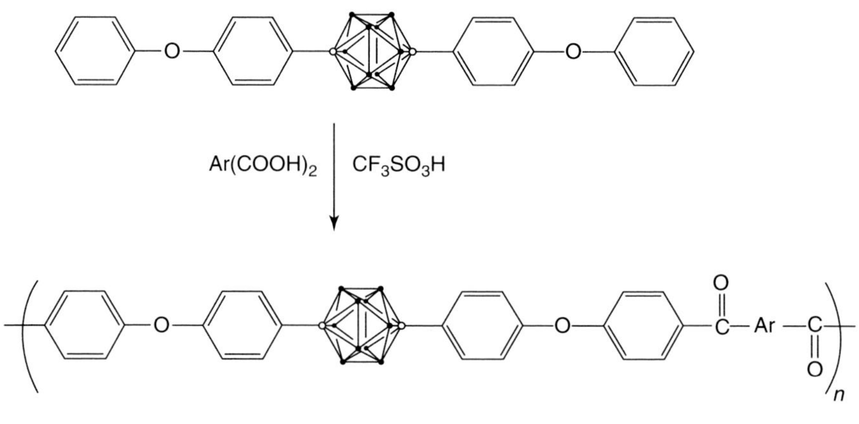

Figure 8.18 Synthesis of polyaryletherketone-*p*-carboranes. Source: Adapted from Colquhoun et al. 1996 [96] with permission.

polybenzoxazines (CCBP) was around 75% at 800 °C in the presence of nitrogen as well under air atmosphere. The superior thermo-oxidative stability of CCBP was also tested by thermal ablation. Polybenzoxazines with (CCBP) or without carborane units (BCBP) were heated for 15 minutes at 500 °C under air atmosphere. Zhang and coworkers investigated the thermal studies on carborane-containing benzoxazine resins from carborane bisphenols [98]. Carborane-bearing benzoxazine precursors were synthesized by Mannich reaction of carborane bisphenol, formaldehyde, and aniline. T_d values for cured carborane-containing benzoxazine and bare non-carborane-containing benzoxazine under nitrogen were 347 and 354 °C, respectively. However, under air conditions these values became 412 and 359 °C, which was much higher than the T_d of bare non-carborane-containing benzoxazine (261 °C). Observed char yield for the carborane-containing benzoxazine resins were 72.6–76.8% at 900 °C under an inert atmosphere, while under an oxidative environment 51.1–77.1% char were obtained. Formation of B–O–B and B–C networks prevents further oxidation thus inhibiting further polymer degradation process. Thermal copolymerization of carborane-containing acetylenic monomer (APCB) and acetylene-terminated imide monomer (*m*-EFDA) was synthesized to improve the thermal and thermo-oxidative stability of the polyimide materials [99]. Addition of carborane-bearing monomer at a loading level of 50 wt% exhibited a high char yield of 91.5% at 1000 °C. The long-term thermo-oxidative stability of the carborane-bearing polymers was also tested after aging in air at 400 and 500 °C for five hours. The formation of a protective layer from the oxidation of carborane units present in the backbone, which was observed by SEM, XPS, and FT-IR, caused a weight retention of 95.6%.

8.3.5.3 Carboranes as Miscellaneous Polymers

Polymers containing carboranyl units that are finally discussed in this chapter are epoxy, polyurethane, polyester, phenolic resins, ROMP-type polymers, etc. Carborane-containing aromatic diamine, 1,2-bis(4-aminophenyl)-1,2-dicarba-close-dodecaborane (HPPA) was synthesized and used in the epoxy curing formulations [100]. The observed char residue for the cured system at a molar ratio of 1.1 was 43.7% and 50.5% at 800 °C under argon and in air atmospheres, respectively. The thermal degradation properties of cured resins were then monitored by TG-FTIR and in situ FTIR to analyze their decomposition profile at different temperatures. The TG-FTIR spectra showed the absorption peaks of volatile pyrolysis products for the cured system at 3652 (υ; O—H), 2967 (υ; saturated C—H), 3024 (υ; unsaturated C—H), 2588 (υ; B—H), 1505 and 1601 (υ; Ar C=C), 1170 (υ; CH_3—C—CH_3), and 1263 (υ; Ar—O) cm^{-1}. The B—H bonds were observed after the temperature reached 415 °C. Incorporation of carborane units in the curing formulation reduced the lower absorption intensity of the volatile compounds and release of pyrolysis gas, forming higher char residue. Furthermore, in situ analysis of FTIR revealed that a new absorption peak at 1336 cm^{-1} corresponding to B—O bonds appeared at around 355 °C in air with a lower temperature than decomposition under argon. Formation of B—O predominantly enhanced under air condition [100]. *O*- or *m*-carborane functional group bearing bisphenols were further utilized as starting materials to synthesize carborane-containing epoxy novolacs and novel carborane-containing poly(hydroxy ethers) [101]. The thermal stability of the resin was further analyzed by TGA operating under air or nitrogen. It was observed that carborane-bearing samples had a great improvement at the initial decomposition temperature and gave 5% weight loss at 395 and 454 °C according to the different quantities of carborane clusters present in the polymer samples. Observed char yield for the cured resin (*o*-carborane substituted novolac) at 900 °C was 91.2% under nitrogen and 93.9% in air, respectively. Carborane cluster is oxidated in the presence of air at elevated temperature and caused the formation of two weight peaks at 450 and 705 °C. The presence of B—H on carborane cage could react with oxygen to form a three-dimensional boron oxide-based insulating layer and thus might block formed radicals and enhance the thermal properties. Qi and coworkers synthesized the carborane-containing polyurethane from hydroxyl-terminated carborane-containing prepolymer as a soft segment and isocyanate-terminated carborane-containing prepolymer [102]. Thermomechanical analysis showed that increasing the carborane biphenol content by 5–10 wt% in the formulation increases the tensile strength up to 20 MPa and carborane group effectively reduces the degradation rate of carborane-containing polyurethane with char yield exceeding 40%. At 600 °C, carborane-free polyurethane matrix had a complete degradation while 20% carbon biphenol bearing polyurethane gave a char up to 62%, indicating that the addition of carborane into the polyurethane backbone dramatically enhanced the degradation temperature and the char yield. A series of *o*-carborane- or *m*-carborane-containing polyesters shown in Figure 8.19 were successfully synthesized by the catalytic polycondensation of carborane diol monomers with carborane diacid chlorides [103]. Following this, the thermal transition of carborane-free polyesters was studied to analyze the thermal resistance of the carborane-containing polymers and to investigate the *structure–thermal property* relationship. The thermal gravimetric analyzer curves showed that the carborane group could effectively enhance the thermal profile and gave a char yield exceeding 64% under air (47% under N_2) at 700 °C.

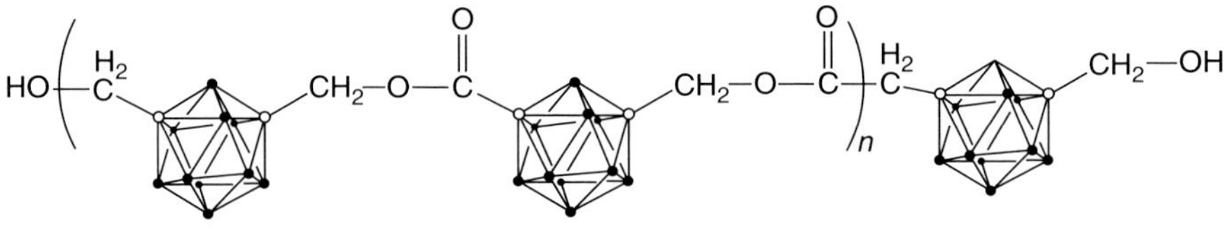

Figure 8.19 *m*-Carborane-containing polyester. Source: Adapted from Li et al. 2016 [103] with permission.

However, carborane-free polyester showed a low char yield of around 0.3% under air (5% under N_2) at the same condition. Meanwhile, it was also claimed that polyesters bearing *m*-carborane group had better thermal stability than the polyester having *o*-carborane functionality.

The hydroxy-terminated polyester containing carborane was prepared via the condensation polymerization of 1,2-bis-(hydroxymethyl) carborane and adipyl chloride and then used in the polyurethane curing formulations [104]. The shear strength and thermal properties of the derived polyurethane adhesives were determined. Adhesives showed higher stability as against nonbearing carborane units and could be used as adhesive joints in the range of 150–200 °C. Ruthenium-catalyzed ring-opening metathesis polymerization (ROMP) of organodecaboranes containing strained-ring cyclic olefinic substituents gave poly(organodecaborane) polymers with higher molecular weights [105]. Studies on ceramic conversion and thermal properties of the resulting polymers were investigated and boron-carbide based ceramic evolution was analyzed using TGA, XRD, DRIFT, Raman, SEM, and elemental analyses as well as density measurements. Table 8.14 shows the properties of pyrolysis end products of poly(organodecaborane).

The ceramic conversions for decaborane-bearing polymers gave char yields of 82% and 76% at 1200 °C. However, when the polymer is heated for 10 hours at 1650 °C under argon gave 51.8% char. The elemental analysis is used to analyze the formed ceramic compositions. It was observed that significant amounts of hydrogen remained in 1000 °C (0.83 wt% or 0.59 equiv per carbon) and 1200 °C (0.45 wt% or 0.34 equiv per carbon). Remaining chemically active species were then lost at 1400 and 1650 °C and resulted in a complete ceramic conversion. Simon and Coughlin synthesized the carborane-containing oxanorbornen

Table 8.14 Results of bulk pyrolyses of poly(organodecaborane) type polymers.

Temperature (°C)	Ceramic/polymer (g/g)	Ceramic yield (%)	*B* (%)	*C* (%)	*H* (%)	*B*:*C*	Nominal composition
1000	0.33/0.59	55.9	59.94	30.56	0.69	2.18	$B_4C/C_{0.82}$
1200	0.32/0.61	52.4	62.42	30.60	0	2.26	$B_4C/C_{0.77}$
1300	0.28/0.52	53.8	65.86	29.73	0.34	2.46	$B_4C/C_{0.63}$
1400	0.38/0.72	52.7	67.76	28.62	0	2.63	$B_4C/C_{0.52}$
1650	0.33/0.65	50.8	68.80	22.90	0	3.30	$B_4C/C_{0.21}$

Source: Wei et al. 2006 [105]. Reproduced with permission of American Chemical Society.

monomer and then derived polyethylene-like materials via ROMP pathway [106]. The inclusion of the silyl-functionalized carborane on the repeating units of polymer backbones' structural and thermal properties was investigated between the range 25—700 °C in nitrogen. It was observed that increasing carborane content increased the char yield. Also, the homopolymer of carborane-bearing oxanorbornene showed the onset of decomposition at 370 °C, which was higher than the monomer itself (onset 148 °C).

8.4 Conclusion

An inorganic polymer with an array of attractive properties and applications comprise one of the most key groups of macromolecules. The properties, characterization, and flame-retardant application of different types of inorganic polymers such as polysiloxane, polyphosphazenes, polysilanes, and carbone are intensively highlighted in this chapter. Some of the interesting characteristics such as flame retardancy and high-temperature stability of these particular polymers are outlined by TGA, UL 94, LOI, and cone calorimeter test strategies. Moreover, properties such as protective layer formation, LOI, reduction in smoke production, and high-temperature stability are reviewed. In this regard, these inorganic polymers can be used as high-performance flame-retardant additives and high-performance polymers in various systems such as thermoplastic, elastomeric, thermosetting matrix.

Acknowledgments

The authors thank the COST Action CM1302 (SIPs): "European Network on Smart Inorganic Polymers."

References

1 Ivings, M.J., Gant, S.E., Saunders, C.J., and Pocock, D.J. (2010). Flammable gas cloud build up in a ventilated enclosure. *J. Hazard. Mater.* https://doi.org/10.1016/j.jhazmat.2010.08.020.
2 Khan, M.M., Tewarson, A., and Chaos, M. (2016). Combustion characteristics of materials and generation of fire products. In: *SFPE Handbook of Fire Protection Engineering*, 5e (ed. M.J. Hurley, D.T. Gottuk, J.R. Hall Jr., et al.), 1143–1232. New York, NY: Springer.
3 Frisch, K.C. and Reegen, S.L. (1975). Relationship between chemical structure and flammability resistance of polyurethanes. In: *Flame-Retardant Polymeric Materials* (ed. E. Pearce), 291–336. Boston, MA: Springer.
4 Bourbigot, S. and Duquesne, S. (2007). Intumescence and nanocomposites: a novel route for flame-retarding polymeric materials. In: *Flame Retardant Polymeric Nanocomposites* (ed. C.A. Wilkie and A.B. Morgan), 131–162.

5 Zhao, B., Hu, Z., Chen, L. et al. (2011). A phosphorus-containing inorganic compound as an effective flame retardant for glass-fiber-reinforced polyamide 6. *J. Appl. Polym. Sci.* https://doi.org/10.1002/app.32860.
6 Qin, Z., Li, D., and Yang, R. (2016). Study on inorganic modified ammonium polyphosphate with precipitation method and its effect in flame retardant polypropylene. *Polym. Degrad. Stab.* 126: 117–124.
7 Cheng, J.J., Shi, B.B., Zhou, F.B., and Chen, X.Y. (2014). Effects of inorganic fillers on the flame-retardant and mechanical properties of rigid polyurethane foams. *J. Appl. Polym. Sci.* https://doi.org/10.1002/app.40253.
8 Dasari, A., Yu, Z.-Z., Cai, G.-P., and Mai, Y.-W. (2013). Author's personal copy recent developments in the fire retardancy of polymeric materials. *Prog. Polym. Sci.*
9 Khobragade, P.S., Hansora, D.P., Naik, J.B., and Chatterjee, A. (2016). Flame retarding performance of elastomeric nanocomposites: a review. *Polym. Degrad. Stab.* https://doi.org/10.1016/j.polymdegradstab.2016.06.001.
10 Martins, J.A. and Silva Dias, M.A.F. (2009). The impact of smoke from forest fires on the spectral dispersion of cloud droplet size distributions in the amazonian region. *Environ. Res. Lett.* https://doi.org/10.1088/1748-9326/4/1/015002.
11 San-Miguel-Ayanz, J., Moreno, J.M., and Camia, A. (2013). Analysis of large fires in European Mediterranean landscapes: lessons learned and perspectives. *For. Ecol. Manage.* https://doi.org/10.1016/j.foreco.2012.10.050.
12 De Lill, D.T. and Carraher, C.E. (2017). Methods for introducing inorganic polymer concepts throughout the undergraduate curriculum. *J. Chem. Educ.* https://doi.org/10.1021/acs.jchemed.7b00028.
13 Salmeia, K.A., Gaan, S., and Malucelli, G. (2016). Recent advances for flame retardancy of textiles based on phosphorus chemistry. *Polymers (Basel)* https://doi.org/10.3390/polym8090319.
14 Horrocks, A.R., Kandola, B.K., Davies, P.J. et al. (2005). Developments in flame retardant textiles – a review. *Polym. Degrad. Stab.* https://doi.org/10.1016/j.polymdegradstab.2003.10.024.
15 Levchik, S.V. (2006). Introduction to flame retardancy and polymer flammability. *Flame Retard. Polym. Nanocomposites* https://doi.org/10.1002/9780470109038.ch1.
16 Kumar Mishra, R., Thomas, S., and Karikal, N. (2017). *Micro and Nano Fibrillar Composites (MFCs and NFCs) from Polymer Blends.* Elsevier https://doi.org/10.1016/B978-0-08-101991-7.09989-1.
17 Garaud, J.L. (2007). *Inorganic Polymers.* Academic Press.
18 Morgan, A.B. and Wilkie, C.A. (2014). *Non-Halogenated Flame Retardant Handbook.* Wiley https://doi.org/10.1002/9781118939239.
19 Loganathan, S., Valapa, R.B., Mishra, R.K. et al. (2017). Thermogravimetric analysis for characterization of nanomaterials. *Therm. Rheol. Meas. Tech. Nanomater. Charact.* 67–108. https://doi.org/10.1016/B978-0-323-46139-9.00004-9.
20 Mishra, R.K., Cherusseri, J., and Joseph, K. (2017). Thermal and crystallization behavior of micro and nano fibrillar in-situ composites. *Polym. Blends* 213–231. https://doi.org/10.1016/B978-0-08-101991-7.00009-1.

21 Bourbigot, S. and Duquesne, S. (2007). Fire retardant polymers: recent developments and opportunities. *J. Mater. Chem.* https://doi.org/10.1039/b702511d.

22 Stojanovic, A., Olveira, S., Fischer, M., and Seeger, S. (2013). Polysiloxane nanotubes. *Chem. Mater.* https://doi.org/10.1021/cm400851k.

23 James, R., Deng, M., Kumbar, S.G., and Laurencin, C.T. (2014). Polyphosphazenes. *Nat. Synth. Biomed. Polym.* https://doi.org/10.1016/B978-0-12-396983-5.00012-0.

24 Naito, M. and Fujiki, M. (2008). Polysilanes on surfaces. *Soft Matter.* https://doi.org/10.1039/B710423E.

25 Cleij, T.J., King, J.K., and Jenneskens, L.W. (2001). Non-ionic water-soluble polysilanes and polysilynes, chameleons in solution polar functionalized polysilanes. In: *Stimuli-Responsive Water Soluble Amphiphilic Polymers.* https://doi.org/10.1021/bk-2001-0780.ch014.

26 Shi, J.B., Zhi, J.G., Bin, T. et al. (2008). Full-conjugated ferrocene-based polymers. *Prog. Chem.*

27 Zaganiaris, E.J., Sperling, L.H., and Tobolsky, A.V. (1967). Elastomeric and mechanical properties of poly-*m*-carboranylenesiloxanes. III. *J. Macromol. Sci. Part A Chem.* https://doi.org/10.1080/10601326708053761.

28 Colquhoun, H.M., Lewis, D.F., Herbertson, P.L., and Wade, K. (1997). Polyetherketones based on para-carborane: synthesis, sulfonation, and membrane-forming characteristics. *Polymer (Guildf).* https://doi.org/10.1016/S0032-3861(96)01016-6.

29 Zhuo, D., Gu, A., Liang, G. et al. (2011). Flame retardancy materials based on a novel fully end-capped hyperbranched polysiloxane and bismaleimide/diallylbisphenol a resin with simultaneously improved integrated performance. *J. Mater. Chem.* https://doi.org/10.1039/c1jm10233h.

30 Ye, J., Liang, G., Gu, A. et al. (2013). Novel phosphorus-containing hyperbranched polysiloxane and its high performance flame retardant cyanate ester resins. *Polym. Degrad. Stab.* https://doi.org/10.1016/j.polymdegradstab.2012.11.015.

31 Hsieh, C.Y., Su, W.C., Wu, C.S. et al. (2013). Benzoxazine-containing branched polysiloxanes: highly efficient reactive-type flame retardants and property enhancement agents for polymers. *Polymer* https://doi.org/10.1016/j.polymer.2013.03.060.

32 Dong, C., Lu, Z., Zhang, F. et al. (2015). Preparation and properties of cotton fabrics treated with a novel polysiloxane water repellent and flame retardant. *Mater. Lett.* https://doi.org/10.1016/j.matlet.2015.03.132.

33 Zhao, Z., Jin, Q., Zhang, N. et al. (2018). Preparation of a novel polysiloxane and its synergistic effect with ammonium polyphosphate on the flame retardancy of polypropylene. *Polym. Degrad. Stab.* https://doi.org/10.1016/j.polymdegradstab.2018.02.007.

34 Zhang, C., Qu, L., Wang, Y. et al. (2018). Thermal insulation and stability of polysiloxane foams containing hydroxyl-terminated polydimethylsiloxanes. *RSC Adv.* 8: 9901–9909. https://doi.org/10.1039/C8RA00222C.

35 Wang, M., Li, B., Wang, J., and Bai, P. (2011). Preparation and properties of polysiloxane grafting multi-walled carbon nanotubes/polycarbonate

nanocomposites. *Polym. Adv. Technol.* 22: 1738–1746. https://doi.org/10.1002/pat.1665.
36 Han, X., Zhao, J., Liu, S., and Yuan, Y. (2014). Flame retardancy mechanism of poly(butylene terephthalate)/aluminum diethylphosphinate composites with an epoxy-functional polysiloxane. *RSC Adv.* 4: 16551–16560. https://doi.org/10.1039/C4RA00515E.
37 Jiang, Z., Yuan, L., Liang, G., and Gu, A. (2015). Unique liquid multi-maleimide terminated branched polysiloxane and its flame retarding bismaleimide resin with outstanding thermal and mechanical properties. *Polym. Degrad. Stab.* 121: 30–41. https://doi.org/10.1016/j.polymdegradstab.2015.08.005.
38 Song, R., Chang, L., and Li, B. (2014). Flame retardancy and thermal properties of carboxyl-containing polysiloxane derivatives in polycarbonate. *J. Appl. Polym. Sci.* 131: n/a–n/a. https://doi.org/10.1002/app.39814.
39 Zhang, L., Wang, Y., and Cai, X. (2016). Effect of a novel polysiloxane-containing nitrogen on the thermal stability and flame retardancy of epoxy resins. *J. Therm. Anal. Calorim.* 124: 791–798. https://doi.org/10.1007/s10973-015-5224-6.
40 Chen, X., Ye, J., Yuan, L. et al. (2014). Multi-functional ladderlike polysiloxane: synthesis, characterization and its high performance flame retarding bismaleimide resins with simultaneously improved thermal resistance, dimensional stability and dielectric properties. *J. Mater. Chem. A.* 2: 7491–7501. https://doi.org/10.1039/C4TA01292E.
41 Dong, C., Lu, Z., Zhu, P. et al. (2015). Combustion behaviors of cotton fabrics treated by a novel guanidyl- and phosphorus-containing polysiloxane flame retardant. *J. Therm. Anal. Calorim.* 119: 349–357. https://doi.org/10.1007/s10973-014-4154-z.
42 Dong, C., Lu, Z., Zhang, F. et al. (2016). Combustion behaviors of cotton fabrics treated by a novel nitrogen- and phosphorus-containing polysiloxane flame retardant. *J. Therm. Anal. Calorim.* 123: 535–544. https://doi.org/10.1007/s10973-015-4914-4.
43 Gao, S., Li, B., Bai, P., and Zhang, S. (2011). Effect of polysiloxane and silane-modified SiO_2 on a novel intumescent flame retardant polypropylene system. *Polym. Adv. Technol.* 22: 2609–2616. https://doi.org/10.1002/pat.1810.
44 Zhou, W., Chen, K., and He, W. (2016). Flame retardancy of polysiloxane solid powder combined with organophosphate in polycarbonate. *MATEC Web Conf.* 67: 6029. https://doi.org/10.1051/matecconf/20166706029.
45 Steiner, A. (2008). Supramolecular structures of cyclotriphosphazenes. *Polyphosphazenes Biomed. Appl.* https://doi.org/10.1002/9780470478882.ch20.
46 Fu, J. and Xu, Q. (2015). Green manufacturing and the application of high-temperature polymer-polyphosphazenes. In: *Green and Sustainable Manufacturing of Advanced Material* (ed. M. Singh, T. Ohji and R. Asthana), 603–646. Elsevier.
47 Chen, D., Cui, Y., Wang, X., and Tang, X. (2002). Polyphosphazene/low-density polyethylene blends: miscibility and flame-retardance studies. *J. Appl. Polym. Sci.* 86: 709–714. https://doi.org/10.1002/app.10970.

48 Reed, C.S., Taylor, J.P., Guigley, K.S. et al. (2000). Polyurethane/poly[bis(carboxylatophenoxy)phosphazene] blends and their potential as flame-retardant materials. *Polym. Eng. Sci.* 40: 465–472. https://doi.org/10.1002/pen.11178.

49 Qiu, S., Wang, X., Yu, B. et al. (2017). Flame-retardant-wrapped polyphosphazene nanotubes: a novel strategy for enhancing the flame retardancy and smoke toxicity suppression of epoxy resins. *J. Hazard. Mater.* 325: 327–339. https://doi.org/10.1016/j.jhazmat.2016.11.057.

50 Qiu, S., Xing, W., Feng, X. et al. (2017). Self-standing cuprous oxide nanoparticles on silica@ polyphosphazene nanospheres: 3D nanostructure for enhancing the flame retardancy and toxic effluents elimination of epoxy resins via synergistic catalytic effect. *Chem. Eng. J.* 309: 802–814. https://doi.org/10.1016/j.cej.2016.10.100.

51 Qiu, S., Ma, C., Wang, X. et al. (2018). Melamine-containing polyphosphazene wrapped ammonium polyphosphate: a novel multifunctional organic–inorganic hybrid flame retardant. *J. Hazard. Mater.* 344: 839–848. https://doi.org/10.1016/j.jhazmat.2017.11.018.

52 Zhang, T., Cai, Q., Wu, D.-Z., and Jin, R.-G. (2005). Phosphazene cyclomatrix network polymers: some aspects of the synthesis, characterization, and flame-retardant mechanisms of polymer. *J. Appl. Polym. Sci.* 95: 880–889. https://doi.org/10.1002/app.21295.

53 Chen, K., Huang, X., Tang, X., and Zhu, L. (2012). Study on the organic–inorganic hybrid polyphosphazene nanotube as a flame retardant for polypropylene. *J. Macromol. Sci. Part B* 51: 269–274. https://doi.org/10.1080/00222348.2010.549435.

54 Liu, H., Wang, X., and Wu, D. (2015). Synthesis of a novel linear polyphosphazene-based epoxy resin and its application in halogen-free flame-resistant thermosetting systems. *Polym. Degrad. Stab.* 118: 45–58. https://doi.org/10.1016/j.polymdegradstab.2015.04.009.

55 Mayer-Gall, T., Knittel, D., Gutmann, J.S., and Opwis, K. (2015). Permanent flame retardant finishing of textiles by allyl-functionalized polyphosphazenes. *ACS Appl. Mater. Interfaces* 7: 9349–9363. https://doi.org/10.1021/acsami.5b02141.

56 Shi, Y., Yu, B., Zheng, Y. et al. (2018). A combination of POSS and polyphosphazene for reducing fire hazards of epoxy resin. *Polym. Adv. Technol.* 29: 1242–1254. https://doi.org/10.1002/pat.4235.

57 Singh, K.P., Mishra, A., Kumar, N. et al. (2018). Evaluation of thermal, morphological and flame-retardant properties of thermoplastic polyurethane/polyphosphazene blends. *Polym. Bull.* 75: 2415–2430. https://doi.org/10.1007/s00289-017-2156-2.

58 Käbisch, B., Fehrenbacher, U., and Kroke, E. (2014). Hexamethoxycyclotriphosphazene as a flame retardant for polyurethane foams. *Fire Mater.* 38: 462–473. https://doi.org/10.1002/fam.2190.

59 Amarnath, N., Appavoo, D., and Lochab, B. (2018). Eco-friendly halogen-free flame retardant cardanol polyphosphazene polybenzoxazine networks. *ACS Sustainable Chem. Eng.* 6: 389–402. https://doi.org/10.1021/acssuschemeng.7b02657.

60 Zhou, X., Qiu, S., Xing, W. et al. (2017). Hierarchical polyphosphazene@molybdenum disulfide hybrid structure for enhancing the flame retardancy and mechanical property of epoxy resins. *ACS Appl. Mater. Interfaces* 9: 29147–29156. https://doi.org/10.1021/acsami.7b08878.

61 Zhao, Z., Zhou, Z., Yu, F., and Ji, J. (2015). Synthesis and flame retardant properties of low density polyethylene/ethylene-vinyl acetate/polyphosphazene derivative composites. *Asian J. Chem.* 27: 919–924. https://doi.org/10.14233/ajchem.2015.17442.

62 Mera, G., Gallei, M., Bernard, S., and Ionescu, E. (2015). Ceramic nanocomposites from tailor-made preceramic polymers. *Nanomaterials* https://doi.org/10.3390/nano5020468.

63 Krempner, C. (2012). Polysilane dendrimers. *Polymers (Basel)* https://doi.org/10.3390/polym4010408.

64 Manners, I. (1996). Polymers and the periodic table: recent developments in inorganic polymer science. *Angew. Chem.* https://doi.org/10.1002/anie.199616021.

65 Shukla, S.K., Tiwari, R.K., Ranjan, A. et al. (2004). Some thermal studies of polysilanes and polycarbosilanes. *Thermochim. Acta* 424: 209–217. https://doi.org/10.1016/j.tca.2004.06.003.

66 Bianconi, P.A. and Weidman, T.W. (1988). Poly(*n*-hexylsilyne): synthesis and properties of the first alkyl silicon $[RSi]_n$ network polymer. *J. Am. Chem. Soc.* 110: 2342–2344. https://doi.org/10.1021/ja00215a077.

67 Ma, N., Yu, Y., Sun, Z., and Huang, S. (2007). Comparative study on the optical properties and stability of linear and branched polysilanes. *J. Lumin.* 126: 827–832. https://doi.org/10.1016/j.jlumin.2006.12.008.

68 Lorkowski, H.J. (1973). Ferrocene polymers. *Polym. Sci. USSR* https://doi.org/10.1016/0032-3950(73)90208-6.

69 Liu, L., Chen, X., and Jiao, C. (2015). Influence of ferrocene on smoke suppression properties and combustion behavior of intumescent flame-retardant epoxy composites. *J. Therm. Anal. Calorim.* https://doi.org/10.1007/s10973-015-4928-y.

70 Carty, P., Grant, J., and Metcalfe, E. (1996). Flame-retardancy and smoke-suppression studies on ferrocene derivatives in PVC. *Appl. Organomet. Chem.* https://doi.org/10.1002/(SICI)1099-0739(199603)10:2<101::AID-AOC484>3.0.CO;2-7.

71 Liao, D.J., Xu, Q.K., McCabe, R.W. et al. (2017). Ferrocene-based nonphosphorus copolymer: synthesis, high-charring mechanism, and its application in fire retardant epoxy resin. *Ind. Eng. Chem. Res.* https://doi.org/10.1021/acs.iecr.7b02980.

72 Mehdipour-Ataei, S. and Babanzadeh, S. (2007). New types of heat-resistant, flame-retardant ferrocene-based polyamides with improved solubility. *React. Funct. Polym.* 67: 883–892. https://doi.org/10.1016/j.reactfunctpolym.2007.05.012.

73 Chen, P., Zhang, F., Li, S., and Cheng, Y. (2016). Smoke suppression properties of epoxy crosslinked structure and intumescent fire retardant in epoxy-based intumescent fire-retardant coating. *J. Appl. Polym. Sci.* https://doi.org/10.1002/app.43912.

74 Plesek, J. (1992). Potential applications of the boron cluster compounds. *Chem. Rev.* https://doi.org/10.1021/cr00010a005.

75 Lu, S.-Y. and Hamerton, I. (2002). Recent developments in the chemistry of halogen-free flame retardant polymers. *Prog. Polym. Sci.* 27: 1661–1712.

76 Grimes, R.N. (2011). Chapter 10 – Icosahedral carboranes: 1,7-$C_2B_{10}H_{12}$ and 1,12-$C_2B_{10}H_{12}$. In: *Carboranes*, 2e. http://dx.doi.org/10.1016/B978-0-12-374170-7.00008-2.

77 Marshall, J., Fei, Z., Yau, C.P. et al. (2014). Incorporation of benzocarborane into conjugated polymer systems: synthesis, characterisation and optoelectronic properties. *J. Mater. Chem. C.* https://doi.org/10.1039/c3tc31663g.

78 Hey-Hawkins, E. and Teixidor, C.V. (2018). *Boron-Based Compounds: Potential and Emerging Applications in Medicine*. Wiley.

79 Apedaile, A., Liggat, J., Parkinson, J. et al. (2012). Lewis acid mediated polymerization of poly(dimethylsiloxane) polymers: investigating reaction kinetics using both NMR spectroscopy and cyclic voltammetry. *J. Appl. Polym. Sci.* https://doi.org/10.1002/app.34449.

80 Henderson, L.J. and Keller, T.M. (1994). Synthesis and characterization of poly(carborane–siloxane–acetylene). *Macromolecules* https://doi.org/10.1021/ma00084a060.

81 Sundar, R.A. and Keller, T.M. (1997). *J. Polym. Sci. Part A Polym. Chem.* 35 (12): 2387–2394.

82 Bucca, D. and Keller, T.M. (1997). *J. Polym. Sci. Part A Polym. Chem.* 35 (6): 1033–1038.

83 Kolel-Veetil, M.K., Dominguez, D.D., Klug, C.A. et al. (2013). Hybrid inorganic-organic poly (carborane-siloxane-arylacetylene) structural isomers with in-chain aromatics: Synthesis and properties. *J. Polym. Sci. Part A Polym. Chem.* 51 (12): 2638–2650.

84 Cheng, S., Zong, L., Yuan, K. et al. (2016). Synthesis and thermal properties of an acetylenic monomer containing boron and silicon. *RSC Adv.* https://doi.org/10.1039/C6RA19410A.

85 Jiang, Y., Li, X., Huang, F. et al. (2015). *O*-Carborane-containing poly(siloxane-arylacetylene)s with thermal and thermo-oxidative stabilities. *J. Macromol. Sci. Part A Pure Appl. Chem.* https://doi.org/10.1080/10601325.2015.1029373.

86 Wang, C., Zhou, Y., Huang, F., and Du, L. (2011). Synthesis and characterization of thermooxidatively stable poly(dimethylsilyleneethynylene-phenyleneethynylene) with *o*-carborane units. *React. Funct. Polym.* https://doi.org/10.1016/j.reactfunctpolym.2011.05.005.

87 Zhang, X.C., Kong, L.H., Dai, L.N. et al. (2011). Synthesis, characterization, and thermal properties of poly(siloxane-carborane)s. *Polymer* 52 (21): 4777–4784.

88 Xie, J., Sun, H., Zhang, X. et al. (2015). Carborane-containing solvent-free liquid silicone resin: synthesis, characterization, and high-temperature adhesive properties. *Phosphorus. Sulfur. Silicon Relat. Elem.* 190: 277–291. https://doi.org/10.1080/10426507.2014.938992.

89 Jiang, Y., Li, J., Huang, F. et al. (2014). Polymer-derived SiC/B 4C/C nanocomposites: structural evolution and crystallization behavior. *J. Am. Ceram. Soc.* 97: 310–315. https://doi.org/10.1111/jace.12670.

90 Wang, C., Huang, F., Jiang, Y. et al. (2012). Oxidation behavior of carbon materials derived from a carborane- and silicon-incorporated polymer. *Ceram. Int.* 38: 3081–3088. https://doi.org/10.1016/j.ceramint.2011.12.007.

91 Lou, P., Zhang, X., Liu, B. et al. (2017). Synthesis and thermal degradation mechanism of polyorganocarboranesiloxanes containing *m*-carboranylmethyl unit. *Polym. Degrad. Stab.* 144: 304–314. https://doi.org/10.1016/j.polymdegradstab.2017.08.029.

92 Packirisamy, S., Schwam, D., and Litt, M.H. (1995). Atomic oxygen resistant coatings for low earth orbit space structures. *J. Mater. Sci.* 30: 308–320. https://doi.org/10.1007/BF00354390.

93 Tong, D., Wang, H., Wang, L. et al. (2017). Coating of poly(carborane-carbosilane-phenylacetylene) on carbon fibers with excellent oxidation protection. *Surf. Coatings Technol.* 319: 335–344. https://doi.org/10.1016/j.surfcoat.2017.04.014.

94 Dash, B.P., Satapathy, R., Maguire, J.A., and Hosmane, N.S. (2011). Polyhedral boron clusters in materials science. *New J. Chem.* 35: 1955. https://doi.org/10.1039/c1nj20228f.

95 Colquhoun, H.M., Lewis, D.F., Anthony Daniels, J. et al. (1997). Polyetherketones from diarylcarboranes: a new approach to semi-inorganic polymers. *Polymer (Guildf).* 38: 2447–2453. https://doi.org/10.1016/S0032-3861(96)00811-7.

96 Colquhoun, H.M., Herbertson, P.L., and Wade, K. (1996). Polyetherketones based onpara-carborane. *J. Polym. Sci. Part A Polym. Chem.* 34: 2521–2523. https://doi.org/10.1002/(SICI)1099-0518(19960915)34:12<2521::AID-POLA27>3.0.CO;2-9.

97 Huang, X., Zhang, Q., Meng, Z. et al. (2015). Greatly enhanced thermo-oxidative stability of polybenzoxazine thermoset by incorporation of *m*-carborane. *J. Polym. Sci. Part A Polym. Chem.* https://doi.org/10.1002/pola.27525.

98 Qi, S.C., Wang, H.R., Han, G. et al. (2016). Synthesis, characterization, and curing behavior of carborane-containing benzoxazine resins with excellent thermal and thermo-oxidative stability. *J. Appl. Polym. Sci.* 133 (23).

99 Cheng, S., Han, J., Wang, X. et al. (2017). Oxidatively stable thermosets derived from thermal copolymerization of acetylene-terminated imide monomer with an acetylenic monomer containing carborane. *Polymer (Guildf).* 115: 96–105. https://doi.org/10.1016/j.polymer.2017.02.070.

100 Men, X., Cheng, Y., Chen, G. et al. (2015). Curing behaviour and thermal properties of epoxy resin cured by aromatic diamine with carborane. *High Perform. Polym.* 27: 497–509. https://doi.org/10.1177/0954008314557049.

101 Han, G., Yang, Z., Yang, X.X. et al. (2016). Synthesis, characterization and thermal stability of novel carborane-containing epoxy novolacs. *Chinese J. Polym. Sci.* 34 (9): 1103–1116.

102 Luo, Y., Lu, Y., Li, N., et al. (2015). Synthesis, structure, and properties of segmented carborane-containing polyurethanes. *J. Appl. Polym. Sci.* https://doi.org/10.1002/APP.42227.

103 Li, N., Zeng, F., Qu, D. et al. (2016). Synthesis and characterization of carborane-containing polyester with excellent thermal and ultrahigh char yield. *J. Appl. Polym. Sci.* 133 (46).

104 Chen, S., Zhao, J., Chen, G., and Huang, P.C. (2011). *Acta Polym. Sin.* (12): 1368–1373

105 Wei, X., Carroll, P.J., and Sneddon, L.G. (2006). Ruthenium-catalyzed ring-opening polymerization syntheses of poly(organodecaboranes): new single-source boron-carbide precursors. *Chem. Mater.* 18: 1113–1123. https://doi.org/10.1021/cm0524603.

106 Simon, Y.C. and Coughlin, E.B. (2010). Ring-opening metathesis copolymerization of cyclooctene and a carborane-containing oxanorbornene. *J. Polym. Sci. Part A Polym. Chem.* 48 (12): 2557–2563.

9

Inorganic Polymers for Potential Medicinal Applications

Andreia Valente[1], Rafaella L. M. Precker[2], and Evamarie Hey-Hawkins[2]

[1] *Faculdade de Ciências da Universidade de Lisboa, Centro de Química Estrutural, Campo Grande, 1749-016 Lisboa, Portugal*
[2] *Leipzig University, Faculty of Chemistry and Mineralogy, Institute of Inorganic Chemistry, Johannisallee 29, 04103 Leipzig, Germany*

9.1 Introduction

Although remarkable progress has been made in understanding the fundamental biology behind many diseases, major advances in their treatment are still to be made. In this context, this chapter gives an overview on the use of inorganic polymers, metal-containing polymers, and metal-organic frameworks (MOFs) in the area of medicinal applications, mainly tissue engineering and drug delivery. With regard to tissue engineering, although significant advances have been made using different biocompatible polymeric materials for orthopedics, dental, or ocular applications, there are still concerns about in vivo toxicity and tolerability prompting further research in this area.

With relevance to drug delivery, current therapeutics are limited by their nonspecific distribution throughout the body leading to high doses, rapid clearance, poor pharmacokinetics, and major side effects. Problems pertaining to instances when drugs must be administered in high concentrations led to an effort to develop systems capable of delivering the drug to a specific target, thus reducing undesirable effects of their actions on other sites.

9.2 Inorganic Polymers and Metal-Containing Polymers for Tissue Engineering and Drug Delivery

9.2.1 Inorganic Polymers

Inorganic polymers are widely known to possess biological activity and research on their use in the field of biomedicine is rapidly growing, motivated by their biocompatibility, bioactivity, and stability [1, 2]. In the next sections, we will discuss some recent examples of applications of polysiloxanes and polyphosphazenes in the realm of tissue engineering and drug delivery.

Smart Inorganic Polymers: Synthesis, Properties, and Emerging Applications in Materials and Life Sciences, First Edition. Edited by Evamarie Hey-Hawkins and Muriel Hissler.

Although polyoxometalates (POMs) are another class of inorganic polymers with potential applications in the medicinal field, this subject will not be addressed in this chapter since several reviews that cover the applications of POMs as antiviral, antitumoral, and anti-HIV agents have already been published during the past 20 years [1, 3–5]. The interest in these early transition metal oxygen anion clusters results from the potential they present to be easily functionalized by tuning their molecular properties. Recently, other applications for POMs are being tested, such as antimicrobial agents [6] and magnetic resonance imaging (MRI) agents [7], or as potential drugs to treat diabetes [8, 9] and Alzheimer's disease [10].

9.2.1.1 Polysiloxanes

Polysiloxane is a biocompatible polymeric material broadly used in the production of biomedical devices due to its biocompatibility, stability, optical transparency, and low toxicity, besides others. However, the hydrophobic surface of polysiloxanes can hinder their application in, for example, long-term cell culture, cell adhesion, or surgical implants. In addition, it can be easily contaminated by proteins and pathogens. To improve polysiloxane hydrophilicity, functionalization of its surface has been attempted using several strategies, both as physical treatments and chemical modifications. In the next paragraphs, some examples of potential medicinal applications of polysiloxanes are discussed.

Polymers that respond to an external stimulus have attracted a lot of attention in the field of drug delivery applications. The encapsulation of model hydrophobic drugs such as nifedipine or ibuprofen has been attempted by using azo-polysiloxane micelles [11, 12] and high-sensitive stimuli-responsive polysiloxanes [13], respectively. Azobenzene responds to light triggering (at 350 nm) due to changes of its stable trans configuration into a metastable cis one. In this context, azo-polysiloxanes modified with quaternary ammonium groups that self-assemble into micelles (70–130 nm) were developed (Figure 9.1a) [11]. The authors showed that the micellar aggregation ($2–8 \times 10^{-3}$ g/l) was influenced only by the hydrophobic/hydrophilic balance and not by the quaternary ammonium groups. The photosensitive micelles with higher azobenzene content were able to better disaggregate after 20 minutes under UV irradiation. The ability of the micelles to encapsulate nifedine was evidenced by UV–visible and dynamic light scattering (DLS) measurements, and the quantity of the drug incorporated was dependent on the polymer's chemical structure as well as on the hydrophilic/hydrophobic segment balance. However, the ability of the micelles to liberate the drug under UV irradiation remains to be shown, though the micelles proved to be stable for at least 21 days after preparation.

Li et al. reported an efficient method for the synthesis of new thermo-, pH-, and salinity-sensitive polysiloxanes with *N*-isopropyl amides as side groups through a catalyst-free aza-Michael addition of poly(aminopropylmethylsiloxane) with *N*-isopropylacrylamide (Figure 9.1b) [13]. After encapsulation, the solubility of ibuprofen was greatly increased from 0.06 to 7.28 mg/ml as the concentration of the new polysiloxanes was increased up to 10 mg/ml, allowing a final drug loading of c. 72–74 wt%. These results indicate that research in this area might allow the future use of these new highly sensitive stimuli-responsive polysiloxanes as drug delivery devices.

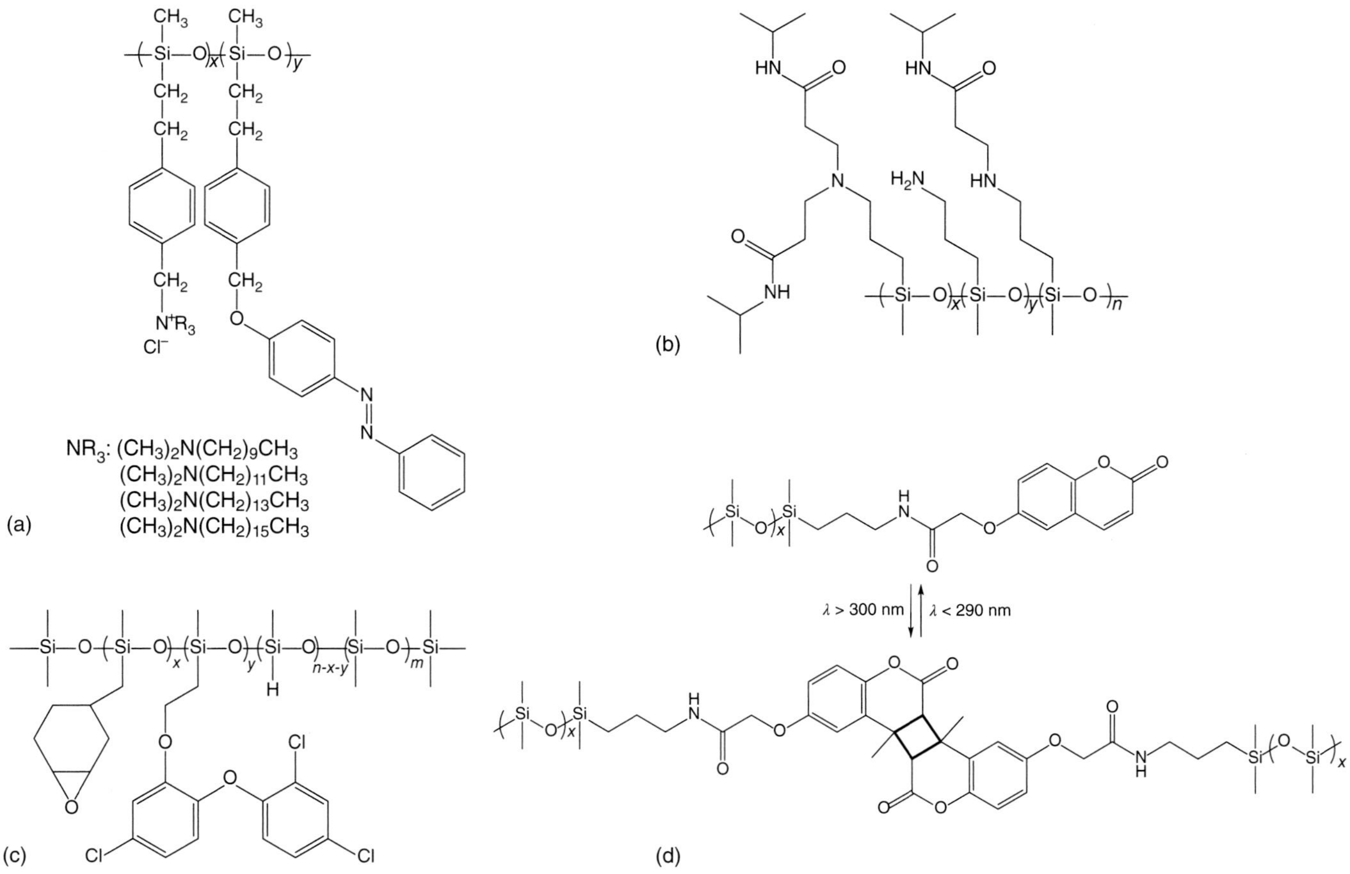

Figure 9.1 Structures of some polysiloxanes with medicinal applications. (a) Azo-polysiloxanes modified with quaternary ammonium groups. (b) Polysiloxanes with *N*-isopropyl amides as side groups. (c) Polysiloxanes functionalized with triclosan (5-chloro-2-(2,4-dichlorophenoxy)phenol). (d) Polydimethylsiloxane functionalized with coumarin.

The integration of antimicrobial activity in materials is important to produce resources for biomedical applications. In this context, Chen et al. reported the synthesis of UV-curable coatings containing polysiloxanes functionalized with the tethered biocide, triclosan (5-chloro-2-(2,4-dichlorophenoxy)phenol) (Figure 9.1c) [14]. The authors varied the content of triclosan, and products ranging from light yellowish (32 wt%) to gray viscous liquids were obtained (47 wt%). Excellent antimicrobial activities were observed against *Staphylococcus epidermidis* and *Escherichia coli*. Interestingly, the authors suggest that the mode of action of these coatings is new with a high specificity to the Gram-positive cell wall, making these polysiloxanes promising candidates for use in infection control.

The fabrication of new corrosion-resistant coatings to be applied in orthopedic and dental applications is a very relevant field. In this context, a novel approach based on the dip coating of polysiloxane/bioactive glass (BG) composite on stainless steel substrates and curing in an oxidative atmosphere (260–500 °C), has been developed for the fabrication of bioactive coatings [15]. Results of in vitro cytotoxicity using osteoblast-like cells (MG-63) confirmed the cyto-compatibility of the polysiloxane/BG composite films.

Bacterial adhesion and consequent biofilm formation can incite severe infections into artificial implants. Thus, a series of polydimethylsiloxane (PDMS) substrates were modified by plasma treatment and the attachment of three opportunistic pathogens that can cause human infections, *E. coli*, *Staphylococcus aureus*, and *Pseudomonas aeruginosa*, was studied [16]. The results showed a decrease in the adhesion of the first two species on the self-assembled monolayers, which was not the case for *P. aeruginosa*. Altogether, the bacterial adhesion could be decreased with the appropriate chemical modifications, showing the applicability of this surface modification.

PDMS is also an interesting material for intraocular lenses due to its transparency. Recently, Jellali et al. reported the synthesis of a PDMS-coumarin (Figure 9.1d) [17]. Under light exposure (>300 nm), these polymers are cured by dimerization of coumarin. Photoreversiblity of the PDMS-coumarin network is achieved via photocleavage of cyclobutane dimers upon irradiation at $\lambda < 290$ nm. The intraocular lens disks made from these materials present high hydrophobicity and good transparency, as required for the specific application. Their biocompatibility was assessed by in vitro cytotoxicity, lens epithelial cell adhesion assays, and rabbit host reaction against implanted disks. Despite good results, decrease in rigidity and suppression of disk opacification at surgery are still challenges that are to be overcome.

Bračič et al. developed a one-step strategy to functionalize PDMS surfaces with chitosan-based bioparticles in order to prevent blood plasma proteins adsorption on the PDMS surface that could lead to infection in a real situation [18]. The best results in terms of stability, improved hydrophilicity, and no adsorption of proteins (tested with bovine serum albumin (BSA)), and enhanced resistance to fibrinogen were obtained for the zwitterionic carboxymethyl chitosan, revealing its potential to be used in the biomedical field.

To improve the surface biocompatibility of PDMS for sustaining long-term cell culture Li et al. modified the surface of PDMS with polydopamine-collagen [19].

This modified PDMS revealed enhanced adhesion and long-term proliferation of L929 fibroblasts and tendon stem cells. Moreover, the cell viability of cells was increased compared to that of native PDMS and apoptosis was decreased, proving that the polydopamine-collagen coating efficiently improved PDMS biocompatibility.

9.2.1.2 Polyphosphazenes

Polyphosphazenes are hybrid inorganic–organic polymers with a phosphorus–nitrogen backbone, substituted with two side groups on the phosphorus atoms. The potential use of polyphosphazenes in the biomedical field is an active area of research, including its use for tissue engineering, vaccine immunoadjuvants, and drug and gene delivery. The interest in this material arises from its degradability, biocompatibility, and biological performance that are easily tunable by introducing different side groups. There are several recent reviews covering their synthesis [20, 21], as well as their degradability and applications [21–26].

Although bone is continuously regenerated during a lifetime, its efficiency is limited by factors such as age, disease, and nutrition. Thus, in some cases, implantation of bone grafts for local bone regeneration needs to be done, although this presents several disadvantages. An alternative is the use of synthetic bone grafts that are able to deliver osteogenic stimulators such as bone morphogenetic proteins (BMPs). In this context, Song and coworkers developed an injectable and thermosensitive hydrogel for the sustained BMP-2 release based on polymeric nanoparticles constituted by a polyphosphazene backbone with hydrophobic isoleucine ethyl ester and hydrophilic polyethylene glycol (Figure 9.2a) [27]. Carboxylic acid moieties were additionally included in the polyphosphazene backbone for ionic interaction with BMP-2. The efficacy of the BMP-2/polymeric nanoparticle hydrogel system was established by in vitro BMP-2 release study and osteogenic activity in mouse myoblast cells (C2C12). In vivo ectopic and orthotopic bone formation studies to prove the ability of BMP-2/polymeric nanoparticles to induce bone generation after a single subcutaneous injection were successful.

Sobhani et al. recently reported another approach using polyphosphazenes for controlled release of BMP-2 for bone tissue engineering [28]. They used a system based on chitosan microspheres from porous scaffolds of calcium phosphate/poly(dimethylaminoethanol)phosphazenes (CaP/PDMAEP) loaded with BMP-2. They evaluated the biodegradability of the resulting systems by SEM, X-ray powder diffraction (XRD), and in vitro degradation. The release rates increased from 20% to 30% with increasing amounts of chitosan microspheres and their bioactivity was observed to accompany the same trend by inducing osteoblast proliferation. In addition, when the system was loaded with a growth factor, the osteogenic differentiation potential of cells increased when compared with that of CaP/PDMAEP alone. All these results validate the hypothesis that this approach might be used in bone tissue engineering.

Teasdale and coworkers combined polyphosphazene with a trifunctional thiol and divinylester in different ratios, followed by thiol-ene photo-polymerization and obtained porous matrices with pore sizes of 100–200 μm (Figure 9.2b) [29]. The degradation rates could be tuned by using different ratios of

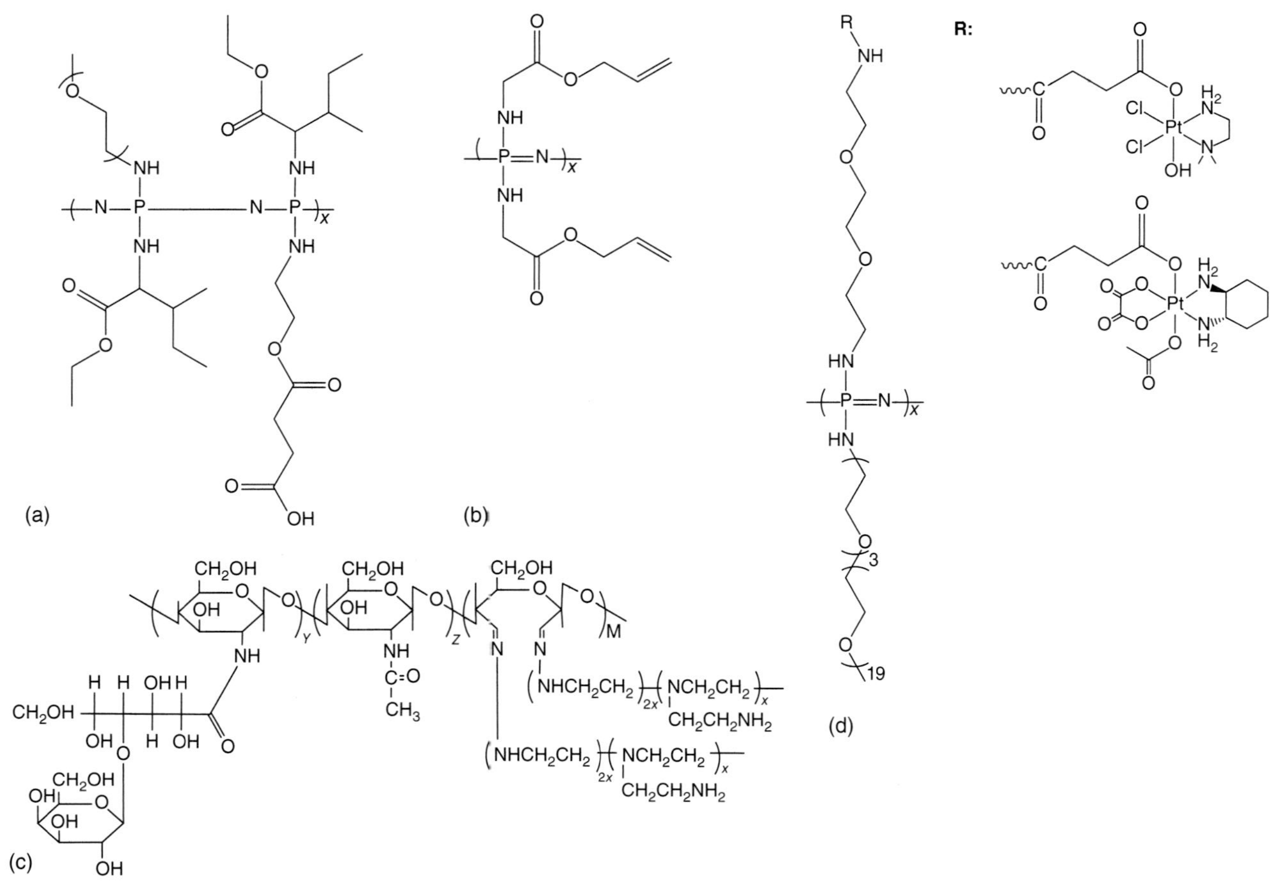

Figure 9.2 Polyphosphazenes with potential medicinal application. (a) Polyphosphazene functionalized with isoleucine ethyl ester and polyethylene glycol. (b) Polyphosphazene functionalized with trifunctional thiol and divinylester. (c) Galactosylated chitosan-graft-polyethylenimine loaded into polyphosphazene hydrogels. (d) Platinum(IV) poly(organo)phosphazenes.

glycine-substituted polyphosphazene and divinyl ester. Preliminary in vitro studies in adipose-derived stem cells confirmed the non-cytotoxicity of the new materials (and their degradation products), as well as cell adhesion and proliferation, confirming the potential of these polymers to be used in the context of tissue engineering.

In the context of drug delivery, there are several formulations of phosphazene functionalizations being tested for gene, antimicrobial, and anticancer agents' delivery.

Ozay et al. reported the synthesis of trimethoprim (antimicrobial) microspheres containing phosphazene as a cross-linker [30]. The microspheres were synthesized using the precipitation polymerization technique by varying the phosphazene ratios. The obtained microsphere sizes ranged between 0.5 and 1.1 μm. The in vitro degradation studies in acidic media revealed that almost full degradation (92%) was accomplished in 50 days. The same microspheres were used for the controlled release of vitamin B_{12} and rhodamine 6G, wherein the release took place during time periods ranging between 26 and 38 hours.

Gene therapy holds significant potential in the treatment of genetic and viral diseases as well as cancer, and its success strongly depends on the vector delivery system. Although there are gene therapy protocols presently in clinical trials there are still safety concerns, which has stimulated the search for new gene delivery systems. Thermosensitive hydrogels have proved to be potential alternatives due to their ability to exhibit a sol–gel transition both at room and elevated temperatures, which induces swelling or de-swelling of the polymer, allowing a controlled release of encapsulated plasmodium DNA, which is dependent on the temperature variations. In this context, Jiang et al. prepared new galactosylated chitosan-graft-polyethylenimine (GC-g-PEI) as gene carriers (Figure 9.2c) that showed low cell toxicity, high transfection efficiency, and high hepatocyte specificity [31]. These GC-g-PEI/DNA complexes were loaded into polyphosphazene hydrogels in order to improve the controlled gene delivery and were tested as hepatocyte targeting gene delivery systems [32]. The results showed a higher transfection efficiency for the loaded polyphosphazenes than for the GC-g-PEI/DNA complexes and good in vitro and in vivo hepatocyte specificity (tested in HepG2 human hepatoblastoma and HeLa human cervix epithelial carcinoma cells, and NCTC3749 murine macrophage and six-week-old female BALB/c mice, respectively).

Selective targeting of tumors and cancer cells might be achieved using several strategies. It is known that macromolecules can selectively accumulate in malignant tissues due to a phenomenon known as the "enhanced permeation and retention effect" (EPR effect) [33]. This happens because the uncontrolled growth of tumors makes them permeable to macromolecules due to defects on their vascular and lymphatic structure. The use of polyphosphazenes as drug delivery systems for anticancer drugs was motivated by the attempts to decrease the side effects of platinum-related anticancer drugs [34–40]. Teasdale and coworkers [38] and Qiu and coworkers [37] have recently reported on this subject. Teasdale prepared novel platinum(IV) poly(organo)phosphazenes macromolecular prodrugs (Figure 9.2d) [38]. The polymers presented relatively narrow dispersities, aqueous solubility, and long-term hydrolytic degradability. The biological assays showed that there was about 30-fold enhancement of Pt accumulation in cancer cells in vitro upon conjugation of the platinum(IV) complexes to the carriers, when compared to the drugs alone. However, the uptake was not directly correlated to the compound's cytotoxicity in the cancer cell lines tested (ovarian A2780 and A2780/cisR, colorectal HCT116 and HCT116/oxR). The study was completed with in vivo evaluation in mice bearing a murine colon cancer cell model (CT-26). The results showed that there was a reduction in the tumor growth on the mice treated with the functionalized poly(organophosphazenes) with platinum(IV) complexes compared with control mice. However, oxaliplatin was much more efficient. It remains to be proved if long-term use will decrease the side effects observed for the low molecular weight platinum-based drugs.

Qiu and coworkers tried to reduce the side effects induced by carboplatin using amino-functionalized polyphosphazene vesicles [37]. For that, polyphosphazenes-containing different primary amine contents were prepared and their ability to encapsulate carboplatin tested. It was verified that the amino content influenced the amount of carboplatin loading. After a series of biological assays, the authors found out that the new vesicles allowed better in vitro cellular uptake, prolonged in vivo circulation on BALB/c mice bearing colon cancer xenografts (CT-26), enhanced decrease in tumor growth, and lessened side effects.

2-Methoxyestradiol (2-ME) has antiangiogenic and antitumor effects on several solid tumors and multiple myeloma. However, it was suspended from clinical trials in the United States due to resulting autoimmune disorders. One of the main problems of 2-ME is its limited solubility which might result in poor bioavailability. In this context, Cho et al. developed a local injectable drug delivery system of 2-ME using poly(organophosphazene) hydrogel aimed to be used in breast cancer therapy [41]. Thus, poly(organophosphazenes) containing L-isoleucine ethyl ester and α-amino-ω-methoxy-poly(ethylene glycol) 550 were prepared. The new thermosensitive polymers, with molecular weights in the 35–38 kDa range and dispersities of 2.38–2.73, sharply enhanced the solubility of 2-ME when compared to phosphate-buffered saline. Successful release of 2-ME from the hydrogel was checked in vitro. In vivo studies carried out in Balb/c nude mice bearing an orthotopic breast tumor (MDAMB-231) model proved that the encapsulation of 2-ME might be advantageous since it improved the antitumor and antiangiogenic activity relative to 2-ME alone. The same strategy, i.e. the use of poly(organophosphazenes) containing α-amino-ω-methoxy-poly(ethylene glycol) as drug delivery systems, was also applied by the same authors for the delivery of docetaxel [42] (an approved anticancer agent) and silibinin [42] (potential anticancer agent) with promising results.

9.2.2 Metal-Containing Polymers

Metal-containing polymers are a unique combination of organic and inorganic/organometallic scaffolds forming a macromolecular system where a synergic effect arising from both components is expected. Research on metal-containing polymers in the field of medicinal applications is growing, mostly inspired by the progress of inorganic and organometallic compounds as anticancer, antibacterial, antifungal, and antivirus agents, as well as biosensors and for bioimaging. The discovery of anticancer properties of cisplatin is certainly the highlight in this field. Thus, this section will be dedicated to metal-containing polymer applications in the area of cancer, stimulated by the successful use of platinum-based anticancer agents in the clinics and the advances on platinum-containing polymers as a less toxic option for the first generation of low molecular weight platinum-based compounds. We will also include examples of other promising elements such as ruthenium and boron.

As discussed in the previous section, the discovery of the EPR effect by Maeda et al. prompted the development of macromolecular assemblies as anticancer agents. Although several developments have been made in this area, the Food and Drug Administration (FDA) has only approved about one dozen of nanotherapeutics for cancer therapy [43].

9.2.2.1 Platinum–Polymer Conjugates

Platinum–polymer conjugates are, by far, the more extensively studied and developed metal-containing polymers. Several formulations based on the combination of different (co)polymers were tested in order to improve stability and the release control of platinum compounds, relative to free drugs [44, 45].

Four platinum–polymer conjugates, namely the polymer–diaminocyclohexane (DACH) platinum conjugate AP 5346 (ProLindac), DACH–platin micelle NC-4016, AP 5280 (Polymer Platinate), and the micellar nanoparticle-encapsulated cisplatin NC-6004, entered clinical trials, although only NC-6004 seems to remain (in the United States and Europe: https://www.cancer.gov/).

NC-6004 is a micellar nanoparticle derived from cisplatin, with ~30 nm, obtained by cross-linking with a polyethylene glycol-*block*-poly glutamic acid copolymer (Figure 9.3a). Early studies showed that this formulation was able to selectively increase accumulation of platinum in tumor, showed lower toxicity than cisplatin, and increased antitumor activity [46]. These features of NC-6004 were demonstrated in both a Phase I study in the United Kingdom, where NC-6004 was administered in monotherapy to patients with solid tumors, and a Phase I/II study in Taiwan and Singapore, where NC-6004 was used in combination with gemcitabine (chemotherapeutic medication) in patients with pancreatic cancer. Then NC-6004 was successfully tested in a Phase I study, also in combination with gemcitabine, in 12 Japanese patients with advanced solid tumors, showing lower acute kidney injury than when treated with cisplatin [47]. The Phase Ib/II study showed that using NC-6004 allowed the delivery of higher equivalent doses of cisplatin without clinically significant neuro-, oto-, or nephrotoxicity [48]. These results showed that the combination of NC-6004 with gemcitabine presents better tolerability and promising activity than cisplatin.

9.2.2.2 Ruthenium–Polymer Conjugates

Ruthenium complexes are currently recognized as alternatives to Pt-based agents. Although there are not yet ruthenium compounds in clinical use, several compounds have passed through clinical trials [49, 50].

Alongside platinum-based anticancer agents, second and third generations of ruthenium-based drugs are being developed in order to overcome issues related

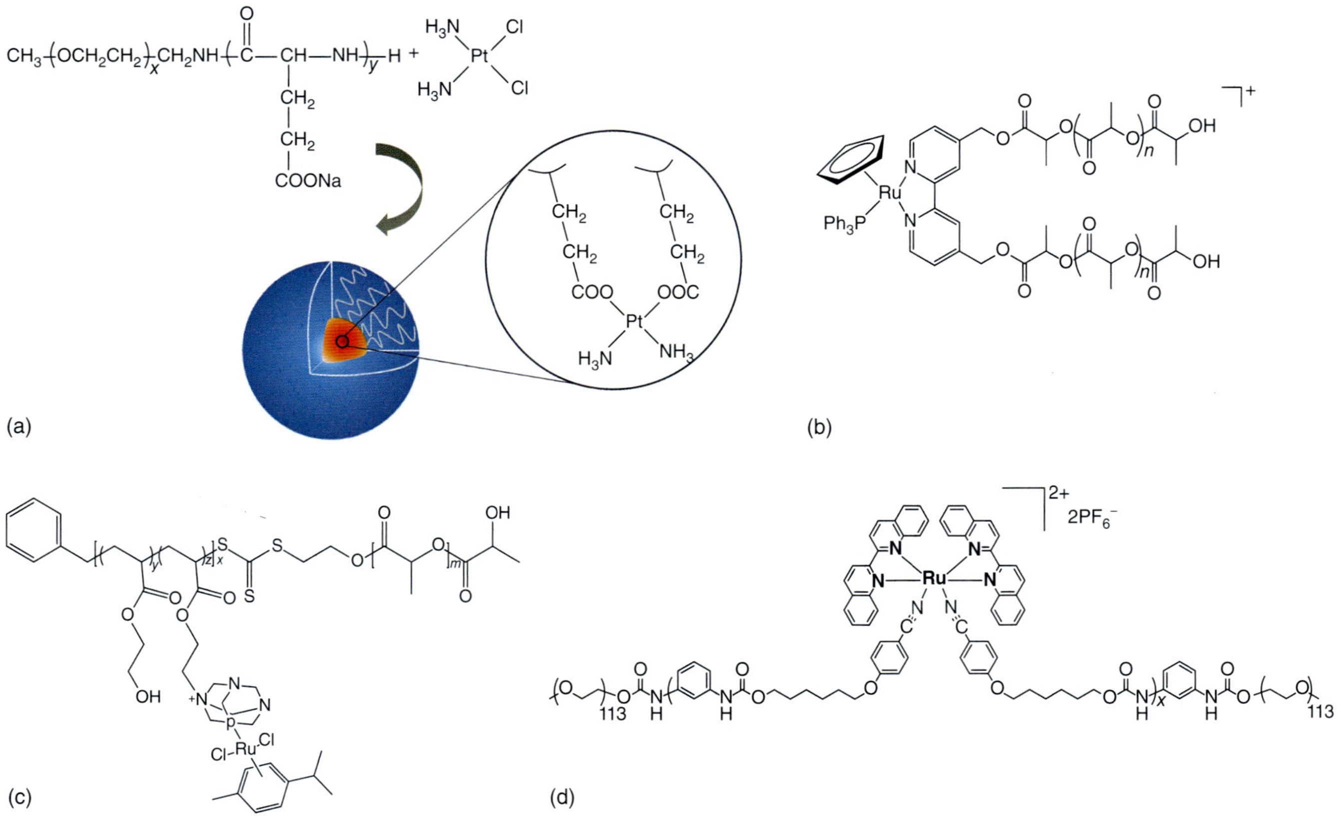

Figure 9.3 Platinum and ruthenium–polymer conjugates with potential medical application. (a) NC-6004. (b) Polylactide–ruthenium cyclopentadienyl conjugate. (c) Copolymer based on polylactide, 2-hydroxyethyl acrylate, and 2-chloroethyl methacrylate conjugated to RAPTA-C. (d) ABC-type triblock copolymer constituted by a photoactive hydrophobic Ru-containing block and two hydrophilic poly(ethylene glycol) blocks.

to in vivo stability and toxicity. The common umbrella is the development of macromolecular assemblies that might benefit from a passive targeting. The strategies involve ruthenium dendrimers [51], ruthenium coordination-cage conjugates [52], ruthenium-HSA conjugates [53] (HSA, human serum albumin), ruthenium nanoparticles [54], or ruthenium–polymer conjugates. We will focus our examples on the last class of compounds.

Based on the very promising results of the first generation of "ruthenium-cyclopentadienyl" ("RuCp") compounds bearing bipyridine-based ligands [55–59], Valente et al. developed a second generation of "RuCp" compounds by incorporating a polylactide polymer at the bipyridine (Figure 9.3b) [60–62]. All compounds were found to be cytotoxic against several cancer cell lines and caused cell death mainly by apoptosis. Results from inductively coupled plasma mass spectrometry (Ru quantification) showed that while the compounds from the first generation were mainly accumulated in the membranes of breast cancer cells [63], probably exerting their action through a cascade of events resulting from this interaction, the macromolecular compounds from the second generation were mainly accumulated inside the cells (nucleus or cytoskeleton) [61]. Owing to this subcellular distribution, in vitro studies with supercoiled plasmid DNA [60] and an F-actin immunofluorescence assay [62] were performed trying to underline the targets for these macromolecular compounds. Results revealed that DNA is not their main target, and that cytoskeleton completely loses its organization leading to cell death.

Another example of the use of the "RuCp" scaffold is the work performed by Romerosa and coworkers. They developed a series of water-soluble organometallic polymers based on the "$RuCp(pta)_2$" fragment (pta = 1,3,5-triaza-7-phosphaadamantane) [64–67]. Given the promising in vitro results obtained for several "RuCp" low molecular weight compounds [55], this family of organometallic polymers might be an interesting alternative to deliver a higher content of compound benefiting from the EPR effect.

Stenzel and coworkers attached RAPTA-C [$RuCl_2$(*p*-cymene)(pta)], a promising anticancer agent [68], to an amphiphilic copolymer based on polylactide, 2-hydroxyethyl acrylate, and 2-chloroethyl methacrylate, the last mentioned being the building block for RAPTA-C conjugation (Figure 9.3c) [69]. The compounds were tested on ovarian cancer cells (A2780, A2780cis, and Ovcar-3) and the micelles were about 10-fold more cytotoxic than RAPTA-C. Confocal microscopy and inductively coupled plasma mass spectrometry studies indicated that micelles' uptake by the cells should proceed via an endocytic pathway and that there was a significant increase on ruthenium uptake for the micelles vs. RAPTA-C. Later, in 2016, the same authors were able to prove the antimetastatic effect of these RAPTA-C macro-assemblies in 2D and 3D in vitro breast tumor cell models [70]. In 2017, they successfully increased the targeting ability of the RAPTA-C micelles by including D-fructose into the polymer backbone in order to benefit from the overexpression of GLUT5 transporter in breast cancer tissues [71].

Another approach to treat cancer is by photodynamic therapy (PDT). Here, the drug is a photosensitizer that, when irradiated at a specific wavelength, produces reactive oxygen species that kill the cells in their surroundings. In this context, photoresponsive ruthenium-containing polymers were developed by Wu and coworkers as potential metallodrugs for PDT. These polymers are constituted of red-light-responsive [Ru^{II}(2,2′:6′,2″-terpyridine)(2,2′-biquinoline)(L)]$^{2+}$, where L is poly(ethylene glycol)-*block*-poly(6-(4-cyanophenoxy)hexyl methacrylate) (PEG-*b*-PCPH) [72]. In aqueous solution, and depending on the polymer molecular weight, different nanostructures such as micelles, vesicles, and large compound micelles are generated. The cellular uptake was found to be dependent on these different morphologies, and the micelles were the structures exhibiting best biocompatibility, enhanced cellular uptake, and better anticancer performance. Red-light irradiation released the anticancer Ru complex [Ru^{II}(2,2′:6′,2″-terpyridine)(2,2′-biquinoline)(H_2O)]$^{2+}$ and generated singlet oxygen ($^1O_2$) inhibiting cancer cells growth.

In 2017, the same authors published another example of this family of compounds based on an ABC-type triblock copolymer constituted by a photoactive hydrophobic Ru-containing block and two hydrophilic poly(ethylene glycol) blocks (Figure 9.3d) [73]. The Ru cytotoxic core was based on $[Ru(Biq)_2(Hob)_2][PF_6]_2$ (Biq = 2,2′-biquinoline, Hob = 4-((6-hydroxyhexyl)oxy) benzonitrile). As in the previous example, these polymers self-assemble into nanoparticles when placed in aqueous solution, are taken up by the cancer cells, and produce $^1O_2$ when irradiated with red light leading to cell death. In vivo assays in mice showed an efficient accumulation of the compounds in the tumor, inhibiting its growth under light irradiation.

9.2.2.3 Carborane–Polymer Conjugates

Dicarba-*closo*-dodecaboranes (abbreviated as carboranes) have a high content of ^{10}B atoms, making them potential candidates for use in boron neutron capture therapy (BNCT). In BNCT, when a tumor that has accumulated sufficient quantities of ^{10}B is irradiated with low-energy thermal neutrons, some ^{10}B atoms can undergo a nuclear reaction to produce α-particles and recoiling ^7Li nuclei. These particles of high energy have a short destructive path length (5–9 μm), making this therapy appealing since, in theory, only those cells with high ^{10}B content will be destroyed. In order to improve ^{10}B accumulation in cancer cells, several strategies based on the EPR effect are being developed using polymers. Some examples of carborane–polymer conjugates are discussed here.

Wu and coworkers synthesized carborane–PEG conjugates, where each carborane was labeled with a fluorescent rhodamine dye (Figure 9.4a) [71]. When placed in water, these polymers self-assembled into spherical vesicles of <200 nm. In vitro assays showed that the vesicles were taken up by the cancer cells. Biodistribution assays in mice using fluorescence microscopy showed that the rhodamine–carborane–PEG had some accumulation in the spleen, but not in other organs.

Sumitani et al. prepared new cross-linked and boron-conjugated micelle by copolymerization of acetal-PEG-*b*-PLA-MA with 1,2-bis(4-vinylbenzyl)-*closo*-carborane ((VB)2-carborane) [70]. These micelles were found to be very stable under physiological conditions and showed good accumulation within the tumors (7.2 ID%/g). Thermal neutron irradiation experiments allowed a significant suppression of tumor growth in mice treated with the micelles. One week after injection the compounds were almost completely excreted from the major organs, except from the tumor.

(a)

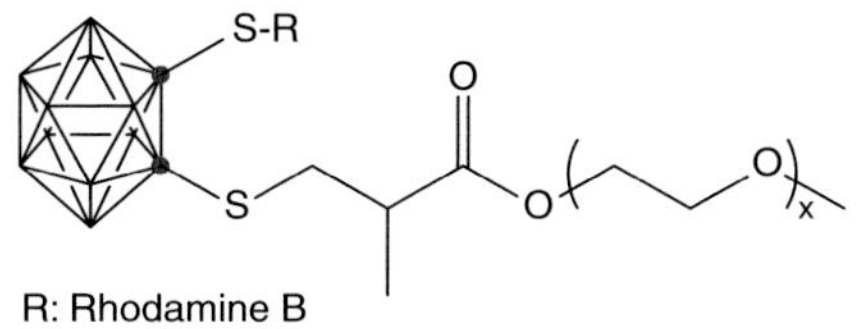

(b)

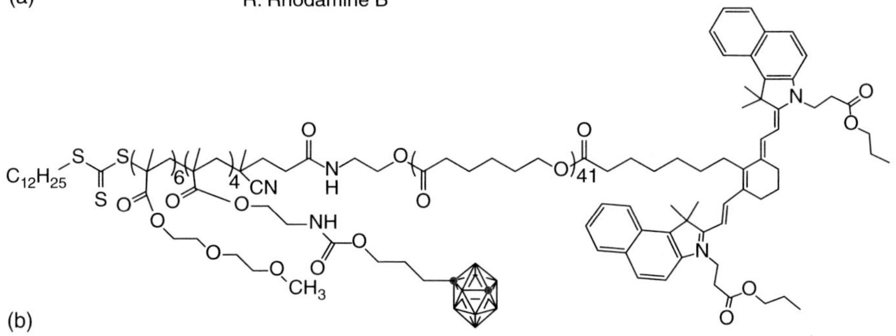

Figure 9.4 Carborane–polymer conjugates with potential medicinal application. (a) Carborane–PEG conjugates labeled with rhodamine. (b) Carborane-containing triblock copolymer based on polycaprolactone, polycarborane, and poly(ethylene glycol)methyl ether methacrylate, conjugated with a near infrared (NIR) fluorescence probe.

Yan and coworkers synthesized a carborane-containing triblock copolymer based on hydrophobic segments of polycaprolactone and polycarborane and hydrophilic segments of poly(ethylene glycol)methyl ether methacrylate, conjugated with a near infrared (NIR) fluorescent probe (Figure 9.4b) [74]. Since these new polymers possess carborane that can be used for BNCT and for an NIR aminocyanine fluorescent probe that can be used for imaging, this new family of micelles might be used as theranostic agents.

The work of Viñas and coworkers has also the potential to be used for theranostics of cancer, by the combined BNCT with the radiolabeling of cobaltabisdicarbollide (COSAN) with radioiodine [75, 76].

In a different approach, Huang and coworkers developed carborane-conjugated amphiphilic copolymer nanoparticles aimed at delivering the known anticancer drug doxorubicin (DOX), leading to a possible combination of chemotherapy and BNCT [77]. Toward this, poly(ethylene glycol)-*b*-poly(L-lactide-*co*-2-methyl-2(2-dicarba-*closo*-dodecarborane)-propyloxycarbonylpropyne carbonate) was synthesized and self-assembled with DOX. The new nanoparticles had prolonged blood circulation time and improved boron and DOX accumulation at the tumor site. The combined therapy allowed the best therapeutic outcome when compared to each of the therapeutics alone.

9.3 Emerging and Potential Applications for Metal-Organic Frameworks for Drug Delivery

MOFs are an emerging class of self-assembled nanoporous materials; their properties can be readily tuned by varying the molecular building blocks. Combining the properties of porous materials with the benefits of nanostructures is expected to improve the performance of classical bulk crystalline MOFs. The design and construction of inorganic–organic hybrid framework materials are currently of great interest in the fields of crystal engineering and supramolecular chemistry, not only for their intriguing topological architectures, but also due to their potential applications in many fields, such as heterogeneous catalysis, drug delivery systems, gas storage, nonlinear optics, magnetism, and luminescent materials [78–84].

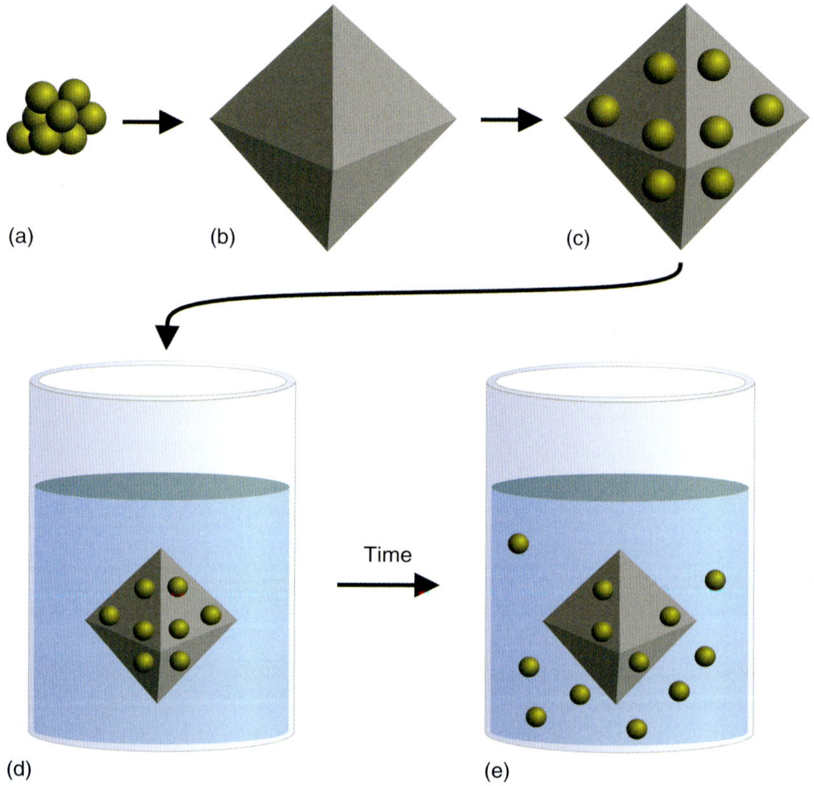

(a) Drug; (b) MOF; (c) loading process; (d) exposure to biological media; (e) release.

In the past decade, interest in ordered porous hybrid materials for medicinal applications has been growing very rapidly, not only as a result of the still increasing number of new structures and compositions, but also due to the significant efforts that have been undertaken to introduce new or optimized functionalities [79, 80]. Furthermore, inorganic and organic nanoparticles with sizes ranging from a few nanometers to several hundred nanometers are promising candidates for biomedical applications, such as active tumor targeting. Advantageous features are their adjustable size, a chemical surface adaptable to biochemical purposes, improvements in terms of biocompatibility, possibility of high drug loading rates, kinetically controlled drug release, enhanced accumulation, and high permeability and retention effect [85, 86].

9.3.1 Metal-Organic Frameworks (MOFs)

MOFs [85] are a class of hybrid materials consisting of metal ions or clusters and bridging organic ligands (linkers) that form highly porous three-dimensional structures (Figure 9.5) [84, 87–89].

The high porosity (up to 90% of the crystal volume) and high surface area (up to several thousand m^2/g) of MOFs are highly interesting from a nanotechnological viewpoint. These highly porous structures with internal diameters of up to 48 Å have promising applications, such as host materials for gas storage (H_2/CH_4) [81, 84, 90–93], adsorption and separation of molecules [93], in catalysis [78, 94], and in the controlled release of drugs [79, 80, 85, 86, 95–99]. As compared to inorganic systems, for example mesoporous silica and silicon that can deliver drugs in a more controlled manner due to their ordered network and porosity but have a limited load capacity, one of the advantages of MOFs is the possibility to adjust pore sizes and functional groups in the frameworks [84, 96–105].

Transition metals are versatile connectors for building MOFs. The transition metals commonly used are Cr^{3+}, Fe^{3+}, Cu^{2+}, Zn^{2+}; some alkaline and alkaline earth ions have also been used [106, 107]. Depending on the metal and the oxidation state, coordination numbers range from 2 to 7, giving rise to various geometries, such as linear or V shaped, T or Y shaped, square planar, tetrahedral, square pyramidal, octahedral, which play an important role in targeting MOF structures [108].

The organic ligands used in the construction of MOFs usually contain a functional group for coordination, such as carboxylate, phosphate, sulfonate, amine, or nitrile. The ligand design can also improve the efficacy of the drug delivery system. Selective binding to targeted tissues can cause a conformational change, resulting in drug release [86].

The secondary building units (SBUs) have intrinsic geometry, which plays an important role in view of the MOF topology. Yaghi et al. described in detail the variety of SBU geometries, which provides ample options for the design of MOF structure [109].

9.3.2 Application of MOFs in Drug Delivery Systems

When developing a controlled release system, one of the challenges is finding nontoxic carriers. Since each drug has a therapeutic range (Figure 9.6), with a

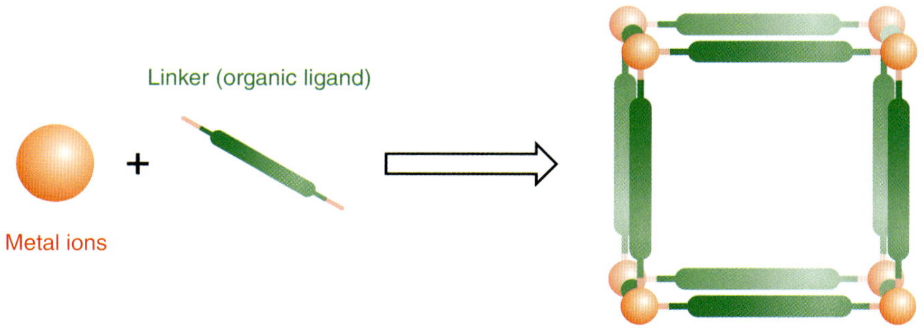

Figure 9.5 Simplified formation of a three-dimensional network through metal ions and organic linkers.

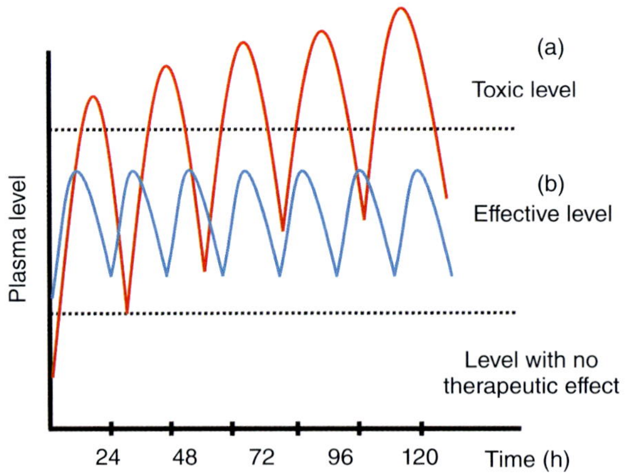

Figure 9.6 Comparison between the variation of drug concentration: (a) when administered by conventional multi-dosage method and (b) the controlled drug release system.

toxic action above and below its ineffectiveness, the plasma levels are dependent on the dosage administered and become more critical if the efficacious dose is close to the toxic dose [105].

In the case of nanocarriers, some of the requirements for an effective therapy are to (i) control the drug release and prevent the "burst effect," (ii) control the matrix degradation, e.g. by using surface modification techniques, (iii) be detectable by various imaging techniques, (vi) efficiently retain drugs with high load capacity [105].

The development of nanomaterials for triggered release through exogenous and/or endogenous activation is a good example of how drug delivery systems can be improved. For an exogenous activation, an external stimulus is needed, for example, a magnetic field, while the endogenous activation takes advantage of chemical differences between the heterogeneous environments within living systems to trigger therapeutic activity [86].

9.3.2.1 Selected Examples of MIL-*n* Frameworks in Drug Delivery

The first example of encapsulation of a drug molecule was reported by Serre, Férey, and coworkers. Ibuprofen (IBU) was used as a model drug in a chromium carboxylate MOF, an example of the cubic (*Fd*3*m*) zeotypic metal carboxylates, named MIL-100 and MIL-101 (MIL = Materials of Institut Lavoisier) [110]. Composed of trimers of metal octahedra and di- or tricarboxylates, the structures exhibit large pores (with diameters of 25–34 Å) and a surface area of 3100–5900 m^2/g. These frameworks exhibited a high loading capacity adsorbing different amounts of ibuprofen, 35 wt% (for MIL-100) and 140 wt% (for MIL-101). The higher drug capacity of MIL-101 is related to its higher surface area, larger pore sizes (Ø: 29 and 34 Å), high pore volumes of 12 700 and 20 600 $Å^3$, and specific interactions between ibuprofen molecules and open metal sites and the organic spacers, while the comparatively lower capacity for MIL-100 (Cr) despite its large pore volumes (8200 and 12 700 $Å^3$) is probably due to its narrow pentagonal windows (5 Å) [110]. With regard to the release, different behaviors were observed for MIL-100 within the first two hours, due to weakly bonded drug molecules probably located on the surface. Release of the

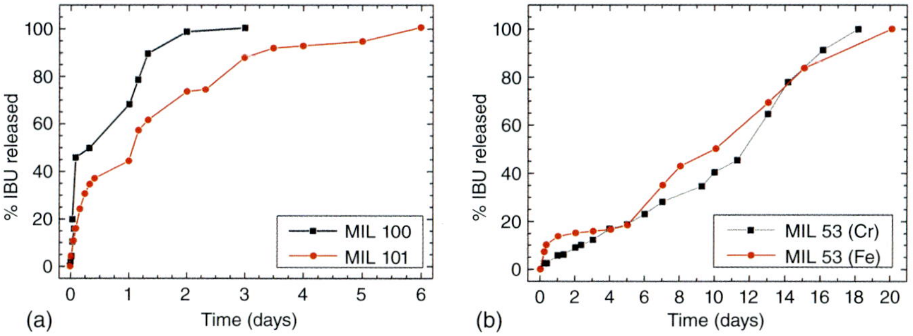

Figure 9.7 (a) Kinetics of delivery of ibuprofen from MIL-100 (Cr) and MIL-101 (Cr) in SBF at 37 °C. (b) Kinetics of delivery of ibuprofen from the MIL-53 (Cr) and MIL-53 (Fe) in SBF at 37 °C. Source: Reprinted with permission from Horcajada 2006 [110], 2008 [111].

entire ibuprofen took place over three days. In the case of MIL-101, the weakly bonded drug molecules were released in the first eight hours, and a complete release took six days, under physiological conditions (simulated body fluid (SBF) at 37 °C) (Figure 9.7a) [110].

Serre, Férey, and coworkers also demonstrated the controlled ibuprofen release in flexible, "breathing" MOFs, such as MIL-53. To circumvent toxicity, chromium was also replaced with iron. The MOFs MIL-53 (Cr) and MIL-53 (Fe) showed a high loading of ibuprofen (around 20 wt%) and a period of release of up to three weeks under physiological conditions (Figure 9.7b) [111].

Singco et al. described the aspirin entrapment in MIL-100 (Fe), in order to avoid the main side effects following oral administration. The loading efficiency after 24 hours was approximately 181% (1806 mg/mg MOF). This loading efficiency was the highest so far for MIL-100 (Fe), which showed a loading efficiency of 35% for ibuprofen and 69% for urea. The release studies revealed a complete release in PBS (phosphate-buffered saline) at 37 °C in 14 days [112].

Other antitumor drugs such as DOX [113, 114] or busulfan [115] were encapsulated in MIL-100 (Fe). In the case of DOX, the drug–MOF interaction occurs through coordination of one or both deprotonated hydroxyl groups of the aglycone moiety of DOX with the unsaturated iron(III) centers, resulting in a highly stable binding, with a loading of 9 wt% and a release of up to two weeks [113]. For busulfan, the encapsulation (25 wt%) and delivery turned out to be a good alternative to avoid side effects, such as accumulation in the liver. However, 61% of the drug was released within the first 30 minutes [115], which could lead to adverse reactions such as the burst effect. In this case, a post-synthetic treatment with poly(ethylene glycol) could be used to modulate the release, therefore avoiding accumulation in liver [115]. Adhikari and Chakraborty also demonstrated a controlled release of DOX from iron- and zinc-based MOFs under different pH, the drug loading was higher for the iron-based MOF due to the stronger binding [114].

Agostoni et al. reported a "green" microwave-assisted hydrothermal synthesis of nontoxic MIL-100 (Fe) with large (29 Å) and small (24 Å) mesoporous cages. The two types of cages are accessible through microporous pentagonal (5.6 Å) or hexagonal windows (8.6 Å) [116]. This group used nanoscale MIL-100 (Fe) to entrap the antiretroviral agent azidothymidine triphosphate (AZT-TP). The interaction was attributed to the efficient coordination of the drug's phosphate groups to unsaturated iron(III) sites within the pores. Different incubation times (5 minutes, 30 minutes, 1 hour, 4 hours, 24 hours) showed that encapsulation efficiencies reached 93.6 ± 8.5% and 98.2 ± 2.4% after 5 and 30 minutes with a drug payload of 7 ± 0.5 wt% and 7.2 ± 0.2 wt%, respectively [117].

It was shown that MIL-100 (Fe) can entrap 5 and 60 times more busulfan and 20 times more azidothymidine triphosphate, when compared to other nanosystems such as polymers or liposomes [95, 99, 115].

Horcajada et al. demonstrated the use of other iron carboxylate MOFs (MIL-*n*), for example, MIL-53, MIL-88A, and MIL-89, besides MIL-100, in the encapsulation of antitumor (busulfan), antiretroviral (cidofovir and azidothymidine triphosphate), and cosmetic agents. A high loading (up to 42 wt%), sustained release (3–14 days) and no burst effect were observed [95].

Lin and coworkers encapsulated the prodrug ethoxysuccinato-cisplatin, *c,c,t*-$[PtCl_2(NH_3)_2(OEt)(O_2CCH_2CH_2CO_2H)]$ (ESCP) in MIL-101 (Fe) by post-synthetic modifications. Incorporation of 2-aminoterephthalic acid (NH_2-BDC, around 17.4 mol%) enables a covalent attachment of ESCP (via formation of an amide bond), in this case, a load of 12.8 wt%, and slow release on MOF degradation (Figure 9.8). A covalent grafting of the fluorescent dye BODIPY (4,4-difluoro-4-bora-3*a*,4*a*-diaza-*s*-indacene) via the amine group showed that both the dye and the prodrug were readily released, due to the instability of the nanosized MIL-101 (Fe). In the corresponding core–shell particles, in which the MOF was covered with a thin silica layer, the release was then slowed to $t_{1/2}$ = 14 hours (BDC-NH-ESCP) and $t_{1/2}$ = 16 hours (BDC-NH-BODIPY) in PBS buffer at 37 °C. To increase the cytotoxicity, the silica shell was further functionalized with a targeting RGD (arginine, glycine, and aspartate) peptide resulting in an IC_{50} value comparable to that of cisplatin. This is a good example of how uncontrolled drug release can be avoided by covalent attachment of active molecules [118].

Related to this work is the development of nanoscale Pt-containing coordination polymers, namely NCP-1, consisting of Tb^{3+} ions and *c,c,t*-(diamminedichlorodisuccinato)platinum(IV), *c,c,t*-$[PtCl_2(NH_3)_2(O_2CCH_2CH_2CO_2H)]$ (DSCP, disuccinatocisplatin), as bridging ligand, which exhibited anticancer activity in vitro as demonstrated by Lin and coworkers. Precipitation from aqueous solution gave nanoparticles with diameters of 58.3 ± 11.3 nm (by DLS measurements). In a similar approach as described above, NCP-1 was encapsulated in shells of amorphous silica with varying thickness and the release of the Pt drug studied. In order to enhance the cellular uptake of NCP-1, the surface here was also modified with cyclo(Arg-Gly-Asp-D-Phe-Lys) [*c*(RGDfK)] (Figure 9.9). These *c*(RGDfk)-functionalized particles showed improved cytotoxicity against human colon adenocarcinoma cells (HT-29), suggesting that the particles were taken up by receptor-mediated endocytosis, followed by reduction to the active Pt^{II} species in the reducing cell environment [119].

The organometallic anticancer drug $[Ru(p\text{-cymene})Cl_2(pta)]$, named RAPTA-C, was encapsulated in MIL-100 (Fe) (loading 42 wt%). The resulting system RAPTA-C@MIL-100 (Fe) combines all advantages described for

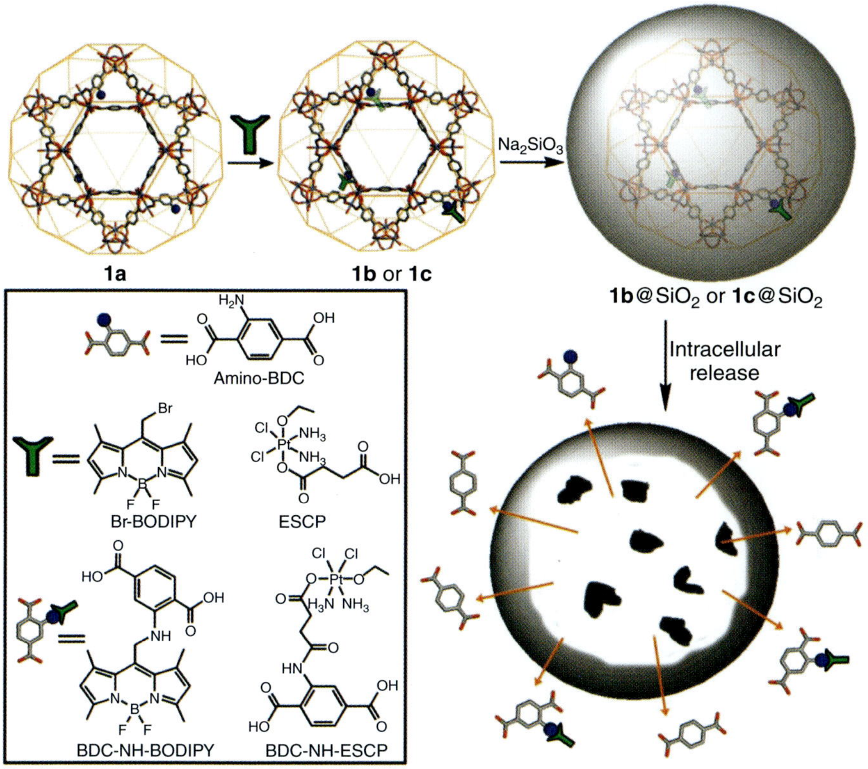

Figure 9.8 Schematic representations of the synthesis, modification, and drug release from amino-functionalized iron MIL-101 (Fe); **1a**: amino-functionalized particles, **1b**: BODIPY-loaded particles, **1c**: ESCP-loaded particles. Source: Reprinted with permission from Taylor-Pashow et al. 2009 [118]. Copyright 2009, American Chemical Society.

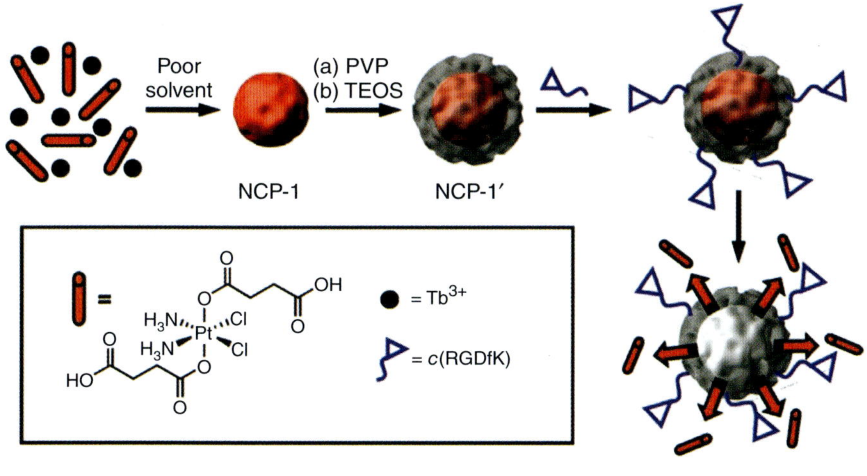

Figure 9.9 Construction of silica-coated NCP-1 particles and further functionalization with *c*(RGDfk). Source: Reprinted with permission from Rieter et al. 2008 [119]. Copyright 2009, American Chemical Society.

MIL-100 (Fe) as a drug delivery system with the antimetastatic activity of RAPTA-C. In SBF, the RAPTA-C delivery process takes place in two steps exhibiting half-life periods of 1.9 and 30 hours. The first step could be related to weakly adsorbed molecules on the external nanoparticles surface as well as partial degradation (10% of benzene-1,3,5-tricarboxylate (BTC) leaching after two days). Especially the long half life of the second step of RAPTA-C delivery offers the possibility of achieving an adequately controlled release of the drug [120].

Wang et al. reported that MIL-101 (Fe) itself has the intrinsic ability to suppress angiogenesis processes in vitro, as well as the potential to effectively distinguish tumor cells, such as SKOV3 (ovary cancer), HeLa (human cervical carcinoma), and A549 (human lung carcinoma) from normal cells, encouraging the use of these MOFs as antitumor agents [119]. Furthermore, MIL-101 (Fe) effectively induced apoptosis of human umbilical vein endothelial cells (HUVECs) through the mitochondrial pathway interfering with the G1/G0 cell division process, and in addition inhibited growth and migration of endothelial cells through processes such as downregulation, i.e. a process by which a cell decreases the amount of a cellular component, in this specific case the expression of the MMP-2/9 (MMP = matrix metalloproteinase) [121].

Recently, Taherzade et al. described two antibiotics with different solubility, tetracycline hydrochloride and doxycycline monohydrate, loaded separately on the nanosized MIL-100 (Fe). Through the release studies, the drug solubility in water was mediated by the MOF; in the case of tetracycline hydrochloride, the release was sustained, and after 48 hours it could be verified that 96% of the drug was released [122].

9.3.2.2 Selected Other Metal-Organic Frameworks Used in Drug Delivery

Hu et al. described the synthesis of MOF-74-Fe(III), prepared through a post-oxidation procedure from neutral MOF-74-Fe(II) [containing $Fe(II)_2(dobdc)$, $dobdc^{4-}$ = 2,5-dioxido-1,4-benzenedicarboxylate, as SBU], which proved to be a good carrier for the selective encapsulation of anionic drugs [123]. This cationic host material exhibits very low cytotoxicity for PC12 cells, which are derived from a pheochromocytoma of the rat adrenal medulla, often used as a model to evaluate the cytotoxicity of nanomaterials. The model drug was ibuprofen and the loading capacity for the corresponding anions was around 0.21 g for 1 g of dried materials, corresponding to ~15.9 wt% in the resulting composite. It was estimated that half the amount of the drug was released in four hours; the process presented different drug delivery rates due to the presence of two distinctive environments of IBU^- anions in host channels, free and coordinated, resulting in the dramatically different release behaviors [123].

Zhu et al. applied the zirconium-based MOF UiO-66 (based on linear 1,4-benzenedicarboxylate (BDC, terephthalate) ligands and $Zr_6O_4(OH)_4$ clusters as 12-connected nodes) for the delivery of alendronate (AL), an amino-bisphosphonate, which is widely used for the treatment of osteoporosis and several other bone diseases, such as solid tumor bone metastases and myeloma bone disease. Owing to poor bioavailability of alendronate, high doses are usually required. Administration of a local, sustained AL delivery system with an enhanced loading capacity and delivery efficiency of AL into cancer cells allows tailoring the controlled capture and release of AL molecules based on the formation of strong Zr—O—P bonds. Thus, it was possible to encapsulate up to 51.4 wt% AL, i.e. 1.06 g of drug per gram of porous nanocarriers, in the UiO-66 nanoparticles. Around 42.7% of AL is released in 60 hours at pH 7.4, whereas more than 59% of the drug is released during the same time interval at pH 5.5, which is useful regarding the sensitivity of AL@UiO-66 in the endosome/lysosome (pH 5) [124].

Filippousi et al. developed a copolymer (poly(ε-caprolactone)–D-α-tocopheryl polyethylene-glycol-succinate (PCL–TPGS)) to modify the surface of two Zr-based MOFs, UiO-66 and UiO-67, to improve drug solubility of hydrophilic cisplatin and hydrophobic taxol. An amount of 140 mg/g taxol could be loaded on the UiO-66 and 100 mg/g on the UiO-67 framework. However, the loading of cisplatin on UiO-66 and UiO-67 (48 and 10 mg/g, respectively) was rather low, due to the hydrophilic nature of the drug as compared to the hydrophobic MOF linkers. It was found that the drug release depends on the MOF–drug interaction while the controlled release rates can be attributed to the microencapsulated formulations [125]. Through molecular dynamics simulations it was possible to visualize the predominant adsorption positions of the adsorbate molecules to get insight into the mobility and adsorption locations. The cisplatin molecules interact with the accessible basic site Zr–O–Zr of the inorganic framework, which is the case for both UiO-66 and UiO-67 [125].

Tamames-Tabar et al. developed a biocompatible and bioactive MOF (BioMOF). In general, Ca^{2+}, Fe^{2+}, Fe^{3+}, Ag^{+}, or Zn^{2+} are used to construct BioMOFs, and the active compounds are delivered via framework degradation. Here, BioMIL-5 (Bioactive Materials from Institut Lavoisier), which is based on therapeutically active azelaic acid (exhibits antibacterial and anti-inflammatory properties) and bioactive Zn^{2+} cations (widely used as a cicatrizing agent, skin moisturizer, and antidandruff astringent) was prepared. In terms of drug delivery, the progressive release of the active Zn^{2+} and azelaic acid from BioMIL-5 in water or bacterial culture media allowed control of the growth of an *S. epidermidis* suspension, a Gram-positive bacterium, over seven days [126].

Another type of BioMOF was reported by Rosi and coworkers. In bio-MOF-1, Zn_8(adeninate)$_4$(biphenyldicarboxylate)$_6$O·2Me_2NH_2, 8DMF, 11H_2O, the anionic nature of the zinc-adeninate structure showed great potential for entrapment of cationic drugs. In this case, procainamide hydrochloride, an antiarrhythmic drug, was encapsulated (0.22 g/g material), and a steady release was observed in PBS buffer (pH 7.4) over 20 hours, and a complete release was reached after 72 hours. The crystalline integrity of the framework was maintained throughout the release process [127].

9.3.3 Toxicity and Stability

9.3.3.1 Toxicity

Recent studies on the toxicity of MOFs have been conducted, mainly because of their potential use as biomaterials. While chromium (LD_{50} 1.8 g/kg), cadmium, nickel (LD_{50} 0.2 g/kg), cobalt (LD_{50} 0.5 g/kg), and other metals are not suitable as metal nodes due to their high toxicity (even though some of them are essential elements, e.g. cobalt), some metals exist in appreciable amounts in the body. For example, iron is an important component of hemoglobin and is approximately 22 μM in blood plasma. The most appropriate metals are Ca, Cu, Mn, Mg, Zn, Fe, Ti, or Zr, but data on their biological fate are limited. Their toxicity is estimated by oral lethal dose 50 (LD_{50}): Ca, 1 g/kg; Cu, 25 μg/kg; Mn, 1.5 g/kg; Mg, 8.1 g/kg; Zn, 350 μg/kg; Fe, 30 g/kg; Ti, 25 g/kg, and Zr, 4.1 g/kg [98].

As organic linkers, carboxylates and phosphonates can be used to build MOFs for biomedical applications. The most commonly used precursors for linkers showed LD_{50} values of 1.13 g/kg (1-methylimidazole), 1.4 g/kg (2-methylimidazole), 1.6 g/kg (5-aminoisophthalic acid), 5 g/kg (terephthalic acid, 2,6-napthalenedicarboxylic acid and isonicotinic acid), and 8.4 g/kg (trimesic acid), indicating that their toxicity is acceptable for bioapplication [98]. Furthermore, some organic linkers might be reused in the body, which would substantially reduce the risk of side effects, for example, amino acid-based MOFs, such as [Cu(L-or D-glutamic acid)(H_2O)]·H_2O [128], or nucleobase-based MOFs, such as Zn_8(adeninate)$_4$(biphenyldicarboxylate)$_6$O·$2Me_2NH_2$, 8DMF, $11H_2O$ (bio-MOF-1) [127].

Horcajada et al. studied the cytotoxic potential of metal ions and organic linkers, and rated MOFs based on iron and zinc as the best candidates for use as biomaterials [95]. Nevertheless, the correlation between physicochemical properties and toxicity is not straightforward; parameters such as colloidal stability, surface charge, shape, size, adsorption behavior, and purity play a role [129].

In vitro and in vivo toxicology studies for some iron carboxylate-based MOFs (iron(III) trimesate MIL-100, iron(III) fumarate MIL-88A and iron(III) tetramethylterephthalate MIL-88B_4CH_3) were conducted [95, 130]. A high dose (up to 220 mg/kg) was administered intravenously in Wistar female rats. Indicators like body and organ weights, absence of immune or inflammatory reactions, animal behavior, and serum parameters were evaluated up to three months after injection and showed no significant differences when compared to the control group [95]. The absorption, distribution, metabolism, and elimination, known as the ADME series, were also evaluated. MIL-100 (Fe) seems to be the most promising with regard to the stability, loading capacities, degradation/release profile, and in vitro and in vivo toxicity. Particle accumulation happened in the liver and spleen but no signs of histological or function alterations were observed [130]. Accordingly, it was concluded that these MOFs are nontoxic [95, 130]. Kundu et al. further confirmed the nontoxic nature of gadolinium(III)-based MOF, Gd–pDBI (pDBI = 1,4-bis(5-carboxy-1*H*-benzimidazole-2-yl)benzene), on leukemia cancer cell line U937 (by 3-(4,5-dimethylthiazol-2-yl)-2,5-diphenyltetrazolium bromide (MTT) assay) [131].

More recently, the zebrafish is being considered as a prominent model as it provides substantial information for the study of chemical toxicity of some nanoMOFs. Maspoch and coworkers reported combined in vitro (HepG2 and MCF-7 cells) and in vivo (zebrafish embryos) studies on the toxicity of 16 archetypical, uncoated nanoMOFs. MOF-74 (Mg), MOF-74 (Co), UiO-66, and UiO-67 presented very low or no toxicity, while MOF-74 (Zn), ZIF-7, and ZIF-8 showed high toxicity values, which is directly correlated with their composition and stability. The Zn^{2+}-based MOF (ZIF-8) showed high toxicity in the breast cancer cell line MCF-7 and human liver cancer cell line HepG2, due to release of Zn^{2+} ions into the cells [132].

9.3.3.2 Stability

Many authors have studied the possibilities for increasing the MOF stability toward hydrolysis [133]. Tan et al. described the hydrolysis mechanisms for the paddle-wheel frameworks M(BDC)(ted)$_{0.5}$ [M = Cu, Zn, Ni, Co; BDC = 1,4-benzenedicarboxylate; ted = triethylenediamine (also known as DABCO, 1,4-diazabicyclo[2.2.2]octane)]. Cu(BDC)(ted)$_{0.5}$ suffers hydrolysis at the Cu–O–C (Cu–BDC) moiety; displacement of the ted linkers by water molecules occurs with Zn(BDC)(ted)$_{0.5}$ and Co(BDC)(ted)$_{0.5}$, while Ni(BDC)(ted)$_{0.5}$ is less susceptible to hydrolysis [133].

Bellido et al. described that MIL-100 (Fe) nanoparticles significantly increased their stability in water after post-synthesis treatment with a KF solution, due to the replacement of pending nonframework carboxylate groups by F^- ions [134]. MOFs can also be stabilized by catenation (interpenetration or interweaving of two or more identical and independent frameworks) [135].

The intrinsic instability of MIL-101 (Fe) toward hydrolysis enables binding of heparin molecules to the framework structure. This composite could serve as a drug-releasing depot for nanofabrication and to introduce anticoagulant activity in medical devices and biocoatings [136].

9.3.4 Biodegradation

Most MOFs exhibit intrinsic biodegradability under physiological conditions (PBS, pH 7.4 at 37 °C, or BSA) as a result of relatively labile metal–ligand bonds. In biological applications, a certain amount of instability is desired in order to avoid endogenous accumulation, and the degradation products can be dealt with by the body's own systems [88].

Studies involving the degradability of MOFs such as MIL-88A (Fe), MIL-100 (Cr), MIL-101 (Cr), and MIL-53 (Fe, Cr), demonstrated that those solids degrade over days or even weeks [95, 112, 113]. The particle size composition and the crystalline MOF structure directly affect the degradation process, which in physiological media can be modulated from a few days up to several weeks.

Recently, Li et al. studied the effect of PBS on nanoMOF degradation (MIL-100 (Fe)) at 37 °C compared to water. After six hours, considerable amounts of the constituting trimesate ligand (34 ± 3 wt%) had been released in PBS [134]. The phosphate concentration directly influenced the nanoMOF degradation mechanism – higher concentration led to faster degradation [134]. Bellido et al. also observed this behavior for MIL-100 (Fe) under similar conditions (incubation in PBS) [137]. Baati et al. observed that MIL-100 (Fe) was degraded in its constitutive components and eliminated in urine and feces after intravenous administration in Wistar female rats [130].

Nevertheless, an appropriate matrix stability is still required in order to maintain the structure before reaching their target. Nanoscale MOFs (nanoMOFs or NMOFs) typically release their drug cargo in less than 24 hours. Surface modifications, e.g. coatings like silica, lipid layer, polymer, oligonucleotides, or nucleic acids, can also slow down the degradation of the nanomaterials [88].

Cyclodextrin molecules were anchored to the nanoMOFs' surface in order to improve the physicochemical properties of drugs (stability, solubility, and bioavailability). The coatings are stable in body fluids and can be further functionalized with targeting ligands, indicating the possibility of a versatile surface modification of nanoMOFs for targeted multifunctional drug delivery and other applications [137].

Lin and coworkers stabilized nanoMOFs by encapsulating them within a silica shell. Silica coatings have been extensively used to modify inorganic and polymer nanoparticles. In order to make this type of coating, the nanomaterials are first coated with a polymer, for example, polyvinylpyrrolidone (PVP), followed by treatment with tetraethyl orthosilicate (TEOS) in basic ethanol to afford silica coatings [88, 138].

The Mn-bisphosphonate NCP developed by Lin and coworkers [119] was coated with lipid and pegylated in order to accomplish controlled release and degradation. These nanoparticles were further functionalized with a targeting group (anisamide) to endow specificity to cancer cells. Release is triggered after being internalized into cells, with the ability to induce apoptosis in both human breast and pancreatic cancer cells [139].

9.4 Final Remarks and Perspectives

The use of (inorganic) polymers in the areas of tissue engineering and drug delivery is a growing area of research due to the versatility that chemical synthesis brings to the conception of new materials. The same backbone can be envisaged to be used for cell adhesion or surgical implants, or, when coated or functionalized with different chemical groups, might act as drug or gene delivery systems. Research in this area is a multidisciplinary challenge and despite all the advances using these (inorganic) polymers there is still room for improvement. For example, there is still the need to develop polymerization strategies allowing the formation of more homogeneous polymers with narrow molecular weight distributions. For this, living polymerizations have been used. Also, the conception of stimuli-responsive polymers is expanding since it leads to lower toxicity of the materials and allows the delivery of higher doses of the active ingredient.

In the area of drug delivery to cancer, the inclusion of receptor-targeting ligands is expected to lead to improved tumor targeting through a combined active and passive targeting (EPR effect), obtained through the inclusion of the receptor-targeting ligand and the polymer backbone, respectively.

Another approach in drug delivery uses metal-organic frameworks (MOFs), especially NMOFs or nanoMOFs, which have been shown to be well suited as nanocarriers for organic drug or metallodrug delivery. Mainly two approaches have been followed: adsorption of drugs in the porous network or use of drug as a MOF constituent. In both cases, tuning the drug release is a major issue. Nontoxic MOFs (so-called BioMOFs) have been developed and their degradation studied. In cases of limited stability, several approaches have been reported, e.g. core-shell particles. In some cases, modification of the shells can also be used to introduce a targeting unit, for example for specific cancer cells. The combination of a suitable drug carrier with a selective targeting unit also seems to be a very promising way here for improving stability as well as selectivity.

Acknowledgments

The authors thank the COST Action CM1302 (SIPs): "European Network on Smart Inorganic Polymers." R. Precker acknowledges funding for a Short Term Scientific Mission from the COST Action CM1302 (SIPs): "European Network on Smart Inorganic Polymers." A. Valente thanks the Investigator FCT2013 Initiative for the project IF/01302/2013 (acknowledging FCT, as well as POPH and FSE – European Social Fund).

References

1 Müller, W.E., Wang, X., and Schröder, H.C. (eds.) (2013). Progress in molecular and subcellular biology. In: *Biomedical Inorganic Polymers*, vol. 54. Berlin/Heidelberg: Springer-Verlag.

2 Mark, J.E., Allcock, H.R., and West, R. (2005). *Inorganic Polymers, 2nd Edition*. New York: Oxford University Press.

3 Hasenknopf, B. (2005). Polyoxometalates: introduction to a class of inorganic compounds and their biomedical applications. *Front. Biosci.* 10: 275–287.

4 Yamase, T. (2005). Anti-tumor, -viral, and -bacterial activities of polyoxometalates for realizing an inorganic drug. *J. Mater. Chem.* 15 (45): 4773.

5 Rhule, J.T., Hill, C.L., Judd, D.A., and Schinazi, R.F. (1998). Polyoxometalates in medicine. *Chem. Rev.* 98 (1): 327–358.

6 Kubo, A.L., Kremer, L., Herrmann, S. et al. (2017). Antimicrobial activity of polyoxometalate ionic liquids against clinically relevant pathogens. *Chempluschem* 82 (6): 867–871.

7 Kong, D.-L., Wei, B., Zhou, S.-Y. et al. (2013). Volume-related efficiency of gadolinium polyoxometalates as MRI contrast agents. *Chem. Res. Chinese Univ.* 29 (6): 1055–1058.

8 Yamase, T., Botar, B., Ishikawa, E., and Fukaya, K. (2001). Chemical structure and intramolecular spin-exchange interaction of $[(VO)_3(SbW_9O_{33})_2]^{12-}$. *Chem. Lett.* 30 (1): 56–57.

9 Bâlici, Ş., Şuşman, S., Rusu, D. et al. (2016). Differentiation of stem cells into insulin-producing cells under the influence of nanostructural polyoxometalates. *J. Appl. Toxicol.* 36 (3): 373–384.

10 Gao, N., Sun, H., Dong, K. et al. (2014). Transition-metal-substituted polyoxometalate derivatives as functional anti-amyloid agents for alzheimer's disease. *Nat. Commun.* 5: 1–9.

11 Prisacaru, A.I., Grama, S., Durdureanu-Angheluta, A.M. et al. (2013). Azo-polysiloxane micelles charged with nifedipine. *Cent. Eur. J. Chem.* 11 (9): 1431–1438.

12 Grama, S., Moleavin, I.-A., Hodorog-Rusu, A. et al. (2013). Photosensitive azo-polysiloxanes for drug delivery applications. *Mater. Plast.* 50 (1): 60–64.
13 Li, S. and Feng, S. (2016). High-sensitivity stimuli-responsive polysiloxane synthesized via catalyst-free aza-Michael addition for ibuprofen loading and controlled release. *RSC Adv.* 6 (101): 99414–99421.
14 Chen, Z., Chisholm, B.J., Stafslien, S. et al. (2010). Novel, UV-curable coatings containing a tethered biocide: synthesis, characterization, and antimicrobial activity. *J. Biomed. Mater. Res. Part A* 95A (2): 486–494.
15 Francis, A., Detsch, R., and Boccaccini, A.R. (2016). Fabrication and cytotoxicity assessment of novel polysiloxane/bioactive glass films for biomedical applications. *Ceram. Int.* 42 (14): 15442–15448.
16 Li, C., Ding, Y., Kuddannaya, S. et al. (2017). Anti-bacterial properties of collagen-coated glass and polydimethylsiloxane substrates. *J. Mater. Sci.* 52 (17): 9963–9978.
17 Jellali, R., Bertrand, V., Alexandre, M. et al. (2017). Photoreversibility and biocompatibility of polydimethylsiloxane-coumarin as adjustable intraocular lens material. *Macromol. Biosci.* 17 (7): 1–12.
18 Bračič, M., Mohan, T., Griesser, T. et al. (2017). One-step noncovalent surface functionalization of PDMS with chitosan-based bioparticles and their protein-repellent properties. *Adv. Mater. Interfaces* 4 (21): 1–11.
19 Li, Q., Sun, L., Zhang, L. et al. (2018). Polydopamine-collagen complex to enhance the biocompatibility of polydimethylsiloxane substrates for sustaining long-term culture of L929 fibroblasts and tendon stem cells. *J. Biomed. Mater. Res. Part A* 106 (2): 408–418.
20 Rothemund, S. and Teasdale, I. (2016). Preparation of polyphosphazenes: a tutorial review. *Chem. Soc. Rev.* 45 (19): 5200–5215.
21 Teasdale, I. and Brüggemann, O. (2013). Polyphosphazenes: multifunctional, biodegradable vehicles for drug and gene delivery. *Polymers (Basel)* 5 (1): 161–187.
22 Allcock, H.R. and Morozowich, N.L. (2012). Bioerodible polyphosphazenes and their medical potential. *Polym. Chem.* 3 (3): 578–590.
23 Andrianov, A.K. (2006). Water-soluble polyphosphazenes for biomedical applications. *J. Inorg. Organomet. Polym. Mater.* 16 (4): 397–406.
24 Baillargeon, A.L. and Mequanint, K. (2014). Biodegradable polyphosphazene biomaterials for tissue engineering and delivery of therapeutics. *Biomed Res. Int.* 8: 16.
25 Ogueri, K.S., Ivirico, J.L.E., and Nair, L.S. (2017). Biodegradable polyphosphazene-based blends for regenerative engineering. *Regener. Eng. Transl. Med.* 3 (1): 15–31.
26 Kumbar, S.G., Bhattacharyya, S., Nukavarapu, S.P. et al. (2006). In vitro and in vivo characterization of biodegradable poly(organophosphazenes) for biomedical applications. *J. Inorg. Organomet. Polym. Mater.* 16 (4): 365–385.
27 Seo, B.B., Choi, H., Koh, J.T., and Song, S.C. (2015). Sustained BMP-2 delivery and injectable bone regeneration using thermosensitive polymeric nanoparticle hydrogel bearing dual interactions with BMP-2. *J. Controlled Release* 209: 67–76.

28 Sobhani, A., Rafienia, M., Ahmadian, M., and Naimi-Jamal, M.-R. (2017). Fabrication and characterization of polyphosphazene/calcium phosphate scaffolds containing chitosan microspheres for sustained release of bone morphogenetic protein 2 in bone tissue engineering. *Tissue Eng. Regen. Med.* 14 (5): 525–538.

29 Rothemund, S., Aigner, T.B., Iturmendi, A. et al. (2015). Degradable glycine-based photo-polymerizable polyphosphazenes for use as scaffolds for tissue regeneration. *Macromol. Biosci.* 15 (3): 351–363.

30 Ozay, H. and Ozay, O. (2014). Synthesis and characterization of drug microspheres containing phosphazene for biomedical applications. *Colloids Surf. A Physicochem. Eng. Asp.* 450 (1): 99–105.

31 Jiang, H.L., Kwon, J.T., Kim, Y.K. et al. (2007). Galactosylated chitosan-graft-polyethylenimine as a gene carrier for hepatocyte targeting. *Gene Ther.* 14 (19): 1389–1398.

32 Jiang, H.L., Kim, Y.K., Lee, S.M. et al. (2010). Galactosylated chitosan-G-PEI/DNA complexes-loaded poly(organophosphazene) hydrogel as a hepatocyte targeting gene delivery system. *Arch. Pharmacal Res.* 33 (4): 551–556.

33 Matsumura, Y. and Maeda, H. (1986). A new concept for macromolecular therapeutics in cancer chemotherapy: mechanism of tumoritropic accumulatio of proteins and the antitumor agents smancs. *Cancer Res.* 46 (12): 6387–6392.

34 Song, R., Yong, J.J., Ju, I.K. et al. (2005). Synthesis, characterization, and tumor selectivity of a polyphosphazene-platinum(II) conjugate. *J. Controlled Release* 105 (1–2): 142–150.

35 Soo, Y., Baek, H., Ha, Y., and Lee, Y. (1997). Synthesis and antitumor activity of novel polyphosphazene-(diamine)platinum(II) conjugates. *Int. J. Pharm.* 153: 79–91.

36 By, H. and Allen, R.W. (1976). Antitumour agents synthesized from K_2PtCl_4 and polymeric or cyclic phosphazenes. *J. Chem. Soc., Chem. Commun.* (1): 717–718.

37 Zhu, X., Peng, Y., and Qiu, L. (2017). Amino-functionalized nano-vesicles for enhanced anticancer efficacy and reduced myelotoxicity of carboplatin. *Colloids Surf. B Biointerfaces* 157: 56–64.

38 Henke, H., Kryeziu, K., Banfi, J. et al. (2016). Macromolecular Pt(IV) prodrugs from poly(organo)phosphazenes. *Macromol. Biosci.* 16 (8): 1239–1249.

39 Allcock, H.R., Allen, R.W., and O'Brien, J.P. (1977). Synthesis of platinum derivatives of polymeric and cyclic phosphazenes. *J. Am. Chem. Soc.* 99 (12): 3984–3987.

40 Jun, Y.J., Kim, J.I., Jun, M.J., and Sohn, Y.S. (2005). Selective tumor targeting by enhanced permeability and retention effect. Synthesis and antitumor activity of polyphosphazene-platinum(II) conjugates. *J. Inorg. Biochem.* 99 (8): 1593–1601.

41 Cho, J.K., Hong, K.Y., Park, J.W. et al. (2011). Injectable delivery system of 2-methoxyestradiol for breast cancer therapy using biodegradable thermosensitive poly(organophosphazene) hydrogel. *J. Drug Targeting* 19 (4): 270–280.

42 Cho, J.K., Hong, J.M., Han, T. et al. (2013). Injectable and biodegradable poly(organophosphazene) hydrogel as a delivery system of docetaxel for cancer treatment. *J. Drug Targeting* 21 (6): 564–573.

43 Caster, J.M., Patel, A.N., Zhang, T., and Wang, A. (2017). Investigational nanomedicines in 2016: a review of nanotherapeutics currently undergoing clinical trials. *WIREs Nanomed. Nanobiotechnol.* 9 (1): e1416, 18 pages.

44 Callari, M., Aldrich-Wright, J.R., De Souza, P.L., and Stenzel, M.H. (2014). Polymers with platinum drugs and other macromolecular metal complexes for cancer treatment. *Prog. Polym. Sci.* 39 (9): 1614–1643.

45 Valente, A. and Zinck, P. (2012). Polymer-metal complexes (PMC) for cancer therapy. *Recent Res. Devel. Polym. Sci.* 11 (2): 99–128.

46 Uchino, H., Matsumura, Y., Negishi, T. et al. (2005). Cisplatin-incorporating polymeric micelles (NC-6004) can reduce nephrotoxicity and neurotoxicity of cisplatin in rats. *Br. J. Cancer* 93 (6): 678–687.

47 Doi, T., Hamaguchi, T., Shitara, K. et al. (2017). NC-6004 phase I study in combination with gemcitabine for advanced solid tumors and population PK/PD analysis. *Cancer Chemother. Pharmacol.* 79 (3): 569–578.

48 Subbiah, V., Grilley-Olson, J.E., Combest, A.J. et al. (2018). Phase Ib/II trial of NC-6004 (nanoparticle cisplatin) plus gemcitabine in patients with advanced solid tumors. *Clin. Cancer Res.* 24 (1): 43–51.

49 Flocke, L.S., Trondl, R., Jakupec, M.A., and Keppler, B.K. (2016). Molecular mode of action of NKP-1339 – a clinically investigated ruthenium-based drug – involves ER- and ROS-related effects in colon carcinoma cell lines. *Invest. New Drugs* 34 (3): 261–268.

50 Alessio, E. (2017). Thirty years of the drug candidate NAMI-A and the myths in the field of ruthenium anticancer compounds: a personal perspective. *Eur. J. Inorg. Chem.* 2017 (12): 1549–1560.

51 Michlewska, S., Ionov, M., Shcharbin, D. et al. (2017). Ruthenium metallodendrimers with anticancer potential in an acute promyelocytic leukemia cell line (HL60). *Eur. Polym. J.* 87: 39–47.

52 Therrien, B. (2015). Biologically relevant arene ruthenium metalla-assemblies. *CrystEngComm* 17 (3): 484–491.

53 Shi, H., Fang, T., Tian, Y. et al. (2016). A dual-fluorescent nano-carrier for delivering photoactive ruthenium polypyridyl complexes. *J. Mater. Chem. B* 4 (27): 4746–4753.

54 Zhao, S., Zhu, X., Cao, C. et al. (2018). Transferrin modified ruthenium nanoparticles with good biocompatibility for photothermal tumor therapy. *J. Colloid Interface Sci.* 511: 325–334.

55 Morais, T.S., Valente, A., Tomaz, A.I. et al. (2016). Tracking antitumor metallodrugs: promising agents with the Ru(II)- and Fe(II)-cyclopentadienyl scaffolds. *Future Med. Chem.* 8 (5): 527–544.

56 Côrte-Real, L., Paula Robalo, M., Marques, F. et al. (2015). The key role of coligands in novel ruthenium(II)-cyclopentadienyl bipyridine derivatives: ranging from non-cytotoxic to highly cytotoxic compounds. *J. Inorg. Biochem.* 150: 148–159.

57 Moreno, V., Font-Bardia, M., Calvet, T. et al. (2011). DNA interaction and cytotoxicity studies of new ruthenium(II) cyclopentadienyl derivative

complexes containing heteroaromatic ligands. *J. Inorg. Biochem.* 105 (2): 241–249.
58 Teixeira, R.G., Brás, A.R., Côrte-Real, L. et al. (2018). Novel ruthenium methylcyclopentadienyl complex bearing a bipyridine perfluorinated ligand shows strong activity towards colorectal cancer cells. *Eur. J. Med. Chem.* 143: 503–514.
59 Côrte-Real, L., Teixeira, R.G., Gírio, P. et al. (2018). Methyl-cyclopentadienyl ruthenium compounds with 2,2′-bipyridine derivatives display strong anticancer activity and multidrug resistance potential. *Inorg. Chem.* 57 (8): 4629–4639.
60 Matos, A., Mendes, F., Valente, A. et al. (2018). Ruthenium-based anticancer compounds: insights into their cellular targeting and mechanism of action. In: *Ruthenium Complexes – Photochemical and Biomedical Applications* (ed. A.A. Holder, J.L. Bullock Jr., M.A. Lawrence, et al.), 201–2019. Wiley-VCH.
61 Valente, A., Garcia, M.H., Marques, F. et al. (2013). First polymer "ruthenium-cyclopentadienyl" complex as potential anticancer agent. *J. Inorg. Biochem.* 127: 79–81.
62 Garcia, M. H.; Valente, A.; Morais, T. S.; Tomaz, A. I. (2016). Macromolecular transition metal complexes for treatment of cancer and process for their preparation. WO 2016/087932, filed 7 December 2015 and issued 9 June 2016.
63 Côrte-Real, L., Matos, A.P., Alho, I. et al. (2013). Cellular uptake mechanisms of an antitumor ruthenium compound: the endosomal/lysosomal system as a target for anticancer metal-based drugs. *Microsc. Microanal.* 19: 1122–1130.
64 Serrano Ruiz, M., Romerosa, A., Sierra-Martin, B., and Fernandez-Barbero, A. (2008). A water soluble diruthenium-gold organometallic microgel. *Angew. Chem. Int. Ed.* 47 (45): 8665–8669.
65 Scalambra, F., Serrano-Ruiz, M., and Romerosa, A. (2015). First water-soluble backbone Ru–Ru–Ni heterometallic organometallic polymer. *Macromol. Rapid Commun.* 36 (7): 689–693.
66 Scalambra, F., Serrano-Ruiz, M., and Romerosa, A. (2018). Water driven formation of channels: unusual solid-state structural transformation of a heterometallic polymer. *Dalton Trans.* 47 (10): 3588–3595.
67 Lidrissi, C., Romerosa, A., Saoud, M. et al. (2005). Stable, water-soluble Pta-based Ru–Ag organometallic polymers. *Angew. Chem. Int. Ed.* 44 (17): 2568–2572.
68 Weiss, A., Berndsen, R.H., Dubois, M. et al. (2014). In vivo anti-tumor activity of the organometallic ruthenium(II)-arene complex [Ru(η^6-*p*-cymene)Cl_2 (Pta)] (RAPTA-C) in human ovarian and colorectal carcinomas. *Chem. Sci.* 5 (12): 4742–4748.
69 Blunden, B.M., Lu, H., and Stenzel, M.H. (2013). Enhanced delivery of the RAPTA-C macromolecular chemotherapeutic by conjugation to degradable polymeric micelles. *Biomacromolecules* 14 (12): 4177–4188.
70 Sumitani, S. and Nagasaki, Y. (2012). Boron neutron capture therapy assisted by boron-conjugated nanoparticles. *Polym. J.* 44 (6): 522–530.

71 Chen, G., Yang, J., Lu, G. et al. (2014). One stone kills three birds: novel boron-containing vesicles for potential BNCT, controlled drug release, and diagnostic imaging. *Mol. Pharmaceutics* 11 (10): 3291–3299.
72 Sun, W., Parowatkin, M., Steffen, W. et al. (2016). Ruthenium-containing block copolymer assemblies: red-light-responsive metallopolymers with tunable nanostructures for enhanced cellular uptake and anticancer phototherapy. *Adv. Healthc. Mater.* 5 (4): 467–473.
73 Sun, W., Li, S., Häupler, B. et al. (2017). An amphiphilic ruthenium polymetallodrug for combined photodynamic therapy and photochemotherapy in vivo. *Adv. Mater.* 29 (6): 1603702, 6 pages.
74 Ruan, Z., Liu, L., Fu, L. et al. (2016). An amphiphilic block copolymer conjugated with carborane and a NIR fluorescent probe for potential imaging-guided BNCT therapy. *Polym. Chem.* 7 (26): 4411–4418.
75 Gona, K.B., Zaulet, A., Gómez-Vallejo, V. et al. (2014). COSAN as a molecular imaging platform: synthesis and "in vivo" imaging. *Chem. Commun.* 50 (77): 11415–11417.
76 Verdiá-Báguena, C., Alcaraz, A., Aguilella, V.M. et al. (2014). Amphiphilic COSAN and I_2-COSAN crossing synthetic lipid membranes: planar bilayers and liposomes. *Chem. Commun.* 50 (51): 6700–6703.
77 Xiong, H., Zhou, D., Qi, Y. et al. (2015). Doxorubicin-loaded carborane-conjugated polymeric nanoparticles as delivery system for combination cancer therapy. *Biomacromolecules* 16 (12): 3980–3988.
78 Lee, J., Farha, O.K., Roberts, J. et al. (2009). Metal-organic framework materials as catalysts. *Chem. Soc. Rev.* 38: 1450–1459.
79 Peer, D., Karp, J.M., Hong, S. et al. (2007). Nanocarriers as an emerging platform for cancer therapy. *Nat. Nanotechnol.* 2: 751–760.
80 Chou, L.Y.T., Ming, K., and Chan, W.C.W. (2011). Strategies for the intracellular delivery of nanoparticles. *Chem. Soc. Rev.* 40: 233–245.
81 Morris, R.E. and Wheatley, P.S. (2008). Gas storage in nanoporous materials. *Angew. Chem. Int. Ed. Engl.* 47: 4966–4981.
82 Wang, P., Wu, G., and Wang, R. (2012). Two d^{10} coordination polymers based on the substitutional aromatic multicarboxylate and N-donor Co-ligands: syntheses, structures and luminescent properties. *J. Inorg. Organomet. Polym.* 22: 1028–1023.
83 Furukawa, H., Cordova, K.E., O'Keeffe, M., and Yaghi, O.M. (2013). The chemistry and applications of metal-organic frameworks. *Science* 341 (6149): 1230444.
84 Kaskel, S. (2016). *The Chemistry of Metal–Organic Frameworks. Synthesis, Characterization, and Applications.* Weinheim: Wiley-VCH.
85 Davis, M.E., Chen, Z., and Shin, D.M. (2008). Nanoparticle therapeutics: an emerging treatment modality for cancer. *Nat. Rev. Drug Discovery* 7: 771–782.
86 Kim, C.S., Duncan, B., Creran, B., and Rotello, V.M. (2013). Triggered nanoparticles as therapeutics. *Nano Today* 8: 439–447.
87 MacGillivray, L.R. (2010). *Metal-Organic Frameworks: Design and Application.* Weinheim: Wiley-VCH.

88 He, C., Liu, D., and Lin, W. (2015). Nanomedicine applications of hybrid nanomaterials built from metal–ligand coordination bonds: nanoscale metal–organic frameworks and nanoscale coordination polymers. *Chem. Rev.* 115: 11079–11108.

89 Wang, L., Zheng, M., and Xie, Z. (2018). Nanoscale metal-organic frameworks for drug delivery: a conventional platform with new promise. *J. Mater. Chem. B.* 6: 707–717.

90 Li, H., Eddaoudi, M., O'Keeffe, M., and Yaghi, O.M. (1999). Design and synthesis of an exceptionally stable and highly porous metal-organic framework. *Nature* 402: 276–279.

91 Eddaoudi, M., Kim, J., Rosi, N. et al. (2002). Systematic design of pore size and functionality in isoreticular MOFs and their application in methane storage. *Science* 295: 469–472.

92 Murray, L.J., Dinca, M., and Long, J.R. (2009). Hydrogen storage in metal–organic frameworks. *Chem. Soc. Rev.* 38: 1294–1314.

93 Li, J.R., Kuppler, R.J., and Zhou, H.C. (2009). Selective gas adsorption and separation in metal–organic frameworks. *Chem. Soc. Rev.* 38: 1477–1504.

94 Ma, L., Abney, C., and Lin, W. (2009). Enantioselective catalysis with homochiral metal–organic frameworks. *Chem. Soc. Rev.* 38: 1248–1256.

95 Horcajada, P., Chalati, T., Serre, C. et al. (2010). Porous metal–organic-framework nanoscale carriers as a potential platform for drug delivery and imaging. *Nat. Mater.* 9: 172–178.

96 McKinlay, A.C., Morris, R.E., Horcajada, P. et al. (2010). BioMOFs: metal–organic frameworks for biological and medical applications. *Angew. Chem. Int. Ed. Engl.* 49: 6260–6266.

97 Wu, M.X. and Yang, Y.W. (2017). Metal–organic framework (MOF)-based drug/cargo delivery and cancer therapy. *Adv. Mater.* 29: 1606134.

98 Horcajada, P., Gref, R., Baati, T. et al. (2012). Metal-organic frameworks in biomedicine. *Chem. Rev.* 112: 1232–1268.

99 Sun, C.-Y., Qin, C., Wang, X.-L., and Su, Z.-M. (2013). Metal-organic frameworks as potential drug delivery systems. *Expert Opin. Drug Deliv.* 10: 89–101.

100 Yaghi, O.M. and Li, H. (1995). Hydrothermal synthesis of a metal-organic framework containing large rectangular channels. *J. Am. Chem. Soc.* 117: 10401–10402.

101 Pettinari, C., Marchetti, F., Mosca, N. et al. (2017). Application of metal–organic frameworks. *Polym. Int.* 66: 731–744.

102 Zhao, Y. (2016). Emerging applications of metal-organic frameworks and covalent organic frameworks. *Chem. Mater.* 28: 8079–8081.

103 Xiao, D.J., Gonzalez, M.I., Darago, L.E. et al. (2016). Selective, tunable O_2 binding in cobalt(II)–triazolate/pyrazolate metal–organic frameworks. *J. Am. Chem. Soc.* 138: 7161–7170.

104 Li, J., Wang, X., Zhao, G. et al. (2018). Metal-organic framework – based materials: superior adsorbents for the capture of toxic and radioactive metal ions. *Chem. Soc. Rev.* 47: 2322–2356.

105 Ward, M.A. and Georgiou, T.K. (2011). Thermoresponsive polymers for biomedical applications. *Polymers* 3: 1215–1242.

106 Wu, H., Zhou, W., and Yildirim, T. (2009). High-capacity methane storage in metal-organic frameworks M_2(dhtp): the important role of open metal sites. *J. Am. Chem. Soc.* 131: 4995–5000.

107 Bertini, I., Gray, H.B., Stiefel, E.I., and Valentine, J.S. (2007). *Biological Inorganic Chemistry: Structure and Reactivity*, Sausalito, CA: University Science Book.

108 Kitagawa, S., Kitaura, R., and Noro, S. (2004). Functional porous coordination polymers. *Angew. Chem. Int. Ed. Engl.* 43: 2334–2375.

109 Tranchemontagne, D.J., Mendoza-Cortes, J.L., O'Keeffe, M., and Yaghi, O.M. (2009). Secondary building units, nets and bonding in the chemistry of metal–organic frameworks. *Chem. Soc. Rev.* 38: 1257–1283.

110 Horcajada, P., Serre, C., Vallet-Regí, M. et al. (2006). Metal–organic frameworks as efficient materials for drug delivery. *Angew. Chem. Int. Ed. Engl.* 45: 5974–5978.

111 Horcajada, P., Serre, C., Maurin, G. et al. (2008). Flexible porous metal-organic frameworks for a controlled drug delivery. *J. Am. Chem. Soc.* 130: 6774–6780.

112 Singco, B., Liu, L., Chen, Y. et al. (2016). Approaches to drug delivery: confinement of aspirin in MIL-100(Fe) and aspirin in the de novo synthesis of metal-organic frameworks. *Microporous Mesoporous Mater.* 223: 254–260.

113 Anand, R., Borghi, F., Manoli, F. et al. (2014). Host–guest interactions in Fe(III)-trimesate MOF nanoparticles loaded with doxorubicin. *J. Phys. Chem. B.* 118: 8532–8539.

114 Adhikari, C. and Chakraborty, A. (2016). Smart approach for in situ one-step encapsulation and controlled delivery of a chemotherapeutic drug using metal-organic framework–drug composites in aqueous media. *ChemPhysChem* 17: 1070–1077.

115 Chalati, T., Horcajada, P., Couvreur, P. et al. (2011). Porous metal organic framework nanoparticles to address the challenges related to busulfan encapsulation. *Nanomedicine* 6: 1683–1695.

116 Agostoni, V., Horcajada, P., Noiray, M. et al. (2015). A "green" strategy to construct non-covalent, stable and bioactive coatings on porous MOF nanoparticles. *Sci. Rep.* 5: 7925.

117 Agostoni, V., Chalati, T., Horcajada, P. et al. (2013). Towards an improved anti-HIV activity of NRTI via metal-organic frameworks nanoparticles. *Adv. Health. Mater.* 2: 1630–1637.

118 Taylor-Pashow, K.M.L., Della Rocca, J., Xie, Z. et al. (2009). Postsynthetic modifications of iron-carboxylate nanoscale metal-organic frameworks for imaging and drug delivery. *J. Am. Chem. Soc.* 131: 14261–14263.

119 Rieter, W.J., Pott, K.M., Taylor, K.M.L., and Lin, W.B. (2008). Nanoscale coordination polymers for platinum-based anticancer drug delivery. *J. Am. Chem. Soc.* 130: 11584–11585.

120 Rojas, S., Carmona, F.J., Maldonado, C.R. et al. (2016). RAPTA-C incorporation and controlled delivery from MIL-100(Fe) nanoparticles. *New J. Chem.* 40: 5690–5694.

121 Wang, J., Chen, D., Li, B. et al. (2016). Fe-MIL-101 exhibits selective cytotoxicity and inhibition of angiogenesis in ovarian cancer cells via downregulation of MMP. *Sci. Rep.* 6: 26126.

122 Taherzade, S.D., Soleimannejad, J., and Tarlani, A. (2017). Application of metal-organic framework nano-MIL-100(Fe) for sustainable release of doxycycline and tetracycline. *Nanomaterials* 7: 215.
123 Hu, Q., Yu, J., Liu, M. et al. (2014). A low cytotoxic cationic metal-organic framework carrier for controllable drug release. *J. Med. Chem.* 57: 5679–5685.
124 Zhu, X., Gu, J., Wang, Y. et al. (2014). Inherent anchorages in UiO-66 nanoparticles for efficient capture of alendronate and its mediated release. *Chem. Commun.* 50: 8779–8782.
125 Filippousi, M., Turner, S., Leus, K. et al. (2016). Biocompatible Zr-based nanoscale MOFs coated with modified poly(ε-caprolactone) as anticancer drug carriers. *Int. J. Pharm.* 509: 208–218.
126 Tamames-Tabar, C., Imbuluzqueta, E., Guillou, N. et al. (2015). A Zn azelate MOF: combining antibacterial effect. *CrystEngComm.* 17: 456–462.
127 An, J., Geib, S.J., and Rosi, N.L. (2009). Cation-triggered drug release from a porous zinc-adeninate metal-organic framework. *J. Am. Chem. Soc.* 131: 8376–8377.
128 Mizutani, M., Maejima, N., Jitsukawa, K. et al. (1998). An infinite chiral single-helical structure formed in Cu(II)-L-/D-glutamic acid system. *Inorg. Chim. Acta* 283: 105–110.
129 Chen, W. and Chunsheng, W. (2018). Synthesis, functionalization, and applications of metal-organic frameworks in biomedicine. *Dalton Trans.* 47: 2114–2133.
130 Baati, T., Njim, L., Neffati, F. et al. (2013). In depth analysis of the in vivo toxicity of nanoparticles of porous iron(III) metal-organic frameworks. *Chem. Sci.* 4: 1597–1607.
131 Kundu, T., Mitra, S., Patra, P. et al. (2014). Mechanical downsizing of a gadolinium(III)-based metal-organic framework for anticancer drug delivery. *Chem. Eur. J.* 20: 10514–10518.
132 Ruyra, A., Yazdi, A., Espin, J. et al. (2015). Synthesis, culture medium stability, and in vitro and in vivo zebrafish embryo toxicity of metal-organic framework nanoparticles. *Chem. Eur. J.* 21: 2508–2518.
133 Tan, K., Nijem, N., Canepa, P. et al. (2012). Stability and hydrolyzation of metal organic frameworks with paddle-wheel SBUs upon hydration. *Chem. Mater.* 24: 3153–3167.
134 Bellido, E., Guillevic, M., Hidalgo, T. et al. (2014). Understanding the colloidal stability of the mesoporous MIL-100(Fe) nanoparticles in physiological media. *J. Am. Chem. Soc.* 30: 5911–5920.
135 Jasuja, H. and Walton, K.S. (2013). Effect of catenation and basicity of pillared ligands on the water stability of MOFs. *Dalton Trans.* 42: 15421–15426.
136 Vinogradov, V.V., Drozdov, A.S., Mingabudinova, L.R. et al. (2018). Composites based on heparin and MIL-101(Fe): the drug releasing depot for anticoagulant therapy and advanced medical nanofabrication. *J. Mater. Chem. B.* 6: 2450–2459.
137 Li, X., Lachmanski, L., Safi, S. et al. (2017). New insights into the degradation mechanism of metal-organic frameworks drug carriers. *Sci. Rep.* 7: 13142.

138 Rieter, W.J., Taylor, K.M.L., and Lin, W. (2007). Surface modification and functionalization of nanoscale metal-organic frameworks for controlled release and luminescence sensing. *J. Am. Chem. Soc.* 129: 9852–9853.

139 Liu, D., He, C., Poon, C., and Lin, W. (2014). Theranostic nanoscale coordination polymers for magnetic resonance imaging and bisphosphonate delivery. *J. Mater. Chem. B.* 2: 8249–8255.

10

Inorganic Dendrimers and Their Applications

Anne-Marie Caminade [1,2]

[1] *Laboratoire de Chimie de Coordination du CNRS, 205 Route de Narbonne, BP 44099, 31077 Toulouse Cedex 4, France*
[2] *Université de Toulouse, LCC-CNRS, CNRS, Toulouse, France*

10.1 Introduction

During the early times of dendrimers, emphasis was on the synthetic aspects and on the methods of characterization and then the interest has moved on to the applications and uses of dendrimers [1]. Indeed, owing to their step-by-step synthesis resulting in perfectly defined structures as well as a large number of terminal functions (groups), dendrimers occupy a special place among all types of polymers and exhibit properties different from those of common polymers. However, the properties of only a few types of dendrimers have been studied in detail. With regard to organic dendrimers, mainly PAMAM (polyamidoamine) [2], and to a lesser extent PPI (polypropyleneimine) [3] and poly(lysine) [4] dendrimers have been studied. Inorganic dendrimers (see Chapter 5) have followed the same trend, and only two families have emerged [5], those based on carbosilane structures [6], and those based on phosphorhydrazone structures [7]. Both families are shown in their linear form in Figure 10.1 (generations 3 in both cases).

This chapter is organized based on the properties observed, and not on the type of inorganic dendrimers. The first type of properties is relative to catalysis, with emphasis on the so-called "dendrimer (or dendritic) effect" [8] and on the recovery and reuse of the dendritic catalyst. The second type is related to the use of dendrimers for affording nanomaterials. This part will include materials exclusively constituted of dendrimers, materials incorporating inorganic dendrimers inside their structure, as well as surfaces of materials modified with either a single layer or with multilayers of inorganic dendrimers, with emphasis on the biological properties of the latter. The third and last type of properties will concern the biological uses of inorganic dendrimers relevant to nanomedicine, highlighting six topics: for bioimaging, for gene therapy, and for treatment of viral infections, brain diseases, cancers, and inflammatory diseases. It must be emphasized that the huge number of related references prevents an exhaustive review of all these fields.

Smart Inorganic Polymers: Synthesis, Properties, and Emerging Applications in Materials and Life Sciences, First Edition. Edited by Evamarie Hey-Hawkins and Muriel Hissler.

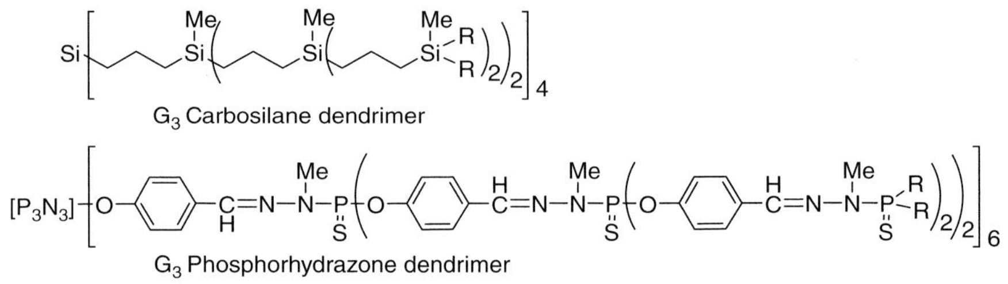

Figure 10.1 Linear representation carbosilane and phosphorhydrazone dendrimers.

10.2 Inorganic Dendrimers as Catalysts

Catalysis is a major topic in chemistry in general, and for dendrimers in particular. Early work in this area has been reviewed early, and fully [9]; more recent reviews [10] have focused with either special emphasis on some particular areas of catalysis [11, 12] or in combination with other topics [13]. Interestingly, the very first examples of the use of any type of dendrimers as catalysts were those of inorganic dendrimers. The first example was a small polyphosphine dendrimer complexing palladium in the full structure. This compound was used for catalyzing the electrochemical reduction $CO_2 \rightarrow CO$ [14]. A few weeks later, a carbosilane dendrimer complexing nickel on its surface was used for catalyzing the Kharasch addition of a polyhalogenoalkane to a C=C double bond [15] (Figure 10.2).

10.2.1 Overview of the Use of Inorganic Dendrimers as Catalysts

Many efforts have been made to try to observe a "dendrimer effect," as will be shown in Section 10.2.2, and also to try to recover and reuse the dendritic catalysts, as will be shown in Section 10.3. In addition to these specific areas, a few salient examples are given in the present section. It must be emphasized that for comparative purposes, the number of catalytic entities (number of metal atoms) is kept constant; for instance, if generation 1 bears six catalytic entities, the efficiency of 1 equiv of generation 1 will be compared to the efficiency of 6 equiv of the monomer. Although in most cases, multiple copies of the catalytic entities are

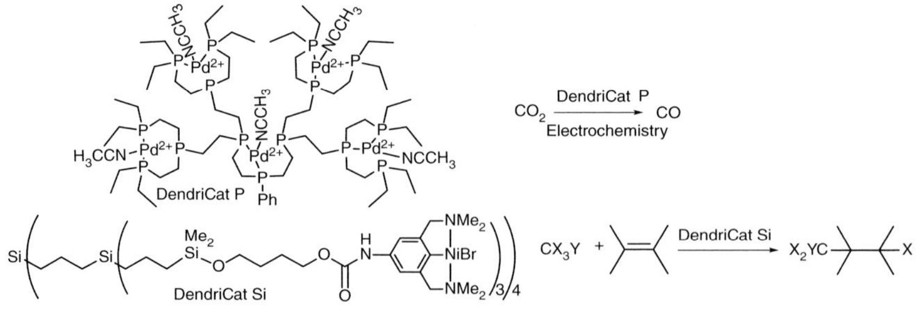

Figure 10.2 The first two types of dendrimers used for catalytic experiments, and the reaction they catalyze.

gathered on the surface of the dendrimers, in some cases, a single catalytic entity is linked to the core. A series of carbosilane dendrimers having a diphosphinoferrocene as the core and allyl terminal groups were synthesized from generation 1 to 3, and compared with dppf (diphenylphosphino ferrocene). The palladium complexes of all these compounds were used for catalyzing the allylic alkylation of 3-phenylallyl acetate with diethyl 2-sodio-2-methyl malonate, affording predominantly the trans product (Figure 10.3a). However, the dppf complex was found to be the most efficient in terms of percentage of conversion (72%, against 17% for generation 3) as well as in terms of selectivity to trans product (90%, against 79% for generation 3) [16]. A diphosphine has been also linked to the core unit of a phosphorhydrazone dendron (dendritic wedge). The third generation has been compared with phosphorhydrazone dendrimers having the same ligands as terminal moieties. In both cases, the corresponding ruthenium hydride complexes were used as catalysts in diasteroselective Michael additions involving ethyl cyanoacetate and diethyl ethylidenemalonate. No difference was observed in terms of percentage of conversion (100%) or diastereoisomeric ratio (72–73% for both dendrimers, both being slightly better than the monometallic species $RuH_2(PPh_3)_4$, 69%) (Figure 10.3b) [17].

The apolar internal structure of carbosilane dendrimers enables their use in supercritical CO_2 ($scCO_2$), which is an alternate friendly reaction medium. In a first experiment, two carbosilane dendrons having a phosphine core unit were associated around $PdCl_2$, and used for catalyzing the Heck coupling reaction of

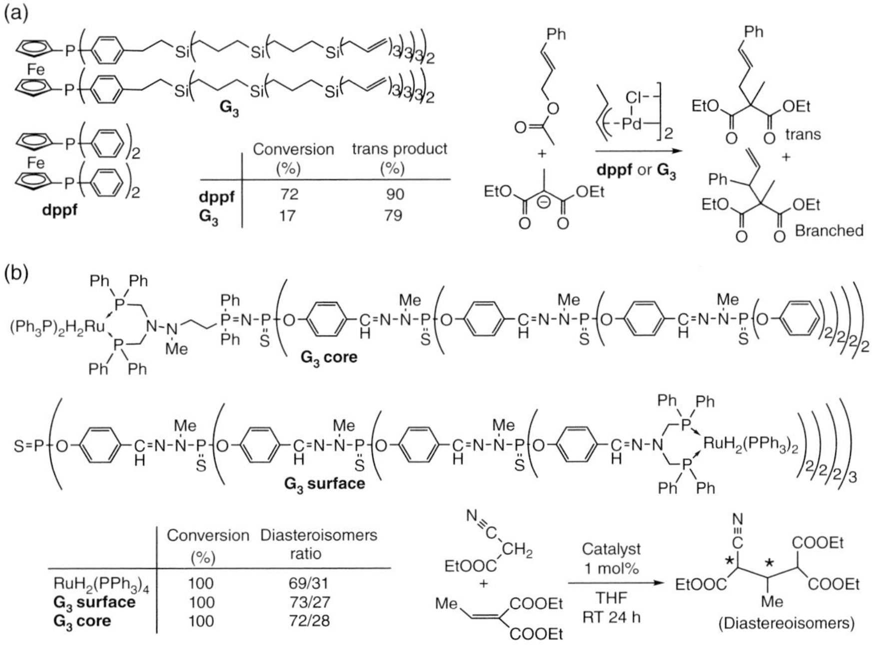

Figure 10.3 Catalytic entities at the core of inorganic dendritic structures. (a) A carbosilane dendrimer for catalyzing allylic alkylations; (b) a phosphorhydrazone dendron and a dendrimer, both used for catalyzing Michael additions.

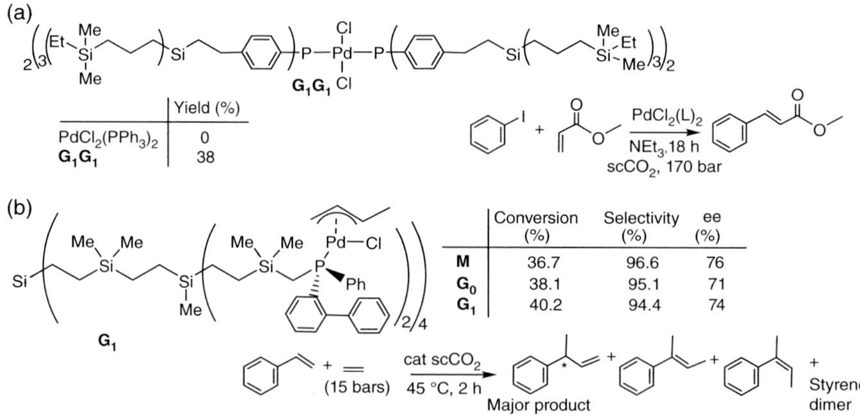

Figure 10.4 Carbosilane dendrimers used as catalysts in $scCO_2$. (a) dendrons used in Heck couplings; (b) dendrimer used for the hydrovinylation of styrene.

iodobenzene in $scCO_2$. The yield was 38% with the dendritic catalyst and 0% with the monomeric $PdCl_2(PPh_3)_2$, illustrating the increased solubility of the dendritic catalyst (Figure 10.4a) [18]. In another experiment, Pd-carbosilane dendrimers containing the P-stereogenic phosphine [P(2-biphenylyl)PhCH_2] as terminal groups were used as catalysts for the hydrovinylation of styrene, in $scCO_2$ medium (Figure 10.4b) [19].

Several carbosilane dendrimers were used in continuous flow membrane reactors, but the turnover numbers are much lower than in batch conditions [20, 21].

An original example of phosphorhydrazone dendrimers concerns the redox-switchable catalysis (RSC), in which oxidation and reduction influence the electron-donating ability of a ligand, resulting in modified activity. Ferrocenyl phosphine ligands complexing Ru(*p*-cym)Cl_2 (cym = cymene) were used as terminal groups and applied as catalysts for the isomerization of 1-octen-3-ol to 3-octanone [22]. The dendrimer was more efficient (turnover frequency, TOF = $600\,h^{-1}$) than the corresponding monomer (TOF = $400\,h^{-1}$). Oxidation of the ferrocene obtained by addition of 1 equiv of [Fe{η^5-C_5H_4C(O)Me}Cp][BF_4] per catalytic function induces a markedly reduced rate of catalysis. Full conversion was observed again with a degree of activity similar to that of the original catalyst when adding [FeCp$^*_2$] (Cp* = C_5Me_5) as reducing agent (Figure 10.5) [23].

10.2.2 The Dendrimer Effect Illustrated with Catalytic Inorganic Dendrimers

The "dendrimer effect" is invoked in the literature to explain phenomena that arise as the generation number of the dendrimers increases [8]. A positive dendrimer effect is observed when the increase of the generation is beneficial for the property studied. Such phenomenon has been mainly observed for catalysis, with a higher efficiency in terms of yield, enantioselectivity, or temperature

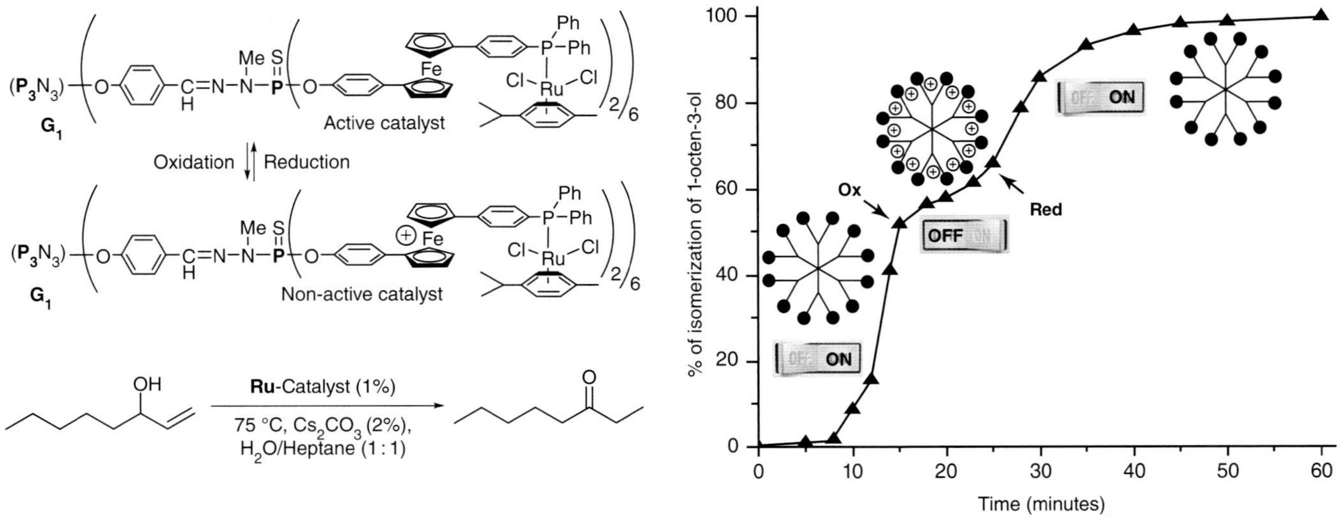

Figure 10.5 Phosphorhydrazone dendrimer for the isomerization of 1-octen-3-ol into 3-octanone. Switch ON/OFF/ON experiments and influence on the catalysis rate.

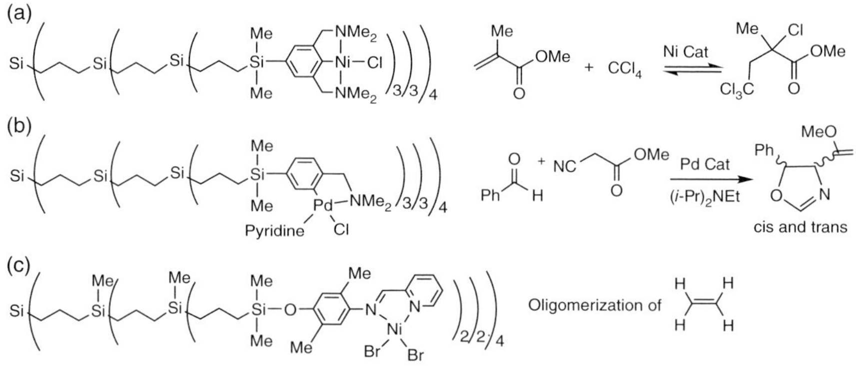

Figure 10.6 Carbosilane dendrimers for which a dendrimer effect has been searched. (a) Negative effect in atom-transfer radical additions; (b) no effect in an aldol condensation reaction; (c) positive effect in the oligomerization of ethylene.

(lower) [24]. However, it must be emphasized that the dendrimer effect can be also negative, with a detrimental effect on the outcome of the catalysis when the generation increases. Such a negative phenomenon was observed very early with carbosilane dendrimers ended by arylnickel complexes, used as catalysts in atom-transfer radical additions, with the TOF (TOF/h and /Ni atom) being 163 for a monomer, 98 for generation 0, 39 for generation 1, and 2 for generation 2 (Figure 10.6a) [25, 26]. In another example, carbosilane dendrimers with aryl palladium terminal groups were used as catalysts in a model aldol condensation reaction. The monomer and all generations up to generation 2 have almost the same selectivity, but the rate of the reaction decreased with increasing steric congestion at the dendrimer periphery (Figure 10.6b) [27]. In fact, positive dendrimer effects have been rarely observed using carbosilane dendrimers as catalysts. One of these rare examples concerns dendrimers containing terminal pyridylimine ligands, complexing nickel, which were used for catalyzing the transformation of ethylene into mixtures of polyethylene and oily oligomers. The TOF values for obtaining the oligomers were 264 for the monomer, 594 for generation 0, 729 for generation 1, and 957 for generation 2 (Figure 10.6c) [28].

Contrarily to carbosilane dendrimers, positive dendrimer effects have been observed frequently with phosphorhydrazone dendrimers. For instance, a series of dendrimers having pyridine-imine ligands complexing copper(I) as terminal groups were used for catalyzing N-arylations. The monomeric complex had practically no activity, whereas all the dendrimers from generations 1 to 3 were highly active, with iodobenzene. When using bromobenzene as arylation agent, the monomer was still non-active, but the dendrimers displayed an increased efficiency when the generation increased, using the same quantity of copper (and ligands) in all cases (Figure 10.7a) [29]. The PTA ligand (PTA = phosphatriazaadamantane) is particularly interesting because it can afford water-soluble complexes. Dendrimers ended by PTA complexes of ruthenium were used as catalysts for the isomerization of allylic alcohols to ketones in

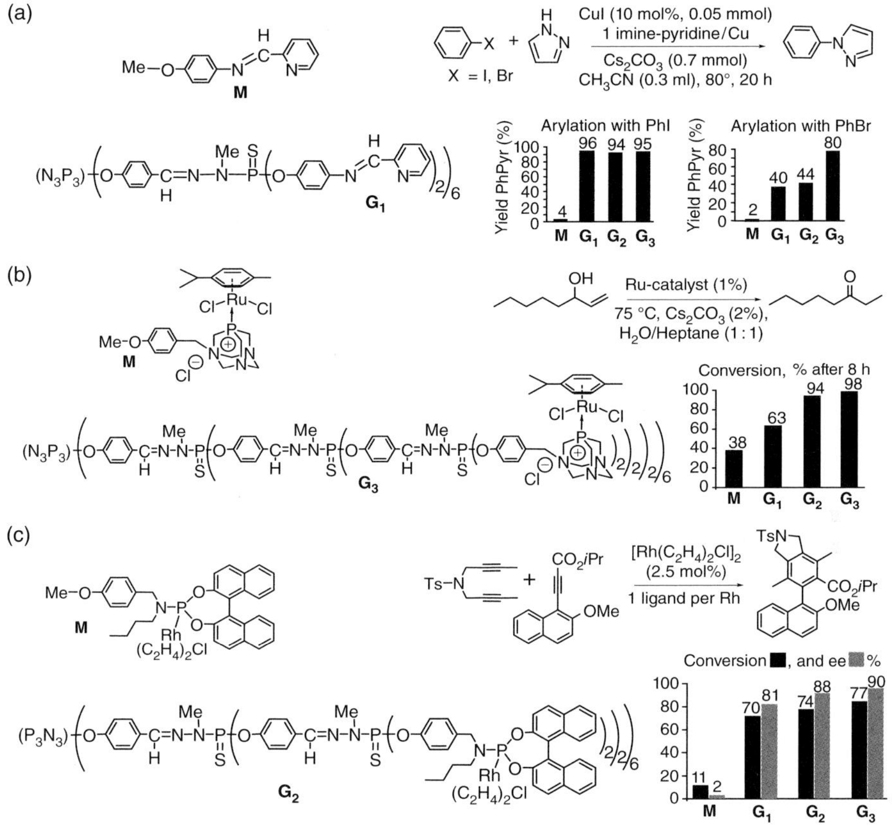

Figure 10.7 Positive dendrimer effects observed with phosphorhydrazone dendrimers as catalysts. (a) Copper-catalyzed arylations; (b) ruthenium-catalyzed isomerization of allylic alcohols; (c) rhodium-catalyzed asymmetric [2+2+2] cycloaddition reactions.

biphasic conditions (water/heptane). A large positive dendrimeric effect was observed when the generation increased (Figure 10.7b) [30].

Another positive dendrimer effect has been observed with phosphorhydrazone dendrimers functionalized with phosphoramidite terminal groups, used for Rh-catalyzed [2+2+2] cycloaddition reactions. The reaction between *N*-tosyl 1,6-diyne and 2-methoxynaphthalene alkynyl derivatives leads to axially chiral diaryl compounds. All the generations of the dendrimers, from 1 to 3, afforded very high yields and enantiomeric excesses, whereas the corresponding monomer induced poor yield and no enantiomeric excess (Figure 10.7c) [31].

10.2.3 The Recovery and Reuse of Catalytic Inorganic Dendrimers

The recovery and reuse of dendritic catalysts has been considered as an important topic since early times [32]. The most common way to recover dendritic catalysts is by precipitation, adding a solvent in which the products, but not the catalyst, are soluble. Such a method is very efficient with phosphorhydrazone dendrimers, which are generally not soluble in ether, whereas most products are. For instance, a generation 4 phosphorhydrazone dendrimer ended by terpyridine–scandium triflate complexes efficiently catalyzed under microwave activation Friedel–Crafts acylations of a wide range of aromatic and heteroaromatic electrophiles. This dendritic catalyst was successfully applied in 12 consecutive runs, using different substrates for each cycle, with recovery and reuse of the dendritic catalyst by precipitation at each end of the cycle (Figure 10.8) [33].

One of the reasons for the successful recovery and reuse of dendritic catalysts is the decrease of metal leaching, which is a major problem encountered in organometallic catalysis, as it is costly, and induces a pollution of the products. Monomeric and phosphorhydrazone dendritic catalysts were applied to Suzuki couplings, with the aim of measuring the influence of both the type of ligand (triphenylphosphine or thiazolyldiphenylphosphine) and the monomer vs. dendrimers, on the leaching of Pd. The leaching of Pd was very large for both monomers (2227 ± 63 ppm for PPh_3, 1432 ± 46 ppm for monomeric thiazolylphosphine), whereas it was reduced with both first generation dendrimers (173 ± 3 ppm for the first generation ended by PPh_3 derivatives, undetectable with the first generation ended by thiazolylphosphine ligands). The metal leaching is responsible for the neat decrease of the catalytic efficiency after recovery, observed with the dendrimer ended by PPh_3 after three runs, whereas the catalytic efficiency remained identical after five runs with the dendrimer ended by thiazolylphosphines (Figure 10.9) [34].

Besides precipitation, the dendritic catalysts can be recovered by other means. A small phosphorhydrazone dendrimer bearing phosphine terminal functions for the complexation of palladium, and a pyrene at the core for interacting by π-stacking with the graphene layers covering the magnetic cobalt nanoparticles, was synthesized. This assembly was used for catalyzing Suzuki couplings, in particular for the synthesis of the anti-inflammatory drug Felbinac. Heating to 60 °C induced desorption of the dendrimers from the surface of the nanoparticles, thus the catalysis occurred in homogeneous conditions. The dendrimers were again completely adsorbed onto the nanoparticles when cooling to room temperature, and the nanoparticles covered by the dendrimers were easily separated from the media, using a magnet. Quantitative yield of Felbinac was obtained from run 1 to 12, demonstrating the efficiency of this recovery method (Figure 10.10) [35].

Relatively few examples of recovery and reuse of carbosilane catalytic dendrimers have been carried out. Bis(oxazoline) ligands were attached to the surface of carbosilane dendrimers, for Cu(II) Lewis acid catalysis of the α-hydrazination of a β-keto ester as well as of the Henry reaction of 2-nitrobenzaldehyde with nitromethane. The dendrimers were in a membrane bag, like in a "tea bag." Seven runs were successfully carried out with only a slight decrease in the efficiency (yield and enantioselectivity) (Figure 10.11) [36].

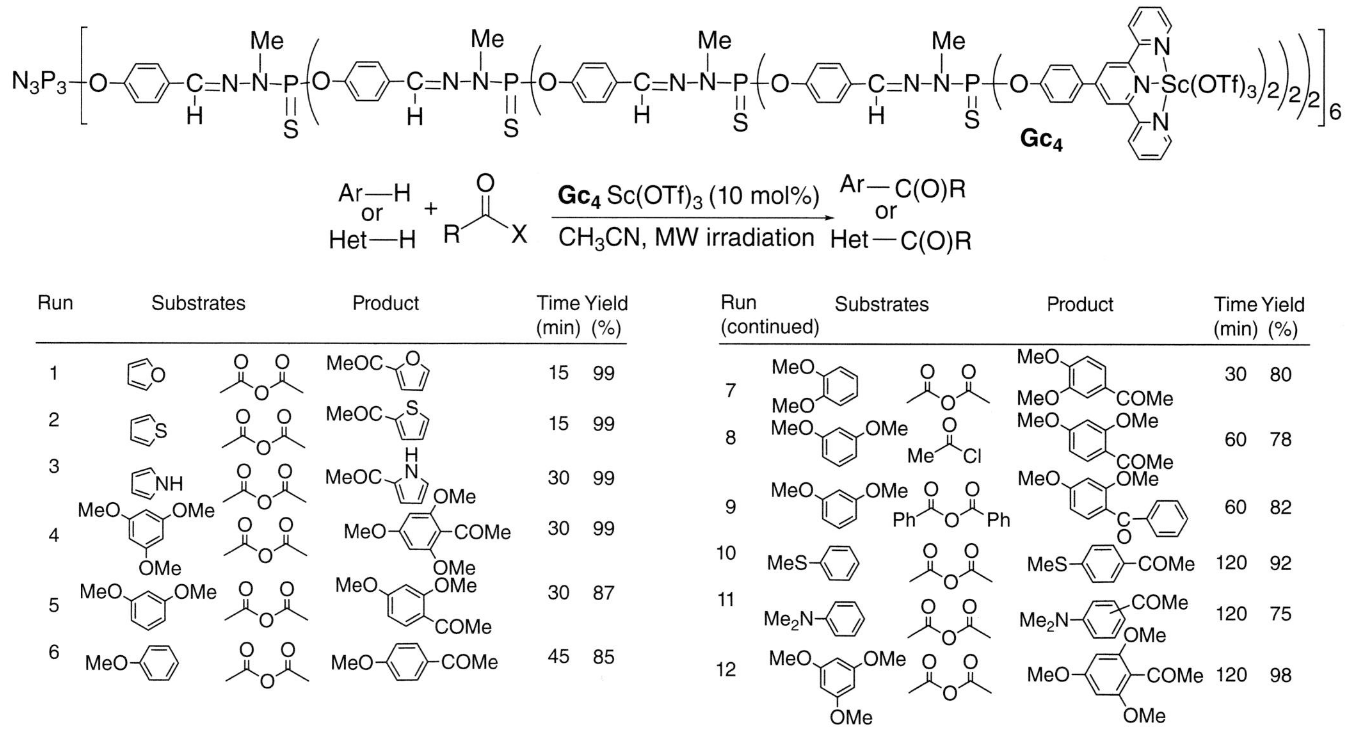

Figure 10.8 Terpyridine complexing scandium on the surface of a phosphorhydrazone dendrimer for catalyzing Friedel–Crafts acylations under microwaves.

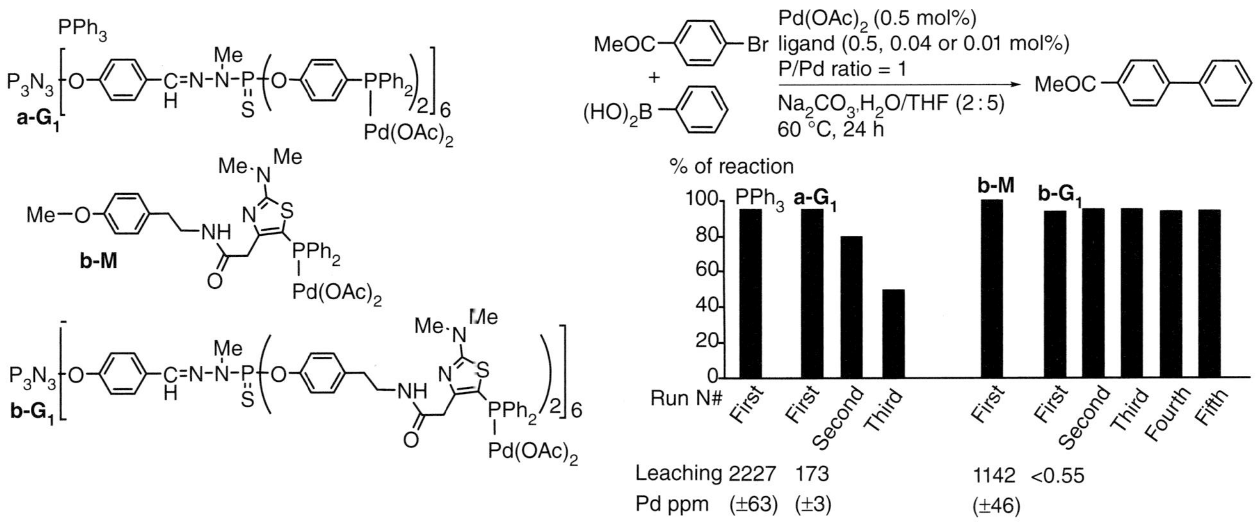

Figure 10.9 Decrease of Pd leaching when using phosphorhydrazone dendrimers in Suzuki couplings on going from monomers to dendrimers, and efficiency of the recovery and reuse depending on the type of ligand.

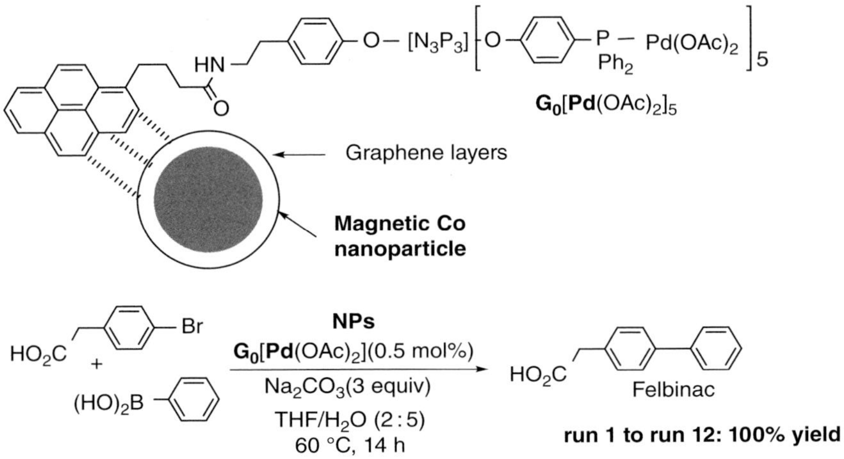

Figure 10.10 Bifunctional phosphorhydrazone dendrimer, interacting by π-stacking with magnetic cobalt nanoparticles protected by graphene layers, and their use for catalyzing the synthesis of the anti-inflammatory drug felbinac by Suzuki couplings.

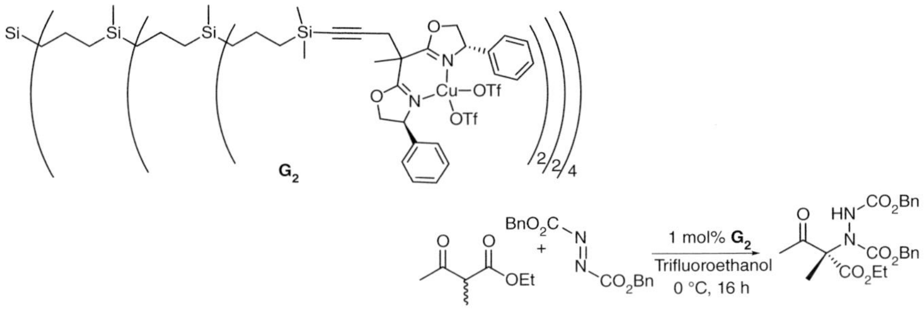

Figure 10.11 Generation 2 carbosilane dendrimer used for catalyzing asymmetric α-hydrazination of a β-keto ester, used in a "tea bag" for an easy recovery.

10.3 Inorganic Dendrimers for Nanomaterials

The use of inorganic dendrimers in the field of materials has been explored very early, in particular with regard to liquid crystals. Introducing mesogenic groups as dendrimer constituents should lead to their self-assembly, which may afford new types of mesophases with uncommon morphologies. The very first examples were of a first generation carbosilane dendrimer functionalized with terminal mesogenic groups based on cyanobiphenyl, methoxyphenyl benzoate, and cholesteryl groups. Different mesophases of the smectic type have been observed, depending on the chemical nature of the mesogenic groups [37]. At the same time, the second generation functionalized with 36 cyanobiphenyl groups was synthesized, which displayed a smectic A phase (Figure 10.12a) [38] as well as the third generation functionalized with 108 mesogenic cholesteryl end-groups (Figure 10.12b) [39].

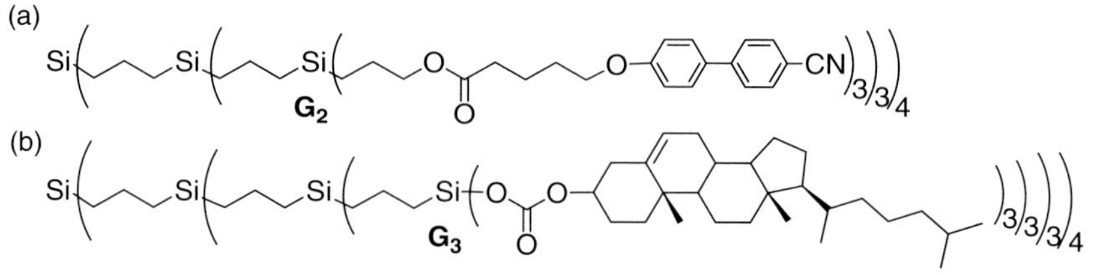

Figure 10.12 Early examples of liquid crystal carbosilane dendrimers, functionalized with either (a) 36 cyanobiphenyl groups, or (b) 108 cholesteryl groups.

Three different types of materials based on inorganic dendrimers have been obtained: (i) materials composed exclusively of dendrimers; this is the case of the liquid crystal dendrimers, and also of some nano-objects; (ii) hybrid materials incorporating dendrimers in their structure, often based on silica; and (iii) surfaces of materials covered at the nanometric scale by dendrimers, often used for biological purposes, in particular, as sensors.

10.3.1 Elaboration of Materials and Nano-objects Exclusively Composed of Inorganic Dendrimers

Only a single example of liquid crystal phosphorhydrazone dendrimers is known to date. This family has been built up to generation 4, bearing 15-membered triolefinic aza-macrocycles containing long hydrocarbon chains of two different lengths (12 and 16 carbon atoms) as terminal functions. These dendrimers adopt an elongated cylindrical shape leading to column mesophases, as opposed to the natural tendency to adopt a globular conformation (Figure 10.13) [40].

Many examples of small carbosilane liquid crystal dendrimers are known [41]. Selected examples of those based on "true" dendrimers (at least of generation 2) are shown in Figure 10.14. Bent-core mesogenic units at the periphery displayed original mesophases, with chiral domains of opposite handedness (Figure 10.14a) [42]. Chiral benzoate derivatives grafted to the surface of a third generation carbosilane dendrimer display smectic C liquid crystal phases (Figure 10.14b) [43]. Dendrimers bearing terminal cyanobiphenyl groups synthesized up to the fifth generation, display two levels of structural organization, i.e. different supramolecular nanostructures of columnar type in addition to smectic-like arrangement of mesogenic groups (Figure 10.14c) [44]. Another series was functionalized with *p*-methoxybenzoic and *p*-butoxybenzoic acid derivatives. All generations from 1

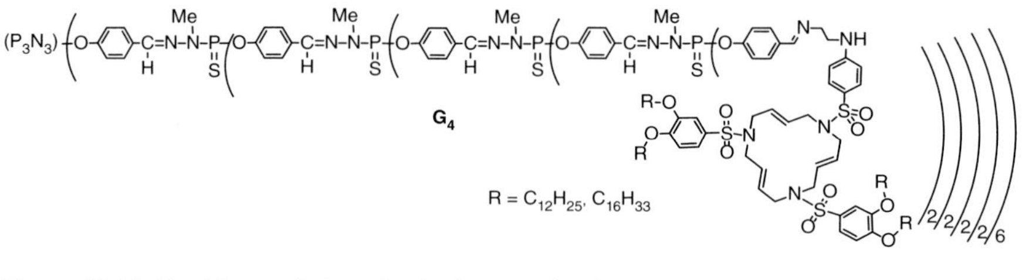

Figure 10.13 Liquid crystal phosphorhydrazone dendrimer.

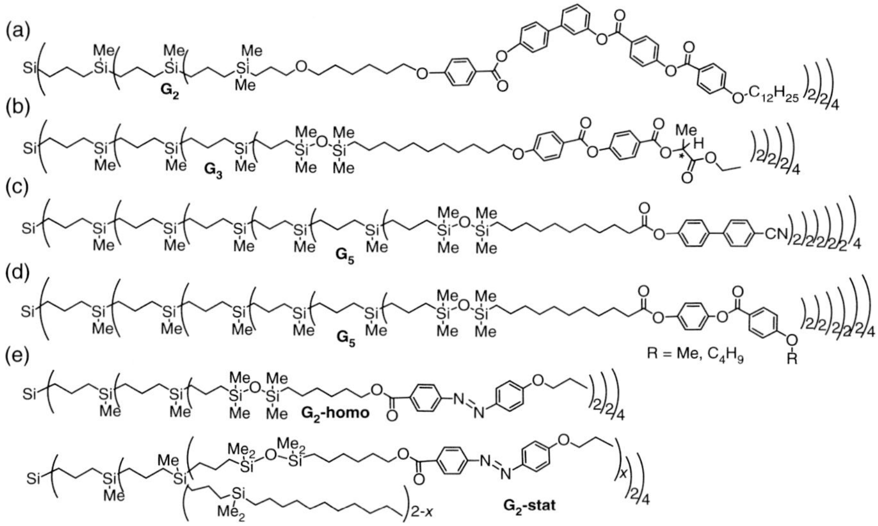

Figure 10.14 Examples of carbosilane dendrimers displaying liquid crystal properties, functionalized with (a) bent-core mesogenic units; (b) chiral benzoate derivatives; (c) cyanobiphenyl groups; (d) *p*-methoxybenzoic and *p*-butoxybenzoic acid derivatives; and (e) azobenzene or a mixture of azobenzene and decyl groups.

to 4 predominantly show smectic phases, whereas the fifth generation forms two different types of columnar phases (Figure 10.14d) [45]. An original series of the second generation carbosilane dendrimers with either azobenzene-containing (photo-isomerizable) mesogenic terminal groups, or decyl terminal groups, or a 1/1 mixture of both, either of the block type (Janus dendrimers) or statistically grafted (random type) was synthesized. All these dendrimers displayed Smectic A phases (Figure 10.14e) [46].

Nano/meso-objects exclusively composed of dendrimers, i.e. nanotubes [47] and microcapsules, have been obtained only with phosphorhydrazone dendrimers, with alternating arrangement of negatively ($\mathbf{G_4}^-$) and positively ($\mathbf{G_4}^+$) charged dendrimers using the electrostatic templating strategy [48] (Figure 10.15a). Nanoporous alumina with holes of 400 nm, was used as template. Twenty bilayers of $\mathbf{G_4}^-$ and $\mathbf{G_4}^+$ dendrimers could be deposited. The alumina template was destroyed using a strong base (KOH), or a strong acid (H_3PO_4), and a strong oxidant (CrO_3), without destroying the assembled dendrimer layers. Original nanotubes exclusively composed of dendrimers, with an external diameter of 400 nm, an inner diameter of 300 nm, and a length of 80 μm were obtained (Figure 10.15b) [49]. If the diameter of the pore is decreased, the penetration of the dendrimers inside the pores is decreased and even forbidden when the ionic strength is increased, even if the diameter of the pores is larger than that of the dendrimers [50]. The same layer-by-layer deposit was carried out on the surface of the microspheres composed of melamine formaldehyde. Better results were obtained using dendrimers and polymers, compared to those obtained with polymers alone or dendrimers alone. Either poly(styrenesulfonate) (PSS) and $\mathbf{G_4}^+$ dendrimers, or poly(allylamine hydrochloride) (PAH) and $\mathbf{G_4}^-$

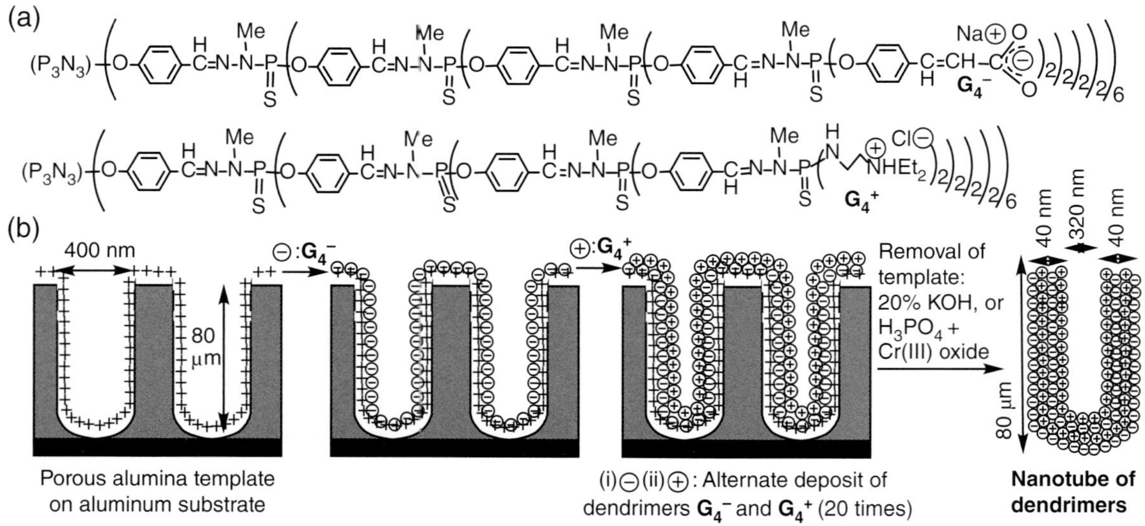

Figure 10.15 (a) Charged phosphorhydrazone dendrimers; (b) synthesis of nanotubes.

dendrimers were used. Removal of the microsphere using HCl generates the microcapsules [51]. Stiffening of these capsules has been observed in the presence of THF [52]. DNA has been used too, instead of the negatively charged polymers [53].

10.3.2 Hybrid Materials Incorporating Inorganic Dendrimers

The very first example of hybrid materials composed of inorganic dendrimers ("soft matter") and "hard matter" concerned carbosilane dendrimers and dendrons, both bearing $Si(OMe)_3$ terminal groups, incorporated inside silica (Figure 10.16a), by a polycondensation sol–gel process, using a stoichiometric amount of water in solvents. Only the second generation afforded xerogels of porous silica [54]. The same process was carried out in benzene with a third generation dendrimer bearing $Si(OEt)_3$ terminal groups (Figure 10.16b), leading to wet monolytic gels with surface area of 490 m^2/g in the first experiment [55], and 800 m^2/g in the second one. After drying, the surface area of the xerogel was found to be 1000 m^2/g. Treatment of this gel with $Ti(O\textit{i}Pr)_4$ afforded an efficient catalyst for epoxidation of cyclohexene [56, 57].

Phosphorhydrazone dendrimers and dendrons have been also incorporated inside silica via sol–gel processes. The first examples were of dendrons from generation 0 to 3 having a single $Si(OEt)_3$ as the core unit (Figure 10.17a). The co-hydrolysis and polycondensation of these dendrons with a defined as well as a varying number of tetraethoxysilane (TEOS) gave rise to dendrimer-silica xerogels. The number of TEOS equivalents needed to obtain the xerogels depends on the generation of the dendrons, from 10 equiv of TEOS for the generation 0 (G_0), to 140 for generation 3 (G_3) [58]. Large dendrimers (generations 4 and 8) bearing ammonium terminal functions (see $\mathbf{G_4}^+$ in Figure 10.15a) were used together with the cationic surfactant CTAB (cetyltrimethylammonium bromide) and TEOS for the elaboration of periodic mesoporous silica of the MCM-41 type. The hexagonal structure MCM-41 is preserved up to about 26% in weight of the dendrimer included in the silica. The cationic surfactant can be selectively removed to liberate the pores while preserving the incorporated dendrimers [59]. More recently, three families of dendritic compounds possessing both $Si(OEt)_3$ group(s) and primary amines were synthesized (Figure 10.17b) [60]. The efficiency of the hybrid hard/soft nanomaterials obtained after the grafting to silica have been tested for trapping CO_2, the most efficient structures being those of the dendrons [61].

Other phosphorhydrazone dendrimers have been embedded in other types of inorganic materials. First generation with alcohol or carboxylic acid terminal

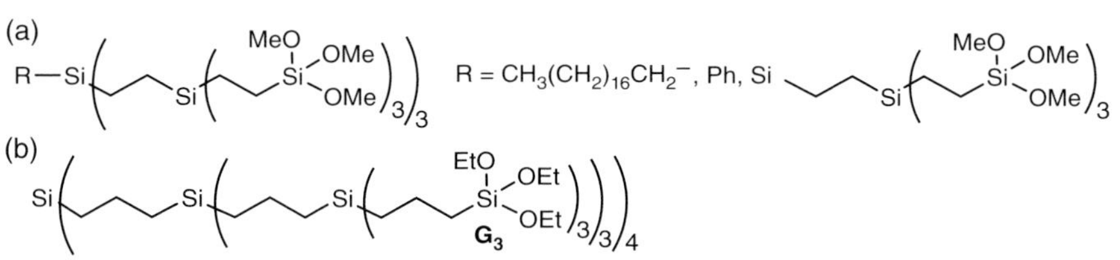

Figure 10.16 Carbosilane dendrons (a) or dendrimers (b) used in sol-gel processes to afford porous silica.

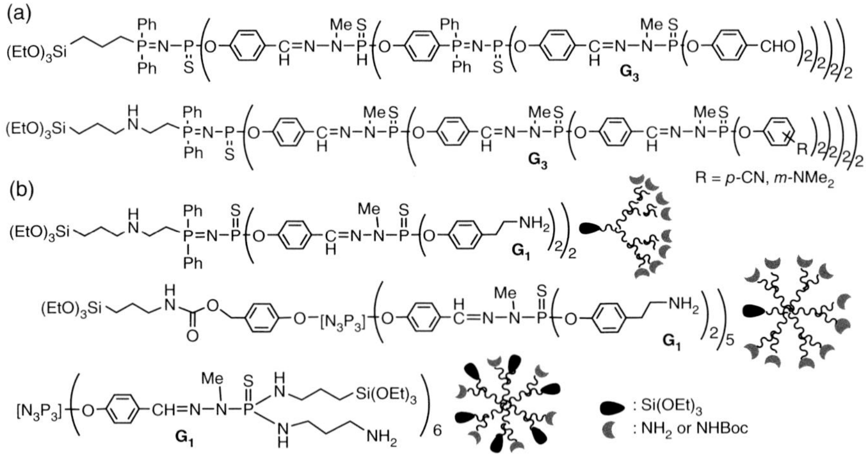

Figure 10.17 Phosphorhydrazone dendrons and dendrimers used for the synthesis of functionalized silica (a), and for grafting to nanoporous silica (b).

functions have been used for the first assembly of nanoscopic building blocks, together with the titanium-oxo cluster $[Ti_{16}O_{16}(OEt)_{32}]$. Mesostructured hybrid materials were obtained, in which the dendrimers dictate a pyramidal local symmetry, which is repeated to form a long-range ordered structure [62]. Dendrimers of generations 5 and 7 bearing carboxylic acid terminal functions (same family as $\mathbf{G_4}^-$ in Figure 10.15a) were mixed in variable ratios in THF with metal alkoxides ($Ce(O\text{-}i\text{Pr})_4$, $Ti(OR)_4$) to afford bicontinuous gels presenting a "sponge-like" structure with pore size ranging from 10 to 30 nm [63].

Three generations of phosphonated dendrimers ($\mathbf{G_4}$ in Figure 10.18a) have been used with $Ti(Oi\text{Pr})_4$ to provide a new family of hierarchically porous dendrimer-bridged titanium dioxide materials, which displayed high porosity with surface areas ranging from 203 to 514 m^2/g [64]. Exploring the scope of this method led to the unexpected discovery of discrete anatase nanocrystals (4.8–5.2 nm in size), in marked contrast to the routinely obtained amorphous titanium dioxide phase. The presence of the dendrimers (Figure 10.18a) prevents the commonly observed anatase-to-rutile phase transformation up to c. 800 °C [65]. The organizational properties of phosphorhydrazone dendrimers has

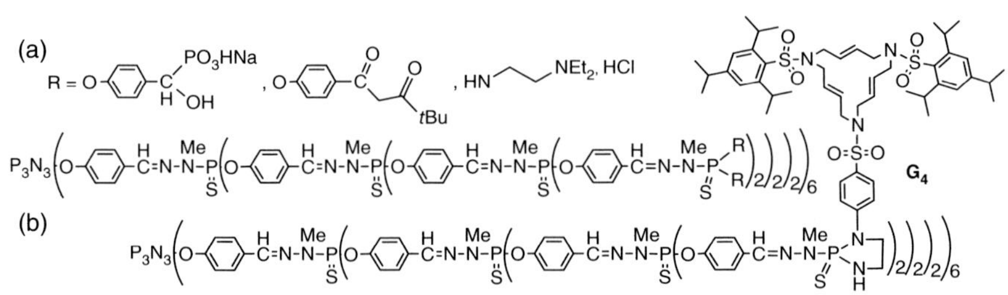

Figure 10.18 Phosphorhydrazone dendrimers used for organizing inorganic materials. (a) Phosphonate, diketone, and ammonium terminal functions. (b) 15-Membered tri-olefinic azamacrocycles terminal functions.

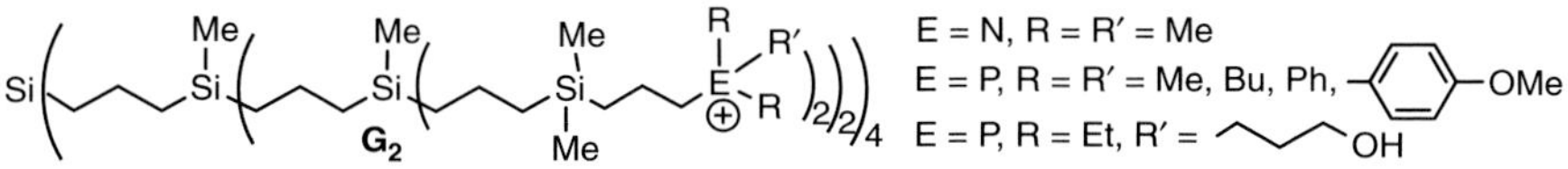

Figure 10.19 Carbosilane dendrimers incorporated in montmorillonite.

been confirmed for dendrimers of the generations 1–4 bearing 15-membered tri-olefinic aza-macrocycles as end-groups (Figure 10.18b), mixed with $Pt_2(dba)_3$ (dba = dibenzylideneacetone). Unprecedented branched supramolecular assemblies composed of dendrimers and coalesced Pt nanoparticles were obtained, which became larger for higher dendrimer generations [66].

Besides synthetic materials, dendrimers have been incorporated also into natural materials, in particular clays. Phosphorhydrazone dendrimers of generations 1 and 2 with ammonium terminal functions (same family as $\mathbf{G_4}^+$ in Figure 10.15a) have been used for cation exchange with montmorillonite, but they behaved differently. Generation 1 formed intercalated nanocomposites, whereas generation 2 resulted in the formation of a mixture of exfoliated and intercalated nanocomposites. These nanocomposites have been used for studying the adsorption of chromate, which is a water pollutant. The maximum uptake of chromate was in the range 23–38 mg/g [67]. The same type of experiment was carried out with the same dendrimers and stevensite. Both types of clay-dendrimers materials have been used for trapping methylene blue, which is another water pollutant [68]. Small carbosilane dendrimers (generations 0 and 1 with NEt_3 end-groups) have been incorporated in montmorillonite as well [69]. Dendrimers with phosphonium groups (up to generation 2, Figure 10.19) have been used recently, and selected samples of the corresponding composite materials were tested as catalysts in the cycloaddition of CO_2 with epoxide (allyl glycidyl ether) [70].

10.3.3 Modification of the Surface of Materials with Inorganic Dendrimers Toward Biological Uses

Biosensors such as DNA microarrays have gained increasing importance in medicinal, forensic, and environmental analyses. The use of dendrimers in such devices induces a higher loading of probes and an easier hybridization, keeping away the biomolecules from the solid support to reduce interferences. The field has been reviewed early [71], and a recent review emphasizes the prominent role played by phosphorhydrazone dendrimers [72]. The first experiment was done with a generation 5 bearing terminal aldehyde groups, grafted to quartz or glass slides pretreated by aminopropyltriethoxysilane [73]. This modified surface was used for the immobilization of a model protein, HSA (human serum albumin) [74]. In a later step, the surface, modified with generation 4 dendrimers, was used for the covalent grafting of amino-modified DNA probes as spots. This "dendrislide" is a sensitive biosensor for the detection of fluorescently labeled complementary oligonucleotide (the target), as illustrated in Figure 10.20 [75]. A detection limit of 1 pM labeled targets was measured, and the "dendrislide" could be reused up to 10 times without any significant change in the fluorescence signal intensity [76]. The process is now developed

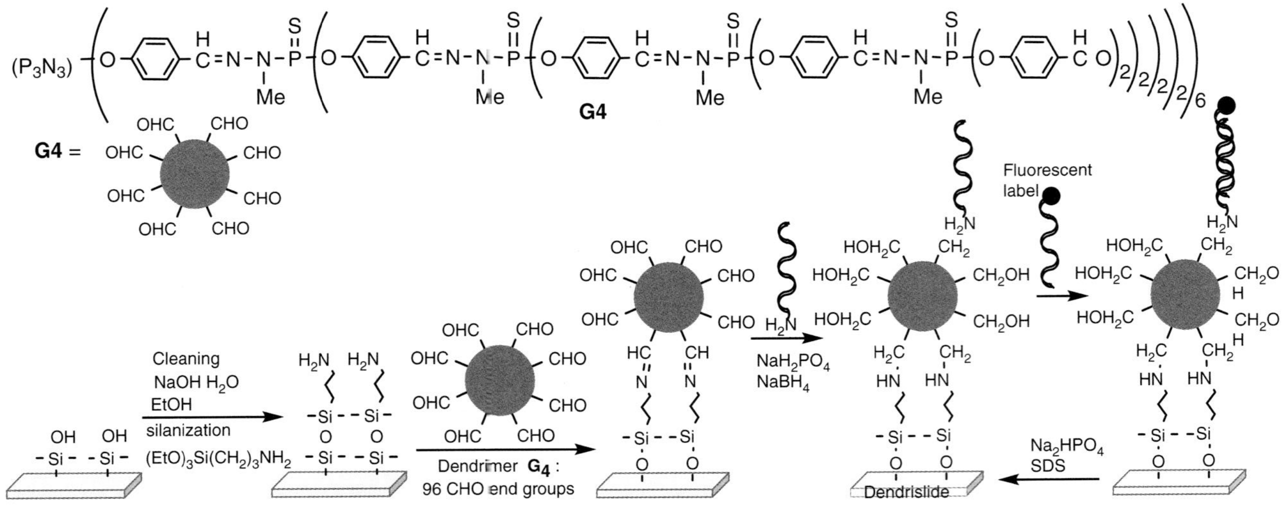

Figure 10.20 The principle of a "dendrislide" (DNA chip) basec on dendrimers.

by the start-up Dendris, in particular for the diagnosis of respiratory infections (http://www.dendris.fr/). These DNA chips have been converted to nanocapsule arrays by grafting on a fluorescent liposome an oligonucleotide complementary to an oligonucleotide bound to the array. This device has potential application for the encapsulation and spotting of proteins [77]. The dendrimer was grafted also onto a piezoelectric membrane and the hybridization was carried out with a complementary oligonucleotide bearing a biotin label, which can be recognized by streptavidin-conjugated gold nanoparticles. The biosensor functionality was validated by measuring the modification of the resonant frequencies of these functionalized piezoelectric membranes [78]. A sequel of this work was conducted for the functionalization of an AFM (atomic force microscopy) tip by a dendrimer, to react further with amino functions of the antibody of GST (glutathione-*S*-transferase). The GST being linked to a dendrimer array, single molecule force interactions between GST and its cognate antibody immobilized on the tip were probed by AFM spectroscopy with high specificity [79].

Several other types of biosensors based on phosphorus dendrimers have been reported. Using the layer-by-layer depositing of charged dendrimers on a flat gold substrate, followed by coupling of probe DNA and hybridization of Cy5 dye labeled complementary target DNA afforded a limit of detection of 30 pM on four bilayers [80]. The principle for the synthesis of nanotubes shown in Figure 10.15b, in which three different luminescent quantum dots have been used instead of the negatively charged dendrimers, offered a graded-bandgap structure. An efficient excitation energy transfer occurred by FRET (fluorescence resonance energy transfer) from the outer to the inner surface of the nanotubes, allowing detection of hybridization with a high sensitivity of 2–5 pM [81, 82]. A further improved sensitivity of detection was obtained using a bilayer formed with a perylene diimide labeled star polymer and a cationic phosphorus dendrimer on a silver substrate. Using surface plasmon enhanced fluorescence spectroscopy technique, a time-resolved ultrasensitive (10^{-18} M) and selective detection of DNA targets was achieved [83]. The only example of carbosilane dendrimers as biosensors is a small (generation 1) dendrimer with ferrocene terminal groups, used for detecting glucose [84].

Surfaces modified by phosphorhydrazone dendrimers can be used as supports for cell cultures. Fetal cortical rat neurons were cultured on films composed of five bilayers of positively and negatively charged dendrimers (see Figure 10.15a). Rat neurons attached preferentially and matured slightly faster on surfaces terminated with positively charged dendrimers than on those that were negatively charged [85]. However, such finding is not a general rule. Indeed, when using a single layer of the first generation dendritic species covalently grafted to a gold surface through a thioctic acid (Figure 10.21), the interaction

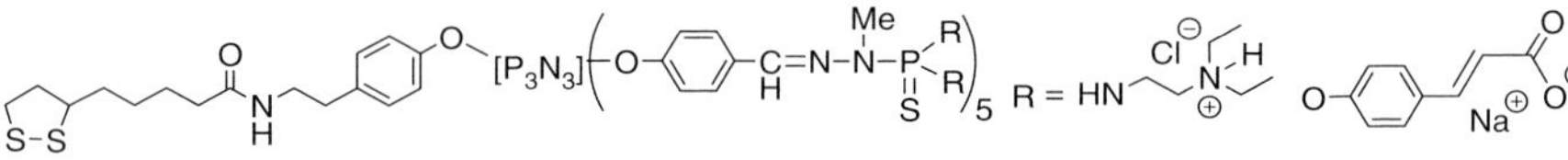

Figure 10.21 Phosphorhydrazone dendrimers with one arm used for the covalent grafting to gold, and 10 positively or negatively charged terminal groups.

with human osteoblast cells depended on the surface coating. Polycationic dendrimers provoke cell apoptosis, whereas negatively charged dendrimers support cell adhesion and proliferation [86]. Silica nanoparticles functionalized by phosphorhydrazone dendrimers, derivatized to phosphonates with lateral poly(ethyleneglycol) (PEG) chains were used for hosting silver nanoparticles, which exhibit antibacterial activity [87].

10.4 Inorganic Dendrimers in Biology/Nanomedicine

The previous paragraphs have already shown the use of inorganic dendrimers for biomaterials. Here, the use of water-soluble inorganic dendrimers for biology will be displayed, with particular reference to bioimaging, gene therapy, for use against viruses, and in the treatment of cancers, brain diseases, and inflammatory diseases.

10.4.1 Inorganic Dendrimers for Bioimaging

The largest part of dendrimers used for bioimaging concerns fluorescent dendrimers [88]. The first examples of fluorescent inorganic dendrimers were a carbosilane dendrimer with one pyrene at the core unit [89], and a phosphorhydrazone dendrimer with 12 pendant pyrenes inside the structure [90]. Fluorescently labeled inorganic dendrimers, especially phosphorhydrazone dendrimers, have been mainly used for imaging their interactions with cells or biological entities. The fluorophore can be at the core, within the branches, or as terminal moieties generally grafted in a stochastic way (a few fluorophores among water-solubilizing functions). A generation 5 dendrimer with a phthalocyanine as core and ammonium terminal groups was used to study its penetration inside human umbilical vein endothelial cells (HUVECs) [91]. A single fluorophore of type maleimide linked to the core of dendrimers with azabisphosphonate terminal groups was used to determine its interaction with human monocytes (Figure 10.22a) [92]. A fluorophore of the julolidine type was also linked to the core of the first generation dendrimer, with azabisphosphonate [93], or the corresponding azabiscarboxylate [94], or other carboxylate terminal groups [95].

An easy way to synthesize fluorescent and water-soluble dendrimers consists in grafting in a stochastic way a small number of fluorophores as terminal functions. This has been done with FITC (fluorescein isothiocyanate) on a dendrimer bearing azabisphosphonate terminal groups [92], and also with a julolidine derivative on a dendrimer bearing mannose derivatives as terminal groups (Figure 10.22b) [96]. Fluorescent entities have been used also as branches (Figure 10.22c), and the corresponding dendrimer has helped in deciphering its anticancer properties [97].

Few examples of fluorescent carbosilane dendrimers have been designed for studying their biological properties. Two examples of stochastic grafting on the surface, the grafting of dansyl groups together with ammonium groups

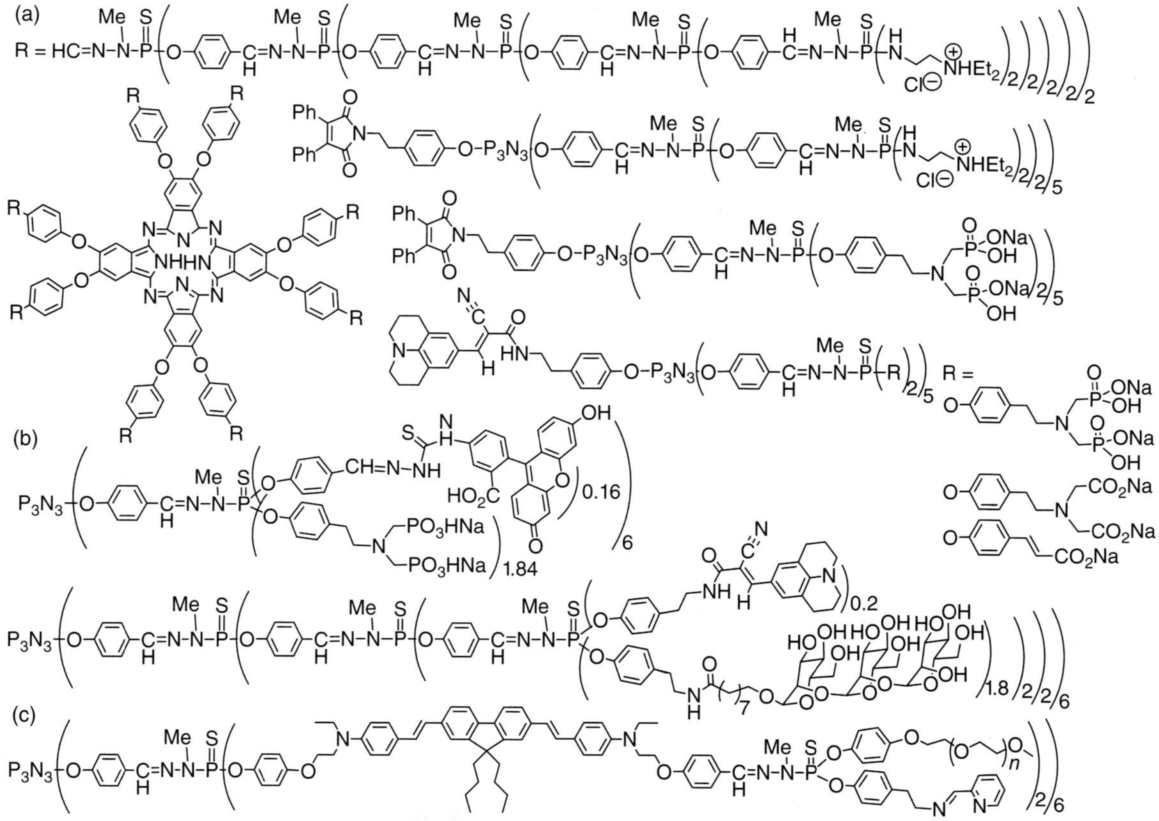

Figure 10.22 Fluorescent phosphorhydrazone dendrimers having the fluorophore, (a) at the core, (b) at a few terminal positions, (c) in the branches.

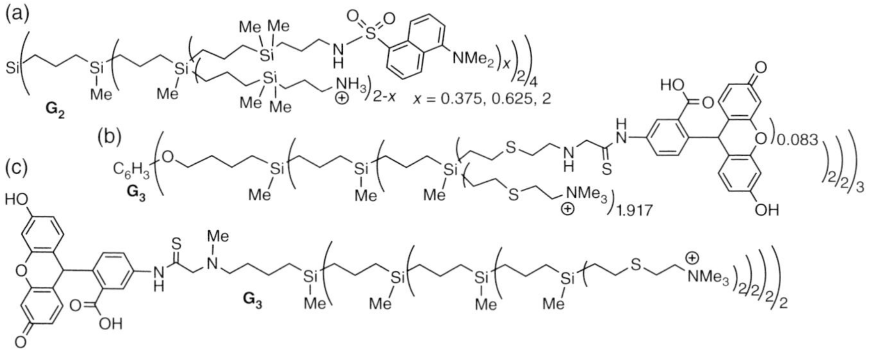

Figure 10.23 Fluorescent carbosilane dendrimers having the fluorophore, (a) and (b) stochastically at a few terminal positions, (c) at the core.

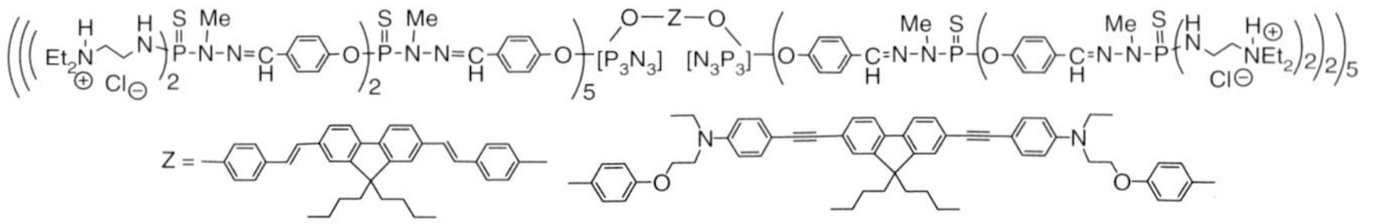

Figure 10.24 Dendrimers having a fluorescent core, for in vivo imaging.

(Figure 10.23a) enabled to study their cellular uptake in peripheral blood mononuclear cells (PBMCs) [98], and the grafting of FITC (Figure 10.23b) enabled to study the penetration inside the brain of mice [99]. A dendron with FITC grafted specifically to the core unit (Figure 10.23c) was used to analyze in vivo bio-distribution in mice [100].

A few phosphorhydrazone dendrimers have been used to image the blood vessels of living animals. These compounds have a special fluorophore core unit (Figure 10.24) that is able to absorb two photons [101, 102], and they have been used in living rats [103] and tadpoles [104].

10.4.2 Inorganic Dendrimers for Gene Therapy

The development of gene therapy in medicine is highly dependent on the transfer efficiency of genes within the desired type of cells (transfection). Viral-based gene delivery is the most efficient method, but immunogenic or potential oncogenic effects constitute important drawbacks and, therefore, alternatives to viruses are investigated. The interaction between positively charged dendrimers and oligonucleotides is called "dendriplex." The first example of inorganic dendrimers forming dendriplexes was the family of positively charged phosphorhydrazone dendrimers shown in Figure 10.15a of the generation from 1 to 5. They were used for transferring the plasmid of the luciferase into mammalian cells. The efficiency increased with the generation, but a plateau was reached for generations 4 and 5, which was comparable to that of the chemical standard in the field, polyethyleneimine [105]. The positively charged dendrimer with a phthalocyanine core (Figure 10.22a) was used as transfection agent for the

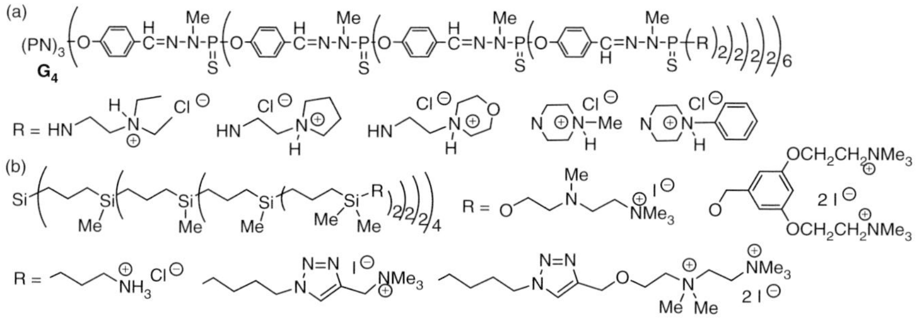

Figure 10.25 Phosphorhydrazone (a) and carbosilane (b) dendrimers used for transfection experiments.

delivery of both fluorescein-labeled oligodeoxyribonucleotide and DNA plasmid containing the functional gene of enhanced green fluorescent protein (EGFP) into HeLa cells [91]. The type of ammonium terminal groups was then varied as shown in Figure 10.25a, and these dendrimers were used to deliver single- and double-stranded DNA into the three cell strains, healthy (HUVEC) or cancerous (human embryonic kidney, HEK 293, HeLa) [106].

Carbosilane dendrimers with terminal ammonium groups linked through Si—O bonds were found to be unstable in water, contrary to those connected through Si—C bonds (Figure 10.25b), but both were used for helping ODN (oligodeoxynucleotide) to penetrate into PBMCs [107]. Other carbosilane dendrimers with ammonium terminal functions were synthesized through "click chemistry" reaction. These dendrimers increased the efficiency of plasmid-mediated gene transfer into HepG2 cells, including in vivo, in mice [108].

10.4.3 Inorganic Dendrimers Against Viruses

The use of inorganic dendrimers against viruses mainly concerns HIV (human immunodeficiency virus) [109], but a few assays with small carbosilane dendrimers have been also carried out against influenza. Suitable terminal groups comprise either one carbohydrate (α-thioglycoside of sialic acid) [110], or three carbohydrates (sialyl α(2→3) lactose (Neu5Acα(2→3)Galβ(1→4)Glcβ1→)) [111], or a sialic acid-mimic peptide (Ala-Arg-Leu-Pro-Arg) [112] (Figure 10.26).

Against HIV, two different strategies have been applied, depending on the charge of the dendrimers. Numerous publications discuss positively charged

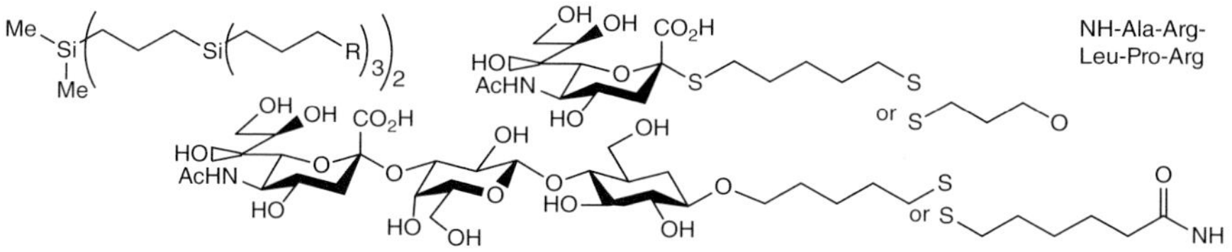

Figure 10.26 Examples of carbosilane dendrimers used against influenza.

carbosilane dendrimers used as carriers of peptides or siRNAs (short interfering RNA), and the field has been reviewed recently [113–115]. The first example is of the dendrimers shown in Figure 10.25b, where R is $O(CH_2)_2NMe(CH_2)_2NMe_3$, and ODN tested in HIV-1 inhibition assays, showing an increase of the protective effect of 25–50% on going from naked ODN to the corresponding dendriplex [116]. The same types of dendrimers were then used for forming dendriplexes with siRNAs, which reduced HIV replication in SupT1 and PBMC [117], and with HIV-derived-peptides (Gp160, P24, and Nef) [118]. The positively charged phosphorhydrazone dendrimer $\mathbf{G_4}^+$ (Figure 10.15a) has been used as well, to form dendriplexes with ODNs and siRNAs, which were found to be efficient to inhibit HIV replication [119], as well as the dendriplexes obtained with HIV-derived-peptides [120].

Negatively charged dendrimers have been used in a different way to fight against HIV [114]. The first examples were of phosphorhydrazone dendrimers with carboxylic acid terminal groups ($\mathbf{G_4}^-$ in Figure 10.15a) associated with galβ1cer analogs, to be used as lure for HIV, as galactosylceramide has a highly specific affinity for the V3 loop region of the gp120 viral envelope protein of HIV-1 [121]. Replacement of the carboxylic acids by different types of phosphonates bearing pendant alkyl chains of different length (Figure 10.27a) [122–124] displayed a similar efficiency for all compounds, due to the partial dissociation of the assembly [125].

Several types of negatively charged carbosilane dendrimers have been used in inhibition experiments of HIV (Figure 10.27b). Dendrimers bearing phosphonate groups have poor anti-HIV capability, carboxylate dendrimers moderately reduce HIV infection, while sulfate and naphthylsulfonate dendrimers are powerful anti-HIV agents [126]. Dendrimers with sulfonate terminal functions achieved 85–90% of HIV inhibition in vivo, without inducing inflammation or vaginal irritation in mice [127]. Another dendrimer with sulfonate terminal functions exerts anti-HIV-1 activity at an early stage of viral replication, and inhibiting the cell-to-cell HIV-1 transmission [128]. Combination of anionic dendrimers with antiretrovirals acting at different stages of HIV replication has

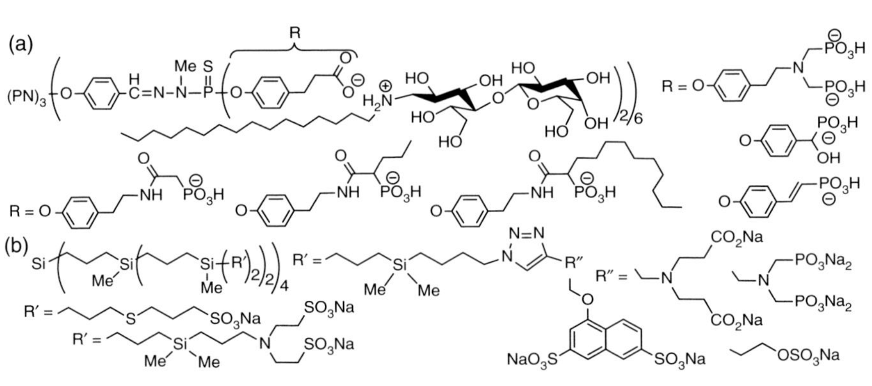

Figure 10.27 Anionic dendrimers used against HIV; (a) phosphorhydrazone dendrimers in combination with galβ1cer analogs; (b) carbosilane dendrimers.

been proposed to decrease HIV infections [129]. A synergistic effect was shown in particular with Tenofovir/Maraviroc [130].

10.4.4 Inorganic Dendrimers in Brain Diseases

Neurodegenerative diseases include the prion diseases due to accumulation of a misfolded protein (Creutzfeldt-Jakob, "mad cow" disease, transmissible spongiform encephalopathies) as well as Alzheimer's and Parkinson's diseases, both caused by accumulation of proteins and peptides to form plaques in the brain. The phosphorhydrazone dendrimer $\mathbf{G_4}^+$ (Figure 10.15a) is the most largely tested dendrimer against such diseases. The first attempt concerned the prion diseases, in which generations 3 and 5 of the same family of dendrimers have been tested. The $\mathbf{G_4}^+$ dendrimer had the highest anti-prion activity, decreasing both PrPSc (scrapie, the infective form of prions) and infectivity in scrapie-infected cells at noncytotoxic doses. It decreases the amount of pre-existing PrPSc from several prion strains, and decreases PrPSc accumulation in the spleen of infected mice by more than 80% [131]. Other experiments have been carried out to study the interaction at the molecular level with heparin [132], and the influence on the aggregation of the prion peptide PrP 185–208 [133]. The influence of the same family of dendrimers has been tested on Alzheimer peptides [134] for preventing their aggregation [135], and also to prevent aggregation of the MAP-Tau protein [136] and of α-synuclein in Parkinson's disease [137]. An overview of the recent findings in the field has been published [138].

10.4.5 Inorganic Dendrimers Against Cancers

A recent review has emphasized the role of dendrimers for the delivery of anticancer drugs [139]. Positively charged PAMAM, carbosilane, and phosphorhydrazone dendrimers have been used for the delivery of anticancer siRNA cocktails. It was shown that the most efficient combinations were based on phosphorhydrazone dendrimers [140]. However, in most cases, the anticancer properties are provided by the dendrimer alone, with suitable terminal groups. Ethacrynic acid (Edecrin) is known to be a poorly potent anticancer agent, but when grafted on phosphorhydrazone dendrimers, it becomes highly effective against human liquid and solid tumor cell lines [141], with an original mechanism of action [142]. The anticancer properties of dendrimers with metal complexes as terminal functions have been tested as well. Phosphorhydrazone dendrimers, which comprise several copper complexes, were tested against a panel of human tumor cell lines, with IC_{50} values (concentration at which 50% of cells died) ranging between 0.3 and 1.6 μM for cancerous cells [143]. The corresponding noncomplexed dendrimers are also efficient, but lesser than the Cu complexes, and both families have a different way of acting inside cells [97]. The same types of dendritic ligands were then used for the complexation of gold instead of copper. An increase in the efficiency was observed, with IC_{50}s being in the low nanomolar range (Figure 10.28a) [144]. Carbosilane dendrimers have also been used for the complexation of copper, and tested in human prostate (PC3) and human cervical (HeLa) cancer cell lines [145]. Ruthenium complexes

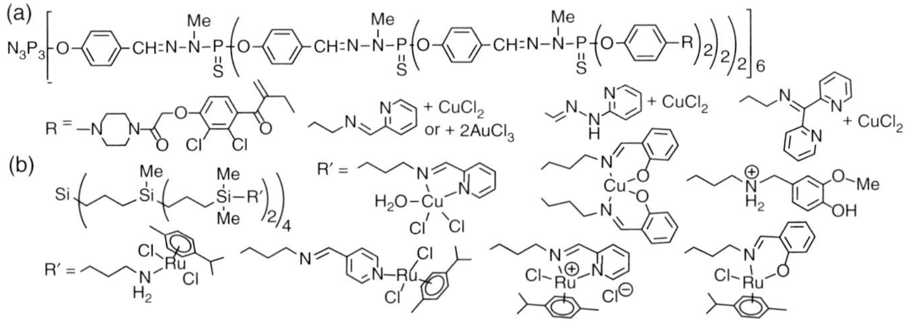

Figure 10.28 Inorganic dendrimers tested against cancer cells, and their terminal functions; (a) for phosphorhydrazone dendrimers; (b) for carbosilane dendrimers.

display cytotoxic activities in the micromolar range in several human cancer cell lines [146]. Carbosilane dendrimers containing vanillin (phenol) on the surface exhibit improved antioxidant efficacy compared to vanillin, and have been studied against HeLa cervical and PC3 human prostate cancer cell lines, with IC_{50}s in the 1–26 μM range (Figure 10.28b) [147].

A small phosphorhydrazone dendrimer functionalized with azabisphosphonate terminal groups (**ABP** Figure 10.29a) is able to induce the multiplication by several hundreds of the number of human natural killer cells (NK). NK cells are a key population with innate immunity, associated with early immune responses against infections, and which play a crucial role in anticancer immunity. The effectiveness of the NK cells produced in this way was assessed against numerous human leukemia and carcinoma cell lines [148]. The efficiency of the process has been also demonstrated with blood from cancer patients with multiple myeloma [149]. The complex mechanism leading to the proliferation of NK cells has been partially elucidated [92, 150].

10.4.6 Inorganic Dendrimers Against Inflammatory Diseases

Acute and chronic inflammatory diseases are often related to the effectiveness of the immune system. Thus, dendrimer **ABP**, which is able to multiply the immune NK cells, may have anti-inflammatory properties. **ABP** is able to activate human monocytes [92], through an anti-inflammatory pathway [151]. It has been successfully tested against several inflammatory diseases of mice, such as rheumatoid arthritis, with **ABP** either injected or even given orally [152], and neuro-inflammation as a model of multiple sclerosis [153]. Seven families of dendrimers bearing the same terminal groups have been tested; only those based on inorganic dendrimers (phosphorus-containing dendrimers, and a carbosilane dendrimer) are active, whereas three families of organic dendrimers (PAMAM, PPI, and lysine) are inactive (Figure 10.29a) [154]. The safety of **ABP** has been demonstrated by repeated injections to nonhuman primates [155]. The start-up IMD-Pharma has been created, with the aim of bringing this dendrimer to Phase 1 clinical trials (www.imd-pharma.com/).

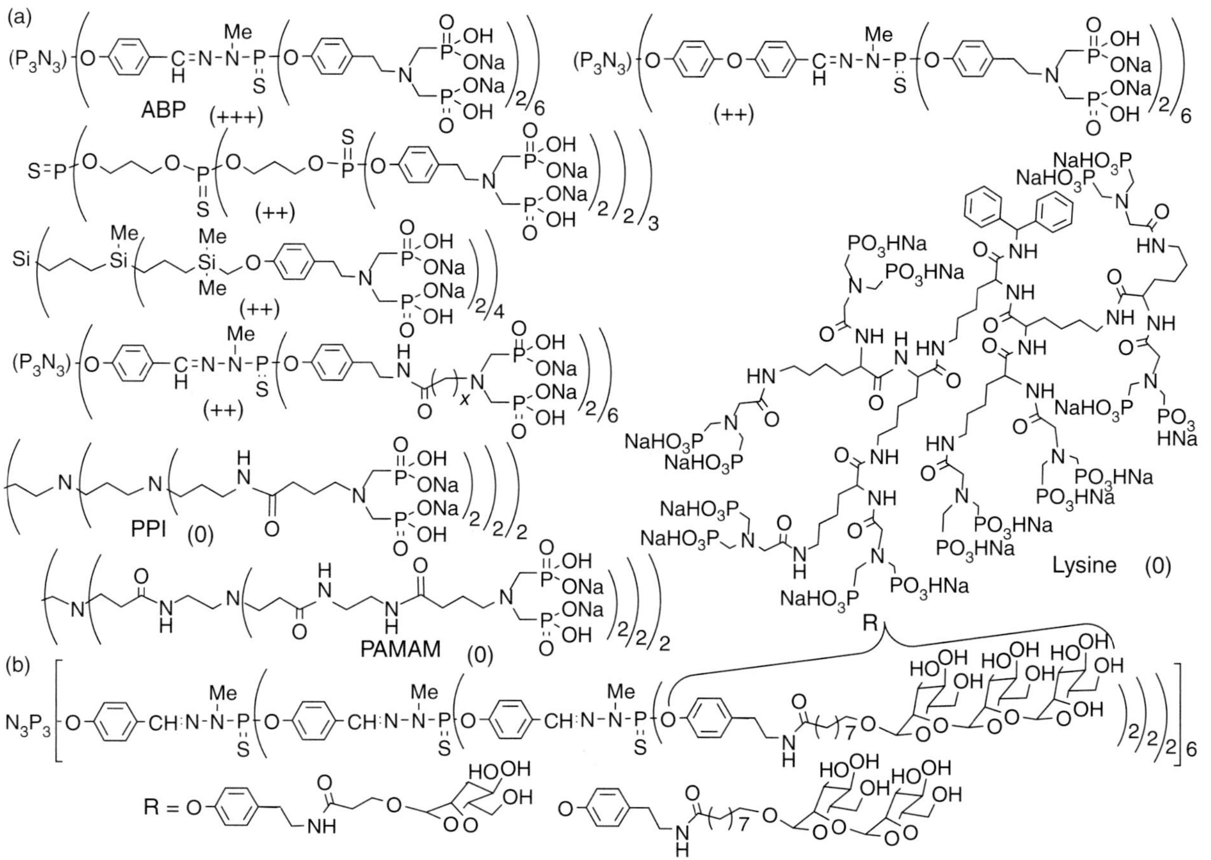

Figure 10.29 (a) Dendrimers used for activating monocytes through an anti-inflammatory pathway, and their efficiency: (+++), the most active; (++), average activity; (0), no activity. (b) Mannoside dendrimers used against acute inflammation.

Besides this work with **ABP**, a series of dendrimers bearing one, two, or three mannose derivatives as terminal groups have been synthesized, to mimic the supramolecular organization of lipoarabinomannans, an inhibitor of pro-inflammatory cytokines in human tuberculosis. The anti-inflammatory efficiency increased both with the generation and the number of mannosides, the tri-mannoside moiety being the most efficient. This dendrimer has been given orally to mice, and was able to protect their lungs against acute inflammation (Figure 10.29b) [96].

10.5 Conclusion and Outlook

The properties of inorganic dendrimers are relevant to many fields, as illustrated in this chapter, but why use inorganic dendrimers? Compared to inorganic polymers, their synthesis is undoubtedly more difficult and more expensive, but their purity (dispersity) and reproducibility of preparing are much better. Although both aspects are not very important for use in materials, they are highly desirable when considering biorelated uses. One could believe that the internal structure of dendrimers is only an innocent scaffold, and that only the terminal moieties are important for the properties of dendrimers. However, the comparison between organic and inorganic dendrimers having the same type of terminal groups shown in Figure 10.29 emphasizes their differences, and demonstrates that the inorganic ones are more efficient and more biocompatible. Considering carbosilane dendrimers and phosphorhydrazone dendrimers, it can be said that both have interesting properties, but in different fields. Catalysis has been studied with both types, but the "dendrimer effect" has been observed mainly with phosphorhydrazone dendrimers. In the field of materials, most liquid crystals are based on carbosilane dendrimers, whereas nano-objects and biosensors have been obtained exclusively with phosphorhydrazone dendrimers. Considering the biological properties, transfection (gene delivery) and anti-HIV properties have been mainly studied with carbosilane dendrimers. On the contrary, phosphorhydrazone dendrimers are the most powerful in the treatment of brain diseases, cancers, and inflammation, including chronic inflammations. The (huge!) work already performed with inorganic dendrimers opens new avenues for future researches, particularly for biology/nanomedicine. Indeed, these regularly branched purely synthetic compounds have no biological equivalent on the molecular scale, even if Nature has privileged the branched structures at the macroscopic and mesoscopic levels, from trees, to the network of lungs, and to the dendritic cells. This fact affords a lot of hope to find new properties, and new ways of action of inorganic dendrimers.

Acknowledgments

The author thanks the CNRS (Centre National de la Recherche Scientifique, France) and the COST Action CM1302 (SIPS): "European Network on Smart Inorganic Polymers."

References

1 Caminade, A.-M., Turrin, C.-O., Laurent, R. et al. (eds.) (2011). *Dendrimers: Towards Catalytic, Material and Biomedical Uses*. Chichester, UK: John Wiley & Sons.

2 Tomalia, D.A., Baker, H., Dewald, J. et al. (1986). Dendritic macromolecules – synthesis of starburst dendrimers. *Macromolecules* 19: 2466–2468.

3 de Brabander-van den Berg, E.M.M. and Meijer, E.W. (1993). Poly(propylene imine) dendrimers – large-scale synthesis by heterogeneously catalyzed hydrogenations. *Angew. Chem. Int. Ed. Engl.* 32: 1308–1311.

4 Aharoni, S.M., Crosby, C.R., and Walsh, E.K. (1982). Size and solution properties of globular tert-butyloxycarbonyl-poly(α,iε-L-lysine). *Macromolecules* 15: 1093–1098.

5 Majoral, J.P. and Caminade, A.M. (1999). Dendrimers containing heteroatoms (Si, P, B, Ge, or Bi). *Chem. Rev.* 99: 845–880.

6 Zhou, L.L. and Roovers, J. (1993). Synthesis of novel carbosilane dendritic macromolecules. *Macromolecules* 26: 963–968.

7 Launay, N., Caminade, A.M., Lahana, R., and Majoral, J.P. (1994). A general synthetic strategy for neutral phosphorus-containing dendrimers. *Angew. Chem. Int. Ed. Engl.* 33: 1589–1592.

8 Caminade, A.M., Ouali, A., Laurent, R. et al. (2015). The dendritic effect illustrated with phosphorus dendrimers. *Chem. Soc. Rev.* 44: 3890–3899.

9 Astruc, D. and Chardac, F. (2001). Dendritic catalysts and dendrimers in catalysis. *Chem. Rev.* 101: 2991–3023.

10 Mery, D. and Astruc, D. (2006). Dendritic catalysis: major concepts and recent progress. *Coord. Chem. Rev.* 250: 1965–1979.

11 Caminade, A.M., Servin, P., Laurent, R., and Majoral, J.P. (2008). Dendrimeric phosphines in asymmetric catalysis. *Chem. Soc. Rev.* 37: 56–67.

12 Caminade, A.M., Ouali, A., Keller, M., and Majoral, J.P. (2012). Organocatalysis with dendrimers. *Chem. Soc. Rev.* 41: 4113–4125.

13 Astruc, D., Boisselier, E., and Ornelas, C. (2010). Dendrimers designed for functions: from physical, photophysical, and supramolecular properties to applications in sensing, catalysis, molecular electronics, photonics, and nanomedicine. *Chem. Rev.* 110: 1857–1959.

14 Miedaner, A., Curtis, C.J., Barkley, R.M., and Dubois, D.L. (1994). Electrochemical reduction of Co_2 catalyzed by small organophosphine dendrimers containing palladium. *Inorg. Chem.* 33: 5482–5490.

15 Knapen, J.W.J., Van der Made, A.W., Dewilde, J.C. et al. (1994). Homogeneous catalysts based on silane dendrimers functionalized with arylnickel(II) complexes. *Nature* 372: 659–663.

16 Oosterom, G.E., van Haaren, R.J., Reek, J.N.H. et al. (1999). Catalysis in the core of a carbosilane dendrimer. *Chem. Commun.* 1119–1120.

17 Maraval, V., Laurent, R., Caminade, A.M., and Majoral, J.P. (2000). Phosphorus-containing dendrimers and their transition metal complexes as

efficient recoverable multicenter homogeneous catalysts in organic synthesis. *Organometallics* 19: 4025–4029.

18 Montilla, F., Galindo, A., Andres, R. et al. (2006). Carbosilane dendrons as solubilizers of metal complexes in supercritical carbon dioxide. *Organometallics* 25: 4138–4143.

19 Rodriguez, L.I., Rossell, O., Seco, M. et al. (2008). Palladocarbosilane dendrimers as catalysts for the asymmetric hydrovinylation of styrene in supercritical carbon dioxide. *J. Organomet. Chem.* 693: 1857–1860.

20 Hovestad, N.J., Eggeling, E.B., Heidbuchel, H.J. et al. (1999). Selective hydrovinylation of styrene in a membrane reactor: use of carbosilane dendrimers with hemilabile P,O ligands. *Angew. Chem. Int. Ed.* 38: 1655–1658.

21 de Groot, D., Eggeling, E.B., de Wilde, J.C. et al. (1999). Palladium complexes of phosphine functionalised carbosilane dendrimers as catalysts in a continuous flow membrane reactor. *Chem. Commun.* 1623–1624.

22 Neumann, P., Dib, H., Sournia-Saquet, A. et al. (2015). Ruthenium complexes with dendritic ferrocenyl phosphanes: synthesis, characterization, and application in the catalytic redox isomerization of allylic alcohols. *Chem. Eur. J.* 21: 6590–6604.

23 Neumann, P., Dib, H., Caminade, A.M., and Hey-Hawkins, E. (2015). Redox control of a dendritic ferrocenyl-based homogeneous catalyst. *Angew. Chem. Int. Ed.* 54: 311–314.

24 Helms, B. and Frechet, J.M.J. (2006). The dendrimer effect in homogeneous catalysis. *Adv. Synth. Catal.* 348: 1125–1148.

25 Kleij, A.W., Gossage, R.A., Jastrzebski, J. et al. (2000). The "dendritic effect" in homogeneous catalysis with carbosilane-supported arylnickel(II) catalysts: observation of active-site proximity effects in atom-transfer radical addition. *Angew. Chem. Int. Ed.* 39: 176–178.

26 Kleij, A.W., Gossage, R.A., Gebbink, R. et al. (2000). A "dendritic effect" in homogeneous catalysis with carbosilane-supported arylnickel(II) catalysts: observation of active-site proximity effects in atom-transfer radical addition. *J. Am. Chem. Soc.* 122: 12112–12124.

27 Kleij, A.W., Gebbink, R., van den Nieuwenhuijzen, P.A.J. et al. (2001). Halide-assisted macrocyclic ring formation in cyclometalated carbosilane dendrimers with 1-[$C_6H_3(CH_2NMe_2)$-4-(PdCl)-3] peripheral groups: application as aldol condensation catalysts. *Organometallics* 20: 634–647.

28 Benito, J.M., de Jesus, E., de la Mata, F.J. et al. (2006). Mononuclear and dendritic nickel(II) complexes containing *N,N′*-iminopyridine chelating ligands: generation effects on the catalytic oligomerization and polymerization of ethylene. *Organometallics* 25: 3876–3887.

29 Ouali, A., Laurent, R., Caminade, A.M. et al. (2006). Enhanced catalytic properties of copper in O- and N-arylation and vinylation reactions, using phosphorus dendrimers as ligands. *J. Am. Chem. Soc.* 128: 15990–15991.

30 Servin, P., Laurent, R., Gonsalvi, L. et al. (2009). Grafting of water-soluble phosphines to dendrimers and their use in catalysis: positive dendritic effects in aqueous media. *Dalton Trans.* 4432–4434.

31 Garcia, L., Roglans, A., Laurent, R. et al. (2012). Dendritic phosphoramidite ligands for Rh-catalyzed [2+2+2] cycloaddition reactions: unprecedented enhancement of enantiodiscrimination. *Chem. Commun.* 48: 9248–9250.
32 van Heerbeek, R., Kamer, P.C.J., van Leeuwen, P., and Reek, J.N.H. (2002). Dendrimers as support for recoverable catalysts and reagents. *Chem. Rev.* 102: 3717–3756.
33 Perrier, A., Keller, M., Caminade, A.M. et al. (2013). Efficient and recyclable rare earth-based catalysts for Friedel–Crafts acylations under microwave heating: dendrimers show the way. *Green Chem.* 15: 2075–2080.
34 Keller, M., Hameau, A., Spataro, G. et al. (2012). An efficient and recyclable dendritic catalyst able to dramatically decrease palladium leaching in Suzuki couplings. *Green Chem.* 14: 2807–2815.
35 Keller, M., Colliere, V., Reiser, O. et al. (2013). Pyrene-tagged dendritic catalysts noncovalently grafted onto magnetic Co/C nanoparticles: an efficient and recyclable system for drug synthesis. *Angew. Chem. Int. Ed.* 52: 3626–3629.
36 Gaab, M., Bellemin-Laponnaz, S., and Gade, L.H. (2009). "Catalysis in a Tea Bag": synthesis, catalytic performance and recycling of dendrimer-immobilised bis- and trisoxazoline copper catalysts. *Chem. Eur. J.* 15: 5450–5462.
37 Ponomarenko, S.A., Rebrov, E.A., Bobrovsky, A.Y. et al. (1996). Liquid crystalline carbosilane dendrimers: first generation. *Liq. Cryst.* 21: 1–12.
38 Lorenz, K., Holter, D., Stuhn, B. et al. (1996). A mesogen-functionalized carbosilane dendrimer: a dendritic liquid crystalline polymer. *Adv. Mater.* 8: 414–416.
39 Coen, M.C., Lorenz, K., Kressler, J. et al. (1996). Mono- and multilayers of mesogen-substituted carbosilane dendrimers on mica. *Macromolecules* 29: 8069–8076.
40 Hincapie, C.A., Sebastian, R.M., Barbera, J. et al. (2014). Supermolecular columnar liquid-crystalline phosphorus dendrimers decorated with sulfonamide. *Chem. Eur. J.* 20: 17047–17058.
41 Dantlgraber, G., Baumeister, U., Diele, S. et al. (2002). Evidence for a new ferroelectric switching liquid crystalline phase formed by a carbosilane based dendrimer with banana-shaped mesogenic units. *J. Am. Chem. Soc.* 124: 14852–14853.
42 Hahn, H., Keith, C., Lang, H. et al. (2006). First example of a third-generation liquid-crystalline carbosilane dendrimer with peripheral bent-core mesogenic units: understanding of "Dark conglomerate phases". *Adv. Mater.* 18: 2629–2633.
43 Boiko, N.I., Lysachkov, A.I., Ponomarenko, S.A. et al. (2005). Synthesis and comparative studies of carbosilane liquid crystalline dendrimers with chiral terminal mesogenic groups. *Colloid. Polym. Sci.* 283: 1155–1162.
44 Ponomarenko, S.A., Boiko, N.I., Shibaev, V.P. et al. (2000). Carbosilane liquid crystalline dendrimers: from molecular architecture to supramolecular nanostructures. *Macromolecules* 33: 5549–5558.

45 Richardson, R.M., Agina, E.V., Boiko, N.I. et al. (2008). Structural investigation of carbosilane liquid crystalline dendrimers. *J. Phys. Chem. B* 112: 16346–16356.
46 Lysachkov, A.I., Boiko, N.I., Rebrov, E.A. et al. (2007). Liquid-crystalline carbosilane dendrimers of different molecular architecture with photochromic mesogenic and aliphatic terminal groups. *Russ. Chem. Bull.* 56: 2407–2413.
47 Caminade, A.M. and Majoral, J.P. (2010). Dendrimers and nanotubes: a fruitful association. *Chem. Soc. Rev.* 39: 2034–2047.
48 Knoll, W., Caminade, A.M., Char, K. et al. (2011). Nanostructuring polymeric materials by templating strategies. *Small* 7: 1384–1391.
49 Kim, D.H., Karan, P., Goring, P. et al. (2005). Formation of dendrimer nanotubes by layer-by-layer deposition. *Small* 1: 99–102.
50 Lazzara, T.D., Lau, K.H.A., Abou-Kandil, A.I. et al. (2010). Polyelectrolyte layer-by-layer deposition in cylindrical nanopores. *ACS Nano* 4: 3909–3920.
51 Kim, B.S., Lebedeva, O.V., Kim, D.H. et al. (2005). Assembly and mechanical properties of phosphorus dendrimer/polyelectrolyte multilayer microcapsules. *Langmuir* 21: 7200–7206.
52 Kim, B.S., Lebedeva, O.V., Park, M.K. et al. (2010). THF-induced stiffening of polyelectrolyte/phosphorus dendrimer multilayer microcapsules. *Polymer* 51: 4525–4529.
53 Kim, B.S., Lebedeva, O.V., Koynov, K. et al. (2006). Effect of dendrimer generation on the assembly and mechanical properties of DNA/phosphorus dendrimer multilayer microcapsules. *Macromolecules* 39: 5479–5483.
54 Boury, B., Corriu, R.J.P., and Nunez, R. (1998). Hybrid xerogels from dendrimers and arborols. *Chem. Mater.* 10: 1795–1804.
55 Kriesel, J.W. and Tilley, T.D. (1999). Dendrimers as building blocks for nanostructured materials: micro- and mesoporosity in dendrimer-based xerogels. *Chem. Mater.* 11: 1190–1193.
56 Kriesel, J.W. and Tilley, T.D. (2000). Synthesis and chemical functionalization of high surface area dendrimer-based xerogels and their use as new catalyst supports. *Chem. Mater.* 12: 1171–1179.
57 Kriesel, J.W. and Tilley, T.D. (2001). Carbosilane dendrimers as nanoscopic building blocks for hybrid organic–inorganic materials and catalyst supports. *Adv. Mater.* 13: 1645–1648.
58 Turrin, C.O., Maraval, V., Caminade, A.M. et al. (2000). Organic–inorganic hybrid materials incorporating phosphorus-containing dendrimers. *Chem. Mater.* 12: 3848–3856.
59 Reinert, P., Chane-Ching, J.Y., Bull, L. et al. (2007). Influence of cationic phosphorus dendrimers on the surfactant-induced synthesis of mesostructured nanoporous silica. *New J. Chem.* 31: 1259–1263.
60 Riegert, D., Pla-Quintana, A., Fuchs, S. et al. (2013). Diversified strategies for the synthesis of bifunctional dendrimeric structures. *Eur. J. Org. Chem.* 5414–5422.
61 Riegert, D., Bareille, L., Laurent, R. et al. (2016). Silica functionalized by bifunctional dendrimers: hybrid nanomaterials for trapping CO_2. *Eur. J. Inorg. Chem.* 3103–3110.

62 Soler-Illia, G., Rozes, L., Boggiano, M.K. et al. (2000). New meso-textured hybrid materials made from assemblies of dendrimers and titanium(IV)-oxo-organo clusters. *Angew. Chem. Int. Ed.* 39: 4250–4254.
63 Bouchara, A., Rozes, L., Soler-Illia, G.J.D. et al. (2003). Use of functional dendritic macromolecules for the design of metal oxo based hybrid materials. *J. Sol-Gel Sci. Technol.* 26: 629–633.
64 Brahmi, Y., Katir, N., Hameau, A. et al. (2011). Hierarchically porous nanostructures through phosphonate-metal alkoxide condensation and growth using functionalized dendrimeric building blocks. *Chem. Commun.* 47: 8626–8628.
65 Brahmi, Y., Katir, N., Ianchuk, M. et al. (2013). Low temperature synthesis of ordered mesoporous stable anatase nanocrystals: the phosphorus dendrimer approach. *Nanoscale* 5: 2850–2856.
66 Franc, G., Badetti, E., Colliere, V. et al. (2009). Dendritic structures within dendritic structures: dendrimer-induced formation and self-assembly of nanoparticle networks. *Nanoscale* 1: 233–237.
67 Beraa, A., Hajjaji, M., Laurent, R. et al. (2016). Removal of chromate from aqueous solutions by dendrimers-clay nanocomposites. *Desalin. Water Treat.* 57: 14290–14303.
68 Beraa, A., Hajjaji, M., Laurent, R., and Caminade, A.M. (2017). Dendrimers-containing organoclays: characterisation and interaction with methylene blue. *Appl. Clay Sci.* 136: 142–151.
69 Strasak, T., Maly, M., Mullerova, M. et al. (2016). Synthesis and characterization of carbosilane dendrimer-sodium montmorillonite clay nanocomposites. Experimental and theoretical studies. *RSC Adv.* 6: 43356–43366.
70 Mullerova, M., Sabata, S., Matousek, J. et al. (2018). Organoclays with carbosilane dendrimers containing ammonium or phosphonium groups. *New J. Chem.* 42: 1187–1196.
71 Caminade, A.M., Padie, C., Laurent, R. et al. (2006). Uses of dendrimers for DNA microarrays. *Sensors* 6: 901–914.
72 Majoral, J.P., François, J.M., Fabre, R. et al. (2018). Multiplexing technology for in vitro diagnosis of pathogens: the key contribution of phosphorus dendrimers. *Sci. China Mater.* 61: 1454–1461.
73 Miksa, B., Slomkowski, S., Chehimi, M.M. et al. (1999). Tailored modification of quartz surfaces by covalent immobilization of small molecules (γ-aminopropyltriethoxysilane), monodisperse macromolecules (dendrimers), and poly(styrene/acrolein/divinylbenzene) microspheres with narrow diameter distribution. *Colloid. Polym. Sci.* 277: 58–65.
74 Slomkowski, S., Miksa, B., Chehimi, M.M. et al. (1999). Inorganic–organic systems with tailored properties controlled on molecular, macromolecular and microscopic level. *React. Funct. Polym.* 41: 45–57.
75 Le Berre, V., Trevisiol, E., Dagkessamanskaia, A. et al. (2003). Dendrimeric coating of glass slides for sensitive DNA microarrays analysis. *Nucleic Acids Res.* 31: 8.
76 Trevisiol, E., Le Berre-Anton, V., Leclaire, J. et al. (2003). Dendrislides, dendrichips: a simple chemical functionalization of glass slides with phosphorus

dendrimers as an effective means for the preparation of biochips. *New J. Chem.* 27: 1713–1719.
77 Chaize, B., Nguyen, M., Ruysschaert, T. et al. (2006). Microstructured liposome array. *Bioconjugate Chem.* 17: 245–247.
78 Nicu, L., Guirardel, M., Chambosse, F. et al. (2005). Resonating piezoelectric membranes for microelectromechanically based bioassay: detection of streptavidin-gold nanoparticles interaction with biotinylated DNA. *Sens. Actuators, B: Chem.* 110: 125–136.
79 Jauvert, E., Dague, E., Severac, M. et al. (2012). Probing single molecule interactions by AFM using bio-functionalized dendritips. *Sens. Actuators, B: Chem.* 168: 436–441.
80 Yu, Y.M., Feng, C.L., Caminade, A.M. et al. (2009). The detection of DNA hybridization on phosphorus dendrimer multilayer films by surface plasmon field enhanced-fluorescence spectroscopy. *Langmuir* 25: 13680–13684.
81 Feng, C.L., Zhong, X.H., Steinhart, M. et al. (2007). Graded-bandgap quantum-dot-modified nanotubes: a sensitive biosensor for enhanced detection of DNA hybridization. *Adv. Mater.* 19: 1933–1936.
82 Feng, C.L., Zhong, X.H., Steinhart, M. et al. (2008). Functional quantum-dot/dendrimer nanotubes for sensitive detection of DNA hybridization. *Small* 4: 566–571.
83 Feng, C.L., Yin, M.Z., Zhang, D. et al. (2011). Fluorescent core-shell star polymers based bioassays for ultrasensitive DNA detection by surface plasmon fluorescence spectroscopy. *Macromol. Rapid Commun.* 32: 679–683.
84 Martinez, M.G., Alonso, B., Casado, C.M. et al. (2011). Multioperational oxidase biosensors based on carbosilane dendrimers with interacting ferrocenes. *Electroanalysis* 23: 2888–2897.
85 Hernandez-Lopez, J.L., Khor, H.L., Caminade, A.M. et al. (2008). Bioactive multilayer thin films of charged N,N-disubstituted hydrazine phosphorus dendrimers fabricated by layer-by-layer self-assembly. *Thin Solid Films* 516: 1256–1264.
86 de Jong, E.R., Deloch, N., Knoll, W. et al. (2015). Synthesis and characterization of bifunctional dendrimers: preliminary use for the coating of gold surfaces and the proliferation of human osteoblasts (HOB). *New J. Chem.* 39: 7194–7205.
87 Hameau, A., Colliere, V., Grimoud, J. et al. (2013). PPH dendrimers grafted on silica nanoparticles: surface chemistry, characterization, silver colloids hosting and antibacterial activity. *RSC Adv.* 3: 19015–19026.
88 Caminade, A.M., Hameau, A., and Majoral, J.P. (2009). Multicharged and/or water-soluble fluorescent dendrimers: properties and uses. *Chem. Eur. J.* 15: 9270–9285.
89 Vitukhnovsky, A.G., Sluch, M.I., Krasovskii, V.G., and Muzafarov, A.M. (1997). Study of polyallycarbosilane dendrimer structure by pyrenyl excimer formation. *Synth. Met.* 91: 375–377.
90 Brauge, L., Caminade, A.M., Majoral, J.P. et al. (2001). Segmental mobility in phosphorus-containing dendrimers. Studies by fluorescent spectroscopy. *Macromolecules* 34: 5599–5606.

91 Maszewska, M., Leclaire, J., Cieslak, M. et al. (2003). Water-soluble polycationic dendrimers with a phosphoramidothioate backbone: preliminary studies of cytotoxicity and oligonucleotide/plasmid delivery in human cell culture. *Oligonucleotides* 13: 193–205.
92 Poupot, M., Griffe, L., Marchand, P. et al. (2006). Design of phosphorylated dendritic architectures to promote human monocyte activation. *FASEB J.* 20: 2339–2351.
93 Portevin, D., Poupot, M., Rolland, O. et al. (2009). Regulatory activity of azabisphosphonate-capped dendrimers on human CD4(+) T cell proliferation enhances ex-vivo expansion of NK cells from PBMCs for immunotherapy. *J. Transl. Med.* 7: 13.
94 Ledall, J., Fruchon, S., Garzoni, M. et al. (2015). Interaction studies reveal specific recognition of an anti-inflammatory polyphosphorhydrazone dendrimer by human monocytes. *Nanoscale* 7: 17672–17684.
95 Ielasi, F., Ledall, J., Anes, A.P. et al. (2016). Influence of PPH dendrimers' surface functions on the activation of human monocytes: a study of their interactions with pure lipid model systems. *Phys. Chem. Chem. Phys.* 18: 21871–21880.
96 Blattes, E., Vercellone, A., Eutamene, H. et al. (2013). Mannodendrimers prevent acute lung inflammation by inhibiting neutrophil recruitment. *Proc. Natl. Acad. Sci. U.S.A.* 110: 8795–8800.
97 Mignani, S., El Brahmi, N., Eloy, L. et al. (2017). Anticancer copper(II) phosphorus dendrimers are potent proapoptotic bax activators. *Eur. J. Med. Chem.* 132: 142–156.
98 Ortega, P., Serramia, M.J., Samaniego, R. et al. (2009). Carbosilane dendrimers peripherally functionalized with dansyl fluorescence tags and their cellular internalization studies. *Org. Biomol. Chem.* 7: 3079–3085.
99 Serramia, M.J., Alvarez, S., Fuentes-Paniagua, E. et al. (2015). In vivo delivery of siRNA to the brain by carbosilane dendrimer. *J. Controlled Release* 200: 60–70.
100 Fuentes-Paniagua, E., Serramia, M.J., Sanchez-Nieves, J. et al. (2015). Fluorescein labelled cationic carbosilane dendritic systems for biological studies. *Eur. Polym. J.* 71: 61–72.
101 Mongin, O., Krishna, T.R., Werts, M.H.V. et al. (2006). A modular approach to two-photon absorbing organic nanodots: brilliant dendrimers as an alternative to semiconductor quantum dots? *Chem. Commun.* 915–917.
102 Terenziani, F., Parthasarathy, V., Pla-Quintana, A. et al. (2009). Cooperative two-photon absorption enhancement by through-space interactions in multichromophoric compounds. *Angew. Chem. Int. Ed.* 48: 8691–8694.
103 Krishna, T.R., Parent, M., Werts, M.H.V. et al. (2006). Water-soluble dendrimeric two-photon tracers for in vivo imaging. *Angew. Chem. Int. Ed.* 45: 4645–4648.
104 Mongin, O., Rouxel, C., Robin, A.C. et al. (2008). Brilliant organic nanodots: novel nano-objects for bionanophotonics. In: *Nanobiosystems: Processing, Characterization, and Applications* (ed. E.M. Heckman, T.B. Singh and J. Yoshida). Bellingham: Spie-Int Soc Optical Engineering.

105 Loup, C., Zanta, M.A., Caminade, A.M. et al. (1999). Preparation of water-soluble cationic phosphorus-containing dendrimers as DNA transfecting agents. *Chem. Eur. J.* 5: 3644–3650.
106 Padie, C., Maszewska, M., Majchrzak, K. et al. (2009). Polycationic phosphorus dendrimers: synthesis, characterization, study of cytotoxicity, complexation of DNA, and transfection experiments. *New J. Chem.* 33: 318–326.
107 Bermejo, J.F., Ortega, P., Chonco, L. et al. (2007). Water-soluble carbosilane dendrimers: synthesis biocompatibility and complexation with oligonucleotides; Evaluation for medical applications. *Chem. Eur. J.* 13: 483–495.
108 Arnaiz, E., Doucede, L.I., Garcia-Gallego, S. et al. (2012). Synthesis of cationic carbosilane dendrimers via click chemistry and their use as effective carriers for DNA transfection into cancerous cells. *Mol. Pharmaceutics* 9: 433–447.
109 Jimenez, J.L., Pion, M., de la Mata, F.J. et al. (2012). Dendrimers as topical microbicides with activity against HIV. *New J. Chem.* 36: 299–309.
110 Sakamoto, J.I., Koyama, T., Miyamoto, D. et al. (2007). Thiosialoside clusters using carbosilane dendrimer core scaffolds as a new class of influenza neuraminidase inhibitors. *Bioorg. Med. Chem. Lett.* 17: 717–721.
111 Oka, H., Onaga, T., Koyama, T. et al. (2008). Sialyl $\alpha(2 \rightarrow 3)$ lactose clusters using carbosilane dendrimer core scaffolds as influenza hemagglutinin blockers. *Bioorg. Med. Chem. Lett.* 18: 4405–4408.
112 Hatano, K., Matsubara, T., Muramatsu, Y. et al. (2014). Synthesis and influenza virus inhibitory activities of carbosilane dendrimers peripherally functionalized with hemagglutinin-binding peptide. *J. Med. Chem.* 57: 8332–8339.
113 Chonco, L., Pion, M., Vacas, E. et al. (2012). Carbosilane dendrimer nanotechnology outlines of the broad HIV blocker profile. *J. Controlled Release* 161: 949–958.
114 Sepulveda-Crespo, D., Gomez, R., De La Mata, F.J. et al. (2015). Polyanionic carbosilane dendrimer-conjugated antiviral drugs as efficient microbicides: recent trends and developments in HIV treatment/therapy. *Nanomedicine* 11: 1481–1498.
115 Perise-Barrios, A.J., Jimenez, J.L., Dominguez-Soto, A. et al. (2014). Carbosilane dendrimers as gene delivery agents for the treatment of HIV infection. *J. Controlled Release* 184: 51–57.
116 Chonco, L., Bermejo-Martin, J.F., Ortega, P. et al. (2007). Water-soluble carbosilane dendrimers protect phosphorothioate oligonucleotides from binding to serum proteins. *Org. Biomol. Chem.* 5: 1886–1893.
117 Weber, N., Ortega, P., Clemente, M.I. et al. (2008). Characterization of carbosilane dendrimers as effective carriers of siRNA to HIV-infected lymphocytes. *J. Controlled Release* 132: 55–64.
118 Ionov, M., Ciepluch, K., Klajnert, B. et al. (2013). Complexation of HIV derived peptides with carbosilane dendrimers. *Colloids Surf., B* 101: 236–242.

119 Briz, V., Serramia, M.J., Madrid, R. et al. (2012). Validation of a generation 4 phosphorus-containing polycationic dendrimer for gene delivery against HIV-1. *Curr. Med. Chem.* 19: 5044–5051.

120 Vacas-Cordoba, E., Bastida, H., Pion, M. et al. (2014). HIV-antigens charged on phosphorus dendrimers as tools for tolerogenic dendritic cells-based immunotherapy. *Curr. Med. Chem.* 21: 1898–1909.

121 Blanzat, M., Turrin, C.O., Aubertin, A.M. et al. (2005). Dendritic catanionic assemblies: in vitro anti-HIV activity of phosphorus-containing dendrimers bearing Gal β(1)cer analogues. *ChemBioChem* 6: 2207–2213.

122 Perez-Anes, A., Spataro, G., Coppel, Y. et al. (2009). Phosphonate terminated PPH dendrimers: influence of pendant alkyl chains on the in vitro anti-HIV-1 properties. *Org. Biomol. Chem.* 7: 3491–3498.

123 Perez-Anes, A., Stefaniu, C., Moog, C. et al. (2010). Multivalent catanionic GalCer analogs derived from first generation dendrimeric phosphonic acids. *Bioorg. Med. Chem.* 18: 242–248.

124 Perez-Anes, A., Rodrigues, F., Caminade, A.M. et al. (2015). Influence of structural parameters on the self-association properties of anti-HIV catanionic dendrimers. *ChemPhysChem* 16: 3433–3437.

125 Perez-Anes, A., Mazeres, S., Caminade, A.M. et al. (2015). Use of a fluorescent aminodeoxylactitol to measure the stability of anti-HIV catanionic dendrimers by spectrofluorimetry. *Tetrahedron Lett.* 56: 1566–1569.

126 Arnaiz, E., Vacas-Cordoba, E., Galan, M. et al. (2014). Synthesis of anionic carbosilane dendrimers via "Click Chemistry" and their antiviral properties against HIV. *J. Polym. Sci., Part A: Polym. Chem.* 52: 1099–1112.

127 Galan, M., Rodriguez, J.S., Jimenez, J.L. et al. (2014). Synthesis of new anionic carbosilane dendrimers via thiol-ene chemistry and their antiviral behaviour. *Org. Biomol. Chem.* 12: 3222–3237.

128 Sepulveda-Crespo, D., Serramia, M.J., Tager, A.M. et al. (2015). Prevention vaginally of HIV-1 transmission in humanized BLT mice and mode of antiviral action of polyanionic carbosilane dendrimer G2-S16. *Nanomedicine* 11: 1299–1308.

129 Sepulveda-Crespo, D., Lorente, R., Leal, M. et al. (2014). Synergistic activity profile of carbosilane dendrimer G2-STE16 in combination with other dendrimers and antiretrovirals as topical anti-HIV-1 microbicide. *Nanomedicine* 10: 609–618.

130 Garcia-Broncano, P., Cena-Diez, R., de la Mata, F.J. et al. (2017). Efficacy of carbosilane dendrimers with an antiretroviral combination against HIV-1 in the presence of semen-derived enhancer of viral infection. *Eur. J. Pharmacol.* 811: 155–163.

131 Solassol, J., Crozet, C., Perrier, V. et al. (2004). Cationic phosphorus-containing dendrimers reduce prion replication both in cell culture and in mice infected with scrapie. *J. Gen. Virol.* 85: 1791–1799.

132 Klajnert, B., Cangiotti, M., Calici, S. et al. (2009). Interactions between dendrimers and heparin and their implications for the anti-prion activity of dendrimers. *New J. Chem.* 33: 1087–1093.

133 Klajnert, B., Cortijo-Arellano, M., Cladera, J. et al. (2007). Influence of phosphorus dendrimers on the aggregation of the prion peptide PrP 185–208. *Biochem. Biophys. Res. Commun.* 364: 20–25.

134 Klajnert, B., Cangiotti, M., Calici, S. et al. (2007). EPR study of the interactions between dendrimers and peptides involved in Alzheimer's and prion diseases. *Macromol. Biosci.* 7: 1065–1074.

135 Ottaviani, M.F., Mazzeo, R., Cangiotti, M. et al. (2010). Time evolution of the aggregation process of peptides involved in neurodegenerative diseases and preventing aggregation effect of phosphorus dendrimers studied by EPR. *Biomacromolecules* 11: 3014–3021.

136 Wasiak, T., Ionov, M., Nieznanski, K. et al. (2012). Phosphorus dendrimers affect alzheimer's (A β(1-28)) peptide and MAP-Tau protein aggregation. *Mol. Pharmaceutics* 9: 458–469.

137 Milowska, K., Gabryelak, T., Bryszewska, M. et al. (2012). Phosphorus-containing dendrimers against alpha-synuclein fibril formation. *Int. J. Biol. Macromol.* 50: 1138–1143.

138 Wasiak, T., Marcinkowska, M., Pieszynski, I. et al. (2015). Cationic phosphorus dendrimers and therapy for Alzheimer's disease. *New J. Chem.* 39: 4852–4859.

139 Mignani, S., Rodrigues, J., Tomas, H. et al. (2018). Dendrimers in combination with natural products and analogues as anti-cancer agents. *Chem. Soc. Rev.* 47: 514–532.

140 Dzmitruk, V., Szulc, A., Shcharbin, D. et al. (2015). Anticancer siRNA cocktails as a novel tool to treat cancer cells. Part (B). Efficiency of pharmacological action. *Int. J. Pharm.* 485: 288–294.

141 El Brahmi, N., Mignani, S.M., Caron, J. et al. (2015). Investigations on dendrimer space reveal solid and liquid tumor growth-inhibition by original phosphorus-based dendrimers and the corresponding monomers and dendrons with ethacrynic acid motifs. *Nanoscale* 7: 3915–3922.

142 Mignani, S., El Brahmi, N., El Kazzouli, S. et al. (2016). A novel class of ethacrynic acid derivatives as promising drug-like potent generation of anticancer agents with established mechanism of action. *Eur. J. Med. Chem.* 122: 656–673.

143 El Brahmi, N., El Kazzouli, S., Mignani, S.M. et al. (2013). Original multivalent copper(II)-conjugated phosphorus dendrimers and corresponding mononuclear copper(II) complexes with antitumoral activities. *Mol. Pharmaceutics* 10: 1459–1464.

144 Mignani, S.M., El Brahmi, N., El Kazzouli, S. et al. (2017). Original multivalent gold(III) and dual gold(III)–copper(II) conjugated phosphorus dendrimers as potent antitumoral and antimicrobial agents. *Mol. Pharmaceutics* 14: 4087–4097.

145 Sanz del Olmo, N., Maroto-Diaz, M., Gomez, R. et al. (2017). Carbosilane metallodendrimers based on copper (II) complexes: synthesis, EPR characterization and anticancer activity. *J. Inorg. Biochem.* 177: 211–218.

146 Maroto-Diaz, M., Elie, B.T., Gomez-Sal, P. et al. (2016). Synthesis and anticancer activity of carbosilane metallodendrimers based on arene ruthenium(II) complexes. *Dalton Trans.* 45: 7049–7066.

147 Mencia, G., Sanz del Olmo, N., Munoz-Moreno, L. et al. (2016). Polyphenolic carbosilane dendrimers as anticancer agents against prostate cancer. *New J. Chem.* 40: 10488–10497.

148 Griffe, L., Poupot, M., Marchand, P. et al. (2007). Multiplication of human natural killer cells by nanosized phosphonate-capped dendrimers. *Angew. Chem. Int. Ed.* 46: 2523–2526.

149 Poupot, M., Turrin, C.O., Caminade, A.M. et al. (2016). Poly(phosphorhydrazone) dendrimers: yin and yang of monocyte activation for human NK cell amplification applied to immunotherapy against multiple myeloma. *Nanomedicine* 12: 2321–2330.

150 Hayder, M., Garzoni, M., Bochicchio, D. et al. (2018). Three-dimensional directionality is a pivotal structural feature for the bioactivity of azabisphosphonate-capped poly(phosphorhydrazone) nanodrug dendrimers. *Biomacromolecules* 19: 712–720.

151 Fruchon, S., Poupot, M., Martinet, L. et al. (2009). Anti-inflammatory and immunosuppressive activation of human monocytes by a bioactive dendrimer. *J. Leukocyte Biol.* 85: 553–562.

152 Hayder, M., Poupot, M., Baron, M. et al. (2011). A phosphorus-based dendrimer targets inflammation and osteoclastogenesis in experimental arthritis. *Sci. Transl. Med.* 3: 11.

153 Hayder, M., Varilh, M., Turrin, C.O. et al. (2015). Phosphorus-based dendrimer ABP treats neuroinflammation by promoting IL-10-producing CD4(+) T cells. *Biomacromolecules* 16: 3425–3433.

154 Caminade, A.M., Fruchon, S., Turrin, C.O. et al. (2015). The key role of the scaffold on the efficiency of dendrimer nanodrugs. *Nat. Commun.* 6: 7722.

155 Fruchon, S., Mouriot, S., Thiollier, T. et al. (2015). Repeated intravenous injections in non-human primates demonstrate preclinical safety of an anti-inflammatory phosphorus-based dendrimer. *Nanotoxicology* 9: 433–441.

11

Other Examples of Inorganic Polymers as Functional Materials

Irene Weymuth and Walter Caseri

ETH Zürich, Multifunctional Materials, Department of Materials, Vladimir-Prelog-Weg 1-5/10, 8093 Zürich, Switzerland

11.1 Introduction

11.1.1 1,2,4-Triazole in Coordination Chemistry

This chapter deals with selected coordination polymers comprising 1,2,4-triazole ligands (Figure 11.1) and some of their materials properties. Note that 4-amino-1,2,4-triazole (NH_2trz) is a particularly nitrogen-rich ligand as it possesses four nitrogen but only two carbon atoms. Basically, 1,2,4-triazole can be substituted at each atom whereby a somewhat special situation is found for N^4 substitution. In such compounds, the N^4 binding site is blocked while the N^1 and N^2 sites remain unaffected in their coordination ability. Accordingly, 1,2,4-triazole ligands contain two (Rtrz) or three (Htrz, NH_2trz) nitrogen atoms that can act as binding sites. Hence, such ligands can adopt various coordination modes, which render them versatile ligands for transition metal complexes. Notably, Htrz can not only exist in different tautomeric forms but also as an anionic ligand (trz^-) and therefore it shows an exceptional variety of complexation possibilities. Basically, coordination by one nitrogen atom only (monodentate coordination) [1, 2], two nitrogen atoms (bridged complexes) [3–5], or even three nitrogen atoms [6] were found when Htrz was employed (Figure 11.2).

It is obvious that a great variety of metal–triazole complexes must exist, differing in their geometry and in the number of coordinated metal atoms. This is evident, for instance, in the variety of complexes with different substituents on the N^4 position of 1,2,4-triazole, most of which are oligo- or polynuclear [4] due to the bridging ability of 1,2,4-triazoles via N^1 and N^2 ligand atoms [7–10]. Besides, coordination by just one nitrogen atom, leading to mononuclear compounds [11, 12], or by three nitrogen atoms, leading to networks [13], also exist in complexes with 1,2,4-triazoles. In general, the diversity of metal complexes with one and the same ligand is particularly remarkable considering the rather small size of (substituted) 1,2,4-triazoles.

Many studies address linear polymers where two adjacent metal atoms are connected by three 1,2,4-triazole molecules, which thus act as bridging ligands (μ-N^1,N^2-triazole bridges), resulting in rigid chains (Figure 11.3) with

Smart Inorganic Polymers: Synthesis, Properties, and Emerging Applications in Materials and Life Sciences, First Edition. Edited by Evamarie Hey-Hawkins and Muriel Hissler.

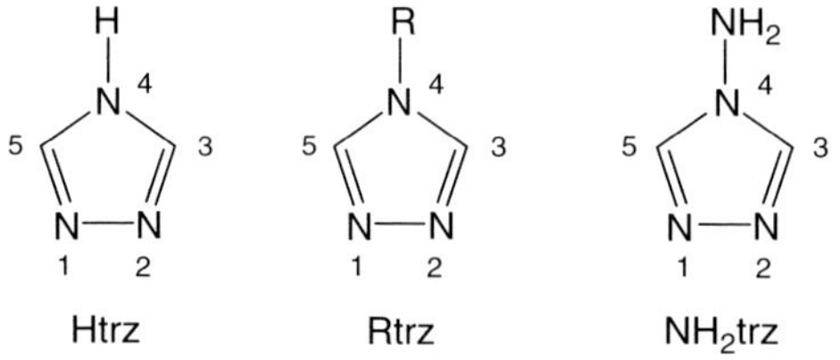

Figure 11.1 Chemical structure of 1,2,4-triazoles on the examples of 4*H*-1,2,4-triazole (Htrz), 4-alkyl-1,2,4-triazole (Rtrz), and 4-amino-1,2,4-triazole (NH_2trz).

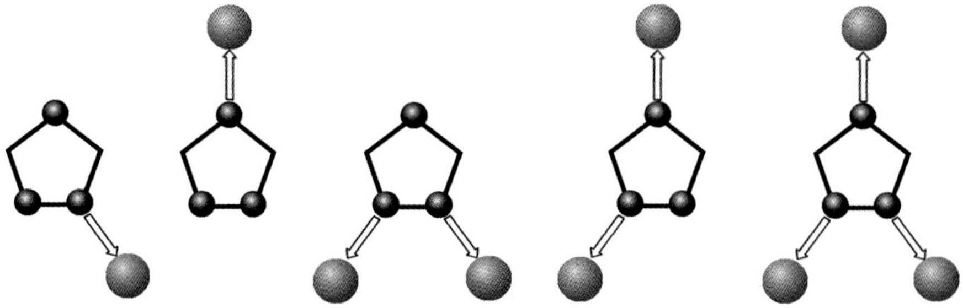

Figure 11.2 Binding modes of 1,2,4-triazole (Htrz, H atoms are omitted for simplicity). Black spheres: nitrogen atoms, gray spheres: metal ions. From left to right: monodentate coordination with N^1/N^2; monodentate coordination with N^4; bridging coordination with N^1 and N^2; bridging coordination with N^1/N^2 and N^4; bridging coordination with N^1, N^2, and N^4.

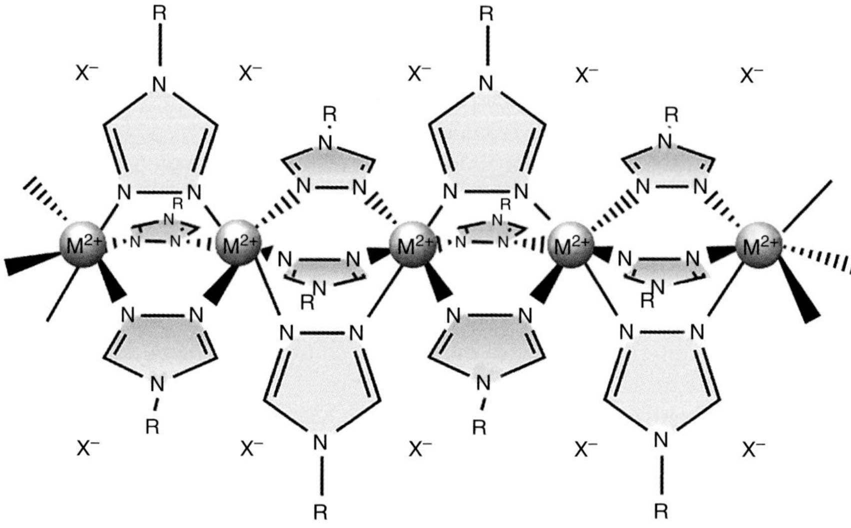

Figure 11.3 Structure of triply bridged metal(II) complexes with 1,2,4-triazoles (the substituent R at N^4 can refer to an alkyl or an amino group) and various counter anions X^-.

metal–metal distances in the order of 3–4 Å [10]. Although these complexes were already prepared in the 1970s [14] and their structure in the solid state was suggested to be linear based on extended X-ray absorption fine structure (EXAFS) [15] and infrared (IR) [14] spectroscopy, a full structural characterization by single crystal X-ray diffraction was only published 20 years later. The first resolved structure was that of the Cu(II) complex $[Cu(hyetrz)_3](ClO_4)_2{\cdot}3H_2O$[1] (hyetrz = 4-(2-hydroxyethyl)-1,2,4-triazole) [10]. The structure of an iron(II) triazole compound, ($[Fe(NH_2trz)_3](NO_3)_2$, NH_2trz = 4-amino-1,2,4-triazole), was presented only recently [8].

11.1.2 Spin-Crossover

Iron(II)–triazole complexes are particularly attractive because they often show spin-crossover behavior, a rare materials property that will be introduced hereafter with ligand field theory [16, 17]. The latter gives a description of the energy splitting of the d-orbitals in transition metal complexes, which is relevant for their chemical and magnetic properties.

For a free transition metal ion all of its five d-orbitals are degenerate and a hypothetical perfectly spherical distribution of ligands around the metal atom would lead to an equal raise in energy for all of them. However, for a discrete number of ligands the degeneracy is broken, as shown in Figure 11.4 for different coordination geometries. Depending on the coordination geometry some of the orbitals rise in energy while others decrease in energy due to the Coulomb repulsion between the electrons of the ligand and those of the metal. Thus, when electrons in d-orbitals are present, the overall energy of the system can be minimized in many cases by population of orbitals at lower energy, depending on the number of d-electrons. In certain cases, however, two possibilities of placing electrons in d-orbitals can arise. For example, in the presence of six d-electrons (d^6 ion), all electrons may occupy the lowest possible energy levels (low-spin state) or they may also occupy higher energy levels (high-spin state) in order to reduce the pairing energy of the electrons (cf. Scheme in Figure 11.4). Thus, the spin state depends on two factors:

- The ligand field splitting energy Δ, which is determined by the ligand field strength of the coordination unit [18];
- The pairing energy of the electrons. This energy is composed of the Coulomb repulsion between the electrons, which increases when an orbital is occupied by two electrons, and the loss of exchange energy between parallel spins. For free Fe^{2+} ions the related values are 89.2 and 139.8 kJ/mol, respectively [18].

If the pairing energy exceeds the ligand field splitting energy, orbitals of higher energy will be occupied in a d^6 atom resulting in a high-spin complex. On the contrary, a low-spin complex will arise if the ligand field splitting energy is larger than the pairing energy. As a matter of fact, in rare cases there exist complexes where

1 N.B. given the polynuclear nature of such complexes the notation $[M(Rtrz)_3]_n\ X_{2n}$ would be more adequate. However, for the sake of simplicity, the empirical formula $[M(Rtrz)_3]X_2$ is used instead throughout this chapter.

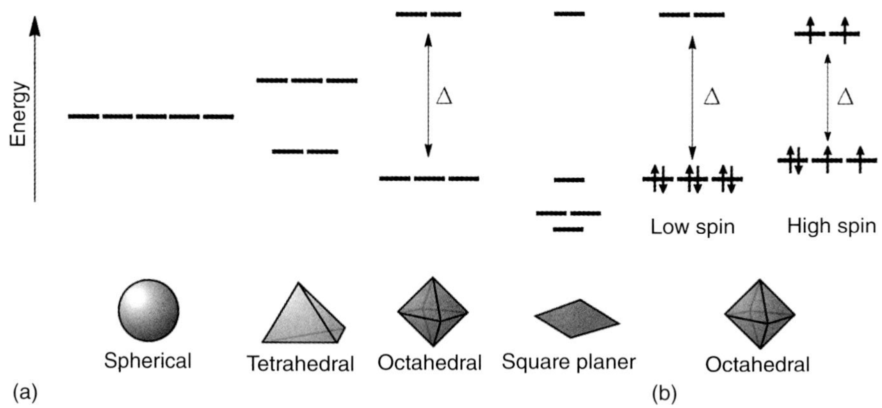

Figure 11.4 (a) Splitting of the energy levels of the d-orbitals in transition metal atoms for different coordination geometries according to crystal field theory. (b) Low-spin and high-spin state of d^6 metal centers in an octahedral coordination geometry. The parameter Δ designates the energy difference between the two sets of d-orbitals (ligand field splitting energy). Source: after Huheey et al. 2003 [18].

the crystal field splitting energy and the pairing energy of the electrons are of nearly equal magnitude between 0 and 100 °C (with a difference in the order of the thermal energy kT, where k is the Boltzmann constant and T is the temperature). For these so-called "spin-crossover" complexes variations in the physical environment suffice to switch between low- and high-spin states. Therefore, external stimuli such as changes in temperature [19–25], pressure [26, 27], light irradiation [28, 29], or strong pulsed magnetic fields [30] can induce spin-crossover.

The first description of this phenomenon traces back to Cambi and Szegö. In 1931, they observed an unexpected susceptibility–temperature relation for some iron(III) dithiocarbamate complexes [31, 32]. After this discovery spin-crossover behavior has also been found in complexes of various other transition metal ions such as iron(II) [33, 34], cobalt(III) [35–37], cobalt(II) [38], chromium(II) [39, 40], manganese(III) [41], and nickel(II) [42]. Particular attention has been paid to spin-crossover in octahedral coordination compounds of d^6 transition metals (e.g. Fe(II) and Co(III)), since this configuration gives rise to a change from a diamagnetic to a paramagnetic state upon spin-crossover. In the low-spin state, the total spin is zero $(^1A_1)$[2] whereas in the high-spin regime a quintet state (total spin of 2, $^5T_2$) is present (Figure 11.5).

Spin-crossover complexes can potentially be used as active components in switchable devices [24, 43, 44]. For this purpose, the compounds should have an abrupt spin-crossover transition around the operation temperature of the device (most often room temperature). For easy detection, a fundamental change in appearance or behavior of the spin-crossover compound should accompany the transition. If the transition proceeds at strikingly different temperatures upon heating and cooling (large hysteresis), there is a region of bistability where the state of the material depends on its thermal history. Therefore, these

2 The Mulliken symbols, $^1A_1$, $^1T_1$, $^1T_2$, $^5T_2$, and ^5E, describe the symmetry and multiplicity of the ground and excited electronic states.

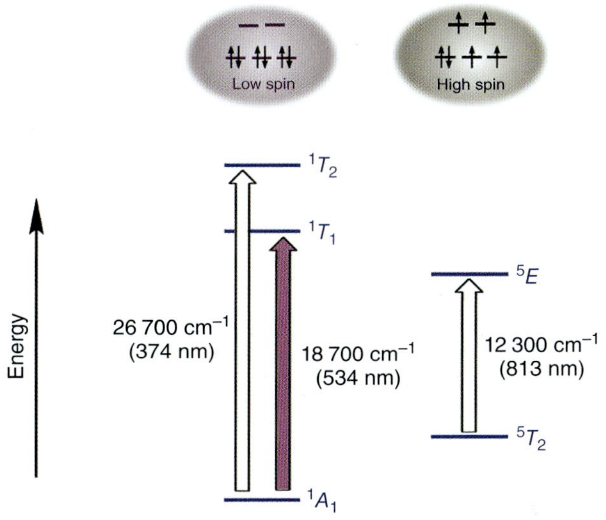

Figure 11.5 Spin-allowed d–d-transitions in polynuclear iron(II)–triazole compounds. The excitations wavelengths correspond to the values reported in ref. [7].

Table 11.1 Examples of polynuclear complexes of iron(II) and 1,2,4-triazole complexes of the type $[Fe(Rtrz)_3]X_2$ (substituent R see table) and their spin-crossover temperature upon heating (T_c↑) and cooling (T_c↓) and the resulting hysteresis width $\Delta T_c = T_c{\uparrow} - T_c{\downarrow}$.

Substituent R at N^4 of 1,2,4-triazole	Counter anion	T_c↑ (°C)	T_c↓ (°C)	ΔT_c (°C)	References
NH_2	NO_3^-	74	41	33	[7]
	1-Naphthalene sulfonate	−38	−44	6	[71]
	2-Naphthalene sulfonate	52	33	19	See Section 11.2.1
3-(Dodecyloxy)propyl	Cl^-	10	10	0	[46]
Octyl	*p*-Toluene sulfonate	32	12	20	[60]
2-Hydroxyethyl	3-Nitrophenyl sulfonate	−158	−173	15	[72]
3-Hydroxypropyl	4-Chloro-3-nitrophenyl sulfonate	−105	−153	48	[73]
H	ClO_4^-	40	23	17	[74]

compounds might be used for data storage given that the kinetic stability is high enough.

The above requirements are met in some linear polymeric iron(II) complexes, e.g. in complexes shown schematically in Figure 11.3. In those and related quasi-one-dimensional compounds, spin-crossover is generally more abrupt with a tendency to larger hysteresis than in mononuclear complexes. Examples are depicted in Table 11.1. Both the substituents at the 1,2,4-triazole and the counter anions influence the spin-crossover temperature (as well as other properties such as the solubility of the materials). Further, the spin-crossover temperature can be influenced by water (even by little amounts, see Section 11.3).

The spin-crossover is associated with a change in the color of the complex. In Fe(II) complexes, the color changes typically from pink to white upon heating.

This thermochromism is caused by the spin-allowed d–d transitions of the iron atoms in the respective spin states (Figure 11.5). The pink color in the low-spin state is caused by the $^1A_1 \rightarrow {}^1T_1$ transition around 18 700 cm^{-1} (534 nm) and another absorption band is found in the ultraviolet region around 26 700 cm^{-1} (374 nm), which originates from the $^1A_1 \rightarrow {}^1T_2$ transition. In the high-spin state there is one band located around 12 300 cm^{-1} (813 nm; $^5T_2 \rightarrow {}^5E$) which is within the near IR range and therefore not perceptible by the human eye [7].

Notably, a major hurdle still to overcome is to find routes for processing of spin-crossover iron(II)–triazole complexes. They usually do not melt prior to decomposition and most of them are considered to be insoluble (e.g. as reported in ref. [45]) unless long-chain aliphatic groups are attached to the ligands [46–48] (certainly at the expense of the percentage of the spin-crossover active iron centers per volume unit of material; see also Section 11.3). Therefore, the possibilities for processing are considerably restricted and only few studies have addressed this issue so far [49–51].

11.2 Coordination Polymers of 4-Aminotriazole and Iron(II)

11.2.1 Solutions

The behavior of spin-crossover coordination polymers in solution has been little addressed so far. Yet, this aspect is of paramount importance for processing of such polymers from solution (note that related polymers commonly do not melt prior to decomposition and hence cannot be processed from melt). Considering the reversible nature of metal–ligand formation, it is not per se obvious that coordination polymers also exist in solution, due to possible equilibrium reactions including species such as free metal ions and free ligands. Hence it is not evident a priori if spin-crossover is manifest also in the dissolved state.

The nature of spin-crossover compounds in solution was studied in detail with the example of $[Fe(NH_2trz)_3](2ns)_2$, with 2ns = 2-naphthalene sulfonate [52]. This compound exhibits spin-crossover in the solid state around room temperature (cf. Table 11.1), i.e. magnetic behavior and color of this compound can be switched by gentle heating and cooling. In most common organic solvents, $[Fe(NH_2trz)_3](2ns)_2$ is either essentially insoluble or it dissociates. One of the rare solvents is *N,N*-dimethylformamide (DMF), and therefore this solvent was selected for investigations in solution.

Accordingly, $[Fe(H_2O)_6](2ns)_2$ was combined with NH_2trz in DMF. Notably, $[Fe(H_2O)_6]^{2+}$ is transformed in DMF to $[Fe(DMF)_6]^{2+}$, whereas DMF appears to coordinate by its oxygen atom [52]. Since $[Fe(DMF)_6]^{2+}$ and NH_2trz hardly absorb light in the region of the maximum absorption wavelength of low-spin iron(II)–NH_2trz complexes (541 nm), formation of coordination compounds of iron(II) and NH_2trz can readily be followed by UV/Vis absorption spectroscopy. By carefully analyzing such spectra, it was found that an excess of NH_2trz (i.e. a ligand-to-metal ratio of 9 : 1) is required to shift the equilibrium in Scheme 11.1 almost completely to the side of fully established coordination polymers (at least

$$[Fe(NH_2trz)_{3-x}]^{2+} + xNH_2trz \rightleftharpoons [Fe(NH_2trz)_3]^{2+}$$

Scheme 11.1 Schematic illustration of equilibrium reactions between iron(II)–NH_2trz complexes and free ligand, where the polymer is represented by a formula unit and x is the average value of NH_2trz molecules per iron(II) ion in the coordination polymers (i.e. x does not have to be an integer but can basically adopt any value between 0 and 3).

at total iron(II) concentrations of 0.02 mol/l or below). At NH_2trz/Fe^{2+} ratios of 6 : 1 and above, the absorbance increases slightly beyond the time expected to attain the plateau value. The associated relatively slow process might be attributed to the formation of aggregates [52]. In general, X-ray absorption spectroscopy and X-ray scattering measurements imply that the low-spin state arises only when the NH_2trz ligand sphere is established to a high extent in polynuclear rigid-rod structures [52]. The Fe–Fe distances and colors depend strikingly on the spin state (around 3.32 Å for the pink low-spin and 3.94 Å for the virtually colorless high-spin states).

Further investigations showed that the formation of such coordination polymers is completely reversible in DMF solutions.

Thus, all in all the results demonstrate that polynuclear coordination compounds of the type $[Fe(NH_2trz)_3]_n^{2+}$ exist in solution, but they are subjected to equilibrium reactions. As a consequence, at an NH_2trz/Fe^{2+} ratio of 3 : 1, corresponding to the stoichiometry of the $[Fe(NH_2trz)_3]^{2+}$ coordination polymers, the polynuclear complexes are not fully developed yet (in the investigated concentration range). Yet, at $NH_2trz/Fe^{2+} = 9 : 1$, the equilibrium is shifted considerably toward the polynuclear rigid-rod entities (Figure 11.6). Importantly, the fact that $[Fe(NH_2trz)_3]^{2+}$ exists in solution paves the way for processing this coordination polymer from DMF.

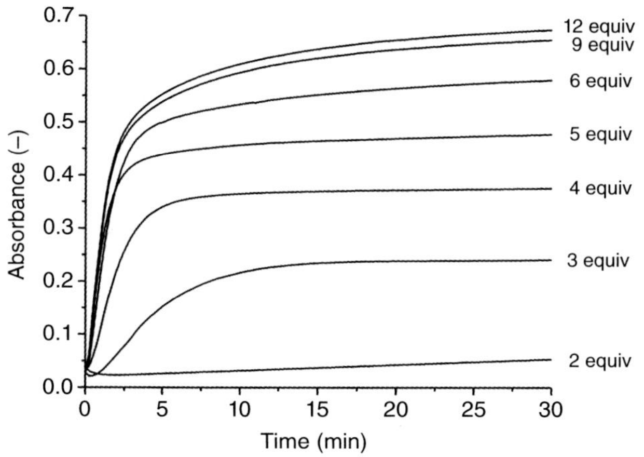

Figure 11.6 Absorbance at 541 nm upon combination of $[Fe(H_2O)_6](2ns)_2$ and NH_2trz in DMF (total Fe^{2+} concentration: 0.019 mol/l) at various molar equivalents of NH_2trz as indicated in the figure. Source: Bräunlich et al. 2014 [52]. Reproduced with permission of American Chemical Society.

11.2.2 Solid State

The packing of the rod-like molecules of coordination polymers with triazole ligands was investigated in detail with $[Fe(NH_2trz)_3](2ns)_2$ (2ns = 2-naphtahlene sulfonate) in various states, besides the as-synthesized material in particular in lyogels, xerogels, and films [51]. The structure of those objects was investigated with X-ray scattering and (cryo-)TEM (transmission electron microscopy).

The spin-crossover temperature of the as-synthesized material depends somewhat on ageing, and it seems that the as-synthesized material is present in a metastable form which transforms by annealing at 250 °C irreversibly to a thermodynamically more stable form of $[Fe(NH_2trz)_3](2ns)_2$. In all cases, however, the spin-crossover temperatures arise at specific temperatures between 22 and 53 °C upon heating (14–35 °C upon cooling).

Dimensionally stable opaque gels (lyogels) are obtained when freshly prepared solutions of $[Fe(NH_2trz)_3](2ns)_2$ in DMF are carefully covered with a layer of toluene [51]. Within one day, lyogels form (Figure 11.7) due to the slow diffusion of DMF into the toluene phase, as the concentration of the coordination polymer increases in the DMF beyond the saturation concentration thus leading to formation of solid structures. In those gels, a columnar hexagonal packing of rigid-rod-like molecules is found. The centers of the rods are separated by a distance of 2.0–2.1 nm and have lengths up to c. 60 nm, corresponding to c. 180 iron atoms in a molecule.

When such gels are dried slowly at reduced pressure, transparent thermochromic films are obtained (Figure 11.8). A color change from pink to colorless arises at about 50 °C (Figure 11.8), reflecting the spin-crossover temperature (cf. Table 11.2). UV/vis analysis of films (Figure 11.8) manifest a similar thermal behavior as solutions of $[Fe(NH_2trz)_3](2ns)_2$ in DMF: below the spin-crossover temperature the absorption band of the low-spin state at 541 nm arises upon cooling and disappears upon heating (Figure 11.8). Remarkably,

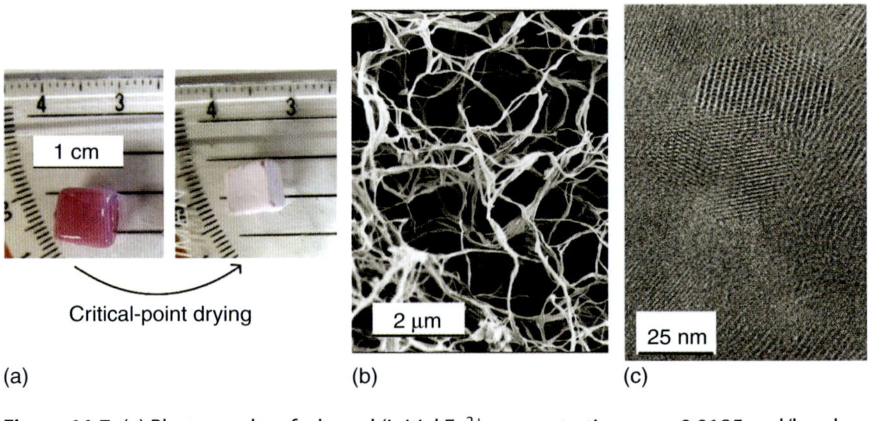

Figure 11.7 (a) Photographs of a lyogel (initial Fe^{2+} concentration c_{Fe} = 0.0185 mol/l and NH_2trz/Fe^{2+} 3/1) before (left) and after (right) critical point drying. (b) SEM image of a dried gel at c_{Fe} = 0.0371 mol/l and NH_2trz/Fe^{2+} 3/1 showing the open cell network. (c) TEM image of a critical point dried gel before (Col_h) heat treatment. Source: Sánchez-Ferrer et al. 2014 [51]. Reproduced with permission of Royal Society of Chemistry.

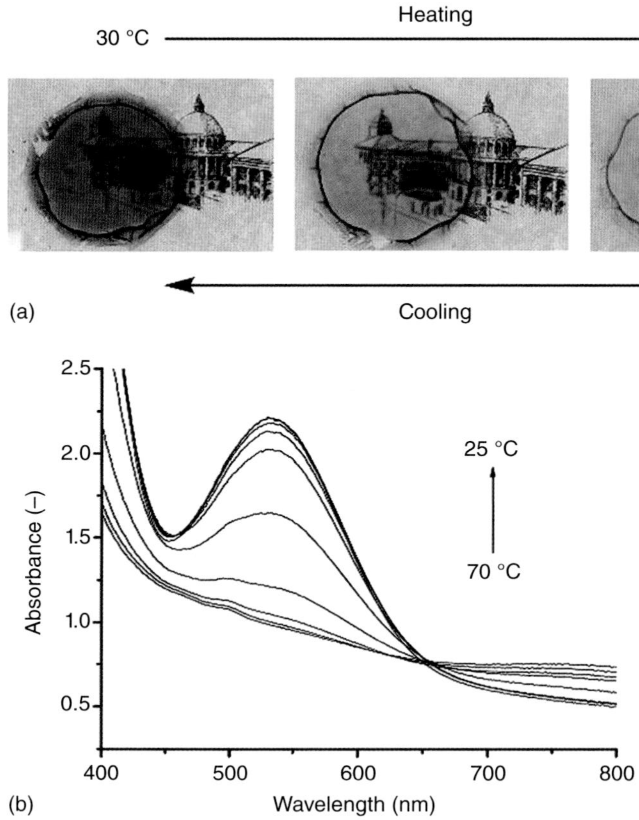

Figure 11.8 (a) Photographs of a thermochromic film of $[Fe(NH_2trz)_3](2ns)_2$ from a gel at $c_{Fe} = 0.0371$ mol/l and NH_2trz/Fe^{2+} 3/1 at 30, 50, and 60 °C. (b) Temperature-dependent UV/Vis spectra (between 25 and 70 °C in steps of 5 °C) of a film (thickness ~240 μm). Source: Sánchez-Ferrer et al. 2014 [51]. Reproduced with permission of Royal Society of Chemistry.

transparency and thermochromic behavior of the films is preserved for at least one year at ambient conditions.

Gels of $[Fe(NH_2trz)_3](2ns)_2$ can also be dried by the CO_2 critical-point drying technique whereby the original shape is essentially preserved although the shrinkage in volume amounts to ~30% of the initial volume [51]. Such materials belong to the class of xerogels. After freeze fracturing, SEM (scanning electron microscopy) images (Figure 11.7) disclose a loose (open cell) network with a mesh size in the micrometer range with density between 0.02 and 0.03 g/cm^3. Cryo-TEM images display a hexagonal packing of rigid rods (Figure 11.7), with similar intermolecular spacings and molecular lengths as noted earlier. Compared to powders of as-synthesized $[Fe(NH_2trz)_3](2ns)_2$ no striking differences arise in spin-crossover behavior (Table 11.2) [51].

Films obtained by drying at elevated temperature and reduced pressure show some opacity. In addition, the spin-crossover transition shifts toward lower temperature (~25 °C upon heating and ~10 °C upon cooling). Further annealing of the sample at 250 °C does not change the spin-crossover temperature to a large extent (Table 11.2). Compared to the clear columnar hexagonal packing (Col_h) of the rod-like molecules in the neat powder and xerogel, films dried in vacuo at ambient temperature consist of a mixture of structures. However, after removing residual DMF from the film the samples exhibit a columnar rectangular packing (Col_r) of the linear stiff rods ($a = 2.01$ nm and $b = 1.08$ nm). This packing is

Table 11.2 Overview of the spin-crossover (SCO) temperature according to DSC measurements: heating (↑) and cooling (↓) and the structural parameters (determined by WAXS, cf. Figure 11.9) of various solid appearances of $[Fe(NH_2trz)_3](2ns)_2$ [51].

Sample	SCO temperature (°C)	Structural parameters
Powder (neat)	↑22–41, ↓14–25	Col_h ($a = 2.24$ nm)
Powder (annealed at 250 °C)	↑~52, ↓~33	Col_r ($a = 2.03$ nm, $b = 1.08$ nm)
Xerogel (neat)	↑~32, ↓~20	Col_h ($a = 2.26$ nm)
Xerogel (annealed at 250 °C)	↑~40, ↓~19	Col_r ($a = 2.10$ nm, $b = 1.11$ nm)
Film (dried at RT, 1–3 mbar, comprises residual DMF)	↑~55, ↓~50	Mixture of different packings
Film (dried at 150 °C, 0.3 mbar)	↑~26, ↓~10	Col_r ($a = 2.01$ nm, $b = 1.08$ nm)
Film (annealed at 250 °C)	↑~29, ↓~17	Col_r ($a = 2.01$ nm, $b = 1.08$ nm)

Source: Sánchez-Ferrer et al. 2014 [51]. Reproduced with permission of Royal Society of Chemistry.

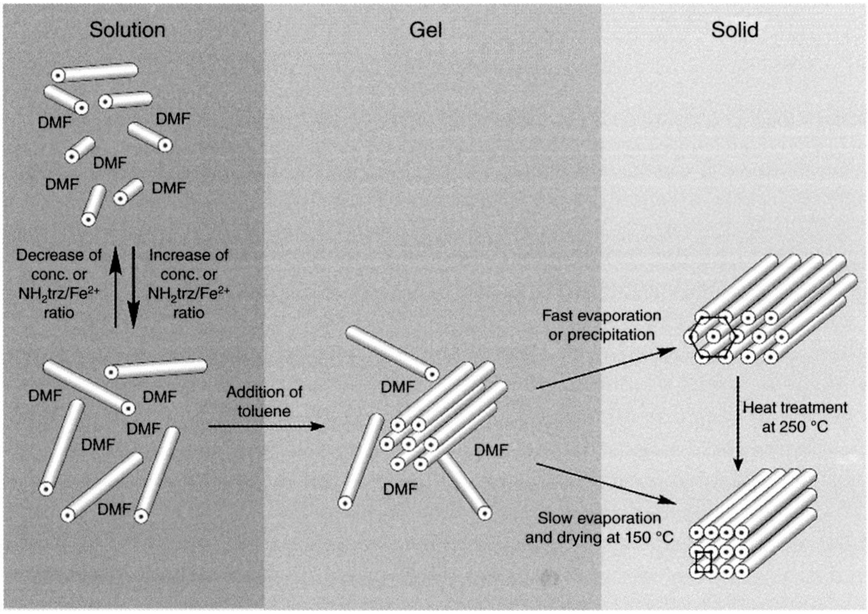

Figure 11.9 Schematic representation of $[Fe(NH_2trz)_3](2ns)_2$ in solution and in the gel and solid state, showing different packings of the rod-like complexes. The hexagonal packing (Col_h) is a metastable state which undergoes a rearrangement toward the more stable rectangular (Col_r) crystalline structure.

essentially preserved upon annealing at 250 °C (Table 11.2). Cryo-TEM images of those films show two periodicities around 2.0–2.1 nm and 0.9–1.0 nm, confirming the columnar rectangular arrangement of the linear stiff rods [51].

As a simple overview, Figure 11.9 displays a schematic representation of the different packings of the rod-like structures in solutions, gels, and solids. The xerogels are constituted by connected struts of the polynuclear linear rigid iron complexes in the metastable columnar hexagonal packing (Col_h) below 200 °C, or the columnar rectangular packing (Col_r) after annealing, similar to the structural behavior of powder samples of $[Fe(NH_2trz)_3](2ns)_2$. The films are the result of the aggregation of a Col_r packing of crystalline domains upon slow evaporation of the residual solvent, which enables the preferred arrangement (Col_r) of the stiff rods. Gels, xerogels as well as films show spin-crossover accompanied by thermochromism in the temperature range of the as-synthesized powders.

11.3 Coordination Polymers of 4-Alkyltriazoles and Iron(II)

In general, coordination polymers of Fe^{2+} and triazole ligands with long alkyl chains (Rtrz) are better soluble in organic solvents than with NH_2trz (complexes with the latter are insoluble in most cases, see earlier discussion). Polymers with various alkyl side chains have been synthesized (Table 11.3), and besides linear

Table 11.3 Overview of coordination polymers of the type $[Fe(Rtrz)_3]X_2$.

R	X
$C_{18}H_{37}$	BF_4^- [55, 60], ClO_4^- [47, 56, 60], $CF_3SO_3^-$ [47, 60], NO_3^- [48, 61], Cl^- [47, 55, 58], *p*tol⁻ [47, 55, 58, 60]
C_nH_{2n+1}, $n = 13, 16$	BF_4^- [55, 60], ClO_4^- [60], $CF_3SO_3^-$ [60], *p*tol⁻ [55, 58, 60], Cl^- [58]
$C_{10}H_{21}$	*p*tol⁻, Cl^- [58]
C_nH_{2n+1}, $n = 4–8$	BF_4^-, ClO_4^-, $CF_3SO_3^-$, *p*tol⁻ [60]
C_3H_7	BF_4^-, $CF_3SO_3^-$ [60]
C_nH_{2n+1}, $n = 1–2$	BF_4^-, ClO_4^- [60]
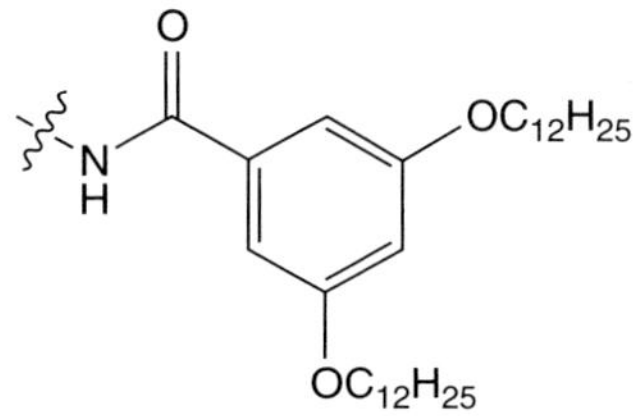	$C_nH_{2n+1}SO_3^-$, $n = 1, 8, 12, 16$, [57] ClO_4^- [62]
4-(3-Dodecyloxy)propyl	Cl^-, 9,10-dimethoxyanthracene-2-sulfonate [63], 4-(3-dodecyloxy)propyl-1-pyrenesulfonate [63]
5-Decylheptadecyl	ClO_4^- [59]

ptol⁻ = p-toluenesulfonate

aliphatic also branched alkyl chains or alkoxy groups were introduced in order to further enhance the solubility. Alkyl-substituted complexes are typically soluble in aprotic solvents, such as toluene, dodecane, diethyl ether, and chloroform. This also allows processing to films by solution casting [46] as well as the preparation of blends with organic polymers [49, 53, 54].

Alkyl-substituted polymers feature a number of properties that have not been found in other Fe^{2+}–triazole complexes:

- They often form gels already at low concentration [55, 56], and in rare cases the spin transition occurs simultaneously with the sol–gel transition [57].
- Many of them are liquid crystalline in the bulk [58, 59] and in solution [57] due to the rod-like shape of the linear Fe^{2+}–triazole backbone.
- Their amphiphilic nature allows preparation of Langmuir–Blodgett films [49, 58].

Polymers of the type $[Fe(Rtrz)_3]X_2$ can undergo spin-crossover in a broad temperature span, ranging from −110 to 70 °C [55–63]. Examples with spin-crossover around room temperature are given in Table 11.4. The spin-crossover is accompanied by a color change from pink to white and an increase of the Fe–N distance (d_{Fe-N}). For example, in $[Fe(C_{18}trz)_3](ptol)_2$ ($ptol^-$ = p-toluenesulfonate) the related values are $d_{Fe-N,LS} = 1.97$ Å in the low-spin state and $d_{Fe-N,HS} = 2.09$ Å in the high-spin state [60]. Of course, as a consequence of the expansion of the Fe—N bond also the distance of two neighboring Fe^{2+} ions increases.

Notably, the spin-crossover temperature in solution is sensitive to the water content of the solvent [64]. Solutions of $[Fe(C_ntrz)_2](DBS)_2$ (n = 14, 16, 18, acronyms see Table 11.4) in dry toluene are colorless at room temperature but turn to pink in the presence of little water (such as ambient humidity), in a reversible process and independent of the length of the alkyl group. This indicates that the Fe^{2+} ions are mainly in the high-spin state in the dry solvent [48] (Figure 11.10 on the example of $[Fe(C_{16}trz)_3](DBS)_2$). The spin-crossover temperature increases with increasing water content which reveals that water

Table 11.4 SCO temperatures of coordination polymers of the generic formula $[Fe(Rtrz)_2]X_2$ during heating (↑) and cooling (↓) [64], with R = octadecyl ($C_{18}trz$), hexadecyl ($C_{16}trz$), and tetradecyl ($C_{14}trz$); and X = 2-naphthalene sulfonate (2ns) and 4-dodecylbenzene sulfonate (DBS).

Compound	SCO temperature (°C)
$[Fe(C_{18}trz)_3](DBS)_2$	↑41, ↓33
$[Fe(C_{16}trz)_3](DBS)_2$	↑38, ↓28
$[Fe(C_{14}trz)_3](DBS)_2$	↑36, ↓27
$[Fe(C_{18}trz)_3](2ns)_2$	↑~15, ↓~12
$[Fe(C_{16}trz)_3](2ns)_2$	↑~20, ↓~15

Source: Bräunlich et al. 2015 [64]. Reproduced with permission of Springer Nature.

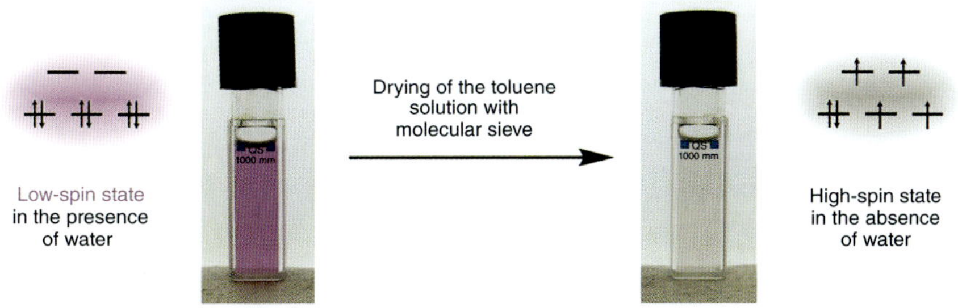

Figure 11.10 Photographs of $[Fe(C_{16}trz)_3](DBS)_2$ solutions (0.020 mol/l in toluene) with 1 equiv H_2O relative to Fe^{2+} (left) and dry (right) at room temperature. Source: Bräunlich et al. 2015 [64]. Reproduced with permission of Springer Nature.

molecules stabilize the low-spin state. It was suggested that water molecules interact with the counterions, thus modifying the crystal field and therewith the spin state [64]. The above findings are of particular interest because the spin-crossover temperature can be tuned around room temperature by the water to $[Fe(C_ntrz)_2](DBS)_2$ ratio – independent of the polymer concentration. Hence, small amounts of water (in the range of parts per million) can be detected in organic solvents with these complexes. Such measurements can be helpful for water determination e.g. in electrically insulating fluids where small amounts of water can be detrimental.

The spin-crossover temperature of $[Fe(C_{18}trz)_3](DBS)_2$ depends on the water content also in the solid state. However, while colorless solids are readily hydrated to the pink form by ambient humidity, the dehydrated (white) form cannot simply be re-established by application of reduced pressure (0.1 mbar). Thus the water molecules seem to be rather strongly bound in the solid material.

In solutions of $[Fe(C_ntrz)_2](DBS)_2$ in toluene an absorption band with a maximum at 536 nm starts to emerge below a temperature of about 45 °C. This band increases fast down to a temperature of ~15 °C and does not increase much anymore at temperatures down to −5 °C [64].

The spin-crossover temperature increases at a higher H_2O/Fe^{2+} ratio (7 °C in the absence of water, 47 °C at H_2O/Fe^{2+} 2 : 1) but is virtually independent of the Fe^{2+} concentration at a given H_2O/Fe^{2+} ratio and length of the alkyl group (from $C_{14}trz$ to $C_{18}trz$). The relationship between water content and spin-crossover temperature is illustrated in Figure 11.11.

The solubility of $[Fe(C_{18}trz)_3](DBS)_2$ allowed processing of the material into films and fibers. Albeit fibers of the pure material could not be prepared, blends with ultra-high molecular weight polyethylene (UHMWPE) as supporting matrix are readily accessible [64]. Fibers with $[Fe(C_{18}trz)_3](DBS)_2$ fractions of 17–80% m/m were thus produced from xylene solution, with fiber thicknesses between 50 and 200 μm. Blending of $[Fe(C_{18}trz)_3](DBS)_2$ with polyethylene does not significantly change the spin-crossover temperature (between 40 and 50 °C as in the neat complex). To visualize the thermochromism associated with spin-crossover, a fiber with 80% m/m $[Fe(C_{18}trz)_3](DBS)_2$ aligned on a hot plate with a temperature gradient between room temperature and ~80 °C is shown in Figure 11.12.

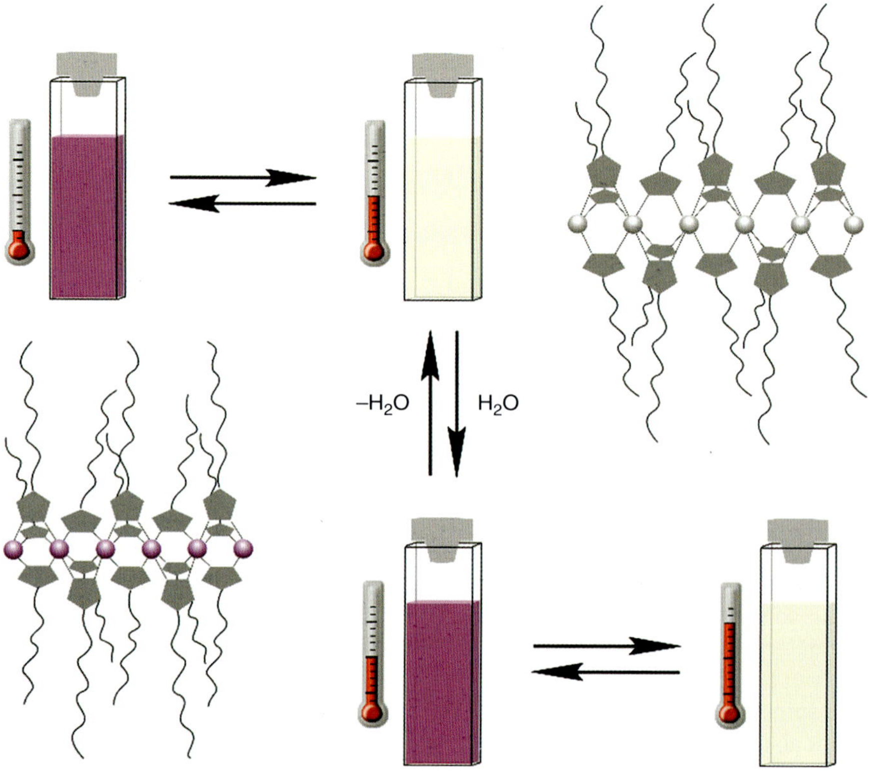

Figure 11.11 Dependence of the spin-crossover temperature of alkyl-substituted Fe^{2+}–triazole complexes in organic solutions on the water content: upon increase of the water content the spin-crossover temperature is raised. It is also indicated in the figure that the Fe–N distances in the low-spin complex (pink) are smaller than in the high-spin complex (white).

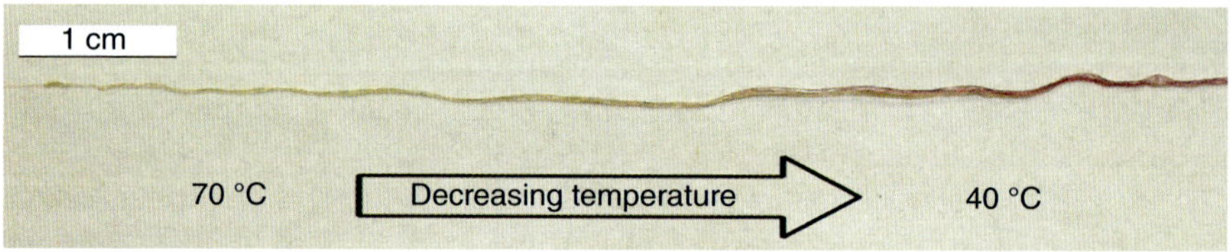

Figure 11.12 Photograph of a fiber consisting of a blend of $[Fe(C_{18}trz)_3](DBS)_2$ (80% m/m) and UHMWPE exposed to a temperature gradient from ~25 to 80 °C. Source: Bräunlich et al. 2015 [64]. Reproduced with permission of Springer Nature.

11.4 Coordination Polymers of 1,2,4-Triazoles and Other Metals

Besides iron(II)–triazole complexes also their cobalt(II) analogs can have varied color upon changing the environment. However, this color change is not reasoned by spin-crossover behavior but arises as a consequence of a change in coordination geometry.

Coordination polymers of cobalt(II) and Rtrz were synthesized with long chains R (e.g. R = alkyl or dialkylether) as they enhance the solubility in organic solvents [48, 61, 65, 66]. Thus, addition of Rtrz to hexaaquacobalt(II) chloride in methanol leads to polymers of the type $[Co(Rtrz)_2]^{2+}$ [66]. The blue precipitate formed after synthesis turns reversibly to pink upon exposure to humidity followed by drying [66]. Therefore, the pink form is associated with a hydrated complex and the blue form with a dehydrated complex (in contrast to earlier assumptions that equilibria comprising both bidentate and monodentate Rtrz ligands are involved in the color change [65]). Accordingly, the polymers also change color reversibly when wet solutions are dried and vice versa. Analysis of UV/vis spectra in solution reveals octahedral coordination geometry for the pink form and tetrahedral coordination geometry for the blue form. Possible structures are shown in Figure 11.13 [66].

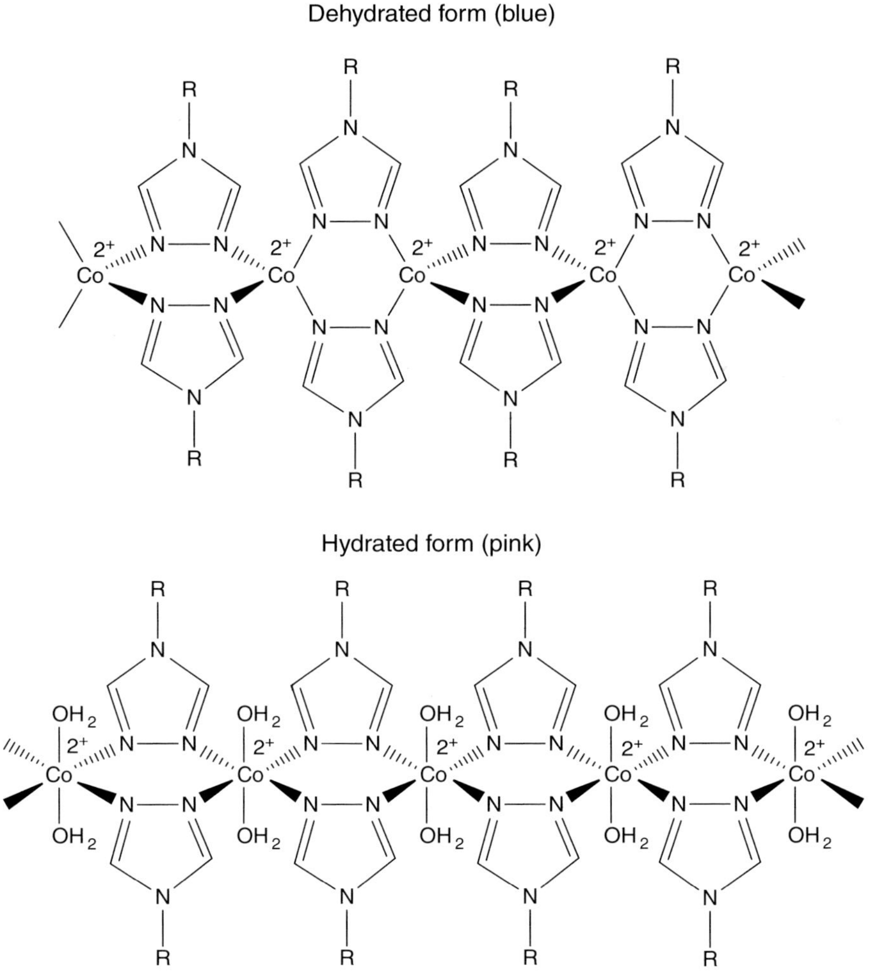

Figure 11.13 Possible structures of the polymer based on $[Co(C_{18}trz)_2]^{2+}$ in the blue and the pink form. The latter structure might also contain units with water molecules in cis position. Source: Bräunlich et al. 2017 [66]. Reproduced with permission of John Wiley & Sons.

In solution, such compounds can show thermochromism in the presence of water. At room temperature, the pink, hydrated form predominates whereas upon heating, the blue, dehydrated form arises reversibly.

When hot solutions of toluene or xylene with Co^{2+} concentrations above c. 0.04 mol/l are cooled down to room temperature gels form [65, 66]. Solutions and gels of previously dried polymers are initially blue but turn slowly (over a period of several days) to pink in contact with ambient humidity, thus indicating that hydration of dehydrated polymers also proceeds in gels.

The large variety of possible coordination compounds with triazole ligands also becomes evident when examining complexes of triazole ligands with other metals. A number of linear NH_2trz – bridged copper(II) coordination polymers are reported, for instance $[Cu(NH_2trz)_3](NO_3)_2{\cdot}H_2O$ [7], $[Cu(NH_2trz)_3](BF_4)_2{\cdot}H_2O$ [67], $[Cu(NH_2trz)_3](ZrF_6){\cdot}H_2O$ [68], $[Cu(NH_2trz)_3]_3(SiF_6)_3{\cdot}8H_2O$ [67], and $[Cu(NH_2trz)_3]_n(BF_4)_n(ZrF_6){\cdot}4H_2O$ [67]. In the case of silver(I), notably complexes with NH_2trz develop structures that can depend critically on the counter ion [69]. For instance, with hexafluorophosphate, zig–zag structures of $[Ag(NH_2trz)_2]^+$ entities with all NH_2trz molecules acting as bridging ligands are formed, while with sulfate a part of the NH_2trz molecules coordinate with only one nitrogen atom of the triazole ring [69]. Complex polymeric structures with $[Ag_4(NH_2trz)_7]^{4+}$ formula units arise in perchlorate, nitrate, and tetrafluoroborate complexes [69, 70]. All NH_2trz molecules coordinate with two nitrogen atoms, and the silver atoms are arranged pairwise in a wave-like manner [69, 70]. With the counter ion tetrafluoroborate also another compound was found, based on the formula unit $[Ag_2(NH_2trz)_3]^{2+}$, where a fraction of the NH_2trz molecules coordinate with three nitrogen atoms (i.e. including the amine group) [69]. Remarkably, all NH_2trz molecules coordinate with three nitrogen atoms in $[Ag(NH_2trz)](CF_3SO_3)$ [70]. Here, the silver atoms and NH_2trz establish three-dimensional coordination polymers comprising "walls" that envelop parallel linear channels that contain the counter ions [70]. Interestingly, several of the silver-NH_2trz polymers are luminescent [69].

11.5 Conclusion and Outlook

Substituted 1,2,4-triazole ligands can establish coordination polymers with a number of metal ions. In solution, an excess of ligand may be required to shift the equilibrium between free ligand and polymeric complexes essentially completely to the side of the final stoichiometry of related solid polymers. A variety of different polymeric structures were established, e.g. with Ag^I and 4-amino-1,2,4-triazole (NH_2trz), due to coordination of NH_2trz with one, two, or three nitrogen atoms. Some of these compounds show luminescence. On the other hand, strictly linear coordination polymers were reported with substituted 1,2,4-triazoles and e.g. Co^{II}, Cu^{II}, and Fe^{II}, where the triazoles act as bridging ligands (coordination of the adjacent nitrogen atoms in the triazole ring). Such polymers can be readily prepared in large quantities. Related compounds with Co^{II} and Fe^{II} show thermochromism, although of different origin: in the case of Co^{II} thermochromism is due to a change in coordination geometry (tetrahedral vs. octahedral), whereas spin-crossover accounts for thermochromism in the Fe^{II} complexes. In the latter case, spin-crossover leads to a transition from diamagnetic to paramagnetic (i.e. a reversible change from a low-spin state to a high-spin state). The color of some of these polymers is also sensitive to small amounts of water, i.e. such polymers act as humidity sensors. Thus, coordination polymers of substituted 1,2,4-triazoles and metal ions can exhibit unusual combinations of properties.

Notably, it was demonstrated on examples of substituted 1,2,4-triazoles and Fe^{II} and partially also of Co^{II} that soluble polymers can be processed to lyogels, xerogels, films, and blended fibers. Spin-crossover and thermochromism in the objects with Fe^{II} is preserved.

So far, coordination polymers with substituted 1,2,4-triazoles have been reported only for a limited number of metal ions, as indicated earlier. Clearly, interesting coordination polymers may be expected also with other metal ions, e.g. Pd^{II} and Pt^{II}, which show square planar coordination geometry. On the other hand, if e.g. application in areas such as humidity or temperature sensors thermochromic fibers or document counterfeiting is envisaged, processing has to be developed in a way that allows upscaling to industrial scales.

Acknowledgments

The authors thank the COST Action CM1302 (SIPs): "European Network on Smart Inorganic Polymers." We cordially thank Paul Smith for his continuous support and C. Mair, S. Lienemann, P. Knüsel, A. Sánchez-Ferrer, R. Mezzenga, M. Bauer, R. Schepper, J. Ruokolainen, and J. Dshemuchadse for their valuable collaboration in the area of metal-triazole coordination polymers.

References

1 Gorter, S. and Engelfriet, D.W. (1981). *Acta Crystallogr. B* 37: 1214.
2 Haasnoot, J.G., de Keyter, G.C.M., and Verschoor, G.C. (1983). *Acta Crystallogr. C* 39: 1207.
3 Englefriet, D.W., den Brinker, W., Verschoor, G.C., and Gorter, S. (1979). *Acta Crystallogr. B* 35: 2922.
4 Jarvis, J.A.J. (1962). *Acta Crystallogr.* 15: 964.
5 Engelfriet, D.W. and Verschoor, G.C. (1981). *Acta Crystallogr. B* 37: 237.
6 Oro, L.A., Pinillos, M.T., Tejel, C. et al. (1984). *J. Chem. Soc., Chem. Commun.* 1687.
7 Dîrtu, M.M., Neuhausen, C., Naik, A.D. et al. (2010). *Inorg. Chem.* 49: 5723.
8 Grosjean, A., Daro, N., Kauffmann, B. et al. (2011). *Chem. Commun.* 47: 12382.
9 Lider, E.V., Peresypkina, E.V., Lavrenova, L.G. et al. (2010). *Russ. J. Coord. Chem.* 36: 337, trans. from Koord. Khim., 2010, 36, 337.
10 Garcia, Y., van Koningsbruggen, P.J., Bravic, G. et al. (1997). *Inorg. Chem.* 36: 6357.

11 Groeneveld, L.R., Vos, G., Gorter, S., and Haasnoot, J.G. (1982). *Acta Crystallogr. B* 38: 2248.
12 Hu, Z.-S., Lin, Y.-H., and Jin, S.-C. (1989). *Acta Crystallogr. C* 45: 1490.
13 Müller-Buschbaum, K., Mokaddem, Y., and Höller, C.J. (2008). *Z. Anorg. Allg. Chem.* 634: 2973.
14 Haasnoot, J.G., Vos, G., and Groeneveld, W.L. (1977). *Z. Naturforsch. B* 32: 1421.
15 Yokoyama, T., Murakami, Y., Kiguchi, M. et al. (1998). *Phys. Rev. B* 58: 14238.
16 Ewald, A.H., Martin, R.L., Ross, I.G., and White, A.H. (1964). *Proc. R. Soc. London A* 280: 235.
17 White, A.H., Roper, R., Kokot, E. et al. (1964). *Aust. J. Chem.* 17: 294.
18 Huheey, J.E., Keiter, E.A., and Keiter, R.L. (2003). *Anorganische Chemie : Prinzipien von Struktur und Reaktivität*, 3ee. Berlin: Walter de Gruyter.
19 Real, J.A., Gaspar, A.B., and Muñoz, M.C. (2005). *Dalton Trans.* 2062.
20 Gütlich, P., van Koningsbruggen, P.J., and Renz, F. (2004). *Struct. Bond.* 107: 27.
21 Gütlich, P., Hauser, A., and Spiering, H. (1994). *Angew. Chem.* 106: 2109.
22 Garcia, Y., Niel, V., Muñoz, M.C., and Real, J.A. (2004). *Top. Curr. Chem.* 233: 229.
23 Gütlich, P. and Goodwin, H.A. (2004). *Top. Curr. Chem.* 233: 1.
24 Barefield, E.K. and Busch, D.H. (1968). *Q. Rev. Chem. Soc.* 22: 457.
25 Goodwin, H.A. (1976). *Coord. Chem. Rev.* 18: 293.
26 Jeftić, J. and Hauser, A. (1997). *J. Phys. Chem. B* 101: 10262.
27 Shepherd, H.J., Rosa, P., Vendier, L. et al. (2012). *Phys. Chem. Chem. Phys.* 14: 5265.
28 Decurtins, S., Gütlich, P., Köhler, C.P., and Spiering, H. (1984). *Chem. Phys. Lett.* 105: 1.
29 Hauser, A. (2004). *Top. Curr. Chem.* 234: 155.
30 Nègre, N., Conséjo, C., Goiran, M. et al. (2001). *Physica B* 294–295: 91.
31 Cambi, L. and Szegö, L. (1931). *Ber. Dtsch. Chem. Ges.* 64: 2591.
32 Cambi, L. and Szegö, L. (1933). *Ber. Dtsch. Chem. Ges.* 66: 656.
33 Baker, W.A. and Bobonich, H.M. Jr., (1964). *Inorg. Chem.* 3: 1184.
34 König, E. and Madeja, K. (1967). *Inorg. Chem.* 6: 48.
35 Kläui, W. (1979). *J. Chem. Soc., Chem. Commun.* 700.
36 Eberspach, W., El Murr, N., and Kläui, W. (1982). *Angew. Chem.* 94: 919.
37 Kläui, W., Eberspach, W., and Gütlich, P. (1987). *Inorg. Chem.* 26: 3977.
38 Zarembowitch, J. and Kahn, O. (1984). *Inorg. Chem.* 23: 589.
39 Halepoto, D., Holt, D.G.L., Larkworthy, L.F. et al. (1989). *J. Chem. Soc., Chem. Commun.* 1322.
40 Hughes, A.K., Murphy, V.J., and O'Hare, D. (1994). *J. Chem. Soc., Chem. Commun.* 163.
41 Liu, Z., Liang, S., Di, X., and Zhang, J. (2008). *Inorg. Chem. Commun.* 783.
42 Brubaker, G.R. and Busch, D.H. (1966). *Inorg. Chem.* 5: 2114.
43 Kahn, O., Kröber, J., and Jay, C. (1992). *Adv. Mater.* 4: 718.
44 Kahn, O. and Jay Martinez, C. (1998). *Science* 279: 44.
45 Lavrenova, L.G., Yudina, N.G., Ikorskii, V.N. et al. (1995). *Polyhedron* 14: 1333.

46 Kuroiwa, K., Shibata, T., Sasaki, S. et al. (2006). *J. Polym. Sci. A* 44: 5192.
47 Armand, F., Badloux, C., Bonville, P. et al. (1995). *Langmuir* 11: 3467.
48 Roubeau, O., Agricole, B., Clérac, R., and Ravaine, S. (2004). *J. Phys. Chem. B* 108: 15110.
49 Rubio, M., Hernández, R., Nogales, A. et al. (2011). *Eur. Polym. J.* 52.
50 Faulmann, C., Chahine, J., Malfant, I. et al. (2011). *Dalton Trans.* 40: 2480.
51 Sánchez-Ferrer, A., Bräunlich, I., Ruokolainen, J. et al. (2014). *RSC Adv.* 4: 60842.
52 Bräunlich, I., Sánchez-Ferrer, A., Bauer, M. et al. (2014). *Inorg. Chem.* 53: 3546.
53 Jeong, S.-H., Kim, K.-N., Kang, J.S. et al. (2007). *Mol. Cryst. Liq. Cryst.* 471: 3.
54 Lee, S.-W., Lee, J.-W., Jeong, S.-H. et al. (2004). *Synth. Met.* 142: 243.
55 Grondin, P., Roubeau, O., Castro, M. et al. (2010). *Langmuir* 26: 5184.
56 Rubio, M. and López, D. (2009). *Eur. Polym. J.* 45: 3339.
57 Fujigaya, T., Jiang, D.-L., and Aida, T. (2007). *Chem. Asian J.* 2: 106.
58 Grondin, P., Siretanu, D., Roubeau, O. et al. (2012). *Inorg. Chem.* 51: 5417.
59 Lee, Y.H., Komatsu, Y., Yamamoto, Y. et al. (2011). *Inorg. Chem. Commun.* 14: 1498.
60 Roubeau, O., Gomez, J.M.A., Balskus, E. et al. (2001). *New J. Chem.* 25: 144.
61 Roubeau, O., Natividad, E., Agricole, B., and Ravaine, S. (2007). *Langmuir* 23: 3110.
62 Fujigaya, T., Jiang, D.-L., and Aida, T. (2003). *J. Am. Chem. Soc.* 125: 14690.
63 Matsukizono, H., Kuroiwa, K., and Kimizuka, N. (2008). *Chem. Lett.* 37: 446.
64 Bräunlich, I., Lienemann, S., Mair, C. et al. (2015). *J. Mater. Sci.* 50: 2355.
65 Kuroiwa, K., Shibata, T., Takada, A. et al. (2004). *J. Am. Chem. Soc.* 126: 2016.
66 Bräunlich, I., Mair, C., Bauer, M., and Caseri, W. (2017). *J. Inorg. Organomet. Polym.* 27: 605.
67 Drabent, K. and Ciunik, Z. (2001). *Chem. Commun.* 37: 1254.
68 Dîrtu, M.M., Rotaru, A., Gillard, D. et al. (2009). *Inorg. Chem.* 48: 7838.
69 Yeh, C.-W., Chang, W.-J., Suen, M.-C. et al. (2013). *Polyhedron* 61: 151.
70 Wang, Q.-L., Xu, H., Hou, H.-W., and Yang, G. (2009). *Z. Naturforsch. B* 64: 1143.
71 van Koningsbruggen, P.J., Garcia, Y., Codjovi, E. et al. (1997). *J. Mater. Chem.* 7: 2069.
72 Garcia, Y., van Koningsbruggen, P.J., Codjovi, E. et al. (1997). *J. Mater. Chem.* 7: 857.
73 Garcia, Y., Moscovici, J., Michalowicz, A. et al. (2002). *Chem. Eur. J.* 8: 4992.
74 Kröber, J., Codjovi, E., Kahn, O. et al. (1993). *J. Am. Chem. Soc.* 115: 9810.

Index

Smart Inorganic Polymers: Synthesis, Properties, and Emerging Applications in Materials and Life Sciences, First Edition. Edited by Evamarie Hey-Hawkins and Muriel Hissler.

d

W

X

Z